Pollen and Spores:
Morphology and Biology

T0141411

Pollen and Spores:
Morphology and Biology

M M Harley, C M Morton and S Blackmore (Editors)

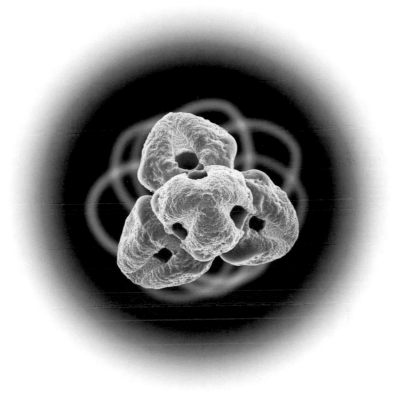

The Royal Botanic Gardens, Kew 2000
In collaboration with The Linnean Society of London,
The Natural History Museum and the Systematics Association

Production Editor: S. Dickerson

Cover design and page make-up by Media Resources,
Information Services Department, Royal Botanic Gardens, Kew

ISBN 1 900347 95 4

Printed in Great Britain by
Whitstable Printers Ltd, Whitstable, Kent

CONTENTS

Pollen and Spores: Morphology and Biology

Contents

KEITH FERGUSON FLS, OBE

We approach the task of writing this dedication to Keith with some trepidation, conscious of the fact that he is very much alive, merely retired. Nevertheless, we wanted to say good things about him while he may enjoy the accolade.

Keith has enjoyed a long and distinguished career during which he has been honoured by a number of learned societies. He is a Fellow of the Linnean Society (FLS), a corresponding member of the Botanical Society of America, and President of the Ray Society. On his retirement he received the accolade of Honorary Research Fellow from the previous Director, Sir Ghillean Prance, and the Board of Directors. Last year he added the unusual, and impressive, achievement of probably being the only palynologist to receive the Order of the British Empire (OBE) from her Majesty, the Queen in the New Year's Honours Lists.

Keith's initial interest in palynology was sparked by Bill Watts at Trinity College, Dublin in the late 1950's, when it became obvious that an investigation of pollen would clarify some of the problems in *Saxifraga*. However, his Professor at Trinity College, David Webb, was not one to rush into print. Time passed and Keith moved on to do a post doc at the Arnold Arboretum and Gray Herbarium at Harvard, working on the flora of the Northeast USA with Carol Wood. Here he made many valuable friends and contacts including James Walker and Duncan Porter.

In 1966 Keith returned to the UK to work on Flora Europaea. He was based at the British Museum (Natural History), as it was then called, and it was there that he met his future wife, Lorna. He persuaded the Keeper, Robert Ross, to let him use the busy Electron Microscopy (EM) Unit to follow up his pollen studies on *Saxifraga*. By the time he joined the Kew staff in 1968, as a Senior Scientific Officer in the Australian section of the Herbarium, he had finally published his first pollen paper (with David Webb): *Pollen morphology in the genus* Saxifraga *and its taxonomic significance.*

Keith is an immensely practical man, with a keen appreciation for new developments in botanical research. With his classical training in plant systematics he was very aware of the significant contribution that pollen morphology could make to systematic botany at Kew. In 1971 he received training in Gunnar Erdtman's laboratory in Stockholm, where he also met John Rowley. Keith and his wife Lorna became good friends of both the Erdtmans and the Rowleys.

Until Kew could be persuaded of the value in acquiring an electron microscope, Keith needed more electron microscope time than the BM(NH) could provide. He approached Don Dring, the Plant Pathologist at the Ministry of Agriculture Fisheries and Food, Pest Infestation Laboratories, Slough, who was able to accommodate him. Keith continued to use the Slough EM facility until 1972 when a small SEM (JSM 1) was installed in the Jodrell laboratory at Kew. At the same time Keith, both enterprising and convincing, had persuaded the Keeper of the Herbarium, Prof. Pat Brenan, to approve the setting up of a small laboratory for pollen studies. SEM work could now be carried out at Kew, but TEM work entailed a journey to the EM unit of Westfield College (University of London), run by Josie Keith Lucas. Although there was little money available for this project Keith, with his tremendously practical approach soon gathered together the necessary ancillary equipment and advice.

Within less than a year he had recruited two students, Oisian MacNamara, the first of a long line of one year sandwich course students, and a young summer vacation student from Reading called Peter Crane. Little did Keith, or Peter, imagine that in 1999 Peter would return to Kew, as its Director. At about the same time MMH was recruited as an Assistant Scientific Officer to assist in the running of the Palynology

Unit. Keith could not have predicted that MMH would be sufficiently enthralled by palynology, that she would stay long enough to succeed him as Head of Palynology at RBG, Kew. Papers on pollen morphology in relation to plant systematics flew from Keith's pen, and have continued to do so ever since. Apart from the scientific integrity of Keith's texts, his beautiful and informative plates of pollen grains, whether SEM, TEM or LM, provide an unmistakable signature.

Keith has also been a great collaborator. A number of his early papers were on Menispermaceae with systematic input from Leonard Forman, and Saxifragaceae and Cornaceae with the late Michel Hideux (C.N.R.S.). There have been many other collaborators including Annick Le Thomas, Philippe Guinet, and Maria Josefa Diez. The very active Kew legume research programme led, until his recent retirement, by Roger Polhill was instrumental in the pollen morphology of the papilionoid and caesalpinioid Leguminosae becoming Keith's major research interest. His most frequent collaborator has been JJS, who he first met in 1972 at a Legume Conference. It was the beginning of a long, enjoyable and fruitful association, and a number of visits to the USA, including six months in 1977 under the auspices of the Missouri Botanic Garden and RBG, Kew, to study techniques and legume pollen. Numerous joint papers on Leguminosae pollen have since been published.

It wasn't long before the research programme soon outgrew the Palynology Unit's nursery quarters. A much larger area became available in the basement of the 1967 Herbarium extension. More space and a solid concrete base provided all the incentive needed for Keith to organise the purchase and installation of in-house EM facilities. First an SEM and later, never one to miss a bargain! he was given a 'JEM 7' TEM by a local pharmaceutical company who were upgrading equipment. In 1981 in collaboration with John Dransfield, Kew's palm systematist, Keith established a long-term programme to study palm pollen. In 1988 Carol Furness joined the Unit, initially to study pollen of Acanthaceae, and Marie Kurmann was also recruited, as a Kew Research Fellow, with a remit to work on the pollen morphology of the conifers.

In spite of an impressive record of research and publication on pollen, Keith's keen eye for detail and his passionate care for the Herbarium, and for the people working within it, led him to apply for one of the three Assistant Keeperships. These positions were set up in 1985 under the Keeper, Gren Lucas. Later, in 1989, Keith became Deputy Keeper. A role he enacted with great verve, not only for Gren Lucas but also for Gren's successor, Simon Owens. Many members of staff and visiting botanists have benefited from Keith's sympathetic and helpful approach to their needs or problems because, although he will be fondly remembered by many as a great talker, he also has the gifts of listening and perception.

Keith is not provincial by temperament, and during the period that he was developing his ideas to put pollen morphology on the systematic research agenda at Kew, he was also visiting a number of other laboratories for discussion, and to accumulate techniques and ideas. In addition to people already mentioned he met, and worked with many well-known palynologists such as, Bernard Lugardon, Jan Muller (Leiden); Siwert Nilsson (Swedish Museum of Natural History), Egon Köhler (Humboldt University, Berlin), Wim Punt (Univ. Utrecht), and Joan Nowicke (Smithsonian Institution, Washington). With all of these people long lasting collaborative, and highly productive professional friendships were founded.

Aware of the need to integrate the palynological fraternity within the UK, and thereby encourage the exchange of ideas, he founded the Linnean Society Palynology Specialist Group (LSPSG) in 1974, and held the first major international meeting devoted primarily to recent pollen. He edited the proceedings with Jan Muller and, in 1985 he co-organised, and edited the proceedings of a second conference this time with another lifelong palynologist, Stephen Blackmore, now Regius Keeper of the

Royal Botanic Garden Edinburgh. As a PhD student applying palynology to classification, Stephen had been greatly helped and encouraged by Keith and, when he was appointed as the palynologist at the BM(NH), he soon sought out his former mentor. At that time Stephen's international reputation in phylogenetic and evolutionary pollen studies was developing rapidly. Stephen Blackmore with Susan Barnes, also edited the third conference proceedings in this series.

The 1998 Kew conference: *Pollen and Spores: Morphology and Biology*, is the fourth in the series and has been dedicated to Keith as a mark of appreciation for his enormous contribution, and his inspiration and encouragement to others, in the field of systematic pollen morphological research.

Madeline M. Harley
Royal Botanic Gardens, Kew

John J. Skvarla
University of Oklahoma

Stephen Blackmore
Royal Botanic Garden,
Edinburgh

[We are indebted to Lorna Ferguson for her enthusiastic help in providing us with additional detail regarding earlier events in Keith's career.]

FOREWORD

It is a great personal pleasure to add a brief comment to this volume in honour of Keith Ferguson. Keith established the Palynology Unit at the Royal Botanic Gardens, Kew in the early 1970s, and under his direction it became a leader in developing the science of palynology. A particular focus has been applying palynological data to problems in the systematics of living plants. These studies, often undertaken in collaboration with colleagues overseas or in the Kew Herbarium, contributed an enormous wealth of valuable new data that had a significant impact on plant systematics, as well as on other botanical subdisciplines. The work also produced some spectacular discoveries as Keith and his colleagues began to chart the extraordinary diversity of angiosperm pollen, of which my particular favourite were the bizarre, enormously elongated pollen grains of *Crossandra stenostachya* (Lindau) C.B. Clarke in the Acanthaceae. But setting aside the specifics of particular palynological discoveries an enduring characteristic of work from the Kew palynology group over the last three decades has been consistent attention to detail in combination with the highest standards of technical excellence and systematic rigour.

In the 30 years of its formal existence the Kew Palynology Unit has been associated with a series of occasional international conferences organised under the auspices of the Linnean Society. The first three of these: *The Evolutionary Significance of the Exine* (1974), *Pollen and Spores. Form and Function* (1985) and *Pollen and Spores: Patterns of Diversification* (1990) have constituted significant landmarks in the development of the field. The present volume *Pollen and Spores: Morphology and Biology* is destined to make a similar mark.

The papers included in this book span an enormous range of subject matter, but all are united by a focus on the diversity of pollen and spores – how that diversity arises in terms of development and evolution, how it is distributed among living and fossil plants, and how it can be utilised - for purposes of systematics and for interpreting vegetational history. The volume is a fitting tribute to the contributions of Keith Ferguson - both to the Royal Botanic Gardens, Kew specifically, and the science of palynology in general. The papers also provide an insight into the current development of spore-pollen studies. In particular, emphasising the integration of data from palynology into a range of botanical subdisciplines including plant phylogenetics, comparative developmental biology, and vegetational history, as well as palaeobotanical studies of the history of plant diversity. In many respects the contributions to this volume reflect the modern maturity of pollen/spore studies as reflected in decreasing emphasis on the distinctiveness of palynology, and increased attention to the extent to which studies of pollen and spores connect to other disciplines: ranging from phylogenetic systematics to pollination biology and plant development.

As Director of the Royal Botanic Gardens, Kew it is also important for me to formally acknowledge the hard work of the Conference Organising Committee comprising Madeline Harley, Paula Rudall, Lisa von Schlippe, Hannah Banks, Cynthia Morton and Steven Blackmore (Natural History Museum, London). We are all especially indebted to Madeline Harley, Cynthia Morton and Stephen Blackmore for their careful editorial work in producing this volume.

Finally, I must take this rare public opportunity to thank Keith Ferguson who was a key figure in the early development of my botanical career. As one of my earliest mentors he helped stimulate my interest in research, and was the first to introduce me to the extraordinary botanical collections and expertise concentrated at the Royal Botanic Gardens, Kew. For both I will always be extremely grateful.

Peter R. Crane
Director
Royal Botanic Gardens, Kew

PREFACE

This volume contains the proceedings of the fourth in an occasional series of international conferences on palynology organised by the Linnean Society Palynology Specialist Group. The first of the conferences, *The Evolutionary Significance of the Exine* was held at the Linnean Society and the Royal Botanic Gardens Kew in 1975. The resulting publication, which was the first of the Society's special volumes, is widely acknowledged to be one of the most important publications on pollen grains and spores. Unlike so many conference volumes it is still routinely cited twenty five years later in current research papers.

The last quarter of a century has seen many exciting developments in the fields of plant reproduction, palaeobotany, systematics, morphology and ultrastructure that directly concern palynology. These have been charted in the subsequent conference volumes. *Pollen and Spores: Form and Function*, published in 1986, reflected the intense interest in functional morphology at that time. As this latest volume shows, interpreting the function of microscopic features continues to be a challenge. *Pollen and Spores: Patterns of Diversification* was published in 1991 when the power of phylogenetic methods was coming into its own in plant systematics. Phylogenetics is now central to an emerging new synthesis in biology that integrates, among other things, systematics, evolution, ecology and developmental biology.

The latest conference, held at the Royal Botanic Gardens Kew and the Natural History Museum from 6th to 9th of July, 1998, captured the continuing excitement that comes from focusing on the microscopic world of pollen grains and spores. 140 delegates representing twelve countries attended the conference and, in additional to the 43 papers that were presented, there were discussion groups and a lively poster session. Two of the conference sessions were dedicated to the memory of two giants of pollen biology: Jack Heslop-Harrison and Bruce Knox who both died in 1998. The achievements of both of these brilliant scientists are well-known and both had contributed to the conferences of the Linnean Society Palynology Specialist Group. Heslop-Harrison's contribution to The Evolutionary Significance of the Exine, was a tour de force of experimental botany that connected plant incompatibility systems with pollen wall ultrastructure. Knox, a student of Heslop-Harrison's, is probably best known for his research into pollen stigma interactions. In the Pollen and Spores: Patterns of Diversification conference he teamed up with Sophie Ducker to provide a superb historical review placing studies of plant sperm cells in their historical context.

The conference was organised by the Linnean Society Palynology Specialist Group, the Systematics Association, the Natural History Museum and the Royal Botanic Gardens Kew. It was timed to coincide with the retirement of Keith Ferguson, the distinguished palynologist who founded both the Palynology Unit at Kew and the Linnean Society Palynology Specialist Group. Keith's many achievements are recognised in the Foreword to this volume by Peter Crane, the Director of Kew and in a short account by Madeline Harley, John Skvarla and Stephen Blackmore.

Many people contributed to the success of the conference and we would particularly like to thank Lisa von Schlippe, Suzy Dickerson, Ann Lucas and members of the Information and Services Division at RBG Kew. Peter Stafford kindly acted as projectionist at the Natural History Museum. The Kew Volunteers organised the conference packs and the refreshments during the conference. Both the Director of the Royal Botanic Gardens, Kew and The Linnean Society generously hosted evening receptions for conference delegates.

Madeline Harley, Royal Botanic Gardens Kew
Cynthia Morton, Freeman Herbarium, Auburn University
Stephen Blackmore, Royal Botanic Garden Edinburgh

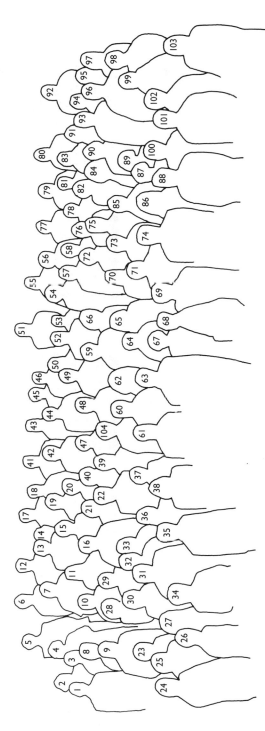

1. Sanna Simán, 2. Satish Srivastava, 3. Deborah Long, 4. Rosalind Srivastava, 5. Favio González, 6. Peter Linder, 7. Michael Zavada, 8. Mary Dettmann, 9. Alan Graham, 10. Rudi Verhoeven, 11. Samantha Cook, 12. Simon Owens, 13. Krysztof Binka, 14. Christa-Charlotte Hofmann, 15. Michael Hesse, 16. Raymond van der Ham, 17. Reinhard Zetter, 18. Paul Rugman-Jones, 19. Peter Martin, 20. Mary Lush, 21. Wan-Pyo Hong, 22. Miklos Kedves, 23. George Callen, 24. Egon Köhler, 25. Laure Civeyrel, 26. Madeline Harley, 27. Agnes Mignot, 28. Shirley Graham, 29. Keith Ferguson, 30. Han van Konijnenberg van Cittert, 31. Siwert Nilsson, 32. Valéry Malecot, 33. Lorna Ferguson, 34. Annick Le Thomas, 35. Ann Cadman, 36. Danielle Lobreau-Callen, 37. Sue de Villiers, 38. Paula Rudall, 39. Marleen Vermoere, 40. Rita Alfaro-Bates, 41. Arjan Stroo, 42. Volker Wilde, 43. Manuel Munuera Giner, 44. Thomas Borsch, 45. Heather Pardoe, 46. Willy Taylor, 47. Rivka Dulberger, 48. Ivan Jaimes, 49. Alan Hemsley, 50. Nancy Kapuskar, 51. Michael Vomberg, 52. Gamal El-Ghazaly, 53. Cynthia Morton, 54. Suzy Huysmans, 55. Charles Wellman, 56. Paul Strother, 57. Harald Schneider, 58. Chih-Hua Tsou, 59. Bob Johns, 60. Hugh Dickinson, 61. Steve Blackmore, 62. Martina Weber, 63. Peter Crane, 64. Giuseppe Colasante, 65. Gerda Horvat, 66. Lara Strittmatter, 67. Maretta Colasante, 68. Else Marie Friis, 69. Kaj Raunsgaard Pedersen, 70. Gerda van Uffelen, 71. Francois Horvat, 72. Elizabeth Moylan, 73. Anna Wróblewska, 74. Ian Glasspool, 75. Peter Hoen, 76. Maria del Carmen Fernandez, 77. Chumpol Khunwasi, 78. Ana Teresa Romero, 79. Elisabeth Lönn, 80. Melanie de Vore, 81. Clare Green, 82. Maria Suárez-Cervera, 83. Kate Stobart, 84. Elsa Arcalis, 85. Elzbieta Weryszko-Chmielewska, 86. Sarah Kreunen, 87. Nina Gabareyeva, 88. Bill Chaloner, 89. Jadwiga Zebrowska, 90. Simon Hiscock, 91. Louise Archer, 92. Barry Tomlinson, 93. Andrei Pozhidaev, 94. Sharon Anuku, 95. Dianne Edwards, 96. Mark Carine, 97. Janine Victor, 98. James Powell, 99. Yiftach Vaknin, 100. Jeffrey Osborn, 101. Philip Ladd, 102. Marzena Masierowska, 103. Ziba Jamzad, 104. Irina Pestova.

xvii

Gabarayeva, N.I. (2000). Principles and recurrent themes in sporoderm development. In: M.M. Harley, C.M. Morton and S. Blackmore (Editors). Pollen and Spores: Morphology and Biology, pp. 1–16. Royal Botanic Gardens, Kew.

PRINCIPLES AND RECURRENT THEMES IN SPORODERM DEVELOPMENT

NINA I. GABARAYEVA

Komarov Botanical Institute, St.-Petersburg, 197376, Russia

Abstract

The formation of the sporoderm is always initiated from a specific cell coating (glycocalyx). Being a groundwork of the exine, the glycocalyx includes two components: a fibrillar, three-dimensional network and spiral elements. The glycocalyx evidently possesses the properties of colloids. The spiral elements are capable of elongating, stretching in width, and multiplying during microspore ontogeny. The diversity of exine patterns rests on the combination of uniformity of exine substructure and variability of the distribution of sporopollenin receptors. The variety of exine pattern is probably determined at the molecular level and effected, to a considerable extent, by self-assembly, as a consequence of simple physical-chemical properties of colloids. Being secreted in monomolecular form, sporopollenin seems to be accumulated through the glycocalyx in the form of micro-globules. The modes of formation of the homologous layers of exines with similar architectures can differ in different taxa. This paper attempts to synthesise a united system of ideas about the formation of sporoderm structure through a discussion of principles and recurrent themes in sporoderm development.

Introduction

The astonishing diversity and exquisite forms of pollen wall, observed in nature, give us an opportunity to try to understand the ways in which their form is determined and the modes by which their structural variety becomes realised. The mature forms keep their secrets, and only by tracing their development can we have a hope to solve an enigma, at least partly. Indeed, only during ontogeny can the transformations between forms be directly observed, an idea emphasised by many investigators (Heslop-Harrison, 1972; Humphries, 1988; Blackmore and Barnes, 1991).

This study has used a combination of the comparative method and ontogenetic approach, in conjunction with electron microscopy. The consideration in parallel of the development of sporoderm layers and activity of cytoplasmic organelles, and the correlation between exine structure and substructure have been the basis of this work. New and previously published results are summarised and considered together with knowledge obtained by other authors, in an attempt to elaborate a united system of ideas about the formation of sporoderm structure. This is presented in the form of several principles or hypothetical suppositions, followed by corresponding discussions.

1

Materials and Methods

Flower buds of 11 species from 5 families of so-called primitive angiosperms (Magnoliidae: Magnoliaceae - *Michelia figo* (Lour.) Spreng., *Manglietia conifera* Dandy, *Magnolia delavayi* Franch., *Liriodendron chinense* (Hemsl.) Sarg.; Annonaceae - *Asimina triloba* (L.) Dunal; *Anaxagorea brevipes* Benth.; Nymphaeaceae - *Nymphaea capensis* Thunb., *Nymphaea colorata* Peter, *Nymphaea mexicana* Zucc.; Illiciaceae - *Illicium floridanum* Ellis; Schisandraceae - *Schisandra chinensis* (Turcz.) Baill.) and of *Borago officinalis* L. (Boraginaceae), and microsporangia of *Stangeria eriopus* (Kunze) Nash (Stangeriaceae) were collected from the Batumi Botanical Garden (Georgia), the Komarov Botanical Institute and the Bergenian Botanical Garden, Stockholm. The material was fixed in Karnovsky (1965) mixture, or in a fixative consisting of 3% glutaraldehyde and 2.5% sucrose in O.1 M phosphate buffer (pH 7.3, 20°C, 24 hr), and post-fixed in 1-2% osmium tetroxide (pH 7.4, 20°C, 1 hr). After acetone dehydration the samples were embedded either in Spurr's hard resin, or in a mixture of Epon and Araldite. Ultrathin sections were contrasted with a saturated solution of uranyl acetate in ethanol and 0.2% lead citrate. Sections were examined with a TEM Hitachi H-600 and with a Zeiss EM-10A. The cytochemical procedures, revealing the chemical compounds of microspore proexine (Gabarayeva, 1992; Gabarayeva and Rowley, 1994), the enzymatic tests on the contents of endoplasmic reticulum cisternae (Gabarayeva, 1990b), and a trace experiment with lanthanum nitrate (Gabarayeva and Rowley, 1994) have been described elsewhere. For stereological analysis of the dynamics of cell organelles in the developing microspore of *Stangeria eriopus* the methods of Weibel (1969) and Steer (1981) were applied.

Principles of sporoderm development

1. The formation of a sporoderm is always initiated from a specific cell coating - the glycocalyx, or primexine matrix

The term "primexine matrix" was introduced to palynology by Heslop-Harrison (1968), adopted by Dickinson (1976a) and intensively used by many developmental palynologists. The more general term of Bennett (1969) "glycocalyx" was preferred by Rowley (1973), and now is employed by other groups of scientists. It seems to me, however, that both terms actually coincide in sense, because Dickinson, as well as Rowley, regarded this initial microspore surface coat as a framework (skeleton, scaffolding or exine template) for sporopollenin accumulation. The term glycocalyx seems preferable, however, because it is more general (applicable both to plant and animal cells) and implies chemical compounds of cell coating.

The appearance of microspore surface coating, or glycocalyx, as the initial structure on the plasma membrane of a young microspore in a tetrad has been demonstrated by most authors tracing sporoderm development. Usually this initial layer is very well pronounced (for example, in gymnosperms - Lugardon, 1995; Audran, 1981; Kurmann, 1989; in so-called primitive angiosperms - Meyer, 1980; Zavada, 1984; Gabarayeva, 1986, 1991a,b; 1995, 1996), but in some cases it is not so evident (Rowley *et al.*, 1992) or is late in development (Waha, 1987; Gabarayeva, 1992). The reasons for the exceptions can be different, from washing out the glycocalyx layer during fixation (at the early tetrad stage the glycocalyx can be easily removed by aqueous fixatives) to weak contrast of the glycocalyx components, or changing of glycocalyx generation by the process of endocytosis. The latter process has been confirmed by tracer experiment with lanthanum nitrate as a part of uptake process during tetrad stage (Gabarayeva and Rowley, 1994).

2. As a framework of the exine, the glycocalyx includes two components: a fibrillar, three-dimensional network and spiral, radially oriented elements, which seem to be inserted into lumina of this network

A fibrillar network in the periplasmic space between plasma membrane and callose envelope has been recognised in ultrathin sections by many investigators. The second glycocalyx component is represented by discrete units, mainly oriented perpendicular to the plasma membrane. In radial sections through a microspore the entities look like spirals, cut longitudinally (Figs. 1-6). These spiral units are small enough to be overlooked even in optimally fixed and stained material. I refer to these two glycocalyx components as exine substructure (because the term "structure" is already defined, Punt *et al.* 1994). Exine substructure is, in essence, a framework supplied with sporopollenin receptors. During development substructure precedes the appearance of structure. The size of the elements of exine substructure units is 10-40 nm, and the size range of the whole substructural unit is about 100-150 nm. It is difficult to observe substructure in its full complexity, but only one or another part or detail in each particular specimen, the rest is the work of spatial imagination. In this case, the unequivocal interpretation of an object is substituted by a concept of *possibly observed.*

The application of different methods of observation inevitably leads to differing interpretations, so there is nothing strange in the appearance of a number of *models* of exine substructure. Each of these models is different, but they share some common features. For example, the polymerisation-promoting surfaces of primexine matrix which serve like a mould (Heslop-Harrison, 1968; Dickinson, 1976a) have much in common with the boundary layer model (Blackmore and Claugher, 1987). The model of Blackmore (1990) suggests also the existence of the second level of substructure: a fine meshwork of equal units about 15 nm in diameter - a point which brings this model closer to substructure interpretation of Nowicke *et al.* (1986). Other methods involving exine degradation reveal ribbed rods (Claugher and Rowley, 1990) and a system of hollow tubules with spirals as walls (Rowley and Claugher, 1996). Fast atom etching reveals, in *Betula* exine, cylindrical structures with a spacing of 100-150 nm and rods which appear to fit into the cylinders (Claugher and Rowley, 1987). The initial model of Rowley and Flynn (1968) and its further versions (Rowley *et al.*, 1981; Rowley, 1990) represent exine substructural units as complex, radially oriented glycocalyx assemblies, consisting of core and binder subunits. This model for organisation has been confirmed with atomic force and scanning tunnelling microscopy, with the exception of *Alnus*, which exhibits clusters of spheroids (Rowley *et al.*, 1995; Wittborn *et al.*, 1996). These two kinds of elements - elongated straight (rod-like) or coiled (helical) ones, and/or spherical or granular elements - are present in other models of exine substructure: fusing and branching threads (Hesse, 1984), interconnected granules (Southworth, 1985, 1986, based on treatment with 2-amino-ethanol), threads and granules (Takahashi, 1993 - freeze-cleavage method), and globule-like unit structures (Kedves, 1986).

The discrete substructure units, observed in our own studies, were spirals or cylinders with spirals as walls and rod-like cores, that is formations most similar to units of Rowley's model. Evidence has been obtained which shows that not only ectexine, but also the endexine develops on the base of radial glycocalyx units (Rowley, 1987-1988, 1995; Gabarayeva and El-Ghazaly, 1997 - see Fig. 3 in this paper). Globules were also observed in exine development in association with spiral elements. The interpretation of these globule elements will be given below.

3. The microspore glycocalyx is not a static formation and it seems to possess the properties of colloids. Its spiral units are capable of growing in height and in width

Cytochemical tests show the presence of polysaccharides in primexine (Dickinson, 1971), of protein and acidic polysaccharides in the microspore glycocalyx before there

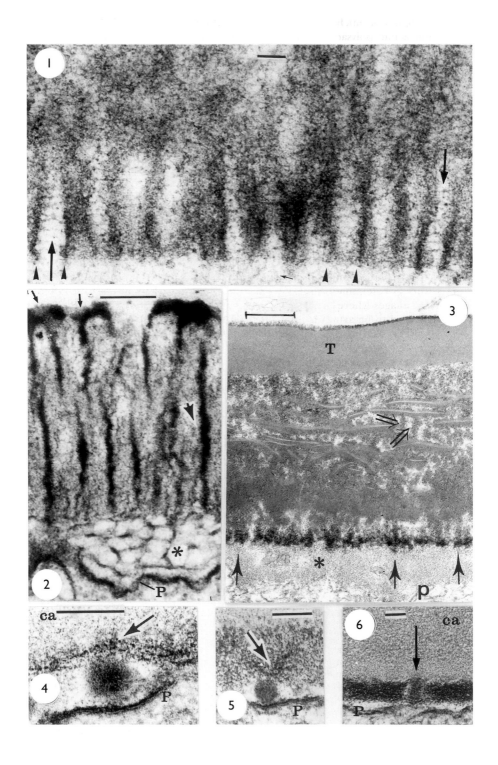

is any, or at least not much, sporopollenin in it (Rowley and Dahl, 1982), and of both neutral and acidic polysaccharides (Southworth, 1990). A number of cytochemical reactions, carried out on sections of *Nymphaea* tetrad microspores, suggest the presence in the glycocalyx and developing exine of protein, acidic polysaccharides, native cations and unsaturated lipids. Lipopolysaccharides embedded within exine at late stages of development were also detected (Rowley, 1975). In terms of chemical compounds of glycocalyx, these results indicate that the glycocalyx is composed of mucopolysaccharides, or glycoproteins, plus lipids in conjunction with protein (lipoproteins) or with polysaccharides (lipopolysaccharides). The supposition was (Gabarayeva, 1990a, 1993), that molecules of these substances form macromolecular aggregates in the hydrophilous medium between the plasma membrane and callose envelope of a tetrad microspore. In other words, the formation of a multiphase system with particles of supramolecular size - of a colloid - takes place in the periplasmic space. The stability of colloids is determined not by chemical interactions between particles, but by physical ones. Then, what we observe as a three-dimensional network and complex spiral units of the glycocalyx, may be, in essence, the phases of a stable colloidal solution.

The difficulty is that radially oriented spiral units of the glycocalyx are mainly seen attached to the plasma membrane, and have been referred to as a part of a single whole — the glycolemma (Abadie *et al*., 1986-1987). Consequently, numerous examples of images where the glycocalyx was detached from the plasma membrane, were considered to be artefacts of fixation until Rowley and Dahl (1977) suggested that glycocalyx units could be separated from the plasma membrane and later reunited to it. Attached and detached portions of the glycocalyx, observed in one and the same microspore, are a common and regularly observed feature in our studies (Gabarayeva, 1986, 1991b, 1995, 1996; Gabarayeva and Rowley, 1994; Gabarayeva and El-Ghazaly, 1997). They may be considered to exhibit the possible mechanism of growth of the glycocalyx units. This mechanism implies periodical movements of the plasma membrane (see below), with corresponding periodical detachment of the glycocalyx units added from the side of the plasma membrane. If so then all the time that the glycocalyx is separated from the plasma membrane, it can be regarded as a colloid. In this case the glycocalyx would have the properties of a colloid, at least in part.

FIG. 1–6. Fig. 1. A fragment of the glycocalyx (cell surface coating) of *Stangeria eriopus* microspore. Close-packed cylindrical glycocalyx units, with spirals as walls, showing loops of spirals as cross striations (arrows). The walls of units are dark contrasted (arrowheads). Tiny arrow points to fine network in the periplasmic space. Bar: 0.1 μm (× 75,000). Fig. 2. Cylinder-like units of *Stangeria eriopus* microspore glycocalyx at the late tetrad stage. The walls of units (arrowhead) and their distal parts (small arrows) are more pronounced and show microglobular nature. The network in the periplasmic space is marked by an asterisk. P (here and in other figures) = plasma membrane. Bar: 0.3 μm (× 45,000). Fig. 3. The spiral radially oriented glycocalyx units (arrows) are still seen in the periplasmic space (asterisk) of *Nymphaea mexicana* microspore at the end of endexine formation (endexine lamellae are shown by outline arrows). T = tectum. (This figure is Fig. 33 from the paper of Gabarayeva and El-Ghazaly, Plant Syst. Evol., 1997, v.204: 1-19). Bar: 0.25 μm (×43,000). Figs. 4–5. Spiral glycocalyx units of a young tetrad microspore of *Nymphaea colorata* (arrows). The proximal parts of spiral units are surrounded by lipoidal globules. Ca = callose. Bars: 0.1 μm (4 - × 180,000; 5 - × 106,000). (These figures are Figs 25A and 26A from the paper of Gabarayeva and Rowley in Nordic Journal of Botany (1994), 14(6): 671-691). Fig. 6. The glycocalyx layer of a young tetrad microspore of *Borago officinalis*. A spiral unit seen as radially oriented cylinder with cross striations is marked by an arrow. Bar: 0.1 μm (× 55,000).

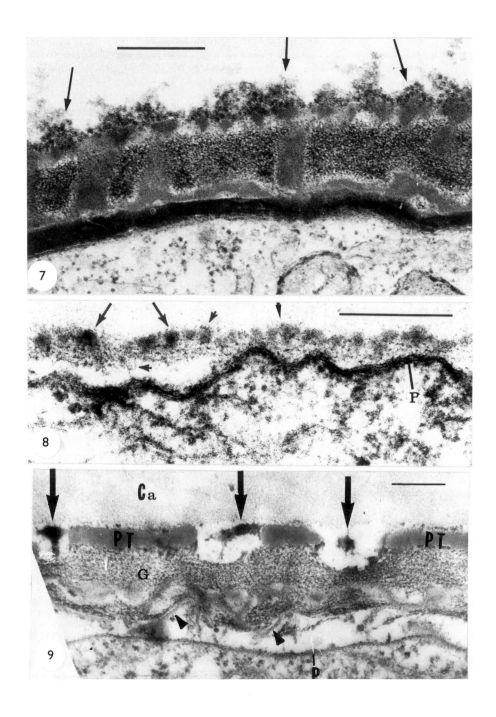

The growth of radially oriented spiral units in height and corresponding thickness of the glycocalyx layer as a whole in the tetrad period is evident. Less evident is the lateral expansion of units. For example, in exine development of *Borago officinalis* it has become clear (Gabarayeva, Rowley and Skvarla, 1998), that funnel-like and cone-like glycocalyx units are not illusions, depending on plane of sections, but are real spatial formations, occurring because the proximal, new-formed part of a unit has undergone less lateral growth than its distal, already expanded part.

4. The diversity of exine patterns rests on the combination of uniformity of exine substructure and variability of the distribution of sporopollenin receptors

The concept "sporopollenin receptor" is rather abstract and can be defined as sites in pro-exine which are specially intended for initial accumulation of sporopollenin. We do not know much about the nature of these receptors, but some suppositions can be put forward. For instance - that these receptors are molecules of enzymes, promoting sporopollenin precursor polymerisation in situ. These enzymes could be separate entities which, at a precise time, become attached to exine substructure units (Gabarayeva, 1990a, 1993), or these receptors might be constituent protein parts (lateral branches) of these units - of complex molecular assemblies (tufts), seen on sections mainly as roundish particles (Skvarla and Rowley, 1987; Rowley, 1990; Rowley, Skvarla and Gabarayeva, in press - see Fig. 7, this paper). These interpretations are very close, if not the same.

The models of exine substructure are different. But, irrespective of what model is closer to reality, the diversity of exine patterns appears not on the level of exine substructure as such, but on the level of selective capacity of this substructure to accumulate sporopollenin, that is, it depends on distribution of sporopollenin receptors through glycocalyx (primexine matrix).

5. The variety of exine pattern is probably determined at the molecular level and carried out, to a considerable extent, by self-assembly, as a consequence of simple physical-chemical properties of colloids

Many studies have been undertaken on spore and pollen walls to elucidate their ultrastructure and development, and much has been achieved recently in terms of exine ultrastructure diversity, but the precise way in which the exine is constructed and how its diversity arises is still obscure. In these circumstances, some new approach might be fruitful, and some hypothetical ways of exine pattern determination should be put forward, even if only in outline. The following approximate way of exine

FIGS. 7–9. Fig. 7. Early free microspore of *Borago officinalis*. There are clusters of dark particles on the tectum which are interpreted as sporopollenin receptors (arrows). Bar: 0.3 μm (× 68,000). Fig. 8. Early tetrad microspore of *Anaxagorea brevipes*. Radially oriented spiral glycocalyx units (arrowheads) are seen in connection with dark contrasted circular spots (arrows), which might be interpreted as lipoidal globules, or as spiral units, seen in top. (This figure is Fig. 13 from Gabarayeva. (1995). Bar: 0.25 μm (× 120,000). Fig. 9. Late tetrad microspore of *Asimina triloba*. The clusters of dark fibrillar substance of tapetal origin (arrows) prevent accumulation of sporopollenin in the places of their localisation (the sites of future exine lumina). In alternative sites, between the clusters, sporopollenin accumulates on the glycocalyx as protectum (PT). G = glycocalyx. Arrowheads point to the first endexine lamellae. Bar: 0.3 μm (× 42,000). (This figure is Fig. 7A Gabarayeva (1992).

structure determination has been proposed (Gabarayeva, 1990a, 1993). The arrangement of a two component (a fibrillar network plus complex spiral units), three-dimensional glycocalyx framework, which is considered to be a colloid, occurs partly by self-assembly, as a result of relatively simple physical-chemical properties of colloids. Sporopollenin receptors (taxon-specific protein molecules, probably - enzymes of sporopollenin polymerisation), being presumably synthesised and released from the rough endoplasmic reticulum of the microspore into the periplasmic space, become inserted into glycocalyx at some definite sites by the way of self-orientation, under the combined influence of all strong and weak intramolecular interactions between protein molecules and macromolecular complexes of the glycocalyx. This self-distribution of sporopollenin receptors through glycocalyx differs in the developing microspore exine of different taxa, as a consequence of different primary, secondary (conformation) and tertiary (rolling up into globule) structure of the protein molecules: a slight difference in primary structure results in considerable difference in protein structure of higher levels. Then, the varying distribution of sporopollenin receptors through glycocalyces causes the diversity in pollen wall structure.

It should be noted that ideas of self-assembly and spontaneously ordered orientation of the units (molecules, microfibrillae, etc.) have been around for a long time (Roland and Vian, 1979; Dickinson and Sheldon, 1986); as far back as 1972 Heslop-Harrison hinted that control must be exerted only at certain strategic points, and physical processes akin to crystallisation must take over to complete the "space-filling" operation. Van Uffelen believes (1991) that spore wall formation is influenced by relatively simple physico-chemical factors, and that a part of the process of spore wall formation is more or less autonomous. It is clear that any ideas on self-assembly should be worked out in co-operation with biochemists, biophysicists, and colloidal chemists. Exactly this approach has been taken by Hemsley and his co-authors (Hemsley *et al*., 1992; Hemsley *et al.*, 1994). In my opinion, their research, including their experimental modelling of exine self-assembly (Hemsley *et al.*, 1996), deserves special attention. Unlike my suppositions (Gabarayeva, 1990a, 1993), referring to the glycocalyx as having the properties of colloids, they (Hemsley *et al.*, 1992) consider sporopollenin to exhibit a colloidal crystal-like organisation in the spore wall. Their main ideas are that sporopollenin is excreted by the tapetum and/or spore cytoplasm in the form of an emulsion. Then all exine patterns can be deposited through self-assembly, by depletion flocculation of regular spherical particles of this emulsion onto long-chained polymer molecules, such as mucopolysaccharides of the spore/microspore surface coat. Mucopolysaccharides or glycoprotein have been reported as a widespread residual component of spore and microspore surfaces (Pettitt and Jermy, 1974; Rowley and Dahl, 1977). Taking this into consideration, the opinion of Hemsley and co-authors (1996) that "...it is the long chain polymer (mucopolysaccharide) chemistry and structure, rather than any characteristics of the sporopollenin, that is responsible for the ultimate arrangement of the exine components" is in accordance with the determinant the role of the glycocalyx and its sporopollenin receptors in exine deposition.

6. Being secreted in monomolecular form, sporopollenin seems to be accumulated through the glycocalyx initially in the form of micro-globules. The receptors of sporopollenin accumulation become distributed through the glycocalyx layer at certain positions that are definable for a species, and the appearance of a taxon-specific pattern occurs. This is probably, by fusion and aggregation of separate sporopollenin micro-globules, localised on glycocalyx units, like a 3-dimensional mosaic

This point is based on numerous observations of lipoidal globules in the periplasmic space of early and mid-tetrad microspores, seen independently (Dickinson, 1976a,b; Hesse, 1985; El-Ghazaly and Jensen, 1986; Gabarayeva, 1986, 1991b) or in connection with spiral glycocalyx units or fibrillar threads (Takahashi, 1993; Gabarayeva and

Rowley, 1994; Gabarayeva, 1995; Gabarayeva and El-Ghazaly, 1997). The diameter of these micro-globules varies between 6-24nm in *Stangeria* (Fig. 2), 24-32nm in *Anaxagorea* (Fig. 8), and 50-80nm in *Nymphaea* (Fig. 4, 5), and might show different degrees of micro-globule coalescence. It should be stressed that we do not know exactly what these lipoidal globules are: they may be sporopollenin, sporopollenin precursors, some lipoproteins, the latter being possible sporopollenin precursors (Gabarayeva, 1990b), or nutritive lipids. Nevertheless, I refer to these micro-globules as the common progenitor of sporopollenin accumulations in exines, since recent data indicate the lipid nature of sporopollenin precursors (Wilmesmeier and Wiermann, 1995). Once they become chemically connected with the glycocalyx receptor sites, they may have become the intrinsic feature of a developing sporoderm. It may be that these initial micro-globules, and the hemispherical surface elements of Hesse (1985), the interconnected granules of Southworth's model (1986), the globular structures of Kedves' model (1986), the granules observed by Takahashi (1993), with the help of freeze-cleavage method, the colloidal sporopollenin spherical dispersion phase of Hemsley and co-authors' model (1994), the pre-granules of Lugardon (1995), and the clusters of spheroids, discovered in *Alnus* exine with atomic force microscopy (Wittborn *et al*., 1996), are essentially the same structures.

7. The modes of the formation of the homologous layers of sporoderm, even with similar architectures, can differ in different taxa. This shows the diversity of modes, by which the morphogenetic tasks are solved in a cell

It is possible to distinguish several morphogenetic modes involved in the process of sporoderm development, among them - the principle of alternativity, the usage of temporal auxiliary structures, and movements of the plasma membrane. The principle of alternativity as a morphogenetic mode is widespread in microspore development. For example, in *Asimina triloba*, in the course of glycocalyx formation, clusters of a substance of tapetal origin take part: where these massulae occur in the glycocalyx layer sporopollenin never accumulates (Fig. 9), and later in development these sites turn to exine lumina (Gabarayeva, 1992). The well-known mode of an aperture development, where the region of future aperture is covered by cisternae of smooth endoplasmic reticulum, preventing exocytosis of Golgi vesicles, and hence the establishment of the glycocalyx layer in this place, is another example of the same principle.

The temporal auxiliary structures could be named "phantom" structures, because they disappear having played their role in morphogenesis. The phantom structures, in the form of strands of a fibrillar substance on the microspore surface of *Illicium floridanum*, exist only in the tetrad period. Sporopollenin accumulates only around them, resulting in the appearance of a system of hollow branching "tunnels" which look like arches in cross sections. Similar temporal auxiliary structures participate in exine ontogeny of *Schisandra chinensis*.

The movements of the plasma membrane within a tetrad microspore were described by many authors, who attached to them a role in the formation of exine architecture (Dickinson, 1971, 1976b; Rowley and Skvarla, 1975; Takahashi and Skvarla, 1991; Takahashi, 1995). These movements are evidently connected with the activity of the cytoskeleton. The microspore plasma membrane movements in different species, or in one and the same species, are probably of various types. The most well-known are those which are connected with the formation of either reticulate patterns of exine, or with the foundation of aperture sites. In both cases these movements, once having taken place, then stop, and the plasma membrane preserves its waving profile for some time, often throughout the tetrad period. The lifting of the formed parts of pro-exine occurs in the post tetrad period, resulting in the appearance of the spatial-convex reticulate pattern which occurs, for example, in *Illicium* and *Schisandra* microspore ontogeny. In the exine ontogeny of *Borago officinalis* retraction of the plasma membrane takes place during the early tetrad stage, determining the sites of the future apertures (Gabarayeva, Rowley and Skvarla, 1998), and leading to spatial angular conformation of the plasma membrane. Such changes in the profile of plasma membrane can be regarded as a result of its macro-movements.

Unlike macro-movements micro-movements, being of small range, probably have a permanent character (the plasma membrane moves all the time, changing the size of the periplasmic space). I suppose that these constant movements of the plasma membrane are necessary in the process of growth of the radial units of the glycocalyx by self-assembly. The images of growing radial glycocalyx units from the side of the plasma membrane are seen during endexine development in *Nymphaea mexicana* (Fig. 3) and in the course of ectexine development in *Stangeria eriopus*. In the case of *Stangeria* the periodic changes in volume to the periplasmic space were confirmed by stereological methods (see following section (8); Grigorjeva and Gabarayeva, 1998).

The modes of intine formation also vary. The most common way of intine deposition is exocytosis of Golgi vesicles with polysaccharide contents. The unusual granular intine of the representatives of Magnoliaceae forms through the activity of specially generated dictyosomes. The widespread channelled intine has always been a subject of special interest (Rowley and Skvarla, 1974), and intine perforations connected with the cytoplasm (Hesse, 1985) or tubular protrusions of the plasma membrane (El-Ghazaly and Jensen, 1987) are similar interpretations. Our understanding of this type of intine is very close; the hypothetical mechanism of its formation was proposed (Gabarayeva, 1995) as a result of parallel activity of Golgi and rough endoplasmic reticulum. The involvement of endoplasmic reticulum in intine construction has been shown for different species (Rodríguez-García and Fernandez, 1990; Hesse, 1991). The similarity of the labyrinth-like intine of *Illicium floridanum* and labyrinths in envelopes of other plant cells (for example, of carnivorous plants - Vassilyev and Muravnik, 1988) is not accidental, but is connected with function (the storage of protein, including enzymes) and the mode of their formation (a direct fusion of endoplasmic reticulum-derived vacuoles, or cisternae, with the plasma membrane).

8. The contribution of endoplasmic reticulum and Golgi to the establishment of the sporoderm layers is considerable but varies depending on developmental stage

The investigations of the role of cytoplasmic organelles in cell wall formation were undertaken by visual estimation and by stereological methods (Weibel, 1969; Steer, 1981). The visual estimation has shown that the involvement of endoplasmic reticulum and Golgi in sporoderm development is especially evident. In the tetrad period smooth endoplasmic reticulum prevails, whereas in free microspore period rough endoplasmic reticulum is predominantly observed in microspore cytoplasm - this seems to be characteristic for many species. It is evidently caused by a different functional role for endoplasmic reticulum in the two periods. In the tetrad period the main events of microspore development are connected with the glycocalyx (surface coat) deposition, and with the initial sporopollenin accumulation through the glycocalyx, resulting in the appearance of pro-exine. The cell organelles most often mentioned in connection with microspore surface coat deposition are dictyosomes and dictyosome-derived vesicles. The latter are indeed plentiful in peripheral cytoplasm. The data received give evidence that various forms of endoplasmic reticulum, among them the unusual aggregates of smooth and rough form as a chain-mail endoplasmic reticulum in young tetrad microspores of *Michelia* and *Manglietia*. A "zebra-endoplasmic reticulum" in the microspores of *Liriodendron* (Gabarayeva, 1991a), are persistently observed in the cytoplasm of the developing microspore, with its cisternae often in close contact with the plasma membrane. However, in the earlier literature, the supposed role of endoplasmic reticulum in the determination of exine structure can scarcely be confirmed. It was shown that smooth endoplasmic reticulum in plant cells participates in synthesis and transport of lipids, terpenoids and steroids (Vassyliev, 1977). My supposition was that smooth endoplasmic reticulum, and its aggregates in tetrad period, synthesised and excreted sporopollenin precursors into the periplasmic space for their initial accumulation through the glycocalyx, and that these substances belonged to the class of lipoprotein. Some evidence was received in favour of this supposition in the course of a cytochemical enzyme experiment (Gabarayeva, 1990b). Rough endoplasmic reticulum prevails, however, in the free microspore period, when formation of intine, and the deposition of self-incompatibility proteins within it, takes place. The activity of rough endoplasmic reticulum in this period is often conjugated with that of Golgi.

Stereological study of the dynamics of the volume density, and absolute volume of 10 microspore components, has confirmed the considerable role of endoplasmic reticulum and Golgi in sporoderm ontogeny (Grigorjeva and Gabarayeva, 1998). The changes of cell components have been studied at four stages of microspore development in *Stangeria eriopus*, from the early mid-tetrad stage to the early free microspore stage. For this period the absolute volume of the developing exine increases three times, and the volume density four times, the growth maximum being observed at the middle tetrad stage — the ontogenetic time of most active increment of the glycocalyx. The absolute volume and volume density of the periplasmic space, which is a centre of all microspore wall growth processes, changes periodically; now increasing, now decreasing. This is in accordance with previously supposed oscillatory movements of the plasma membrane (see section 7). The absolute volume and volume density of dictyosomes has increased 1.5 times by the end of tetrad period and decreased abruptly in the beginning of post tetrad period, when the number of dictyosomes per cell has decreased by 6 times. This confirms the general knowledge about participation of Golgi in formation of the glycocalyx and pro-exine, based on visual estimation. By the completion of pro-exine deposition, when the formation of intine is still not initiated, the presence of Golgi elements has become inconsiderable.

The volume density (about 15% per cytoplasm) and absolute volume of smooth endoplasmic reticulum in *Stangeria* microspore cytoplasm is maximal at the end of mid-tetrad stage, this is in accordance with our opinion on the role of smooth endoplasmic reticulum as a site of synthesis of sporopollenin precursors, the latter accumulate in the glycocalyx exactly at this ontogenetic time.

The quantitative (morphometric) data give evidence of the participation of endoplasmic reticulum and Golgi in exine formation of *Stangeria eriopus*.

Conclusions

To conclude, in brief, the glycocalyx being a framework of exine, includes two components: fibrillar three-dimensional network and spiral elements. The glycocalyx evidently possesses the properties of colloids. Spiral elements are capable of elongating and growing in width. The diversity of exine patterns rests on the combination of uniformity of exine substructure and variability of the distribution of sporopollenin receptors. The variety of exine pattern is probably determined at the molecular level and carried out, to a considerable extent by self-assembly, as a consequence of the simple physical-chemical properties of colloids. Being secreted in monomolecular form sporopollenin seems to be accumulated through glycocalyx, initially in the form of microglobules. The principle of alternativity, and movements of the plasma membrane, are involved in the process of sporoderm development. Stereometrical study of the dynamics of volume density and the absolute volume of the microspore components has confirmed the previously supposed movements of the plasma membrane and suggested the considerable role of endoplasmic reticulum and Golgi in the process of exine development.

Acknowledgements

(This work was partly supported by grants from the Swedish Natural Science Research Council, the Swedish Institute and the Russian Foundation of Fundamental Investigations, No. 96-04-50732).

References

Abadie, M., Hideux, M. and Rowley, J.R. (1986-1987). Ultrastructural cytology of the anther. II. Proposal for a model of exine considering a dynamic connection between cytoskeleton, glycolemma and sporopollenin-synthesis. *Annales des Sciences Naturelles Botanique Paris.* Ser. 13, 8: 1–16.

Audran, J. (1981). Pollen and tapetum development in *Ceratozamia mexicana* (Cycadaceae): sporal origin of the exinic sporopollenin in cycads. *Review of Palaeobotany and Palynology* 33: 315–346.

Bennett, H.S. (1969). Cell surface: components and configurations, and movements and recombinations. In: A. Lima-de-Faria (editor). Handbook of molecular cytology, pp. 1261–1306. North-Holland Publishing Co., Amsterdam.

Blackmore, S. (1990). Sporoderm homologies and morphogenesis in land plants, with a discussion on *Echinops sphaerocephala* (Compositae). *Plant Systematics and Evolution* 5: 1–12.

Blackmore, S. and Barnes, S.H. (1991). Palynological diversity. In: S. Blackmore and S.H. Barnes (editors). Pollen and spores: patterns of diversification, Clarendon Press, pp. 1–8. Oxford.

Blackmore, S. and Claugher, D. (1987). Observations an the substructural organization of the exine in *Fagus sylvatica* L. (Fagaceae) and *Scorzonera hispanica* L. (Compositae: Lactuceae). *Review of Palaeobotany and Palynology* 53: 175–184.

Claugher, D. and Rowley, J.R. (1987). *Betula* pollen grain substructure revealed by fast atom etching. *Pollen et Spores* 29: 5–20.

Claugher, D. and Rowley, J.R. (1990). Pollen exine substructure in *Fagus* (Fagaceae): role of tufts in exine expansion. *Canadian Journal of Botany* 68: 2195–2200.

Dickinson, H.G. (1971). The role played by sporopollenin in the development of pollen in *Pinus banksiana.* In: J. Brooks, P.R. Grant, M. Muir, P. Van Gijzel, G. Shaw (editors). Sporopollenin, pp. 31–67. London.

Dickinson, H.G. (1976a). The deposition of acetolysis-resistant polymers during the formation of pollen. *Pollen et Spores* 18: 321–334.

Dickinson, H.G. (1976b). Common factors in exine deposition. In: I.K. Ferguson and J. Muller (editors). The evolutionary significance of the exine, pp. 67–89. London.

Dickinson, H.G. and Sheldon, J.M. (1986). The generation of patterning at the plasma membrane of the young microspore of *Lilium.* In: S. Blackmore and I.K. Ferguson (editors). Pollen and spores: form and function, Academic Press, London. pp. 1–17.

El-Ghazaly, G. and Jensen, W. (1986). Studies of the development of wheat (*Triticum aestivum*) pollen: formation of the pollen aperture. *Canadian Journal of Botany* 64: 3141–3154.

El-Ghazaly, G. and Jensen, W.A. (1987). Development of wheat (*Triticum aestivum*) pollen. II. Histochemical differentiation of wall and Ubisch bodies during development. *American Journal of Botany* 74: 1396–1418.

Gabarayeva, N.I. (1986). The development of exine in *Michelia fuscata* (Magnoliaceae) in connection with changes of the cytoplasmic organelles of microspore and tapetum. *Botanicheski Zhurnal* (Leningrad), 71: 311–322 (In Russian, with English summary).

Gabarayeva, N.I. (1990a). Hypothetical ways of exine structure determination. *Botanicheski Zhurna* (Leningrad) 75: 1353-1362 (In Russian, with English summary).

Gabarayeva, N.I. (1990b). On the sites of sporopollenin precursor synthesis in developing pollen grains of the representatives of Magnoliaceae. *Botanicheski Zhurnal* (Leningrad), 75: 783–791 (In Russian, with English summary).

Gabarayeva N.I. (1991a). Patterns of development in primitive angiosperm pollen. In: S. Blackmore and S.H. Barnes (editors). Pollen and spores: patterns of diversification, pp. 257–268. Clarendon Press, Oxford.

Gabarayeva, N.I. (1991b). Ultrastructure and development of exine and orbicules in *Magnolia delavayi* (Magnoliaceae) in tetrad and at the beginning of post-tetrad periods. *Botanicheski Zhurnal* (Leningrad) 76: 10–19 (In Russian, with English summary).

Gabarayeva N.I. (1992). Sporoderm development in *Asimina triloba* (Annonaceae). I. The developmental events before callose dissolution. *Grana* 31: 213–222.

Gabarayeva, N.I. (1993). Hypothetical ways of exine pattern determination. *Grana* 33 (suppl. 2): 54–59.

Gabarayeva, N.I. (1995). Pollen wall and tapetum development in *Anaxagorea brevipes* (Annonaceae): sporoderm substructure, cytoskeleton, sporopollenin precursor particles, and the endexine problem. *Review of Palaeobotany and Palynolo* 85: 123–152.

Gabarayeva, N.I. (1996). Sporoderm development in *Liriodendron chinense* (Magnoliaceae): a probable role of the endoplasmic reticulum. *Nordic Journal of Botany* 16: 1–17.

Gabarayeva, N.I. and El-Ghazaly, G. (1997). Sporoderm development in *Nymphaea mexicana* (Nymphaeaceae). *Plant Systematics and Evolution* 204: 1–19.

Gabarayeva, N.I. and Rowley, J.R. (1994). Exine development in *Nymphaea colorata* (Nymphaeaceae). *Nordic Journal of Botany* 14: 671–691.

Gabarayeva, N.I., Rowley, J.R. and Skvarla, J.J. (1998). Exine development in *Borago* (Boraginaceae). 1. Microspore tetrad period. Taiwania 43: 203–214.

Grigorjeva, V.V. and Gabarayeva, N.I. (1998). Dynamics of cell components in developing microspores of *Stangeria eriopus* (Stangeriaceae) in the course of exine formation: a stereological study. *Botanicheski Zhurnal* (St.-Petersburg) 83: 1–11.

Hemsley, A.R., Collinson, M.E. and Brain A.P.R. (1992). Colloidal crystal-like structure of sporopollenin in the megaspore walls of recent *Selaginella* and similar fossil spores. *Botanical Journal of the Linnean Society* 108: 307–320.

Hemsley, A.R., Collinson, M.E., Kovach, W.L., Vincent, B. and Williams, T. (1994). The role of self-assembly in biological systems: evidence from iridescent colloidal sporopollenin in *Selaginella* megaspore walls. *Philosophical Transactions of the Royal Society of London B* 345: 163–173.

Hemsley, A.R., Jenkins, P.D., Collinson, M.E. and Vincent, B. (1996). Experimental modelling of exine self-assembly. *Botanical Journal of the Linnean Society* 121: 177–187.

Heslop-Harrison, J. (1968). Tapetal origin of pollen coat substances in *Lilium. New Phytologist* 67: 779–786.

Heslop-Harrison, J. (1971). Wall pattern formation in angiosperm microsporogenesis. Control mechanism of growth and differentiation: *Symposium of the Society of Experimental Biology*, pp. 277–300. Cambridge.

Heslop-Harrison, J. (1972). Pattern in plant cell wall: morphogenesis in miniature. *Proceedings of the Royal Institution of Great Britain* 45: 335–351.

Hesse, M. (1984). An exine architecture model for viscin threads. *Grana* 23: 69–75.

Hesse, M. (1985). Hemispheric surface elements of exine and orbicules in *Calluna* (Ericaceae). *Grana* 24: 93–98.

Hesse, M. (1991). Cytology and morphogenesis of pollen and spores. In: H.-D. Behnke (editor). Progress in botany, vol. 52, pp. 19–34. Springer Verlag, Berlin, Heidelberg.

Humphries C.J. (1988). Ontogeny and systematics. Columbia University Press, New York.

Karnovsky, M.J. (1965). A formaldehyde-glutaraldehyde fixative of high osmolarity for use in electron microscopy. *Journal of Cell Biology* 27: 137A.

Kedves, M. (1986). In vitro destruction of the exine of recent palynomorphs I. *Acta Biologica.* Szeged 32: 49–60.

Kurmann, M.H. (1989). Pollen wall formation in *Abies concolor* and a discussion on wall layer homologies. *Canadian Journal of Botany* 67: 2489–2504.

Lugardon, B. (1995). Exine formation in *Chamaecyparis lawsoniana* (Cupressaceae) and a discussion on pteridophyte exospore and gymnosperm exine ontogeny. *Review of Palaeobotany and Palynology* 85: 35–51.

Meyer, N.R. (1980). Probable trends of sporoderm evolution in gymnosperms and some angiosperms. In: S.G. Zhilin (editor). Systematics and evolution of higher plants, pp. 86–92. Nauka, Leningrad.

Nowicke, J.W., Bittner J.L. and Skvarla J.J. (1986). *Paeonia*, exine substructure and plasma ashing. In: S. Blackmore and I.K. Ferguson (editors). Pollen and spores: form and function, pp. 81–95. London.

Pettitt, J.M. and Jermy, A.C. (1974). The surface coat on spores. *Biological Journal of the Linnean Society* 6: 245–257.

Punt, W., Blackmore, S., Nilsson, S., Le Thomas, A. (1994). Glossary of pollen and spore terminology. Laboratory of Palaeobotany and Palynology Foundation, Utrecht, Heidelberglaan, The Netherlands.

Rodríguez-García, M.I. and Fernandez, M.C. (1990). Ultrastructural evidence of endoplasmic reticulum changes during the maturation of the olive pollen grain (*Olea europaea* L., Oleaceae). *Plant Systematics and Evolution* 171: 221–231.

Roland, J.C. and Vian, B. (1979). The wall of the growing plant cell: its three-dimensional organization. *International Review of Cytology* 61: 129–167.

Rowley, J.R. (1973). Formation of pollen exine bacules and microchannels on a glycocalyx. *Grana* 13: 129–138.

Rowley, J.R. (1975). Lipopolysaccharide embedded within the exine of pollen grains. 33rd Annual proceedings of the Electron Microscopy Society of America. Las Vegas. Nevada, p. 572.

Rowley, J.R. (1987-1988). Substructure within the endexine, an interpretation. *Journal of Palynology* 23–24: 29–42.

Rowley, J.R. (1990). The fundamental structure of the pollen exine. *Plant Systematics and Evolution* 5: 13–29.

Rowley, J.R. (1995). Are the endexines of pteridophytes, gymnosperms and angiosperms structurally equivalent? *Review of Palaeobotany and Palynology* 85: 13–34.

Rowley, J.R. and Claugher, D. (1996).Structure of the exine of *Epilobium angustifolium* (Onagraceae). *Grana* 35: 79–86.

Rowley, J.R. and Dahl, A.O. (1977). Pollen development in *Artemisia vulgaris* with special reference to glycocalyx material. *Pollen et Spores* 19: 169–297.

Rowley, J.R. and Dahl, A.O. (1982). A similar structure for tapetal surface and exine "tuft"-units. *Pollen et Spores* 24: 5–8.

Rowley, J.R., Dahl, A.O., Sengupta, S. and Rowley, J.S. (1981). A model of exine substructure based on dissection of pollen and spore exines. *Palynology* 5: 107–152.

Rowley, J.R. and Flynn, J.J. (1968). Tubular fibrils and the ontogeny of the yellow water lily pollen grain. *Journal of Cell Biology* 39: 159.

Rowley, J.R., Flynn J.J. and Takahashi, M. (1995). Atomic force microscope information on pollen exine substructure in *Nuphar*. *Botanica Acta* 108: 300–308.

Rowley, J.R. and Skvarla, J.J. (1974). Origin of the inner intine in pollen of *Canna*. 32nd Annual Proceedings of the Electron Microscopy. Society of America, St. Louis, Missouri: Arceneaux, pp. 84–85.

Rowley, J.R. and Skvarla, J.J. (1975). The glycocalyx and initiation of exine spinules in microspores of *Canna*. *American Journal of Botany* 62: 479–485.

Rowley, J.R., Skvarla, J.J. and Gabarayeva, N.I. (1999). Exine development in *Borago* (Boraginaceae). 2. Free microspore stages. Taiwania 44: 212–229.

Rowley, J.R., Skvarla, J.J. and Petitt, J.M. (1992). Pollen wall development in *Eucommia ulmoides* (Eucommiaceae). *Review of Palaeobotany and Palynology* 70: 297–323.

Skvarla, J.J. and Rowley, J.R. (1987). Ontogeny of pollen in *Poinciana* (Leguminosae). I. Development of exine template. *Review of Palaeobotany and Palynology* 50: 313–331.

Southworth, D. (1985). Pollen exine substructure. 1. *Lilium longiflorum*. *American Journal of Botany* 72: 1274–1283.

Southworth, D. (1986). Substructural organization of pollen exines. In: S. Blackmore and I.K. Ferguson (editors). Pollen and spores: form and function, Academic Press, London. pp. 61–69.

Southworth, D. (1990). Exine biochemistry. In: S. Blackmore and R.B. Knox (editors). Microspores. evolution and ontogeny, pp. 193–212. Academic Press, London.

Steer, M.W. (1981). Understanding Cell Structure. Cambridge University Press.

Takahashi, M. (1993). Exine initiation and substructure in pollen of *Caesalpinia japonica* (Leguminosae: Caesalpinioideae). *American Journal of Botany* 80: 192–197.

Takahashi, M. (1995). Exine development in *Aucuba japonica* Thunberg (Cornaceae). *Review Palaeobotany and Palynology* 85: 199–205.

Takahashi, M. and Skvarla, J.J. (1991). Exine pattern formation by plasma membrane in *Bougainvillea spectabilis* Willd. (Nyctaginaceae). *American Journal of Botany* 78: 1063–1069.

Van Uffelen, G.A. (1991). The control of spore wall formation. In: S. Blackmore and S.H. Barnes (editors). Pollen and spores: patterns of diversification, pp. 89–102. Clarendon Press, Oxford.

Vassilyev, A.E. (1977). Functional Morphology of the Secretory Plant Cells. Nauka, Leningrad.

Vassilyev, A.E. and Muravnik, L.E. (1988). The ultrastructure of the digestive glands in *Pinguicula vulgaris* L. (Lentibulariaceae) relative to their function. II. The changes on stimulation. *Annals of Botany* 62: 343–351.

Waha, M. (1987). Sporoderm development of pollen tetrads in *Asimina triloba* (Annonaceae). *Pollen et Spores* 29: 31–44.

Weibel, E.R. (1969). Stereological principles for morphometry in electron microscopic cytology. *International Review of Cytology* 26: 235–302.

Wilmesmeier, S. and Wiermann, R. (1995). Influence of EPTC (s-ethyl-dipropyl-thiocarbamate) on the composition of surface waxes and sporopollenin structure in *Zea mays*. *Journal of Plant Physiology* 146: 22–28.

Wittborn, J., Rao, K.V., El-Ghazaly, G. and Rowley, J.R. (1996). Substructure of spore and pollen grain exines in *Lycopodium, Alnus, Betula, Fagus* and *Rhododendron. Grana* 35: 185–198.

Zavada, M.S. (1984). Pollen wall development of *Austrobaileya maculata. Botanical Gazette* 145: 11–21.

El-Ghazaly, G., Moate, R., Huysmans, S., Skvarla, J. and Rowley, J. (2000). Selected stages in pollen wall development in *Echinodorus, Magnolia, Betula, Rondeletia, Borago* and *Matricaria*. In: M.M. Harley, C.M. Morton and S. Blackmore (Editors). Pollen and Spores: Morphology and Biology, pp. 17–29. Royal Botanic Gardens, Kew.

SELECTED STAGES IN POLLEN WALL DEVELOPMENT IN *ECHINODORUS, MAGNOLIA, BETULA, RONDELETIA, BORAGO* AND *MATRICARIA*

GAMAL EL-GHAZALY[1], ROY MOATE[2], SUZY HUYSMANS[3], JOHN SKVARLA[4] AND JOHN ROWLEY[5]

[1]Swedish Museum of Natural History, Palynological Laboratory, SE-104 05 Stockholm, Sweden
[2]University of Plymouth, Faculty of Science, PL4 8AA Plymouth, UK
[3]Katholieke Universiteit Leuven, Botanical Institute, Laboratory of Plant Systematics, B-3001 Heverlee, Belgium
[4]University of Oklahoma, OK 73019 NORMAN
[5]University of Stockholm, Department of Botany, SE-106 91 Stockholm, Sweden

In *Echinodorus cordifolius, Magnolia grandiflora, Betula pendula* and *Rondeletia odorata*, in tetrad period different patterned arrays of tubular probaculae condense around the protrusions of the plasmalemma. Oblique sections of the exine of young free microspores show that the tectum consists of the distal portions of close-packed bacular units that appear as a honeycomb pattern. When the foot layer forms in *Echinodorus* and *Magnolia*, after the microspores are released from callose envelopment, there is a white line at its inner surface. As the early exine thickened white lines were seen at the inner surface of the foot layer and still later in microspore development between the foot layer and a zone which we called transitory or rudimentary endexine. In the dicotyledon species the endexine is well-developed and consists of several lamellae.

Cryo-SEM study of *Betula* microspores and pollen grains shows a well-organised network of strands connecting the tapetal cells with the exine. The filaments are generally connected to the microchannels of the exine. Since the plasmalemma is apparently protruded and lining the microchannels, we then assume that the network of filaments are connecting the plasmalemma of the microspores through the microchannels with the plasmalemma of the tapetal cells.

Introduction

The development of pollen wall, origin and control of the wall pattern and the source and nature of sporopollenin, are among the subjects that have attracted attention in the past and continued to do so. The diversity in exine ontogeny indicates the potential of information on development for research on phylogeny and evolution.

In early tetrad stages of microspores, when the tectum is first evident, the exine units show a honeycomb pattern resulting from close packing and interdigitation of the units. El-Ghazaly and Jensen (1985, 1986a) show the occurrence of such a honey comb arrangement in *Triticum* (Poaceae) in early stages of microspore development. Huysmans *et al.* (1998) observed a similar pattern in *Rondeletia odorata* (Rubiaceae).

Wodehouse (1935) pointed out that a reticulate pattern is a common theme in nature, and is formed where even shrinkage occurs within a uniform matrix.

Patterned arrays of tubular probaculae then condense within the primexine and these subsequently develop into mature wall elements by the accumulation of sporopollenin. This sequence gives the idea that the primexine contains receptors that promote sporopollenin polymerisation (Dickinson 1970; Dickinson and Sheldon 1986; Blackmore 1990; Blackmore and Barnes 1990; Heslop-Harrison 1962; Heslop-Harrison 1968a, b, c; Rowley 1973; Rowley and Dahl 1982).

In the microspore tetrad of *Echinodorus* (Alismataceae), during the callose period, the exine begins as rods originating from the plasmalemma. These rods are exine units that upon further development become baculae as well as part of the tectum and foot layer. Oblique sections of the early exine show that the tectum consists of the distal portions of close-packed exine units.

The patterned sexine (tectum, baculae and foot layer) is formed within the primexine. A reticulum formed of hexagonal units is one of the most common patterns encountered in biological systems. This pattern also occurs in the pollen wall generation mechanism, for example, in *Lilium* (Dickinson 1970), in *Triticum* (El-Ghazaly and Jensen 1985), and in *Echinodorus* (El-Ghazaly and Rowley 1999).

In our work on the development of selected stages of microspores and pollen grains we include *Echinodorus* (Alismataceae), *Magnolia* (Magnoliaceae), *Betula* (Betulaceae), *Rondeletia* (Rubiaceae), *Borago* (Boraginaceae) and *Matricaria* (Asteraceae). This work is an attempt to present different ideas and interpretation of pollen wall development and substructure.

Material and Methods

For SEM a few anthers of *Magnolia grandiflora* were fixed in 1% GA for 5 minutes, dehydrated in an alcohol series, frozen with liquid nitrogen, fractured, sputter coated with gold for 1 minute, and then examined with a JEOL 6300.

For cryo-SEM pollen grains from *Betula pendula* (syn. *B. verrucosa*) were collected at the beginning of April until the end of 1998 in the Stockholm area. Some anthers at free microspore stage were prepared for Cryo-SEM technique. The anthers were treated with 1% glutaraldehyde, or with 0.05 M sodium cacodylate buffer at pH 7.5 (Maldonado *et al.* 1986) or deposited on wet filter paper. They were then placed on SEM stubs, attached to the stub by silver paint, frozen in liquid nitrogen and transferred to a cryo unit where they kept at –180°C, fractured, etched, coated with gold and examined with SEM and photographed at under –140°C.

For TEM, fresh anthers from different species, at different developmental stages, were treated with a mixture of 1% paraformaldehyde and 3% glutaraldehyde in 0.05 M Na-cacodylate buffer, pH 7.4, at room temperature, for 24 hours, postfixed with 2% OsO_4 for 24 hours. Anthers were then dehydrated in a graded ethanol series, embedded in Spurr's medium (Spurr 1969) and sectioned. Sections were treated with 1% aqueous uranyl acetate for 5 minutes, washed with distilled water, and floated on lead citrate stain for 5 minutes. Sections were observed with a Zeiss EM 906 at the Botany Department, Stockholm University, Sweden.

Results

Borago, Matricaria and *Echinodorus*

In early stages the exine-units (columellae) have a light contrasted core zone and dark binder zone (for example, in *Borago*, Figs. 1 and 2) preceded or followed by a dark core and less dark binder zone (for example, *Matricaria*, Figs. 3 and 4). When tectal components are initiated at the beginning of exine development they are close packed. Consequently the junction between tectal components is hexagonal or polygonal (Figs. 2 and 4).

The columellae and primexine matrix are ca 100 nm high in interapertural sectors (for example, *Echinodorus*, Fig. 5). They taper down to < 25 nm at the margin of apertures.

Magnolia

The young microspore of *Magnolia* in Fig. 6 is in a stage prior to the vacuolate period. The plasmalemma and cytoplasm protrude across a pericytoplasmic space and contact the exine. We consider these cytoplasmic protrusions to be involved with uptake of nutrients and other material from the loculus. It is evident that, where the foot layer is periodically thick and thin, the thin regions are underlain by glycocalyx.

The late free microspore of *Magnolia* in Fig. 7 is in a stage at the beginning of the vacuolate period. Vesicular-fibrillar material is present in the periplasmodial space. There is evidence that this material is attached to the plasmalemma and may occur in conjunction with sites where cisternae of the ER "contact" the plasmalemma. In many sites the "vesicles" are adjacent to one another or joined to show as an elongate section of a conduit form. There is considerable variation in the thickness of the foot layer and endexine, as at the younger stage in the above figure.

The decrease in thickness of the endexine where the foot layer is thick is prominent in the young pollen grain of *M. grandiflora* in Fig. 8. The membranous boundary of the vesicular-fibrillar components is more apparent here than in Fig. 7. These components in the zone below the exine and above the intine in Fig. 8 are probably in transition between the exine and cytoplasm and are not properly assigned to either endexine or intine. The spheroidal inclusions within the intine are also considered to represent material in transition between the loculus, exine and cytoplasm.

Rondeletia

Cytoplasmic protrusions extending out to the exine are numerous in late microspores. In micrographs of such stages there are apparent contacts between the ER cisternae and these cytoplasmic protrusions (Figs. 9 and 10). Later in microspore development cytoplasmic protrusions remain numerous but their continuity to the endexine and foot layer is not so evident as earlier (Fig. 11).

In mature pollen of *R. odorata* the endexine components splay out over the pore (Fig. 12). In such regions the endexine can be seen to consist of tubular components with a core that is low in contrast (Fig. 13).

Strands extending between tapetal cells and microspores in *Magnolia* and *Betula*

Figs. 14, 15 and 16 illustrate strands extending from tapetal cells to the surface of microspores in locules. The diameter of such strands is much greater in the cryo prepared material of *Betula* (Figs. 15 and 16) than in the chemically prepared material of *Magnolia* (Fig. 14). In Fig. 16 it is evident that the tapetal to microspore strands do not occur on the fractured surfaces of microspores or tapetal cells.

Pollen and Spores: Morphology and Biology

PLATE 1

FIGS 1–2. Mid-microspore tetrad stage in *Borago officinalis* L. The columellae in Fig. 1 are sectioned radially. The exine-units that make-up the columellae have a low contrasted core (arrowheads) inside a strongly contrasted binder zone at this stage of development. The oblique section in Fig. 2 shows that the distal (tectal) portions of these exine-units are close packed, forming a honey-comb or hexagonal pattern. Fig. 1, Bar = 0.5 μm; Fig. 2, Bar = 1 μm. (Fig. 1=20,000×; Fig. 2= 50,000×). FIGS 3–4. Early microspore tetrad stage in *Matricaria inodora* L. The closely spaced exine units will become columellae. Fig. 3 shows the units in longitudinal section while the units were sectioned transversely in Fig. 4. The core zone of units is darkly contrasted and can be seen as circular in transverse section and rod-shaped in longitudinal sections. In Fig. 4 several units are obliquely sectioned and show as rods (arrowhead) as in Fig. 3. For both figures the Bar = 1 μm. (22,000×). FIG. 5. Young microspore tetrad in *Echinodorus cordifolius* (L) Griseb. enveloped in callose (C). The tapetal cell (T) is rich with organelles. Exine units can be seen to be circular in the microspore (M-1) because of the oblique plane of section. In the transverse section of the microspore (M-2) exine units can be seen to be rod-shaped (arrow). Both sections include an aperture (A). Bar = 1 μm. (18,000×).

PLATE 2

FIG. 6. Free microspore of *Magnolia grandiflora* prior to vacuolation. Protrusions of the plasmalemma are distinct (arrows); these protrusions contact the exine. There is great variation in the thickness and shape of the foot layer (F) and endexine (arrowheads). There is glycocalyx material (asterisks) in the columellar arcade and evaginations of the foot layer/endexine. The cytoplasm contains large vesicles filled by granules apparently made up of fibrillar material. Bar = 1 μm (56,000×). FIG. 7. Late free microspore (early vacuolate) stage of *M. grandiflora* showing many coarse vesicular-fibrillar components in the periplasmic space (asterisk) between the plasmalemma and the thin endexine. Circular frame marks a site where there is contact between the vesicular components and the plasmalemma. The endexine is separated from the foot layer (F) by a white line lamellation (arrowhead). There are several cisternae of the ER that appear to be in contact with the plasmalemma (arrow). Bar = 1 μm (32,000×).

PLATE 3

FIG. 8. Young pollen grain of *Magnolia grandiflora* showing the great enlargment of the zone of vesicular-fibrillar material (arrowheads) and development of the intine (I). The membranous, and tubular boundary, of the vesicular-fibrillar material is more evident than earlier in Fig. 7. The inclusions (arrow) within the intine appear to be spheroidal in sections. There is variation in the thickness of foot layer (F) and endexine (E). The cytoplasm contains many mitochondria (star). Bar = μm (33,000×).

PLATE 4

FIG. 9. Late microspore of *Rondeletia odorata* with a great many plasmalemma protrusions extending outward to the exine. Bar = 1 μm (33,000×). FIG. 10. An enlargement of the periplasmodial space in Fig. 9 showing ER contacts with plasmalemma protrusions (arrows). Bar = 0.5 μm (68,000×). FIG. 11. Early pollen grain stage of *R. odorata* with beginning of intine development (I). Plasmalemma protrusions are reduced or masked after intine development but in this micrograph the protrusions are evident, as is the proximity of ER cisternae to them (arrowhead). Bar = 1 μm (17,500×).

PLATE 5

FIG. 12. Pollen grain stage in *Rondeletia odorata* with the generative cell and nucleus (N) in the plane of section. The medial section of the aperture marked (A) passes through the pore and shows the somewhat separated elements of the endexine. Bar = 1 μm (5,000×). FIG. 13. Section of an aperture of *R. odorata* similar to the one marked (A) in Fig. 14. The endexine elements show a medial white line (arrows) and a circular profile with a low contrasted core (arrowheads) in cross-sections. Bar = 1μm (40,000×).

PLATE 6

FIG. 14. A section of *Magnolia grandiflora* microspores (M) adjacent to tapetal cells (T). There are many strands between tapetal cells and the microspores and around microspores. Compare the results of this chemical fixation of *Magnolia* with the low temperature stablisation and SEM micrographs in microspores of *Betula* in the following figures. Bar = 10 μm (2,300×).

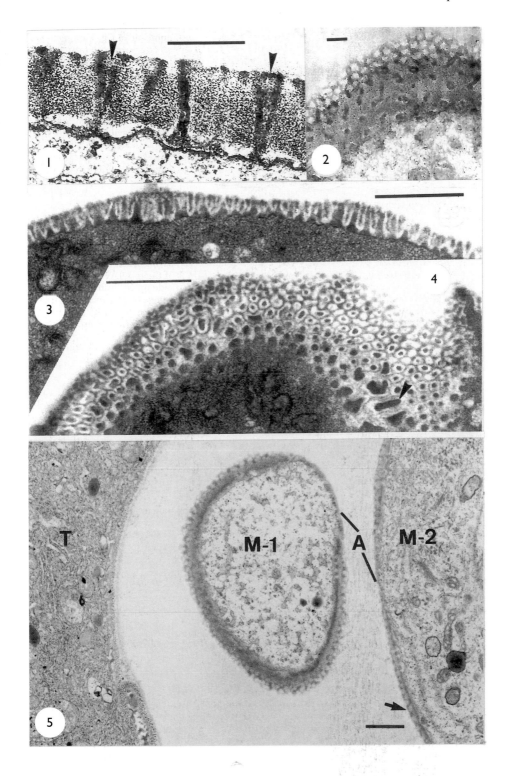

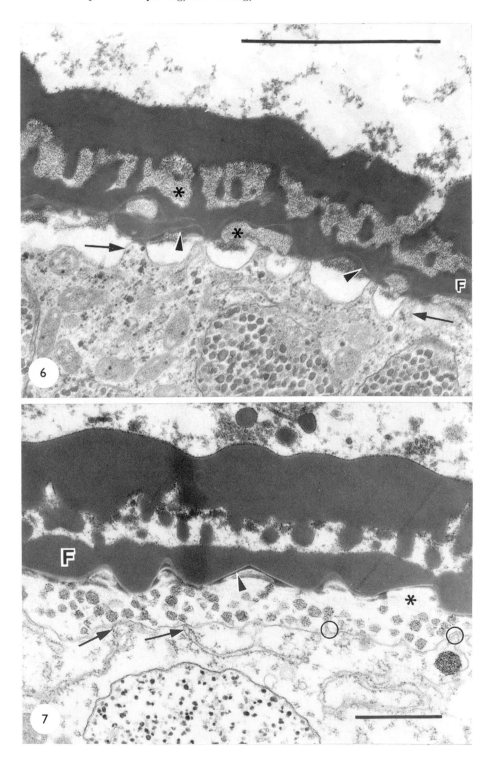

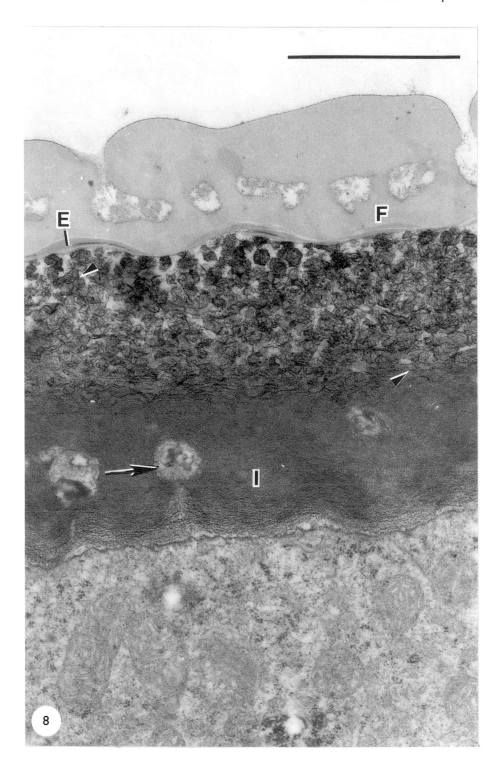

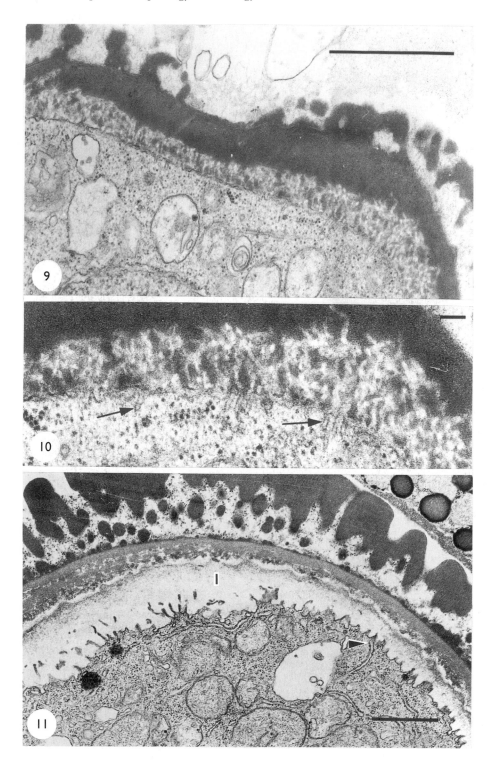

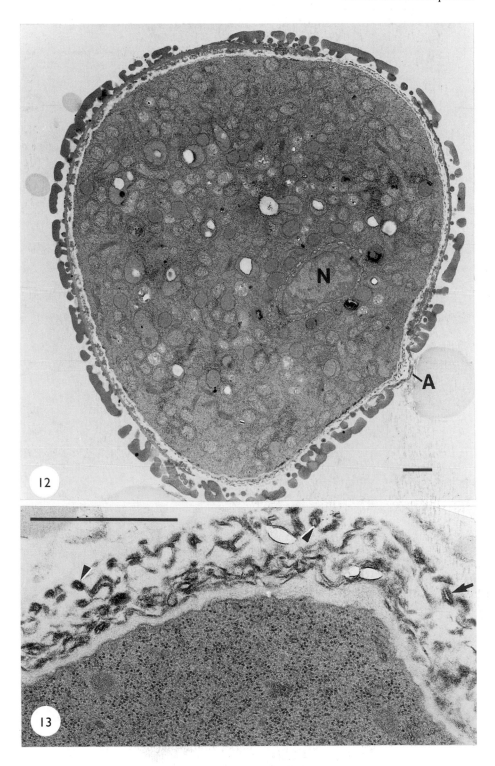

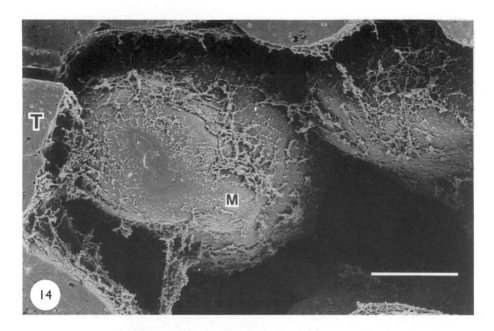

Discussion

Blackmore and Claugher (1984, 1987) used fast atom bombardment in studying the exines of *Fagus* (Fagaceae) and *Scorzonera* (Compositae) and found that the exines were composed of hollow tubes. Blackmore (1990) in a developmental study of pollen of *Echinops* (Compositae) found that exine processes appeared to be hollow during early stages.

In our work with *Echinodorus* we found that there were cylindrical units in the exine having a low contrast core zone (El-Ghazaly and Rowley 1999). Units with a low contrast core in *Borago* are seen in Figs. 1 and 2. Nowicke *et al.* (1986) used plasma-ashing on many species of *Paeonia* (Paeoniaceae) and found that rod-shaped substructures were evident in pollen whereas they have not been observed before ashing.

Figs. 9 and 10 show apparent contacts between plasmalemma protrusions and ER and the exine. Heslop-Harrison (1963) showed the presence of cisternae of the endoplasmic reticulum under the plasmalemma in connections with the conduits of columellae. Similar observations were reported by Skvarla and Larson (1966) as part of their ontogenetic study of *Zea mays* L. The interpretation then was that there might be a connection between sites of columellar initiation and ER cisternae. Our interpretaton now is that early columellae are part of a transfer system. The ER connection with early columellae is seen as part of an uptake problem.

The localisation of initiation sites of columellae in the plasmalemma would seem to be a problem requiring information about early exine development. In the three examples cited here units of the early exine are close-packed, i.e., in *Borago*, *Matricaria* and *Echinodorus*. For these taxa what is important for final exine form is columellar location with respect to exine pattern, sequences of middle and later development rather than the initiation of exine-units.

In the endexine of mature pollen grains of *Rondeletia* the endexine of apertures shows lamellations that have white line centres in longitudinal profile, and a low contrast central zone in cross sections. In these apertural zones the endexine components are tubular with a low contrast core. In the *Poinciana* (Leguminosae)

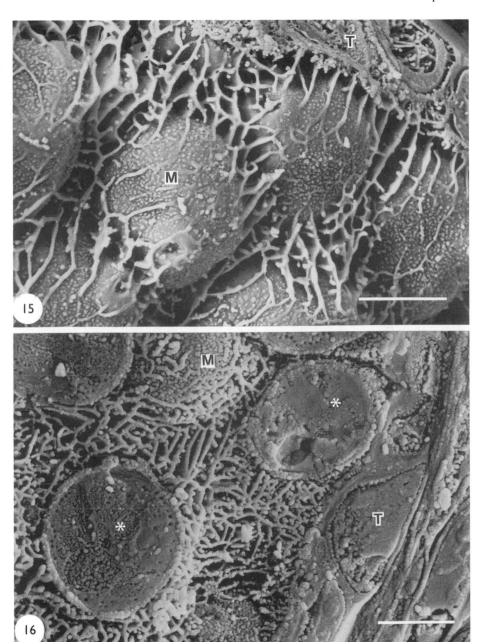

FIGS. 15–16. Fig. 15. Low temperature (-180°C) preservation of living *Betula* microspores followed by cryo-SEM (180°C). The tapetal cells (T) have been sectioned but the microspores (M) show intact exine surfaces. Strands from the surface of tapetal cells connect to surface of microspores (presumably to microchannels). Bar = 10 μm (3,000×). Fig. 16. Preservation and microspore stage the same as in Fig. 15. Both anther wall and tapetal (T) cells are sectioned. The micrograph shows fractured microspores (asterisks) and intact exine surfaces (M). The strands between tapetal cells and microspores are the loculus and surfaces of microspores and tapetal cells. They do not occur on the cut cytoplasm of either tapetal cells or microspores. General organelles profiles are apparent in the cut surface of cytoplasm. Bar = 10 μm (2,300×).

Rowley and Skvarla (1987) showed that endexine components in apertural regions were tubular with a low contrast central core. In interapertural regions, where white lines in the endexine are commonly believed to represent lamellae, a compromise interpretation might be fascicles of tubular form.

Strands between tapetal cells have had a limited reception except for the viscin threads of Onagraceae and Ericaceae (Skvarla et al. 1978). Hesse (1981) has reviewed reports of viscin thread-like strands in several other taxa. Rowley (1963; see also Fig. 12 in Erdtman 1969) reported strands between tapetal cells, Ubisch bodies and spinules for *Poa annua* L. microspores.

The strands shown in chemically fixed microspores of *Magnolia* are greatly shrunken and entangled compared with those of the cryo prepared and cryo-SEM examined *Betula* in Figs. 15 and 16. The results from quite different fixations for these distantly related taxa are, however, complementary with respect to the common occurrence of strands connecting tapetal cells and microspores. The function of such strands requires work with tracers although transport of nutrients is a reasonable supposition.

Acknowledgements

This research was partially supported by a grant from the Swedish Royal Academy of Science to G. El-Ghazaly.

References

Blackmore, S. (1990). Sporoderm homologies and morphogenesis in land plants, with a discussion of *Echinops sphaerocephala* (Compositae). *Plant Systematics and Evolution* (Supplementum 5): 1–12.

Blackmore, S. and Barnes, S.H. (1990). Pollen wall development in angiosperms. In: Blackmore S, Knox RB (eds). Microspores: evolution and ontogeny. Academic Press, London, pp. 173–192.

Blackmore, S. and Claugher, D.(1984). Ion beam etching in palynology. *Grana* 23: 85–89.

Blackmore, S. and Claugher, D. (1987). Observations on the substructural organisation of the exine in *Fagus sylvatica* L. (Fagaceae) and *Scorzonera hispanica* L. (Compositae: Lactuceae). *Review of Palaeobotany and Palynology* 53: 175–184.

Claugher, D. and Rowley, J.R. (1987). *Betula* pollen grain substructure revealed by fast atom etching. *Pollen et Spores* 29: 5–20.

Dickinson, H.G. (1970). Ultrastructural aspects of primexine formation in the microspore tetrad of *Lilium longiflorum*. *Cytobiology* 1: 437–449.

Dickinson, H.G. and Sheldon, J.M. (1986). The generation of patterning at the plasma membrane of the young microspore of *Lilium*. In: Blackmore S, Ferguson I.K. (eds). Pollen and spores: form and function. Academic Press, London, pp. 1–17.

El-Ghazaly, G. and Jensen, W. (1985). Studies of the development of wheat (*Triticum aestivum*) pollen: III. Formation of microchannels in the exine. *Pollen et Spores* 27: 5–14.

El-Ghazaly, G. and Jensen, W. (1986a). Studies of the development of wheat (*Triticum aestivum*) pollen: I. Formation of the pollen wall and Ubisch bodies. *Grana* 25: 1–29.

El-Ghazaly, G. and Rowley, J. R. (1999). Microspore and tapetal devlopment in *Echinodorus cordifolius* (Alismataceae). *Nordic Journal of Botany* 19(1): 101–120.

Erdtman, G. (1969). Handbook of Palynology. Munksgaard, Copenhagen.

Heslop-Harrison, J. (1962). Origin of exine. *Nature* 195: 1069–1071.

Heslop-Harrison, J. (1963). An ultrastructural study of pollen wall ontogeny in *Silene pendula. Grana Palynologica* 4: 7–24.

Heslop-Harrison, J. (1968a). Tapetal origin of pollen-coat substances in *Lilium. New Phytology* 67: 779–786.

Heslop-Harrison, J. (1968b). Pollen wall development. *Science* 16: 230–237.

Heslop-Harrison, J. (1968c). Wall development within the microspore tetrad of *Lilium longiflorum. Canadian Journal of Botany* 46: 1185–1192.

Hesse, M. (1981). Viscinfäden bei Angiospermen - homologe oder analoge Gebilde? *Mikroskopie* (Wien) 38: 85–89.

Huysmans, S., El.Ghazaly, G. & Smets, E. (1998). Pollen wall, tapetum and orbicule development in *Rondeletia odorata* (Rubiaceae). In: Pollen and spores: morphology and biology, London 1998. Abstracts. P.11. Royal Botanic Gardens, Kew and Natural History Museum, London.

Nowicke, J.W., Bittner, J.L. & Skvarla, J.J. (1986) *Paeonia*, exine substructure and plasma ashing. In: S. Blackmore & I.K. Ferguson, Pollen and spores: form and function (eds), pp. 81–95. Linnean Society Symposia Series 12, Academic Press, London.

Rowley, J. R. 1963. Nonhomogeneous sporopollenin in microspores of *Poa annua* L. *Grana Palynology* 3(3): 3–20.

Rowley, J.R. (1973). Formation of pollen exine bacules and microchannels on a glycocalyx. *Grana* 13: 129–138.

Rowley, J.R. and Dahl, A.O. (1982). A similar substructure for tapetal surface and exine tuft-units. *Pollen et Spores* 24: 5–8.

Rowley, J. R. and Skvarla, J. J. (1987). Ontogeny of pollen in *Poinciana* (Leguminosae). II. Microspore and pollen grain periods. *Review of Palaeobotany and Palynology* 50: 313–331.

Skvarla, J. J. and Larson, D. A. (1966). Fine structural studies of *Zea mays* pollen. I. Cell membrane in exine ontogeny. *American Journal of Botany* 52: 1112–1125.

Skvarla, J. J., Raven, P. H., Chissoe, W. F. and Sharp, M. (1978). An ultrastructural study of viscin threads in Onagraceae pollen. *Pollen et Spores* 20: 5–143.

Spurr, A.R. (1969) A low-viscosity epoxy resin embedding medium for electron microscopy. *Journal of Ultrastructural Research* 26: 31–43

Wodehouse, R.P. (1935). Pollen grains. Their structure, identification and significance in science and medicine. McGraw-Hill, New York.

Hemsley, A.R., Collinson, M.E., Vincent, B., Griffiths, P.C. and Jenkins, P.D. (2000). Self-assembly of colloidal units in exine development. In: M.M. Harley, C.M. Morton and S. Blackmore (Editors). Pollen and Spores: Morphology and Biology, pp. 31–44. Royal Botanic Gardens, Kew.

SELF-ASSEMBLY OF COLLOIDAL UNITS IN EXINE DEVELOPMENT

[1]ALAN R. HEMSLEY, [2]MARGARET E. COLLINSON, [3]BRIAN VINCENT,
[4]PETER C. GRIFFITHS AND [5]PAUL D. JENKINS

[1]Department of Earth Sciences, Cardiff University, PO Box 914, Cardiff,
CF1 3YE, Wales, UK
[2]Geology Department, Royal Holloway University of London, Egham,
Surrey TW20 0EX, UK
[3]School of Chemistry, University of Bristol, Cantock's Close, Bristol BS8 1TS, UK
[4]Department of Chemistry, University of Wales Cardiff, PO Box 912,
Cardiff, CF1 3TB, Wales, UK
[5]Ian Wark Research Institute, University of South Australia, Warrendi Road,
The Levels, South Australia 5095, Australia

Abstract

Spore and pollen walls commonly exhibit complex internal ultrastructure and surface patterning. The processes governing the formation of these features remain obscure despite considerable study, although it has been postulated that part of construction is self-assembling. Our investigations of exines in a number of lycophyte megaspores have shown that aqueous mixed colloidal systems containing polymers, lipids and colloidal particles are a major component in spore wall development within this, and probably other, groups of plants. Colloidal simulation of both mega- and microspore wall construction using a model system has enhanced our understanding of possible mechanisms of spore wall generation and has defined the role of self-assembly within exine construction.

Introduction

Two problems confront the palynologist who seeks to explain the diversity and complexity of surface pattern and internal ultrastructure prevelant in both living and fossil spores and pollen. One is the nature of the controlling mechanism that gives rise to pattern. The other concerns the composition of the material used by plants to build these patterns. In the past, these two problems have been addressed independently by those who consider the developmental ultrastructure of spore and pollen wall formation (for example, Rowley and Southworth, 1967; Sheldon and Dickinson, 1983, 1986; Blackmore and Barnes, 1987; Rowley and Dunbar, 1990; Rowley and Claugher, 1991) and those interested in the chemical nature of the highly intractible biomacromolecule, sporopollenin (Brooks, 1971; Shaw, 1971; Brooks and Shaw, 1977; Schulze Osthoff and Wiermann, 1987; Guilford et al., 1988; Herminghaus et al., 1988; Wehling et al., 1989; Gubatz et al., 1993; Hemsley et al., 1993; van Bergen et al., 1995). Our investigation of the development of heterosporous lycophyte spore walls has sought to combine what is known about the causes of pattern formation and the interaction of the latter within the chemical and physical environment of the sporangium, an approach considered by few in the past (for example, van Uffelen, 1991; Gabarayeva, 1993; Godward and Pell; 1994, Scott, 1994).

It has been assumed that some form of template must be present upon (or within) the cellulosic sporocyte wall to act as a trigger/guide for pattern formation (Scott, 1994). The presence, but not the exact nature, of the template was demonstrated by Sheldon and Dickinson (1986) using centrifugation to cause disruption to the exine unit distribution. More recently, analyses of mutant *Arabidopsis* pollen development has shown that micro-undulations of the sporocyte surface play a role in pattern development since where these are absent, exines with the expected structure are not formed (Takahashi, 1989; Takahashi and Skvarla, 1991; Paxson-Sowders *et al.* 1997). Sporocytic templates are clearly genetically governed (Haque and Godward, 1983) and must, in many spores and pollen, direct the principle distribution of exine components such as spines or columellae/muri. These templates, however, are restricted in their influence on the development of exines of great thickness and/or radially varying ultrastructure. This is especially so where the exine appears to contain no system by which a pattern might be transmitted through already-deposited wall material to regions of peripheral development (for example, Fig. 1, but see Morbelli and Rowley, 1993). Similarly, it is difficult to see how 'exine' structure can develop upon ubisch bodies/orbicules (Hesse, 1986) which presumably lack an internal template. Clearly an understanding of the mode of deposition of wall construction materials is critical in these cases (i.e. identification of the physical and chemical nature of the materials as well as the sequence of deposition in respect to the template).

Exine chemistry

Sporopollenin, the principal exine component of most archegoniates, has been much studied. Despite this, it has failed to reveal its precise chemical nature although it is established that the principal components are C, H, and O in an approximate ratio of 4:6:1 (Shaw, 1971). The main analytical procedures currently in use (pyrolysis GCMS and solid state NMR) provide results which confirm the presence of aliphatic (fatty acid), aromatic (probably a cinnamic acid) and minimal carboxylic acid components (Guilford *et al.*, 1988; Wehling *et al.*, 1989; van Bergen *et al.*, 1993, 1995; Kawase and Takahashi, 1995; Hemsley *et al.*, 1996b). They do not, however, suggest consistency of this ratio between different plant groups, implying that sporopollenin varies in its proportional composition, but not in the actual components. Sporopollenin may be considered therefore as a randomly cross-linked biomacromolecule without a repetitive large-scale structure, a characteristic which would inherently make this material resistent to enzymic attack, and to many laboratory procedures designed to reduce/return it to its principal components.

Exine formation

The bulk of sporopollenin involved in exine production is believed to be derived in some way from the tapetum in most groups (Pacini and Franchi, 1991). It seems unlikely that sporopollenin itself is a tapetal product, more likely that the tapetum is responsible for the production of the components of sporopollenin (monomers) which subsequently combine in some way to form the polymer, possibly upon the introduction of some catalyst (enzymic (Sakagami *et al.*, 1991) or metallic (Frausto Da Silva and Williams, 1991)). If the above view of sporopollenin composition is correct, it is also difficult to envisage enzymic catalysis of either polymerisation, or structure formation since both would require at least some repetitive molecular structure for active site involvement. It is with these points in mind that the larger scale (colloidal) properties of polymers should be considered since their interaction within the sporangial environment could give rise to both ultrastructure and surface patterning. Indeed, our preliminary investigations of both fossil and living *Selaginella* megaspores (Hemsley *et al.*, 1992; Collinson *et al.*, 1993; Hemsley *et al.*, 1994, 1996a), showed that some structures present within these complex exines (for example, Fig. 1) were easily simulated by colloidal flocculation of polystyrene latex dispersions (Figs. 2 and 3). Structure develops as a consequence of the initial components within the system and their relative instantaneous concentrations, the patterns produced are therefore essentially self-assembling.

Further supporting evidence of colloidal construction was obtained from observation of developing megaspores and the structure of the components involved. Wall development has been investigated by Taylor (1991) and Morbelli and Rowley (1993). Both show that aggregation of particles (and laminae) is rapid. By chance, a developing megaspore (*Selaginella laevigata*, Fig. 4) at a critical stage was extracted from a sporangium and analysed using SEM. This has demonstrated the existence of three-dimensional raspberry-like structures of around 5-10 μm diameter (Fig. 5). These occur as a stage in exine construction and are known to be a typical form of colloidal particle aggregation in mixed particle systems (Maeda and Armes, 1993). Normally such structures are obscured by subsequent exine development, but they can also occur as separate ubisch-type aggregates, commonly upon the surface of mature megaspore walls (Fig. 6). The presence of lipids/oils within *Selaginella* megasporangia has been established (Morbelli and Rowley, 1993) and lipids would appear to be important in orbicule formation (Hesse, 1986). The appearance of the rapidly-flocculated sporopollenin on the surface of the developing spore (Fig. 4) is very similar to a floc of polystyrene particles grown in our apparatus (Figs. 7 and 8 respectively). This reinforces our view that destabilising colloidal polystyrene latex dispersions can mimic some aspects of colloidal sporopollenin self-assembly and demonstrates that, although of different composition, these organic colloids probably follow an analogous mechanism (see also Lin *et al.*, 1989). It would seem that lipid components play an important role in shaping the developing exine by constraining the distribution of the colloidal sporopollenin particles and aggregates. The following results are derived from the introduction of comparable components into our simulated system.

Materials and Methods

Polystyrene latex dispersions were used in our simulations of exine development since they have some similarities with sporopollenin (for example, density, refractive index) but more significantly because there is a wide range of literature covering the behaviour of colloidal polystyrene particles, their preparation and behaviour, under differing conditions (Pieranski, 1983; Shih and Stroud, 1984; Everett, 1988; Hemsley *et al.*, 1994, 1996; Adams *et al.*, 1998). Most of our simulations entail the production of a polystyrene latex, essentially an aqueous suspension of colloidal particles, typically spherical in shape (polyballs) with a diameter of around 0.2-0.5 μm. Details of the preparation of such a latex are given by Goodwin *et al.* (1973), Hemsley *et al.* (1996a, 1998), utilising styrene monomer, sodium chloride (as an electrolyte to enhance colloidal interaction), and ammonium persulphate (as an initiator of the styrene monomer which comprise the colloidal particles). Carboxymethylcellulose was used in some simulations to provide a long-chain free polymer in solution that would cause flocculation. Some lattices were manufactured with a proportion of cyclohexane present forming an emulsion. In others, a proprietary rapeseed oil replaced the

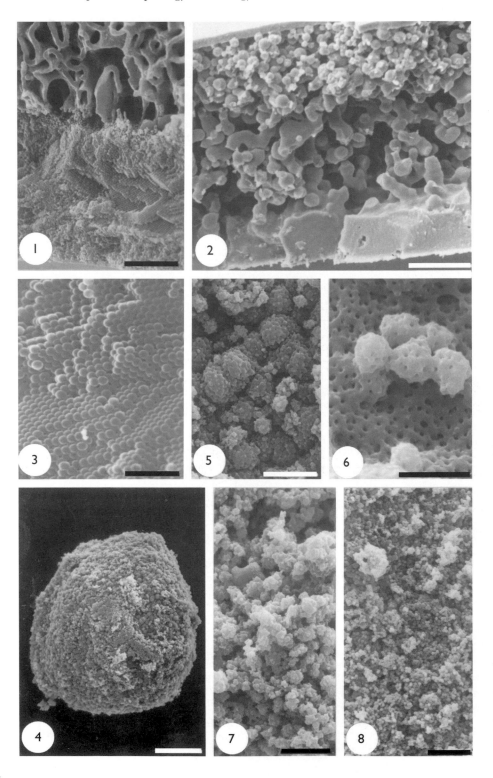

cyclohexane. Details of the proportions involved are given in Table 1. A comparison of the living and experimental systems is given in Table 2.

TABLE 1. The proportions of components used to produce the structures shown in the indicated figures.

	Fig. 2	Figs. 3 and 10	Fig. 8	Figs. 9, 12-15	Figs.16 and 17	Figs. 18-22
Distilled water	335 ml	670 ml	670 ml	670 ml	670 ml	670 ml
Styrene	50 ml	80 ml	50 ml	50 ml	50 ml	50 ml
Cyclohexane	335 ml	-	50 ml	100 ml	50 ml	10 ml
Rapeseed oil	-	-	-	-	30 ml	30 ml
Carboxymethyl-cellulose (in 20 ml distilled water)	5 g	5 g	-	-	-	-
Sodium chloride (10% solution)	1.25 ml	1.25 ml	1.25 ml	1.25 ml	1.25 ml	1.25 ml
Ammonium persulphate (3% solution)	25 ml	25 ml	20 ml	20 ml	20 ml	20 ml

FIGS. 1–8. All SEM micrographs. Fig. 1. A broken megaspore wall of *Selaginella myosurus* illustrating (from bottom) layers containing irregular particles, close-packed particles forming a colloidal crystal, and exinous ridges consisting of convoluted laminae. Scale bar = 10 μm. Fig. 2. A flocculation of polydisperse (variable size) polystyrene particles utilising carboxymethyl-cellulose as the flocculant. Large particles appear to have linked (possibly by bridging flocculation) before smaller particles (deposited by depletion flocculation). Such ordering is not simply a gravitational effect. Residual carboxymethylcellulose has formed an intine-like layer at the base of the simulation. Scale bar = 50 μm. Fig. 3. A flocculation of monodisperse (same size) particles again utilising carboxymethylcellulose. Here, due to the regularity of size, particles have become close-packed (by depletion flocculation) and have formed a colloidal crystal (compare with middle layer in Fig. 1). Scale bar = 1 μm. Fig. 4. A megaspore of *Selaginella laevigata* extracted during wall development. Rapid flocculation of colloidal sporopollenin has occurred over much of the surface obscuring the developing structure. Scale bar = 50 μm. Fig. 5. Detail top right of 4 reveals the presence of raspberry-like particle aggregates forming a stage in the colloidal system of wall construction. Scale bar = 10 μm. Fig. 6. Raspberry-like ubisch bodies adhering to the surface of a megaspore of *Selaginella selaginoides*. Normally occurring as a stage in wall development, occasionally these fail to become incorporated within the final exine as shown here. Scale bar = 5 μm. Fig. 7. Detail of the flocculated sporopollenin covering the bulk of the spore shown in 4. Flocculation was probably triggered (upon removal from the sporangium) by contact with air and subsequent dehydration. Scale bar = 5 μm. Fig. 8. An air-dried flocculation of polydisperse polystyrene particles, appearing similar in general features to that shown in 7. Scale bar = 2 μm.

TABLE 2. Comparison of the components of the living and simulated systems.

	SPORANGIUM	SIMULATION
Supporting fluid	Water	Water
Polymer latex	Sporopollenin	Polystyrene
Flocculant	Mucopolysaccharides	Carboxymethylcellulose/ air drying
Hydrocarbon	Lipids (Fatty acids)	Cyclohexane/ Rapeseed oil
Electrolyte	Various salts	Sodium chloride
Initiator	Enzyme? UV? O_2? Cobalt?	Ammonium persulphate

Lattices were air dried (during which aggregation usually occurred if it had not already done so) within glass phials. In some cases, samples were dried directly upon SEM stubs or mounted from a distilled water droplet onto negative film following the method used for modern pollen (Moore *et al.*, 1991). Dried-down lattices within phials were broken up (using a mounted needle) and fragments placed on SEM stubs upon which a sticky tab had already been placed. Samples were sputter coated with gold/palladium and observed using a Cambridge Instruments Stereoscan 360 SEM (15kV) or coated with gold in a Philips 501 B SEM (30 kV).

FIGS. 9–15. All SEM micrographs. Fig. 9. 'Stringy' particle aggregates demonstrating fusion of particles which can occur when cyclohexane is present as droplets (and monomer solvent) within the reaction chamber. Strings result from the aggregation of 'sticky' particles. Scale bar = 10 μm. Fig. 10. Structures resulting only from the gradual accumulation of polystyrene upon peripheral parts of our apparatus during the production of a colloidal latex demonstrates how readily complexity may be developed from simple components. Scale bar = 20 μm. Fig. 11. Part of the megaspore wall of a fossil *Salvinia cerebrata* (Miocene of the Czech Republic) showing a remarkable similarity of structure with the artificial polystyrene system shown in 10. Scale bar = 5 μm. Fig. 12. Polystyrene particles and particle aggregates forming a coating around droplets of cyclohexane. Scale bar = 10 μm. Fig. 13. A larger droplet of cyclohexane sports an almost reticulate ornament of particle aggregates. Scale bar = 10 μm. Fig. 14. detail of part of the surface of 13 showing the fusion of many of the particles comprising the wall of this artificial spore. Scale bar = 2 μm. Fig. 15. A broken wall of an artificial spore like that shown in 13 reveals the way in which particles have accumulated at the surface of the central droplet of cyclohexane. Scale bar = 2 μm.

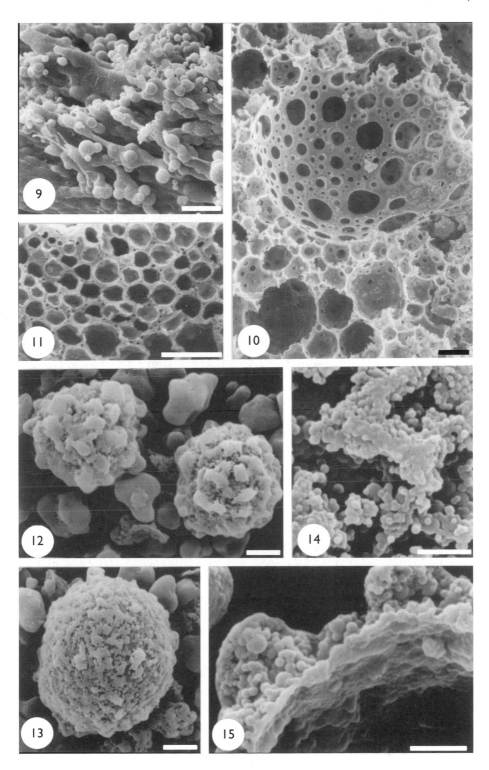

Results and Discussion

Mixed systems containing discrete colloidal particles, their aggregates, and lipid droplets clearly offer the prospect of a wide variety of interactions dependent upon the relative proportions of each component. In addition, our reaction chamber (due to its size) contains a gradation of environments ranging from highly agitated (adjacent to the rotor) to relatively still (close to the glass wall of the vessel). From most preparations we are therefore able to extract at least two distinct products formed under these two extremes. Figures 9 and 10 illustrate the polymer structures formed at the base of the reaction vessel, i.e. from an environment in which greater fusion of particles might be expected. These are complex structures which tend to show considerable linkage between constituent particles and particle aggregates. These illustrations serve principally to show the degree of complexity of structure that may be obtained from such simple mixtures, however, there are clear similarities in structure between these and features found in lycopsid and fern (Fig. 11, *Salvinia*) megaspore walls (Kempf, 1973; Friis, 1977; van Bergen *et al.*, 1993) indeed, wherever the volume of sporopollenin involved in wall production permits this scale of self-assembled structure.

Figures 12-15 depict structures obtained by aggregation of the bulk of the agitated colloidal material from a reaction containing cyclohexane as the hydrocarbon component. Here, droplets of cyclohexane have acted as adsorbing surfaces for polystyrene particles (due to a hydrophobic reaction), thus forming aggregates. Most droplets are very small, resulting in 2-10 μm aggregates. Some droplets, however, are larger and have acted as adsorbing surfaces not just for particles but also of particle aggregates. One might envisage the sporocyte wall acting in a similar manner within a sporangium. Were such a wall to have irregularities (a template) these might affect the distribution of particles and aggregates. In our simulation, the cyclohexane droplets have no irregularities and yet, despite this, a surface ornament has developed resulting from the hydrophilic/hydrophobic and surface charge interactions between particles, particle aggregates and lipid droplets. The similarity of ornament with that shown by many fern spores is striking (see Tryon and Lugardon, 1991, and Figs. 3.7, 87.1, and 113.4 therein). This result, however, is significant in that it clearly demonstrates that surface ornament can result without the aid of a template.

The incorporation of rapeseed oil (in addition to cyclohexane) into the mixed colloidal system results in larger droplets, principally due to the greater viscosity and hydrophobicity of the oil. Reaction rates are generally slower and individual particles

FIGS. 16–23. All SEM micrographs. Fig. 16. A spherical particle of polystyrene formed among dried 'tapetal debris' of rapeseed oil and polystyrene latex. Individual polystyrene particles are adherent to the surface. Scale bar = 10 μm. Fig. 17. A similar, but larger sphere to that shown in 16. Here, the surface has accumulated many adherent polystyrene particles which are self-ordered in a surface ornament. Scale bar = 5 μm. Fig. 18. Large aggregates of polystyrene particles and rapeseed oil/polystyrene latex illustrate a consistent surface ornament despite variation in size. Scale bar = 500 μm. Fig. 19. Detail of a spherical aggregate shown in 18. Scale bar = 200 μm. Fig. 20. Small aggregates of less spherical shape may be formed among larger aggregates. Commonly these have a more pronounced ornament. Scale bar = 500 μm. Fig. 21. Detail of the surface ornament of a large spherical aggregate as shown in 18. Scale bar = 200 μm. Fig. 22. Wall composition of spherical aggregates is obscured by the high proportion of rapeseed oil/polystyrene latex, however, some evidence of its construction from numerous small aggregates is apparent. Scale bar = 250 μm. Fig. 23. *Selaginella selaginoides* megaspore showing a similar ornament to the spherical polystyrene aggregates (Figs. 18–22) and under attack from fungal hyphae.

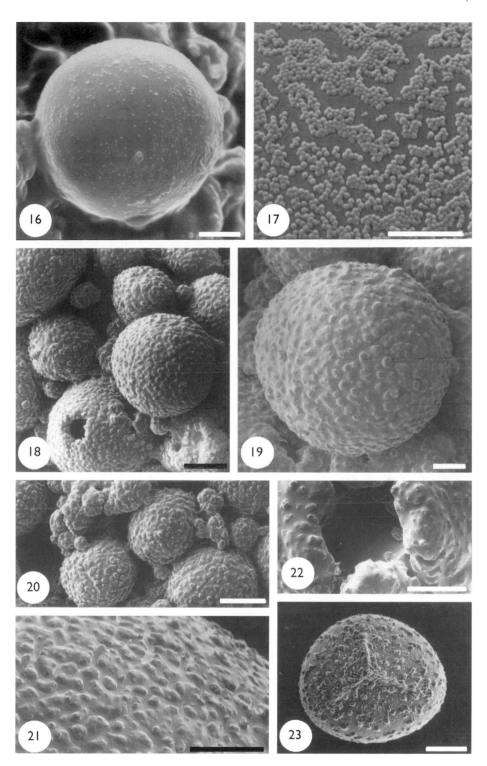

can be much larger. Figure 16 shows a large polystyrene particle formed in such a reaction. It is surrounded by the dried-down mixture of polystyrene and rapeseed oil which lacks any structure. Some small polystyrene particles are adherent to the surface of the large particle. Figure 17 shows a very large polystyrene particle from such a reaction on which the interaction of surface charge and stickyness between the large particle and numerous small particles has given rise to distinct (but varying) surface patterning (see also figure 3.13 in Hunter, 1993; Yeh *et al.*, 1997).

Further reduction of cyclohexane, but retention of rapeseed oil yields particularly large droplets (of megaspore dimensions) which accumulate particles and particle aggregates (Figs. 18-22) in much the same way as the smaller cyclohexane droplets in Figs. 12 and 13. In this case, however, droplets become filled with particles (Fig. 22) and particle aggregates possibly due to the greater cohesiveness and decreased miscibility of the oil. Again, one might substitute a lipid droplet for an 'oily' megaspore sporocyte wall upon which particles and particle aggregates (such as those shown in Figs. 5 and 6) might coalesce to form a robust ornament. These artificial megaspores constructed from polystyrene particles, and rapeseed oil in an aqueous system show a remarkable resemblance of surface structure to the spores of certain *Selaginella* species (Fig. 23; and Minaki, 1984, pl. 7, (2); pl. 10 (8); pl. 19 (2) therein).

These results have important implications with respect to the use of spore morphology and ultrastructure in taxonomy. It is clear from our work that small changes in the components of this self-assembly system produce often very different final structures. Similarly, large changes in components can produce relatively small effect in structure (components used to produce the structures shown in Figs. 8 and 14 both produce a polydisperse latex with particles 0.2 to 0.4 μm). Chaotic systems such as these do not possess a direct mapping between the proportions of components and the ultimate structure. Likewise, one cannot therefore expect a direct mapping between genetic information and ultimate structure and for this reason, where self-assembly mechanisms are involved in development, one might expect discrepancies between taxonomies based on morphology and gene sequences (Hemsley, 1998). In addition, groupings of spores based on similarity of structure may not reflect underlying genetic diversity, nor would one expect the plants producing such groupings to show an eqivalent degree of relatedness based on overall morphology. That spores (and pollen) *can* be used as taxonomic adjuncts demonstrates that although self-assembly mechanisms in wall development may be important, there is enough genetic control (in most cases) for different species to produce different spore (or pollen) types. However, the physical and chemical constraints of such systems and their utilisation of similar components probably accounts for the widespread occurrence of reticulate and polygonal patterns (Scott, 1994), and perhaps more prominent surface features such as spines.

Conclusions

Following initial investigations focussing on colloidal crystal structure within *Selaginella* megaspore walls (Taylor and Taylor, 1987; Taylor 1991; Collinson *et al.*, 1993; Hemsley *et al.*, 1994, 1996, 1997, 1998) it has become clear that colloidal construction of walls plays an important role in the self-assembly of these particularly thick exines. The results presented here serve to further demonstrate the diversity of comparable structure that may be obtained from simple mixed colloidal systems. These structures, particularly those associated with, and forming around, droplets of hydrocarbon, strongly suggest that pattern formation over distal surfaces and contact faces of many spores, could well proceed principally by self-assembling colloidal 'flocculations. Features such as laesurae (Fig. 23), and various forms of aperture in pollen, would result from interference in the flocculation process caused by factors acting from within, or at the surface of, the sporocyte. Where exines are thin, a physical or chemical template may provide a cause of variation in flocculation resulting in the the initiation of a (distantly) genetically controlled pattern, however, it seems likely that subsequent wall development is self-assembling.

These simulations have yet to produce the complexity of internal structure of many spores and pollen. The rapid radial changes in component form in pollen (for example, laminae in the endexine, columellae and tectal units in the ectexine) may result from rapid changes in the chemical/physical environment within the sporangium (anther) to which the self-assembly mechanism responds by production of differing sequential layers. This may be difficult to simulate in our current apparatus but could be approached by an attempt to achieve significant changes in local concentrations in a smaller reaction vessel. Furthermore, a move toward more 'natural' components may be necessary (for example, p-coumaric acid polymers (Wehling et al., 1989), fatty acids (Guilford, 1988), and mucopolysaccharides (Morbelli and Rowley, 1993)) in order to achieve more complex, comparable structures. None the less, this work serves to demonstrate the potential significance of colloidal polymer latex/water/lipid (hydrocarbon) mixtures in the creation of spore and pollen walls, and shows that the self-assembly of such systems, even without a template, can produce ornamentation (upon an artificial spherical base) which is comparable to that of real spores.

Acknowledgements

We thank A. Oldroyd and P. Fisher (both Cardiff, Earth Sciences) for technical assistance and J. Crawley and V. Williams (Cardiff, Biology) for photographic work. The support of a Royal Society University Research Fellowship (to ARH) is gratefully acknowledged.

References

Adams, M., Dogic, Z., Keller, S.L. and Fraden, S. (1998). Entropically driven microphase transitions in mixtures of colloidal rods and spheres. *Nature* 393: 349–352.

Blackmore, S. and Barnes, S.H. (1987). Embryophyte spore walls: origin, development, and homologies. *Cladistics* 3: 185–195.

Brooks, J. (1971). Some chemical and geochemical studies on sporopollenin. In: J. Brooks, P.R. Grant, M. Muir, P. van Gijzel and G. Shaw (editors). Sporopollenin, pp. 351–407. Academic Press, London.

Brooks, J. and Shaw, G. (1977). Recent advances in the chemistry and geochemistry of pollen and spore walls. *Transactions of the Bose Research Institute* 40: 19–38.

Collinson, M.E., Hemsley, A.R. and Taylor, W.A. (1993). Sporopollenin exhibiting colloidal organization in spore walls. *Grana* Supplement 1: 31–39.

Everett, D.H. (1988). Basic principles of colloid science. Royal Society of Chemistry, London.

Frausto Da Silva, J.J.R. and Williams, R.J.P. (1991). The biological chemistry of the elements: The inorganic chemistry of life. Clarendon Press, Oxford. 561pp.

Friis, E.M. (1977). EM-studies on Salviniaceae megaspores from the Middle Miocene Fasterholt flora, Denmark. *Grana* 16: 113–128.

Gabarayeva, N.I. (1993). Hypothetical ways of exine structure determination. *Grana* Supplement 2: 54–59.

Godward, M.B.E. and Pell, K. (1994). Inheritance of exine pattern in *Nicotiana* × *sanderae* (Solanaceae). *Botanical Journal of the Linnean Society* 115: 145–159.

Goodwin, J.W., Hearn, J., Ho, C.C. and Ottewill, R.H. (1973). The preparation and characterisation of polymer lattices formed in the absence of surface active agents. *British Polymer Journal* 5: 347–362.

Gubatz, S., Rittscher, M., Menter, A., Nagler, A. and Wiermann, R. (1993). Tracer experiments on sporopollenin biosynthesis. An overview. *Grana* Supplement 1: 12–17.

Guilford, W.J., Schneider, D.M., Labovitz, J. and Opella, S.J. (1988). High resolution solid state ^{13}C NMR. Investigation of sporopollenins from different plant classes. *Plant Physiology* 86: 134–136.

Haque, M.Z. and Godward, M.B.E. (1983). Evidence of genetic control of exine pattern from irradiation studies in Lactuceae. *Pollen et Spores* 25: 421–436.

Hemsley, A.R. (1998). Non-linear variation in simulated complex pattern development. *Journal of Theoretical Biology* 192: 73–79.

Hemsley, A.R., Barrie, P.J., Chaloner, W.G. and Scott, A.C. (1993). The composition of sporopollenin and its use in living and fossil plant systematics. *Grana* supplement 1: 2–11.

Hemsley, A.R., Collinson, M.E. and Brain, A.P.R. (1992). Colloidal crystal-like structure of sporopollenin in the megaspore walls of Recent *Selaginella* and similar fossil spores. *Botanical Journal of the Linnean Society* 108: 307–320.

Hemsley, A.R., Collinson, M.E., Kovach, W.L., Vincent, B. and Williams, T. (1994). The role of self-assembly in biological systems: evidence from iridescent colloidal sporopollenin in *Selaginella* megaspore walls. *Philosophical Transactions of the Royal Society, London B* 345: 163–173.

Hemsley, A.R., Griffiths, P., Vincent, B. and Collinson, M.E.C. (1997). Advances in the understanding of complex pattern formation in spore and pollen walls. In: G. Jeronimidis and J.F.V. Vincent (editors). Plant biomechanics, pp. 96–102. University of Reading.

Hemsley, A.R., Jenkins, P.D., Collinson, M.E. and Vincent, B. (1996a). Experimental modelling of exine self-assembly. *Botanical Journal of the Linnean Society* 121: 177–187.

Hemsley, A.R., Scott, A.C., Barrie, P.J. and Chaloner, W.G. (1996b). Studies of fossil and modern spore wall biomacromolecules using 13C solid state NMR. *Annals of Botany* 78: 83–94.

Hemsley, A.R., Vincent, B., Collinson, M.E. and Griffiths, P.C. (1998). Simulated self-assembly of spore exines. *Annals of Botany* 82: 105–109.

Herminghaus, S., Gubatz, S., Arendt, S. and Wiermann, R. (1988). The occurrence of phenols as degradation products of natural sporopollenin - a comparison with "synthetic sporopollenin". *Z. Naturforsch.* 43c: 491–500.

Hesse, M. (1986). Orbicules and the ektexine are homologous sporopollenin concretions in spermatophyta. *Plant Systematics and Evolution* 153: 37–48.

Hunter, R.J. (1993). Introduction to modern colloid science. Oxford University Press, Oxford.

Kawase, M. and Takahashi, M. (1995). Chemical composition of sporopollenin in *Magnolia grandiflora* (Magnoliaceae) and *Hibiscus syriacus* (Malvaceae). *Grana* 34: 242–245.

Kempf, E.K. (1973). Transmission electron microscopy of fossil spores. *Palaeontology* 16: 787–797.

Lin, M.Y., Lindsay, H.M., Weitz, D.A., Ball, R.C., Klein, R. and Meakin P. (1989). Universality in colloidal aggregation. *Nature* 339: 360–362.

Maeda, S. and Armes, S.P. (1993). Preparation of novel polypyrrole-silica colloidal composites. *Journal of Colloid and Interface Science* 159: 257–259.

Minaki, M. (1984). Macrospore morphology and taxonomy of *Selaginella* (Selaginellaceae). *Pollen et Spores* 26: 421–480.

Moore, P.D., Webb, J.A. and Collinson, M.E. (1991). Pollen analysis. 2nd edition. Blackwell Scientific Publishers, Oxford.

Morbelli, M.A. and Rowley, J.R. (1993). Megaspore development in *Selaginella*. 1. "Wicks", their presence, ultrastructure, and presumed function. *Sexual Plant Reproduction* 6: 98–107.

Pacini, E. and Franchi, G.G. (1991). Diversification and evolution of the tapetum. In: S. Blackmore and S.H. Barnes (editors). Pollen and spores: patterns of diversification, pp. 301–316. Clarendon Press, Oxford.

Paxson-Sowders, D.M., Owen, H.A. and Makaroff, C.A. (1997). A comparative ultrastructural analysis of exine pattern development in wild-type *Arabidopsis* and a mutant defective in pattern formation. *Protoplasma* 198: 53–65.

Pieranski, P. (1983). Colloidal crystals. *Contemporary Physics* 24: 25–73.

Rowley, J.R. and Claugher, D. (1991). Receptor independent sporopollenin. *Botanica Acta* 104: 316–323.

Rowley, J.R. and Dunbar, A. (1990). Outward extension of spinules in exine of *Centrolepis aristata* (Centrolepidaceae). *Botanica Acta* 103: 355–359.

Rowley, J.R. and Southworth, D. (1967). Deposition of sporopollenin on laminae of unit membrane dimensions. *Nature* 213: 703–704.

Sakagami, H., Oh-hara, T., Kohda, K. and Kawazoe, Y. (1991). Lignified materials as a potential medicinal resource IV. Dehydrogenation polymers of some phenylpropenoids and their capacity to stimulate polymorphonuclear cell iodination. *Chemical and Pharmacological Bulletin* 34: 950–955.

Schulze Osthoff, K. and Wiermann, R. (1987). Phenols as integrated compounds of sporopollenin from *Pinus* pollen. *Journal of Plant Physiology* 131: 5–15.

Scott, R.J. (1994). Pollen exine - the sporopollenin enigma and the physics of pattern. In: R.J. Scott and M.A. Stead (editors). Molecular and cellular aspects of plant reproduction, pp. 49–81. Cambridge University Press, Cambridge.

Shaw, G. (1971). The chemistry of sporopollenin. In: J. Brooks, P.R. Grant, M. Muir, P. van Gijzel and G. Shaw (editors). Sporopollenin, pp. 305–348. Academic Press, London.

Sheldon, J.M. and Dickinson, H.G. (1983). Determination of patterning in the pollen wall of *Lilium henryi*. *Journal of Cell Science* 63: 191–208.

Sheldon, J.M. and Dickinson, H.G. (1986). Pollen wall formation in *Lilium*: The effect of chaotropic agents, and the organisation of the microtubular cytoskeleton during pattern development. *Planta* 168: 11–23.

Shih, W-H. and Stroud, D. (1984). Theoretical study of miscibility and glass-forming trends in mixtures of polystyrene spheres. *Journal of Chemistry and Physics* 79: 6254–6260.

Takahashi, M. (1989). Pattern determination of the exine of *Caesalpinia japonica* (Leguminosae: Caesalpinioideae). *American Journal of Botany* 76: 1615–1626.

Takahashi, M. and Skvarla, J.J. (1991). Exine pattern formation by plasma membrane in *Bougainvillea spectabilis* Willd. (Nyctaginaceae). *American Journal of Botany* 78: 1063–1626.

Taylor, W.A. (1991). Ultrastructural analysis of sporoderm development in megaspores of *Selaginella galeottii* (Lycophyta). *Plant Systematics and Evolution* 174: 171–182.

Taylor, W.A. and Taylor, T.N. (1987). Subunit construction of the spore wall in fossil and living lycopods. *Pollen et Spores* 29: 241–248.

Tryon, A.F. and Lugardon, B. (1991). Spores of the Pteridophyta. Springer-Verlag, New York.

van Bergen, P.F., Collinson, M.E., Briggs, D.E.G., de Leeuw, J.W., Scott, A.C., Evershed, R.P. and Finch, P. (1995). Resistent biomacromolecules in the fossil record. *Acta Botanica Neerlandica* 44: 319–342.

van Bergen, P.F., Collinson, M.E. and de Leeuw, J.W. (1993). Chemical composition and ultrastructure of fossil and extant salvinialean microspore massulae and megaspores. *Grana* Supplement 1: 18–30.

van Uffelen, G.A. (1991). The control of spore wall formation. In: S. Blackmore and S.H. Barnes (editors). Pollen and spores: patterns of diversification, pp. 89–102. Clarendon Press, Oxford.

Wehling, K., Niester, Ch., Boon, J.J., Willemse, M.T.M. and Wiermann, R. (1989). *p*-Coumaric acid - a monomer in the sporopollenin skeleton. *Planta* 179: 376–380.

Yeh, S-R., Seul, M. and Shraiman, B.I. (1997). Assembly of ordered colloidal aggregates by electric-field-induced fluid flow. *Nature* 386: 57–59.

Romero, A.T. and Fernández, M.C. (2000). Development of exine and apertures in *Fumaria densiflora* DC from the tetrad stage to maturity. In: M.M. Harley, C.M. Morton and S. Blackmore (Editors). Pollen and Spores: Morphology and Biology, pp. 45–56. Royal Botanic Gardens, Kew.

DEVELOPMENT OF EXINE AND APERTURES IN *FUMARIA DENSIFLORA* DC FROM THE TETRAD STAGE TO MATURITY

ANA TERESA ROMERO[1] AND MARÍA DEL CARMEN FERNÁNDEZ[2]

[1]Departamento de Biología Vegetal, Facultad de Ciencias, Universidad de Granada, 18001 Granada, Spain
[2]Estación Experimental del Zaidín (CSIC), Profesor Albareda 1, 18008 Granada, Spain

Abstract

A study of the pollen wall and apertures of *Fumaria densiflora* DC from the end of the tetrad period to the bicellular stage has been made. During the initial ontogenetic stages white-line-centred lamellae appear in the nexine stratum, differentiated into two layers during the young and vacuolate microspore stages, with the outer layer having white-line-centred lamellae and the inner layer only white lines in the area where both are joined. Following cytochemical analysis the latter proved to be an endexine layer. The apertures are plugged by a fluffy material juxtaposed alongside the endexine and intine layers. The exine structuring is discussed, as is the presence of the fluffy material in the apertures.

Introduction

Our interest in the family Fumariaceae goes back over 10 years, to when we first studied the pollen morphology of the order Papaverales, among other characters (Morales Torres *et al.*, 1988), in species of the south-eastern Iberian Peninsula. Other authors have described various aspects of the pollen of this family (for example, Candau, 1987; Blackmore *et al.*, 1995).

The most abundant genus in terms of the number of member species (over 200) is *Corydalis*, this genus is represented only by 3 species distributed in the western Iberian Peninsula, whereas the genus *Fumaria* is represented by 24 of the total of 50 species found world-wide (Lidén, 1986a).

Fumaria presents pantoporate pollen, of which no ultrastructural or ontogenic studies have been made with transmission electron microscopy. Since the genus *Fumaria* is the most species-rich genus of the family found in Spain, and given the ease of collecting anthers from wild populations of the genus, we decided to study one of its species, *Fumaria densiflora* DC.

Fumatory taxa are annual plants with a transverse-zygomorphic spurred corolla, with two tripartite stamens, the upper stamen with one basal nectarium penetrating into the spur. They are basically spring-flowering, although they may flower at any time of year and in many cases are cleistogamic and self-compatible (Lidén, 1986b).

The purpose of this paper was to study the ultrastructure of the mature pollen of *F. densiflora*, although the particular features found in this model led us to investigate its ontogeny.

Material and Methods

Anthers of *Fumaria densiflora* DC collected in wild populations from Gójar (Granada, Spain) were used at different stages of development. For both scanning and transmission electron microscope studies, the samples were fixed in 3% glutaraldehyde and post-fixed with a mixture of 1% osmium tetroxide and 2% potassium ferrocyanide. For the cytochemical studies a mixture of 1% glutaraldehyde and 2% paraform-aldehyde was used. In all cases the samples were dehydrated and embedded in Spurr's resin. The techniques used for the cytochemical analysis were PTA acetone for proteins and periodic acid-thiosemicarbazide-silver proteinate for neutral polysaccharides (Thiéry 1967). The treatments and viewing were performed at the Scientific Instruments Centre of the University of Granada and in the Pollen Biology Group laboratory at the "Estación Experimental del Zaidín", also in Granada.

Results

The pollen wall of *Fumaria densiflora* is composed of a very thick tectum with perforations of highly electron-dense material, in certain cases crossing from the inside to the outside, Fig. 1. There is no columellate infratectum or any intercolumellate spaces, but rather a narrow band of fibrous-granular material. The tectum surface is covered with a loosely arranged fibrous material. Underneath the tectum and the narrow granular region, all that can be distinguished is a layer that is three times thinner than the tectum, uneven in thickness, which covers the intine. After staining with osmium tetroxide and potassium ferrocyanide, there is no clearly discernible difference between the foot layer and endexine, Fig. 1. The intine, which is extremely homogeneous in terms of its thickness and structure, presents no special features.

The apertures are of the pore type and are surrounded by an aspis (a special form of annulus, see Punt *et al.* 1994), Fig. 2. The pores are blocked by an external plug composed of a fluffy material, made of highly organised and densely packed lamellated structures which, when viewed at high magnification, are found to be either tri- or penta-lamellated (dark-white-dark or dark-white-dark-white-dark), Fig. 3. This plug is also located within the area that corresponds to the infratectum, described above as a narrow band of fibrous-granular material, (Fig. 2).

Ontogeny of the wall

At the end of the tetrad period in very young microspores, when there are still remnants of callose between the microspores, the ectexine presents a thick tectum, perforated perpendicularly by abundant, evident perforations, Fig. 4. During this period, unlike the mature grain, we can start to refer to an infratectum, although it is narrow and has very few columellae, which are short and barely visible, resting on a layer that some authors have called "nexine". The nexine has a large quantity of highly visible white-line-centred lamellae, which led us to believe at first that it was endexine, Fig. 4. The infratectal gap, the inside of the perforations of the tectum and the microspore surface are filled with a fibrous-granular material.

In the young microspore the nexine thickens noticeably, Fig. 5, such that two nexine strata (foot layer and endexine) can now start to be distinguished beneath the infratectum. The innermost layer of the new formation has a few white-line-centered lamellae, Fig. 5. We had never observed this characteristic in the other pollen types we had studied previously. The microspores touch the surface of the tapetum or the many dense globules that fill the locule after being secreted.

In the vacuolate microspore the wall thickness is practically consolidated, Fig. 6. The fibrous-granular material that covered the exine in previous states is now a more highly organised outer covering. This material also fills the tectum perforations. The infratectum becomes less noticeable as the size of the other layers (tectum and nexine) increases. The intine, which may be observed for the first time during this stage, forms a uniform layer. The dense globules disappear from the locule of the anther (compare Figs. 5 and 6).

In bicellular pollen the exine shows a change in its osmiophilic affinity in comparison with earlier stages, mainly related to the material filling the infratectal areas and perforations of the tectum. During this stage, the two nexine layers (foot layer–endexine) can no longer clearly be distinguished. The outer covering described for the previous stage is no longer present, although the locule and granum surface are once again occupied by another type of loose, electron-dense material, which may be the result of the process of degeneration of the tapetum which occurs during this stage, compare Figs. 1 and 6.

In the samples fixed only with glutaraldehyde, Fig. 7, the inner nexine layer is seen to be almost white. When the PTA acetone technique is used, Fig. 8, both this nexine layer and the outer exine covering are highly positive, for which reason it should be attributed to the endexine. The endexine appears as isolated amorphous material with white line centered lamellae on the outer edge, Fig. 9. With periodic acid-thioscmicarbazide-silver proteinate, the only positive reaction in the wall is observed at the intine and exine covering level (data not shown).

Ontogeny of the aperture

At the end of the tetrad stage, the tectal region around the aperture thickens, forming the aspis. The apertures are composed of a thickening of the fibrous-granular material filling the infratectum and some white-line-centred lamellae of nexine origin. Because of its location and structure, this thickening corresponds to the exine oncus (EO), Fig. 10.

In the young microspore stage, dense globules from the tapetum are present in the locule of the anther, Fig. 11, although the tapetum remains highly active in vacuolate microspores, as shown by its ultrastructure and the abundance of cytoplasmic prolongations on its cell surface, Fig. 12.

Three strata can now be distinguished in the young and vacuolate microspores, Figs. 11 and 12:

1. The outermost layer is still composed of fibrous-granular material, a continuation of that filling the infratectum.
2. The middle layer is the nexine, now thickened, with its two distinguishable strata, Fig. 13, the outer one presents a larger number of free white line centered lamellae, while the inner one is continuous and stretches from the intra-apertural areas.
3. The intine shows no apertural differentiation in intinous oncus.

In bicellular pollen the outermost layer has modified its structure and the fluffy fibrous material, Fig. 14, described for mature pollen, Fig. 2, appears here for the first time.

Figure abreviations :

A- Aspis.	C- Callose debris.
EO- Exinous oncus.	EN- Endexine.
FGM- Fibrillar-granular material.	FL- Foot layer.
FM- Fluffy material.	G- Globules.
I- Intine.	N- Nexine (foot layer-endexine).
IF- Infratectum.	WL- White-line-centered-lamellae.
T- Tectum.	

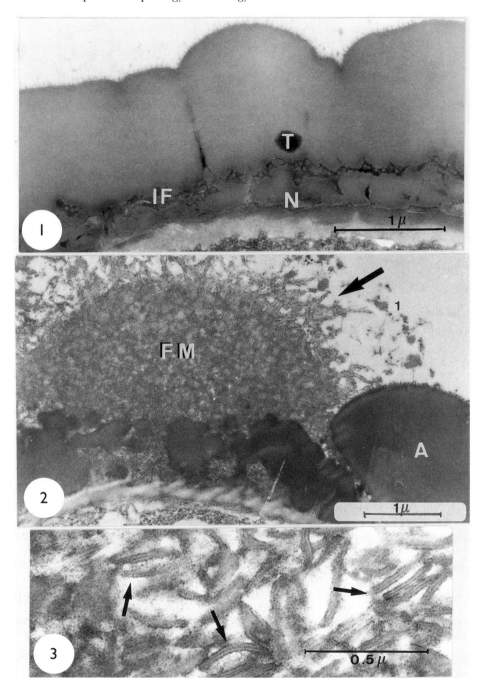

FIGS. 1–3. Wall and aperture of the mature pollen grain of *Fumaria densiflora*. Fig. 1. Wall showing the different strata (tectum, infratectum and foot layer–endexine). Fig. 2. Detail of the apertural area plugged by the fluffy material. Fig. 3. High-magnification detail of the fluffy material showing its penta-lamellated structure.

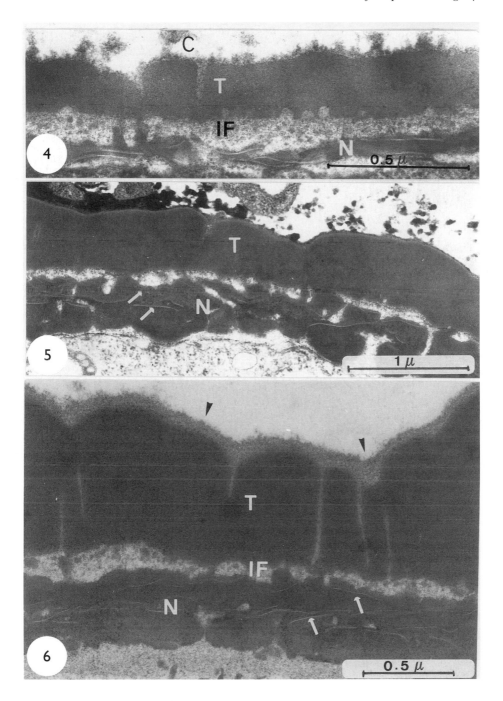

FIGS. 4–6. Ontogenic development of the wall of *F. densiflora.* Fig. 4. Microspore recently released from the callose after the tetrad period. Fig. 5. Young microspore. Fig. 6. Vacuolate stage.

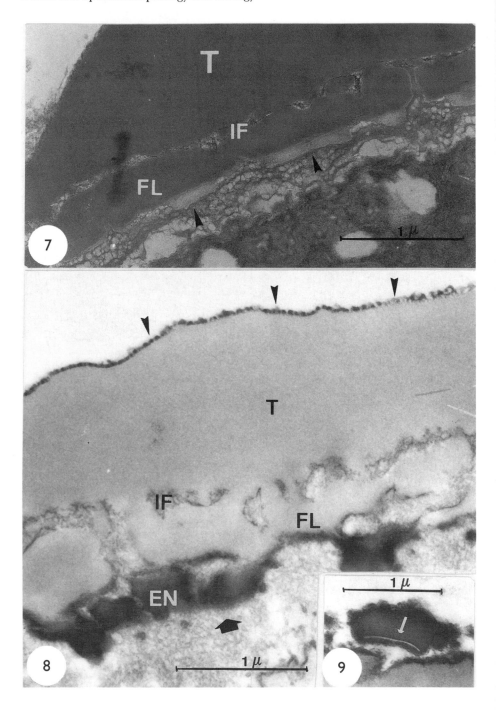

FIGS. 7–9. Pollen wall of *F. densiflora* during the mature stage. Fig. 7. Fixing only with glutaraldehyde: the inner nexine layer appears electron lucent (endexine). Fig. 8. PTA acetone: the inner nexine layer is highly electron-dense (endexine). Fig. 9. Detail of endexine showing white-line-centred lamellae in the outer area.

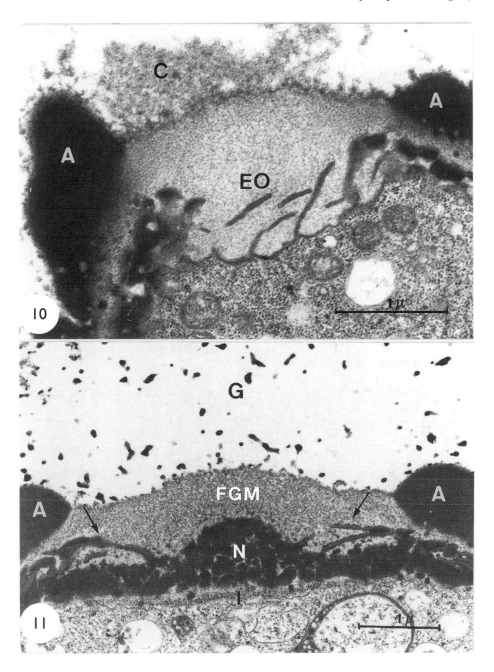

FIGS. 10–11. Ontogenic development of the aperture of *F. densiflora*. Fig. 10. Microspore recently released from the callose after the tetrad period. Fig. 11. Young microspore.

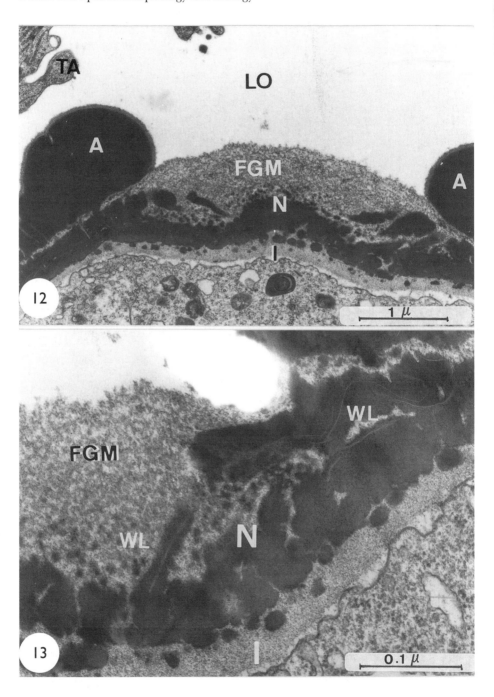

FIGS. 12–13. Apertural area in vacuolate microspore of *F. densiflora*. Fig. 12. General view. Fig. 13. Detail of nexine in this area.

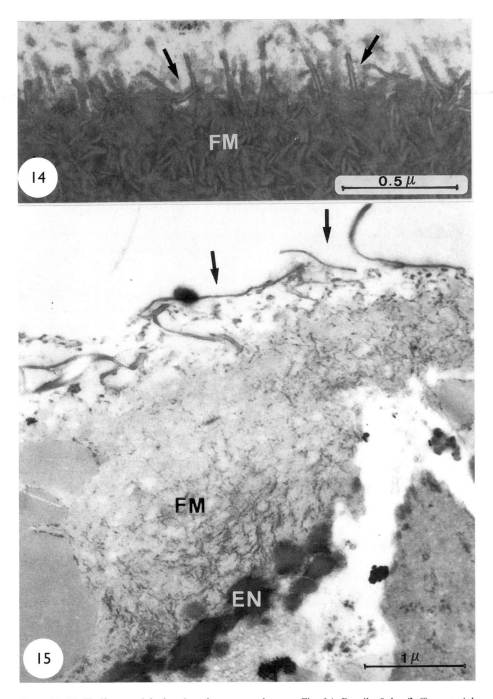

FIGS. 14–15. Fluffy material plugging the apertural areas. Fig. 14. Detail of the fluffy material under normal conditions. Fig. 15. Fluffy material after treatment with PTA acetone.

A positive response for the nature of the fluffy-type material of the aperture was only obtained with the PTA acetone technique, Fig. 15.

Discussion

Ontogenetic criteria provide the best available guide to distinguishing between the different strata of the pollen wall (Blackmore and Crane, 1988; Fernández and Rodríguez-García, 1988; Fernández et al., 1992; and other authors). These criteria were also of use in this case to distinguish between ectexine and endexine.

In the mature pollen of Fumaria densiflora the stratification of the wall is not sufficiently clear, since it all has the same density when observed under a transmission electron microscope, as occurs with Krameria (Simpson and Skvarla, 1981) and other species. In such cases it is extremely difficult to locate and identify the different layers correctly. Some authors, including Dahl and Rowley (1991) and Dunbar and Rowley (1984), have denominated the layer below the infratectum in microspores recently released from the callose as "nexine" (Erdtman, 1952). Both these criteria have led us to refer to this stratum as nexine. This term, which includes the foot layer and endexine, has been used more extensively in optical microscopy (for terminological details see Punt et al., 1994). Following the work of Faegri and Iversen (1975), the structuring of the exine appears to have been clarified, and its ontogeny has been widely discussed with respect to the lamellated arrangement of the endexine and/or foot layer (Dickinson and Heslop-Harrison, 1968; Le Thomas and Lugardon, 1972; Le Thomas, 1980-81; Guédès, 1982; Rowley, 1987-1988; Blackmore and Crane, 1988; Blackmore, 1990). In the opinion of Guédès (1982) and Blackmore and Crane (1988), however, a true foot layer is not lamellated, even during development. Thus, these authors consider that taxa in which lamellation appears immediately below the infratectum lack a foot layer as such. However, in general terms, arrangements on circumferentially white-line-centred lamellae are normally considered endexinous. In fumatory taxa the nexine has circumferentially white-line-centred lamellae until the mature-pollen stage, when they are no longer observed. Thus, after our initial analysis we believed that the stratum we called nexine was a single endexinous stratum composed of two layers: an outer one with white lines throughout its thickness and an inner one with white lines only in the area where both are joined, as described for other species such as Olea europaea L. (Fernández and Rodríguez-García, 1988). However, in pollen fixed only with glutaraldehyde and those treated with PTA acetone the inner layer (which lacks white-line-centered lamellae) appears white or electron-dense, respectively. This led us to reconsider the term "nexine" and the presence of endexine, and we therefore assigned the classical terms "endexine" for the inner layer (non-lamellated and composed of amorphous islets) and "ectexine foot layer" to the outer, lamellated layer.

In our opinion, the fact that endexine is the layer structured from white lines need not necessarily always hold true for all the taxa, and we agree with Rowley (1987–88) that these structures may also be involved in ectexine deposition. Thus, the deposition model would depend on the evolution and phylogenetic position of the taxa. For these reasons future ontogenic studies, which are done less and less frequently, will gradually provide more information as the number of species studied increases.

Another of the more striking characteristics of this pollen type is the presence of the fluffy material located between the apertural ring which plugs the apertures already formed. To date, this type of material, made of perfectly organised lamellated structures, has been described only as "surface fuzz" for *Rumex acetosa*, as "tubules" for *Aegiceras corniculatum* (Rowley, 1981) or as a "fuzzy surface coating" for *Canna* microspores (Rowley and Skvarla, 1986). In these cases the lamellated structures described are tri-lamellated and cover the surface of the tectum, and as far we know they have not previously been described in the apertures or with a penta-lamellated structure, as described here for *F. densiflora*. The authors, who have found similar types of structures, suggest no specific role for this material. However, in all cases it is material that covers the outside of the pollen. It seems reasonable to assume that this plug made of material with a complex composition and structure plays a protective role in *Fumaria* against desiccation and other damaging effects, since this pollen has 6–8 large-sized pores.

We believe that this material may well be the result of the degradation of the tapetum (although at no time have we seen it inside the locule of the anther), which is highly active in this species, as shown by the constant changes in material observed on the surface of the wall throughout ontogeny (loose fibrous material, dense globules, well-organised material, etc.).

As occurs with other species (Fernández *et al.*, 1992 ; Fernández and Rodríguez García, 1995), the harmomegathy process of this pollen does not centre only upon the apertures; the wall also seems to play a key role. Thus, although the tectum is very thick, it is micro-perforated, whereas the infratectum is barely present, the foot layer is lamellated and the endexine is "fragmented" into amorphous islets. As a result, the wall offers little resistance to changes in volume.

Acknowledgements

The authors express their thanks to Dr M.I. Rodríguez-García for all her comments and observations on this paper, as well as for kindly arranging for us to have access to the use of her laboratory. To Mohammed M'rani Alaoui for his technical assistance and to Ross Howard for revising the English-language version of the manuscript.

References

Blackmore, S. (1990). Sporoderm homologies and morphogenesis in land plants, with discussion of *Echinops sphaerocephala* (Compositae). *Plant Systematics and Evolution* (Suppl. 5) : 1–12 .

Blackmore, S. and Crane, P.C. (1988). The systematic implications of pollen and spore ontogeny. In Humphries, C.J. (ed.): Ontogeny and systematics, pp. 83–115. New York. Columbia University Press, and London. British Museum (Natural History).

Blackmore, S., Stafford, P. and Persson V. (1995). Palynology and systematics of Ranunculiflorae. In Jensen, U. and Kadereid, J.W. (eds.) Systematics and Evolution of Ranunculiflorae. *Plant Systematics and Evolution* (Suppl.) 9: 71–82.

Candau, P. (1987). Fumariaceae. In Valdés, B., Díez, M.J. and Fernández, Y. (De.) Atlas polínico de Andalucía occidental 87–89. Sevilla.

Dahl, A.O. and Rowley, J.R. (1991). Microspore development in *Calluna* (Ericaceae). Exine formation. *Annales Sciences Naturelles, Bot. Paris Serie 13* 11: 155–176.

Dickinson, H.G. and Heslop-Harrison, J. (1968). Common mode of deposition for the sporopollenin of sexine and nexine. *Nature* 220: 926–927.

Dunbar, A. and Rowley, J.R. (1984). *Betula* pollen development before and after dormancy: exine and intine. *Pollen et Spores* 26(3–4): 299–338.

Erdtman, G. (1952). Pollen Morphology and Plant Taxonomy. Angiosperms. Almqvist and Wiksell, Stockholm 539 pp.

Faegri, K. and Iversen, J. (1975). Textbook of Pollen Analysis. 3rd edition. New York. Hafner.

Fernández, M.C. and Rodríguez-García, M.I. (1988). Pollen wall development in *Olea europaea* L. *New Phytologist* 108 : 91–99.

Fernández, M.C., Romero, A.T. and Rodríguez-García, M.I. (1992). Aperture structure and function in *Lycopersicum esculentum* Miller (Solanaceae) pollen grain. *Review of Palaeobotany and Palynology* 72 : 41–48.

Fernández, M.C. & Rodríguez García, M.I. (1995). Pollen grain apertures in *Olea europaea* L. (Oleaceae). *Review of Palaeobotany and Palynology*, 85 : 99–109.

Guédès, M. (1982). Exine stratification, ectexine structure and angiosperm evolution. *Grana* 21 : 161–170.

Le Thomas, A. (1980-1981). Ultrastructural characters of pollen grains of African Annonaceae and their significance for the phylogeny of primitive Angiosperms. *Pollen et Spores* 22: 267–342, 23: 5–36.

Le Thomas, A. and Lugardon, B. (1972). Sur la structure fine des tétrades de deux Annonacées (*Asteranthe asterias* et *Hexalobus monopetalus*). CRS *Academie Scientifique Paris, ser D* 275: 1749–1752.

Lidén, M. (1986a). Synopsis of Fumarioideae (Papaveraceae) with a monograph of the tribe Fumarieae. *Opera Botanica* 88: 5–115.

Lidén, M: (1986b). *Fumaria* L. In Castroviejo, S. *et al.* (Eds.) Flora Ibérica 1: 447–467.

Morales Torres, C., Mendoza, R. and Romero, A.T. (1988). La posición sistemática de *Papaver argemone* L. interés evolutivo del Orden Papaverales. *Lagascalia* 15: 181–189.

Punt, W.S., Blackmore, S., Nilsson, S and Le Thomas, A. (1994). Glossary of pollen and spore terminology. LPP contribution series no. 1. Utrecht, 106 pp.

Rowley, J.R. (1981). Pollen wall characters with emphasis upon applicability. *Nordic Journal of Botany* 1: 357–380.

Rowley, J.R. (1987-1988). Substructure within the endexine, an interpretation. *Journal of Palynology* 23–24: 29–42.

Rowley, J.R. and Skvarla, J.J. (1986). Development of the pollen grain wall in *Canna*. *Nordic Journal of Botany* 6: 39–65.

Simpson, B.B. and Skvarla, J.J. (1981). Pollen morphology and ultrastructure of *Krameria* (Krameriaceae): utility in question of intrafamilial and interfamilial classification. *American Journal of Botany* 68: 277–294.

Thiéry, J.P. (1967). Mise en évidence des polysaccharides sur coupes fines en microscopie. *Journal Microscopy* 6: 987–1018.

Suárez-Cervera, M., Le Thomas, A., Goldblatt, P., Márquez, J., Seoane-Camba, J.A. (2000). The channelled intine of *Aristea major*: ultrastructural modifications during development, activation and germination. In: M.M. Harley, C.M. Morton and S. Blackmore (Editors). Pollen and Spores: Morphology and Biology, pp. 57–71. Royal Botanic Gardens, Kew.

THE CHANNELLED INTINE OF *ARISTEA MAJOR*: ULTRASTRUCTURAL MODIFICATIONS DURING DEVELOPMENT, ACTIVATION AND GERMINATION

[*1]MARÍA SUÁREZ-CERVERA, [2]ANNICK LE THOMAS, [3]PETER GOLDBLATT, [1]JESÚS MÁRQUEZ, [1]JUAN A. SEOANE-CAMBA

[1]Department of Botany, Faculty of Pharmacy, University of Barcelona,
Av. Diagonal s/n, 08028 Barcelona, Spain;
[2]Laboratoire de Phanérogamie, Museum national d'Histoire naturelle,
16 rue Buffon, F-75005 Paris, France;
[3]Missouri Botanical Garden, P.O. Box 229, St. Louis, Missouri 63166-0229, USA

Abstract

Ultrastructural modifications of the thick and channelled intine during the processes of maturation, activation and germination are described for the monosulcate pollen grains of *Aristea major* (Iridaceae). The intine of this species is three-layered. At the aperture the middle intine is very thick and conspicuously channelled. At the time of channelled layer formation the plasmalemma forms long prolongations (microvilli); when the channelled intine is already thick, Golgi vesicles appear to enter the microvilli and seem to be related to enlargement of the channels. At the time of activation before the germination process, the apertural exintine is ruptured, the material situated in the channels is released, and the endintine undergoes strong hydration. The pollen tube develops near one end of the sulcus. In this region the channelled intine layer undergoes progressive gelation and eventually disappears. The role of the channelled layer is discussed, and it is compared with other analogous channelled layers in monocotyledons.

Introduction

The Afro-Madagascan genus *Aristea* (Iridaceae), with about 50 species, is remarkably diverse in pollen aperture variations (Goldblatt, 1990). A preliminary cladistic analysis suggests that the genus originated in the Cape region (Goldblatt and Le Thomas, 1997). The pollen grains are simple, heteropolar, and monosulcate, with a wide distal aperture (Fig. 2). The exine is thin and the intine is three-layered, with a middle intine that is very thick and conspicuously channelled at the aperture (Le Thomas *et al.*, 1996, 1997, and Fig. 1). This study focuses on formation of the channelled intine layer and seeks to determine the role of this layer in the germination process of *Aristea major* pollen grains.

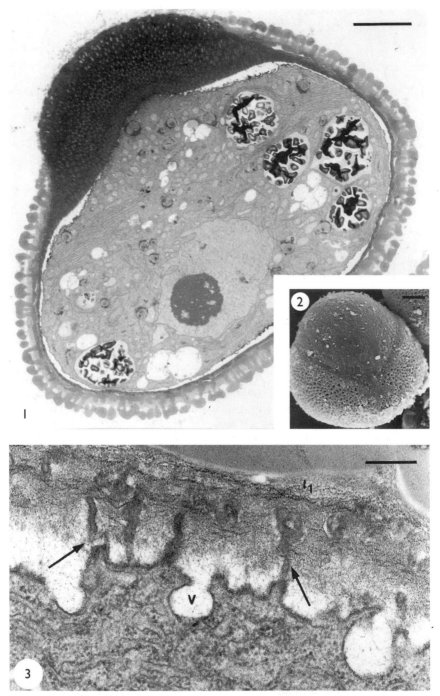

FIGS. 1–3. Fig. 1. Cross section through the aperture. Fig. 2. SEM. General view. Fig. 3. Young pollen grain. Deposition of the middle intine (or channelled intine) on the non-apertural region. Long prolongations of plasmalemma (arrows) penetrating into the matrix, large vesicles (V) with fibrillar material fused with the plasmalemma. Note the exintine (11). Bars: 5 μm (1, 2), 0.25 μm (3).

Materials and Methods

Young and mature anthers from *Aristea major* Andrews (Cape, P. Goldblatt) were used. In vitro germination: pollen was collected from freshly collected anthers and sprinkled onto a 0.7% agar medium with 0.35 M sucrose, 300 mg/l $CaCl2$ and 100 mg/l boric acid, for 10, 15 or 30 minutes and 2–4 hours at room temperature. After these times the process was interrupted and pollen grains were removed and fixed. Fixation and embedding: the samples were fixed in 2% paraformaldehyde and 2.5% glutaraldehyde solution in 0.1 M cacodylate buffer for 12 hours at room temperature. They were washed in 0.1 M cacodylate buffer and post-fixed with 1% osmium tetroxide in 0.8% phosphate buffered $K_3Fe(CN)_6$ for 7 hours at 4°C (Weber, 1992). This was followed by dehydration in acetone and embedding in Spurr's resin. For localisation of neutral polysaccharides, sections were treated with the periodic acid thiocarbohydrazide silver protein (PA-TCH-SP) method (Thiéry, 1967). To detect total lipids we used the osmium-thiocarbohydrazide-osmium (OTO) technique (Seligman *et al.*, 1966). The Thiéry test was used without periodic acid oxidation for the detection of unsaturated lipids (Rowley and Dahl, 1977). Phosphotungstic acid (PTA) in acetone was used for detection of glycoproteins (Marinozzi, 1968). Finally, the sections were stained with uranyl acetate for 15 min. and lead citrate for 5 min. and observed in a Philips EM301.

Results

Channelled intine deposition

The deposition of the middle intine (or channelled intine) takes place at the free microspore stage and the end of vacuolate phase. In the cytoplasm, mitochondria, numerous stacks of Golgi apparatus with electronlucent vesicles, and endoplasmic reticulum with thick lumina are observed. Close to the wall, large vesicles with fibrillar material have fused with the plasmalemma (Fig. 3). The plasmalemma forms prolongations or microvilli, which penetrate into the matrix of the middle intine (Figs. 3, 4). Later, these microvilli seem to become channels. At this time, the endoplasmic reticulum is situated near the plasmalemma, contributing to the constitutive secretion of the microvilli (Fig. 4). Near the end of channel deposition, the Golgi apparatus is located near the plasmalemma. When the channelled intine is already thick, abundant Golgi vesicles seem to enter the microvilli (Fig. 5). These vesicles and lamellae seem to be related to enlargement of the channels. Polysaccharides, lipids and glycoproteins are detected in the plasmalemma and in the microvilli (Figs. 4–6).

In the next stage, the endintine is deposited (Fig. 8), and the three-layered intine is completed (Figs. 8, 9). In the non-apertural region the exintine is irregular, skirting the inner part of the foot layer (Fig. 9), the middle intine shows a small number of channels, and the endintine is fibrillar and homogeneous. At the aperture the exintine is reduced to just a thin layer, the endintine is thin, and the middle intine is very thick and channelled (Fig. 10). The thick channels are observed to be empty, with remnants of the lamellae, or with amorphous or granular material (Figs. 8–11). The mature apertural channelled intine shows a characteristic polygonal distribution of its channels in tangential sections (Fig. 11). Proteins are stained in the outer part, and in a thin layer of the inner part, of the channelled intine. The endintine is PTA negative. No proteinaceous material has been detected in the channels (Fig. 7). The Thiéry test for polysaccharides and the test for lipids show a slight reaction in the channelled intine.

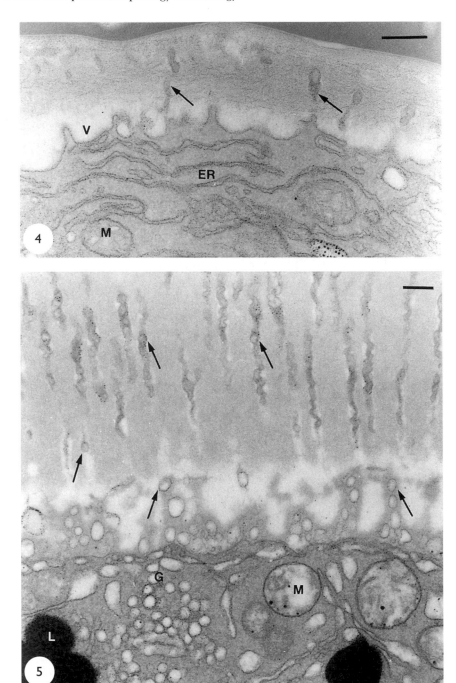

FIGS. 4–5. Young pollen grains. Fig. 4. Thiéry test. The endoplasmic reticulum (ER) is situated near the plasmalemma. Mitochondria (M), vesicles (V), microvilli (arrows). Fig. 5. Test for unsaturated lipids. The channelled middle intine layer is already thick. Note abundant Golgi vesicles near the plasmalemma (G) entering the channels (arrows). Mitochondria (M), lipids (L). Bars: 0.25 μm.

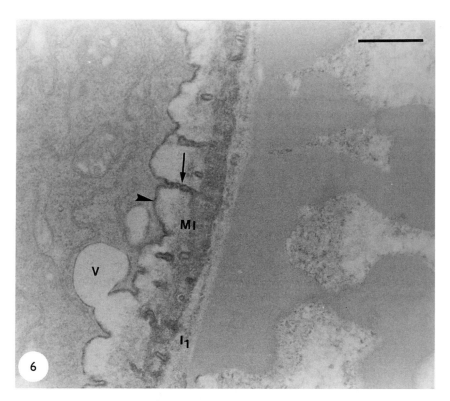

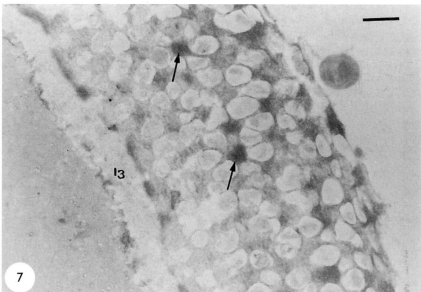

FIGS. 6–7. Young pollen grains. Fig. 6. PTA test. Microvilli (arrow), plasmalemma (arrow head) and outer part of the matrix of middle intine (MI) show a positive reaction for glycoproteins. Vesicles (V) and exintine (11) is PTA negative. Fig. 7. Mature pollen grains, oblique section. PTA test. The glycoproteins are detected in the middle intine (arrows) and are missing in the endintine (13). Bars: 0.5 μm.

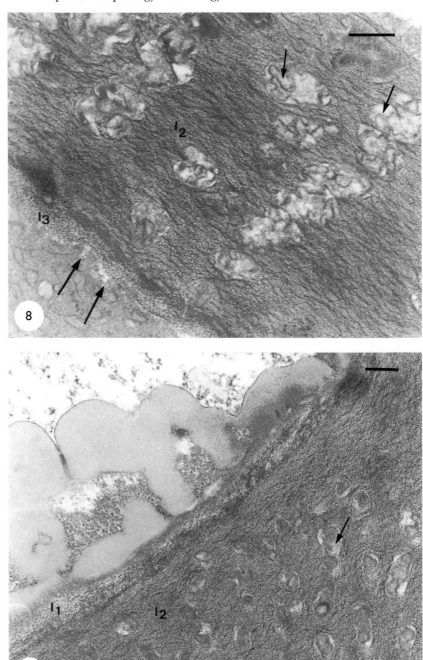

FIGS. 8–9. Young pollen grains. Fig. 8. The endintine (I3) is forming (long arrows). Note the lamellae, possible remnants of Golgi vesicles in the channels (short arrows) of middle intine (I2). Fig. 9. Mature pollen grain. Exintine (I1) near the aperture and middle intine (I2) with amorphous material (arrow). Bars: 0.25 μm.

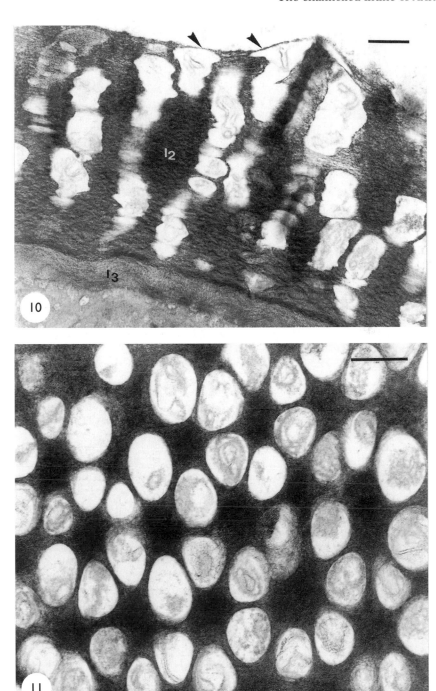

FIGS. 10–11. Apertural region. Fig. 10. Mature pollen grain. The three-layered intine: reduced exintine (arrow heads), large and channelled middle intine (12), and fibrillar endintine (13). Fig. 11. Tangential section. Apertural channelled intine with a characteristic polygonal distribution of its channels. Bars: 0.25 µm.

Activation and germination process

The apertural intine undergoes great modification during activation and germination. At the time of activation (5, 10, 30 min.), hydration is considerable in the endintine, the channelled middle intine shows open channels, because the thin exintine is ruptured, and abundant material is released through the channels. During germination (2–4 hours), at the end of the sulcus where the pollen tube develops, the channelled layer of the intine becomes gelatinous and undergoes progressive disintegration (Figs. 12, 14, 15). However, the channelled intine remains unaltered in the region furthest away from the pollen tube, but the channels appear empty (Figs. 12, 13). When the pollen tube is formed, the channelled intine is very reduced near the tube, and a thick layer, probably of callose material, is situated around the base of the pollen tube (Figs. 16, 17).

Discussion

Intine stratification

The unusual channelled stratum below the exine in many families of monocotyledons was interpreted in the past as: intine (Knox, 1971; Knox and Heslop-Harrison, 1971) in *Gladiolus, Iris* and *Crocus;* middle pectic layer of intine (Heslop-Harrison, 1975, 1979, 1987; Heslop-Harrison and Heslop-Harrison, 1991); intine 2 (Nakamura, 1979) in *Lilium;* outer intine (exintine) (Kress and Stone, 1982; Hesse and Waha, 1983; Kress, 1986; Stone, 1987; Theilade and Theilade, 1996) in Zingiberales; channelled oncus-like zone (Rowley and Skvarla, 1986) in *Canna;* onciform zone (Rowley and Dunbar, 1996; Rowley *et al.*, 1997) in Centrolepidaceae and Strelitziaceae; massive non-acetolysis resistant channelled zone (B-layer) (Kronestedt-Robards and Rowley, 1989) in Strelitziaceae. In our opinion the intine of *Aristea major* is three-layered, in agreement with the structural model proposed by Heslop-Harrison and Heslop-Harrison (1991): the exintine is a well differentiated pectic layer in the non-apertural region. However, it is very thin at the aperture, in contrast to the situation in other monocotyledons such as Poaceae, where the apertural exintine forms a strong *Zwischenkörper* (Heslop-Harrison and Heslop-Harrison, 1980; Márquez *et al.*, 1997a). The pectic-proteinic middle intine has a channelled structure, which is very conspicuous in the apertural region and, as seen in tangential sections, its channels are arranged in a characteristic polygonal structure. The endintine is a fibrillar cellulosic layer.

To confirm that this channelled layer is part of the intine the following considerations can be cited: in *Aristea* the development of the channelled layer is initiated at the free microspore phase, after the vacuolate stage, and when the exintine is completely developed. In *Canna* (Rowley and Skvarla, 1986) and in *Zingiber* (Theilade and Theilade, 1996) the channelled zone develops during the microspore tetrad period and no exintine is described. In other taxa initiation of the intine occurs after microspore mitosis (Blackmore and Barnes, 1990; Rowley *et al.*, 1997), but in some taxa it is present soon after release of the microspores, as in *Gladiolus* (Knox, 1971), *Lilium* (Nakamura, 1979) or *Cucurbita* (Nepi *et al.*, 1995). This thick channelled layer at the aperture is a continuation of the thin channelled layer of the middle intine in the non-apertural region. The strong digitation of the plasmalemma and the intervention of endoplasmic reticulum and Golgi vesicles in the development of the channelled layer are very similar to the observed deposition of other intines as, for example, in *Triticum* (El-Ghazaly and Jensen, 1986, 1987) or *Lolium* (Márquez *et al.*, 1997 a, b), and correspond to the usual localisation of cell wall polymers, which are synthesised in the endoplasmic reticulum and Golgi apparatus (Carpita and Gibeaut,

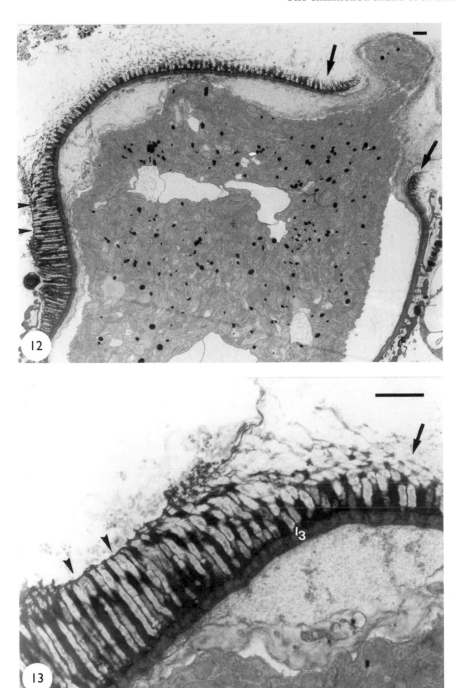

FIGS. 12–13. Germination (2 hours). Fig. 12. Pollen tube growing at the end of the sulcus. The channelled layer of the intine becomes gelatinous and is disintegrated (arrows) near the pollen tube and stays unaltered in the region in some distance of the pollen tube (arrows heads). Fig. 13. Detail of the unaltered channels (arrows heads) and the disintegrated zone (arrow). Note the hydrated endintine (I3). Bars: 1 μm.

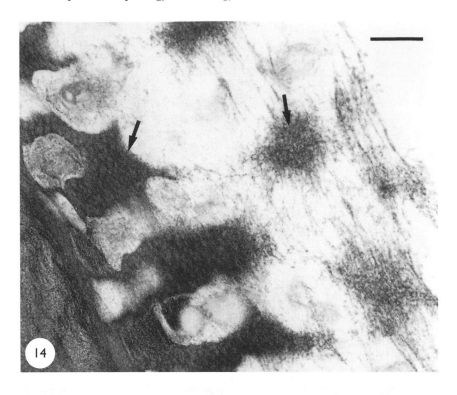

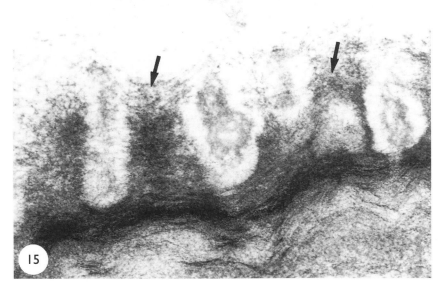

Figs. 14–15. Germination (2 hours). Progressive disintegration of the channelled middle intine. Note the loose structure of the matrix (arrows). Bars: 0.25 μm.

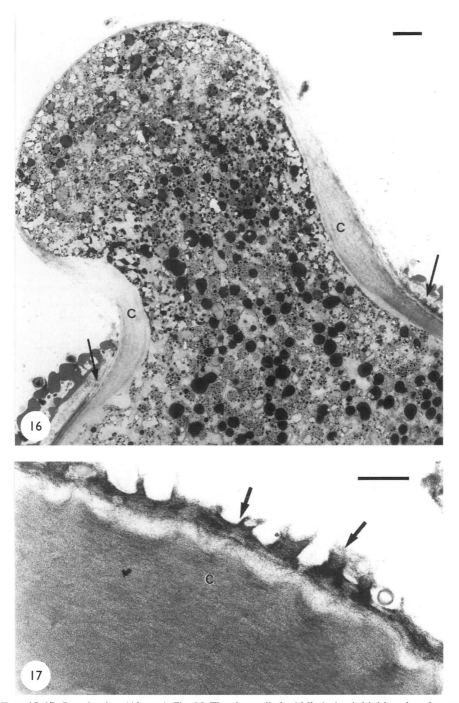

FIGS. 16–17. Germination (4 hours). Fig. 16. The channelled middle intine is highly reduced near the tube (arrows). Note a thick layer, probably of callose material, around the tube (C). Fig. 17. Remnants of middle intine (arrows) and thick layer, probably of callose material (C). Bars: 2 μm (16), 0.5 μm (17).

1993). In *Aristea* reaction to the Thiéry test is low, due possibly to the fact that some cellulose may be present, although pectins seem to represent the main component; however, its reactions to saturated and unsaturated lipids and glycoproteins are similar to those of other intines (Heslop-Harrison, 1987; El-Ghazaly and Grafström, 1995). In intines of both ungerminated and germinated pollen of *Aristea*, a radial gradient of methyl esterified pectins was observed, pectins are less esterified with increasing distance from the plasma membrane and also as activation and germination times advance (Suárez-Cervera *et al.*, in preparation); these observations are similar to those described by Geitmann *et al.* (1995) in *Brugmansia*, and by Li *et al.* (1995) in *Nicotiana*. The cytochemical staining reactions of the channelled layer in *Aristea* show a typical middle pectic layer of the intine (Heslop-Harrison, 1987). A similar disintegration of the middle intine layer of *Aristea* was also observed in the region around the pore, during germination of *Zea mays* (Márquez *et al.*, 1997b).

Functions of the channelled intine

The channels are arranged in a characteristic polygonal structure, which suggests an organised mechanism for protection of the pollen grain at the wide aperture. The partial gelation of the channelled layer in the germination process is a newly recognised system in the pathway of pollen tube growth. The organised and channelled apertural intine may have several important roles in the germination process: to prevent desiccation of the gametophyte during pollination; to maintain the shape of the pollen grain in the hydration process prior to pollen tube growth; to release germinating exudates through the channels; to mark out routes of water and solute ingress during hydration; to control enlargement of the apertural region, by means of the unmodified region of apertural intine, which seems to take on the role of an apertural endexine, which is generally rare in the monocotyledons (Kress and Stone, 1982; Zavada, 1983); to facilitate pollen tube growth, since the channelled layer becomes gelatinous only in the zone where the pollen tube grows, near one end of colpus. This narrow zone is equivalent to the outer intine layer observed in other Iridaceae, with pectins ready for rapid gelation (Knox and Heslop-Harrison, 1971; Heslop-Harrison, 1977). It is also similar to the pathway of the thick exintine of Poaceae (Heslop-Harrison and Heslop-Harrison, 1980, 1991; Márquez *et al.*, 1997a). In *Aristea* the apertural exintine is very thin; its role may be to facilitate the release of the materials located in the channels, which are extruded at hydration and activation times. Two preferred sites for tube emergence, one at each end of the sulcus, were observed in the monosulcate pollen grains of *Narcissus*, where, at the beginning of hydration, changes in the cytoplasm associated with the evolution of the actin cytoskeleton facilitate pollen tube emergence only in the narrow area at the end of the sulcus (Heslop-Harrison and Heslop-Harrison, 1992). The preferred locations in the monosulcate grain are at one or other end of the apertural slit, where the intine microfibrils are shorter and the texture looser, rather than in the central zone (Heslop-Harrison and Heslop-Harrison, 1992). In *Aristea* the developed channelled intine is thinner at the ends of the sulcus and possibly the intine microfibrils are loose there, but it still has a complex structure and its gelation may be the key factor in facilitating pollen tube growth.

Conclusion

The thickened, channelled layer of the apertural intine of *Aristea major* is observed in other species of this genus (Le Thomas *et al.*, 1996, 1997) and other genera of Iridaceae such as *Crocus* (Knox and Heslop-Harrison, 1971; Heslop-Harrison, 1975, 1987),

Gladiolus (Knox, 1971), *Iris* (Knox, 1971; Halbritter and Hesse, 1995), *Micranthus, Moraea, Watsonia* and *Witsenia* (J. Manning, Cape Town, personal communication) and also in *Bobartia, Homeria* and *Romulea* (personal observations). In our opinion, the mechanism of germination observed in *Aristea major* may be extended to other Iridaceae with a thick, channelled intine, and shows a high degree of evolutionary adaptation in the apertural system.

Acknowledgements

We thank Prof. James A. Doyle, University of California Davis, who was generous with his time for helpful discussions with us. The authors are also very grateful to the Scientific Technical Services at the University of Barcelona, for their careful preparation of the sections for TEM used in this study. This work was supported by Grant DGES PB96-0393, Spain.

References

Blackmore, S. and Barnes, S. H. (1990). Pollen wall development in angiosperms. In: S. Blackmore and R. B. Knox (editors). Microspores. Evolution and ontogeny, pp. 173–192. Academic Press, London.

Carpita, N.C. and Gibeaut, D.M. (1993). Structural models of primary cell walls in flowering plants: consistency of molecular structure with the physical properties of the walls during growth. *The Plant Journal* 3: 1–30.

El-Ghazaly, G. and Grafström, E. (1995). Morphological and histochemical differentiation of the pollen wall of *Betula pendula* Roth. during dormancy up to anthesis. *Protoplasma* 187: 88–102.

El-Ghazaly, G. and Jensen, W.A. (1986). Studies of the development of wheat (*Triticum aestivum*) pollen. I. Formation of the pollen wall and Ubisch bodies. *Grana* 25: 1–29.

El-Ghazaly, G. and Jensen, W.A. (1987). Development of wheat (*Triticum aestivum*) pollen. II. Histochemical differentiation of wall and Ubisch bodies during development. *American Journal of Botany* 74: 1396–1418.

Geitmann, A., Hudak, J., Vennigerholz, F. and Walles, B. (1995). Immunogold localisation of pectin and callose in pollen grains and pollen tubes of *Brugmansia suaveolens*. Implications for the self-incompatibility reaction. *Journal of Plant Physiology* 147: 225–235.

Goldblatt, P. (1990). Phylogeny and classification of Iridaceae. *Annals of the Missouri Botanical Garden* 77: 607–627.

Goldblatt, P. and Le Thomas, A. (1997). Palynology, phylogenetic reconstruction, and classification of the Afro-Madagascan genus *Aristea* (Iridaceae). *Annals of the Missouri Botanical Garden* 84: 263–284.

Halbritter, H. and Hesse, M. (1995). The convergent evolution of exine shields in angiosperm pollen. *Grana* 34: 108–199.

Heslop-Harrison, J. (1975). The physiology of the pollen grain surface. *Proceedings of the Royal Society of London* 199: 275–299.

Heslop-Harrison, J. (1979). Pollen walls as adaptive systems. *Annals of the Missouri Botanical Garden* 66: 813–829.

Heslop-Harrison, J. (1987). Pollen germination and pollen tube growth. *International Review of Cytology* 107: 1–78.

Heslop-Harrison, J. and Heslop-Harrison, Y. (1980). Cytochemistry and function of the Zwischenkörper in grass pollens. *Pollen et Spores* 22: 5–10.

Heslop-Harrison, J. and Heslop-Harrison, Y. (1991). Structural and functional variation in pollen intines. In: S. Blackmore and S.H. Barnes (editors). Pollen and spores: patterns of diversification, Systematics Association Special Volume No 44, pp. 331–343. Clarendon Press, Oxford.

Heslop-Harrison, Y. (1977). The pollen-stigma interaction: pollen tube penetration in *Crocus*. *Annals of Botany* 41: 913–922.

Heslop-Harrison, Y. and Heslop-Harrison, J. (1992). Germination of monocolpate angiosperm pollen: evolution of the actin cytoskeleton and wall during hydration, activation and tube emergence. *Annals of Botany* 69: 385–394.

Hesse, M. and Waha, M. (1983). The fine structure of the pollen wall in *Strelitzia reginae* (Musaceae). *Plant Systematics and Evolution* 141:285–298.

Knox, R.B. (1971). Pollen-wall proteins: localisation, enzimic and antigenic activity during development in *Gladiolus* (Iridaceae). *Journal of Cell Science* 9: 209–237.

Knox, R.B. and Heslop-Harrison, J. (1971). Pollen-wall proteins: electron microscopic localisation of acid phosphatase in the intine of *Crocus vernus*. *Journal of Cell Science* 8: 727–733.

Kress, W.J. (1986). Exineless pollen structure and pollination systems of tropical *Heliconia* (Heliconiaceae). In: S. Blackmore and I.K. Ferguson (editors). Pollen and spores: form and function. pp. 329–345. Linnean Society Symposium Series 12, Academic Press, London.

Kress, W.J. and Stone, D.E. (1982). Nature of the sporoderm in monocotyledons, with special reference to the pollen grains of *Canna* and *Heliconia*. *Grana* 21: 129–148.

Kronestedt-Robards, E.C. and Rowley, J.R. (1989). Pollen grain development and tapetal changes in *Strelitzia reginae* (Strelitziaceae). *American Journal of Botany* 76: 856–870.

Le Thomas, A., Suárez-Cervera, M. and Goldblatt, P. (1996). Deux types polliniques originaux dans le genre *Aristea* (Iridaceae-Nivenioideae): Implications phylogéniques. *Grana* 35: 87–96.

Le Thomas, A., Suárez-Cervera, M. and Goldblatt, P. (1997). Pollen ultrastructure of some South-African species of *Aristea* (Iridaceae). 3rd Symposium of African Palynology, pp. 25 (Abstracts).

Li, Y.Q., Faleri, C., Geitmann, A., Zhang, H.Q. and Cresti, M. (1995). Immunogold localisation of arabinogalactan proteins, unesterified and esterified pectins in pollen grains and pollen tubes of *Nicotiana tabacum* L. *Protoplasma* 189: 26–36.

Márquez, J., Seoane-Camba, J.A. and Suárez-Cervera, M. (1997a). Allergenic and antigenic proteins released in the apertural sporoderm during the activation process in grass pollen grains. *Sexual Plant Reproduction* 10: 269–278.

Márquez, J., Seoane-Camba, J.A. and Suárez-Cervera, M. (1997b). The role of the intine and cytoplasm in the activation and germination processes of Poaceae pollen grains. *Grana* 36: 328–342.

Marinozzi, V. (1968). Phosphotungstic acid (PTA) as a stain for polysaccharides and glycoproteins in electron microscopy. In: Proceedings 4th European Regional Conference on Electron Microscopy, pp. 55–56. Rome.

Nakamura, S. (1979). Development of the pollen grain wall in *Lilium longiflorum*. *Journal of Electron Microscopy* 28: 275–284.

Nepi, M., Ciampolini, F. and Pacini, E. (1995). Development of *Cucurbita pepo* pollen: ultrastructure and histochemistry of the sporoderm. *Canadian Journal of Botany* 73: 1046–1057.

Rowley, J. R. and Dahl, A.O. (1977). Pollen development in *Artemisia vulgaris* with special reference to the glycocalyx material. *Pollen et Spores* 19: 169–284.

Rowley J.R. and Dunbar, A. (1996). Pollen development in *Centrolepis aristata* (Centrolepidaceae). *Grana* 35: 1–15.

Rowley J.R. and Skvarla, J.J. (1986). Development of the pollen grain wall in *Canna*. *Nordic Journal of Botany* 6: 39–65.

Rowley J. R., Skvarla, J.J. and Chissoe, W.F. (1997). Exine, onciform zone and intine structure in *Ravenala* and *Phenakospermum* and early wall development in *Strelitzia* and *Phenakospermum* (Strelitziaceae) based on aborted microspores. *Review of Palaeobotany and Palynology* 98: 293–301.

Seligman, A.M., Wasserkrug, H.D. and Hanker, J.S. (1966). A new staining method (OTO) for enhancing contrast of lipid containing membranes and droplets in osmium tetroxide fixed tissue with osmiophilic thiocarbohydrazide (TCH). *Journal of Cell Biology* 30: 424–432.

Stone, D.E. (1987). Developmental evidence for the convergence of *Sassafras* (Laurales) and *Heliconia* (Zingiberales) pollen. *Grana* 26: 179–191.

Theilade, I. and Theilade, J. (1996). Ontogeny of pollen grains in *Zingiber spectabile* (Zingiberaceae). *Grana* 35: 162–170.

Thiéry, J.P. (1967). Mise en évidence des polysaccharides sur coupes fines en microscope électronique. *Journal de Microscopie* 6: 987–1018.

Weber, M. (1992). The formation of pollenkitt in *Apium nodiflorum* (Apiaceae). *Annals of Botany* 70: 573–577.

Zavada, M.S. (1983). Comparative morphology of monocot pollen and evolutionary trends of apertures and wall structures. *Botanical Review* 49: 331–379.

Linder, H.P. (2000). Pollen morphology and wind pollination in angiosperms. In: M.M. Harley, C.M. Morton and S. Blackmore (Editors). Pollen and Spores: Morphology and Biology, pp. 73–88. Royal Botanic Gardens, Kew.

POLLEN MORPHOLOGY AND WIND POLLINATION IN ANGIOSPERMS

H.P. LINDER

Bolus Herbarium, University of Cape Town, Rondebosch 7700, South Africa

Abstract

Pollen characteristics are critical in pollination syndromes. Yet most of the attention in searching for the attributes of the syndromes has been directed at floral morphology, and little attention has been paid to the pollen morphological attributes. Standard pollination biology textbooks list smooth pollen ornamentation and a size range of 20-50 μm as correlated attributes of wind pollination, and the functional value of these attributes has been explored in theoretical terms. However, the correlation between aperture type and wind pollination has not been explored. I show that there is a correlation between these two traits. Mapping them onto phylogenies indicates a pattern where the plesiomorphic, elongate aperture type persists in most anemophilous lineages, but that in the most diverse anemophilous lineages (Cyperales-Poales, Urticales, and the Betulaceae – Juglandaceae) the pollen apertures are circular. Statistical testing indicates that circular apertures evolved more commonly in anemophilous than in biotically-pollinated families, and in addition that anemophilous lineages with circular pollen apertures are more speciose than anemophilous lineages with elongate pollen apertures. However, the fossil record does not support the hypothesis that lineages with circular aperturate pollen grains might persist longer, and this implies that circular pollen apertures might somehow be linked to more rapid speciation. The functional advantage of circular apertures in wind dispersed pollen is unknown.

Introduction

Anemophily has received substantial interest over the past couple of decades, from many different aspects. The efficacy with which various structures can trap pollen has been explored in a series of wind-tunnel experiments (Buchmann *et al.*, 1989; Camazine and Niklas, 1984; Niklas, 1981; 1985; 1987; Niklas and Buchmann, 1985; 1987). This work showed that pollen-collecting structures, such as gymnosperm cones and grass inflorescences, might vary in their ability to trap pollen of different densities. The bulk of the attention in these studies has been on the collection organs, rather than on the pollen morphology.

The importance of pollen morphology in pollination has been investigated in some detail in biotically-pollinated plants. In 1982 Ferguson and Skvarla described correlations between pollination vector and pollen surface morphology which cut across the taxonomic relationships of the plants, leading them to suggest that these ornamentation differences might be adaptations to the pollination mode, rather than reflective of the phylogenetic history of the plants. Hemsley and Ferguson (1985) extended these correlations to hummingbird pollination, showing that hummingbird pollinated flowers have drier pollen than passerine pollinated flowers. In 1986 Ferguson and Pearce, in a study of the pollen of *Bauhinia* which shows some correlations between pollen morphology and pollination type, also note "It may be postulated that failure to directly correlate pollen characters with pollinator can be the results of secondary adaptation from one pollination mode or pollen vector with the concept of constraints limiting further associated exine modification.".

Hesse conducted similar work on the pollen morphological attributes of anemophilous pollen. In a series of excellent papers he compared closely related anemophilous and entomophilous European species of the Ranunculaceae, Hamamelidaceae, Platanaceae, Fagaceae (Hesse, 1978), Oleaceae, Scrophulariaceae, Plantaginaceae, Asteraceae (Hesse, 1979c), Salicaceae, Tiliaceae, Ericaceae (Hesse, 1979b), and Polygonaceae (Hesse, 1979a). This work demonstrated the importance of the pollenkitt volumes to wind pollination, and related the changing stickiness of the pollen to changes in the ornamentation of the pollen grains. This provided a functional link between two of the commonly cited attributes of anemophilous pollen: smooth ornamentation (Bolick, 1990; Faegri and Van der Pijl, 1979; Proctor *et al.*, 1996; Whitehead, 1983) and dry grains.

The aerodynamic implications of different pollen morphologies have been evaluated in terms of the principles of fluid mechanics, and particularly in terms of the effect different pollen morphologies would have on the Reynolds Number of the grains (Bolick, 1990; Crane, 1986; Whitehead, 1969). Pollen ornamentation may have significant effects on the aerodynamic behaviour of the grains. Medium sized pollen, from 20-40 μm (Whitehead, 1983) - compared to 10-300 μm for angiosperms as a whole - may also be related to aerodynamic properties of the grains, as it would increase the volume to surface ratios.

In this paper I explore the possibility that the aperture type might be linked to wind pollination, and use a range of statistical methods to search for generalisations about the relationship between pollen aperture type and pollination mode. An initial evaluation between a morphological attribute and its putative functional value is based on the correlation between the presence of morphology and its function. This type of analysis was used by Lloyd and Webb (1992) to explore the relationship between floral morphological features and heterostyly. This approach, however, cannot evaluate the causative relationships, as it cannot distinguish between the persistence of plesiomorphic conditions and the joint origination of structure and function. To use the terminology of Gould and Vrba (1982), it cannot distinguish between exaptation and adaptations. It may indicate current utility of a morphology, although this should really be supported by experimental evidence (Baum and Larson, 1991). In addition, care has to be taken with any interpretation of correlation analyses, as they may over-estimate the statistical significance of the relationship, due to pseudo-replication (Harvey *et al.*, 1995).

Almost all the work in searching for a relationship between pollen attributes and wind pollination to date has been based on correlating changes within families (for example, Hesse, 1978; 1979a; 1979b; 1979c). These methods do not fall into the pseudo-replication trap, as they investigate changes in close phylogenetic proximity to each other. However, this means that they cannot be used to explore the consequences of the persistence of these functional attributes (i.e., anemophily), as they investigate only the changes at the origination of the function. They are also reliant on the inclusion of closely related species with different pollination modes. This reduces the possibility of detecting adaptations to wind pollination that may happen only after the evolution of wind pollination, and will find only those attributes which evolve at or very soon after the establishment of wind pollination.

The shortcomings of the methods used by Hesse and by Bolick can be avoided by mapping the attributes over the cladograms to indicate if there are any patterns. Such visual inspection is still widely used as the most sophisticated investigation technique (for example, Carpenter, 1989; Coddington, 1988; Donoghue, 1989; Kohn *et al.*, 1996; Les *et al.*, 1997). It allows the simultaneous comparison of two variables over many lineages, and has the advantage of both mapping transitions from one state to the other, as well as displaying the variation within a large clade that may be consistent for one variable, but not the the other. This has been termed the "homology method" by Coddington (1994). The results are sensitive to character optimisation used, and both accelerated and delayed optimisation methods should be employed to ensure that all logically possible scenarios are investigated. However, there has been criticism of these methods used in the absence of statistical testing (Harvey and Pagel, 1991; Maddison, 1990). Appropriate statistical tests are difficult to design, since all tests are sensitive to biases in the sampling of terminals (Síllen-Tullberg, 1993). However, in the absence of testing it is not possible to make statements about any correlations of less than 100 percent. Statements like "this feature is usually associated with that phenomenon" need to have some probability statement attached to them to make them meaningful.

In this paper I seek to establish whether circular aperturate pollen is more frequently found in wind-pollinated families than in biotically-pollinated families, whether circular aperturate pollen evolved more frequently in wind-pollinated lineages than in other lineages, and finally whether the possession of circular aperturate pollen confers some form of advantage to wind-pollinated lineages.

Methods

Sources of pollen and pollination data

The basic data on pollination mode and pollen morphology were obtained from the interactive database on the Intkey platform (Dallwitz, 1980; Watson and Dallwitz, 1991). These data were checked against the family descriptions published by Cronquist (1981), and in some detailed cases against the palynological literature. Pollination literature is in general scattered, and it is difficult to compile a detailed list of the anemophilous species. The situation is further confused by the occurrence in several groups of ambophilous species.

The pollen terminology used in this paper follows Kremp (1965). I have attempted to simplify the pollen apertures to two functional types: those with circular apertures, and those with elongate apertures. Included under the circular aperture types are ulcerate, porate and foraminate apertures. I therefore did not take into account the number of apertures. Elongate apertures include sulcate, colpate, colporate, rugate, sulculate, spiraperturate and zoniaperturate apertures. This is a very wide group, and clearly includes very disparate elements, with the only uniting theme being that they form slits rather than circular apertures. These functional types would obviously include non-homologous structures, but hopefully they all have the same functional constraints.

Taxonomic units used

The terminal units used are families, rather than species or genera. There are a number of problems with using families as terminal units. Firstly, there is a degree of arbitrariness in the ranking and delimitation of families. Secondly, several of the families used may well not be monophyletic, or may be embedded within other families, this might apply particularly to the anemophilous families with reduced flowers embedded within larger-flowered entomophilous families. However, data are readily available at family rank, while descriptive data at generic rank are largely scattered in the literature and not easily accessible. The phylogenetic relationships among families, although still poorly known, are substantially better researched than relationships among all the genera - for the majority of families these data are simply not available. One of the consequences of using families as terminals is that they are often polymorphic for the interesting characters. This needs to be taken into account when the tests are designed, to ensure that the interpretation will err towards being conservative. Ideally lineages, and not families, should be used, but for this much more substantial phylogenetic information needs to be available.

Phylogeny

Phylogenetic relationships among the monocot families were based on the combined analyses of morphological and *rbcL* data of Linder and Kellogg (1995) and Chase *et al.* (1995). For the dicot families no combined analyses are as yet available, and the phylogenetic relationships based on the *rbcL* data set assembled by Chase *et al.* (1993) and re-analysed by Rice *et al.* (1997) were used.

One of the critical areas concerns the relationships among the "Higher Hamamelid" families: Fagaceae, Betulaceae, Myricaceae, Nothofagaceae, Juglandaceae and Casuarinaceae. The *rbcL* sequence data (Chase, *et al.*, 1993; Rice, *et al.*, 1997), does not provide adequate resolution in this clade. Earlier, morphological, studies indicate that Fagaceae are the sister-group to the rest of the higher Hamamelidae (Nixon, 1989). Restriction site analysis of the inverted repeat in the chloroplast genome indicated that Nothofagaceae is the sister to Fagaceae and a group made up of Casuarinaceae, Juglandaceae, Myricaceae and Betulaceae (Manos *et al.*, 1993). The relationships among the families were finally established by a combined analysis of *matK* and *rbcL* sequence data (Manos and Steele, 1997), and this is the system which is followed in this paper.

Several purely wind-pollinated families are difficult to place. Barbeyaceae, a monotypic family from East Africa and Arabia, is generally considered to be closely related to the Urticales. Its colporate pollen is unique in the order, which has ulcerate pollen, and Cronquist (1981) notes: "The more or less distinct carpels and primitive phloem of *Barbeya* indicate a position fairly low on the evolutionary tree of angiosperms . . ." (p. 189). Myrothamnaceae is generally regarded as being closely related to Hamamelidaceae (Cronquist, 1981), but Endress (1989) argues for a position in the Trochodendrales, near the family Cercidiphyllaceae. However, on *rbcL* sequence data, Cercidiphyllaceae is not considered to be a member of the Trochodendrales. There is clearly no current consensus on the affinities of this strange family of resurrection plants, and so it has been left out of the analysis. Plantaginaceae is not included in the the *rbcL* analysis, but I have followed Cronquist (1981) and Reeves and Olmstead (1998) in placing the family near Scrophulariaceae.

Statistical testing

Character optimisation procedures were achieved using MacClade (Maddison and Maddison, 1993), using both ACCTRAN and DELTRAN optimisations to plot the pollen types and pollination modes over the available cladograms.

Statistical tests were done using Statistica version 5.1 (StatSoft Inc., Tulsa, Oklahoma), employing the Fisher Exact Test. For correlation analyses, the distribution of the two pollen apertures among wind- and biotically-pollinated families was investigated. The same test was used to evaluate transitions from elongate to circular pollen apertures by comparing the number of transitions from circular to elongate and circular to circular (i.e., no change) apertures for both wind and biotically-pollinated families.

Results and Discussion

Correlation analyses

In an initial analysis I searched for a correlation between aperture type (specifically circular apertures versus elongate apertures) and pollination mode. In this analysis I used only families which are monomorphic for pollination type, thus excluding families like Palmae and Moraceae, which include substantial numbers of both wind- and biotically-pollinated species. Where families are polymorphic for aperture types (which is common) they have been scored as separate instances. For example, a family that contains sulcate and sulculate pollen, and is biotically dispersed, would be scored as both sulcate and sulculate, and would therefore have two entries. This may lead to artifacts in the analysis, as it effectively subdivides the unit that is being used in the analysis from a family to a sub-unit of a family. These problems are likely to recur irrespective of the initial taxonomic level chosen, as polymorphism is found at all levels, including the species. This is one of the weaknesses of the correlation approach when studying the relationship between two variables at supraspecific level.

The results (Table 1) indicate that porate and ulcerate apertures are more common in wind-pollinated families than in biotically-pollinated families, while colporate, sulcate and colpate pollen apertures are most common in biotically-pollinated families. To test for significance, the pollen types were simplified to circular (porate, ulcerate and foraminate) and elongate (the remainder, including colporate, sulcate, colpate, rugate, sulculate, spiraperturate and zoniaperturate) apertures (Table 2). Using the Fisher Exact Test indicates that these are significantly different: circular apertures are therefore much more likely to be found in wind-pollinated taxa, than in biotically-pollinated taxa. This may be due either to historical (phylogenetic) factors, or to circular apertures being functionally superior in wind pollination to elongate apertures. A correlation analysis cannot dissect out these two variables (for example, Gould and Lewontin, 1979; Wanntorp, 1983).

TABLE 1. Distribution of aperture types among the wind and biotic pollinated families. Only families which are monomorphic for pollination type are used, but all occurrences of aperture types are recorded.

	Wind		Biotic	
	total	percent	total	percent
Porate, ulcerate	18	44	24	14
Colporate	8	20	56	33
Sulcate, colpate	10	24	58	34
Foraminate	2	5	15	9
Rugate	3	7	0	0
Sulculate	0	0	9	5
Spiraperturate	0	0	1	1
Zoniaperturate	0	0	4	2

TABLE 2. Simplified aperture types in those families that are totally wind or biotically-pollinated. Ulcerate and porate are treated as "circular" aperture types, while all the others are included as "elongate" apertures. The Fisher Exact Test provides a Yates Corrected Chi-test of p = 0.0001.

	Circular apertures	Elongate apertures
Wind pollination	18	23
Biotic pollination	24	143

Mapping aperture types on phylogenies

Mapping pollen aperture types and pollination mode onto a phylogeny provides a graphical impression of the nature of the association between these two features. It is not necessary to include all the families in this approach. Consequently, families which are biotically-pollinated, have elongate apertures, and are not sister-families of anemophilous and/or circular aperturate families were left out because they do not add to the information in the analysis. This is convenient, as it would be difficult to map all families on one figure.

In the monocots ulcerate/porate apertures evolved five times:

1. In the large wind-pollinated clade, consisting of Cyperales, Poales and Typhales. This clade is retrieved by both *rbcL* data (Duvall *et al.*, 1993) and morphological data (Linder and Kellogg, 1995), although Dahlgren *et al.* (1985) did not retrieve it.
2. In Burmanniales, which include Burmanniaceae, Corsiaceae and Thismiaceae (Dahlgren, *et al.*, 1985). This clade was also retrieved by Stevenson and Loconte (1995) and Chase *et al.* (1995). These families are all biotically-pollinated.
3. In *Cyclanthus bipartitus*. This is a single species in the Cyclanthaceae, a family of *ca.* 180 species, and is therefore not mapped.
4. In Lemnaceae. These peculiar water plants may be water-pollinated or biotically-pollinated and derived from the biotically-pollinated Araceae (French *et al.*, 1995; Mayo *et al.*, 1995; 1997), which almost never have ulcerate pollen.
5. In Alismataceae and Limnocharitaceae. Alismataceae are described as polyforaminate (Haynes *et al.*, 1998a) and Limnocharitaceae pantoporate or inaperturate (Haynes *et al.*, 1998b). Since these two families are sisters (Les, *et al.*, 1997) in the Alismatales, circular apertures probably evolved only once in the Alismatales. Both families are fly pollinated, but wind pollination occurs sometimes in Alismataceae (Haynes, *et al.*, 1998a).

The mapping of the pollen aperture types on a reduced cladogram of the monocots (Fig. 1) shows that the link between pollen type and anemophily is rather complex. The Alismatales were not included in this analysis, as they show a complex mixture of pollination types. Circular apertures evolved in two lineages not associated with anemophily, but the large clade of Poales – Cyperales – Typhales – Restionales all have porate pollen, and are all wind-pollinated.

Circular aperture types appear to have evolved many times in the dicots, and a search through the Watson and Dallwitz (1991) database indicates that 101 families have at least some species with ulcerate, porate or foraminate pollen apertures, while relatively few families have only circular apertures. This indicates that the evolution of circular apertures happens frequently within families, but rarely below the origination of the families.

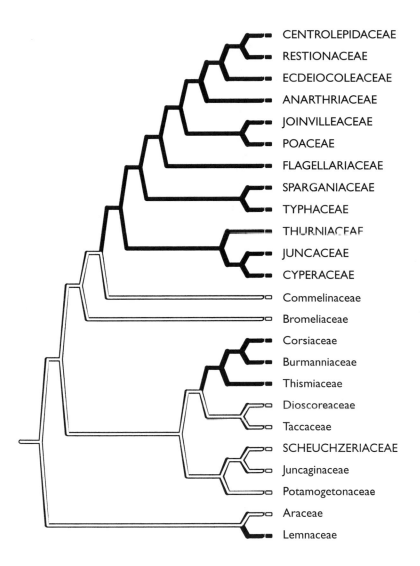

FIG. 1. Phylogeny of monocots, using the families which are completely anemophilous or with circular apertures, as well as their immediate sister families. Lineages with circular apertures are mapped in solid lines, lineages with elongate apertures are mapped with hollow lines, all with DELTRAN optimisations. Anemophilous families are given in capitals, biotically-pollinated families in lower case.

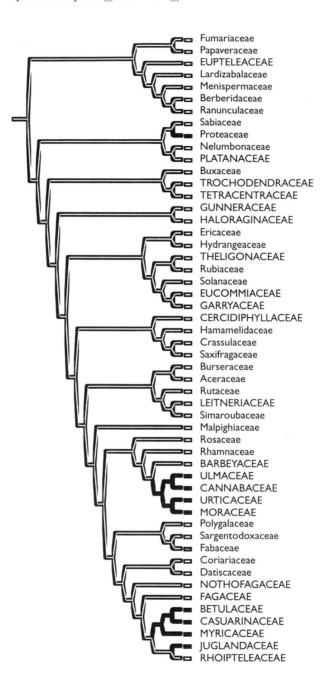

FIG. 2. Phylogeny of the rosids, using the anemophilous families and their closest relatives, based on the *rbc*L phylogeny of Rice *et al.* (1997), with some modifications as indicated in the text. Lineages with circular apertures are mapped in solid lines, lineages with elongate apertures are mapped with hollow lines, all with DELTRAN optimisations. Anemophilous families are given in capitals, biotically-pollinated families in lower case.

Below the family level, the pattern in the dicots (Fig. 2) is similar to the monocots, with the large clades of Urticales and the Betulaceae – Casuarinaceae – Myricaceae with porate pollen. If optimisation is done using ACCTRAN, then this latter clade is larger, including Juglandaceae, with a secondary reversal to elongate pollen in Rhoipteleaceae. In general, in most single family originations of anemophily the families retained the ancestral (elongate) aperture type. In addition, in the Proteaceae, which are largely biotically-pollinated, circular pollen apertures predominate.

Evolution of circular apertures

Mapping pollen types and pollination mode on a phylogeny suggests an association between these traits, but by itself does not provide an indication of whether this association is statistically significant or not.

Any statistical analysis would need to be based on the proportion of possible changes from elongate to circular aperturate pollen that occur in both wind and biotically-pollinated lineages, compared to the number of actual changes observed. This methodology follows that suggested by Síllen-Tullberg (1993). Families polymorphic for pollination mode or pollen aperture type were scored as the optimised basal condition: this was ambiguous only in the higher Hamamelid clade. Using the optimised basal conditions for each family means that the variation within families is ignored, and so the polymorphism in pollen aperture type is conveniently ignored. This means that a great deal of potentially interesting information is lost. Síllen-Tullberg (1993) circumvented the problem of polymorphism by lowering the level at which she worked in polymorphic taxa until the polymorphism was lost. This was not possible in this study, as the data were not available. All families included within the Eudicots in the Rice et al. (1997) analysis were included in the analysis. The pollen aperture types were drawn in, and these optimised manually to the nodes below. Where a terminal was polymorphic, the node below was used to assign a single state to the family. Similarly, pollination mode for polymorphic families was inferred in the same way. The Betulaceae clade was resolved as in the mapping approach. In addition, the Plantaginaceae was included.

This shows that circular aperturate pollen evolves significantly more frequently in wind-pollinated families than in biotically-pollinated families, in both the monocots (Table 3) and the dicots (Table 4). Combining the data sets for all angiosperms will only strengthen this relationship. However, since only transitions at family level were taken into account in this analysis, this is not informative on the patterns within the families. Most transitions to wind pollination occur within families (unpublished data), but there appears to be no obvious reason why these should differ from the patterns above family level.

TABLE 3. Number of transitions from elongate apertures (0) to circular apertures (1) and reversals, and lack of transitions, for monocots. Since it was difficult to resolve Alismatales (which contain a mixture of anemophilous and hydrophilous species) they have been excluded. The Fisher Exact Test for columns 1 and 2 has a significance level of $p = 0.0426$.

	0 -> 0	0 ->1	1-> 0	1->1
Wind-pollinated	0	1	0	13
Biotic pollination	45	1	0	2

TABLE 4. Number of transitions from elongate apertures (0) to circular apertures (1) and reversals, and lack of transitions, for the Eudicots. The Fisher Exact Test for columns 1 and 2 has a significance level p = 0.0013.

	0 -> 0	0 ->1	1-> 0	1->1
Wind-pollinated	17	4	0	3
Biotic pollination	128	2	0	0

Evolutionary implications of circular pollen apertures

Pollen with circular apertures is found more commonly in anemophilous families than in biotically-pollinated families. It has evolved more frequently in these lineages, and is often associated with speciose anemophilous lineages, such as Urticales, the Betulaceae clade, and the Poales – Cyperales clade. None of the isolated anemophilous families have circular aperturate pollen.

There are two possible explanations for this pattern. The first is that circular pollen apertures may enable anemophilous lineages to survive longer. Anemophilous lineages which have circular aperturate pollen, by this argument, would have a lower likelihood of extinction than anemophilous lineages with elongate pollen. The persistence hypothesis is easy to test. Anemophilous pollen is well-represented in the fossil record, as anemophilous plants produce large quantities of pollen and the grains are widely dispersed. Much of the fossil pollen can be identified to family, or at least to groups of families, and it is therefore possible to get reasonable estimates of the minimum age of the families. We obviously need to assume that families or groups of families that are now totally wind-pollinated, would also have been so in the Cretaceous. This is probably a reasonable assumption, as the presence of the families from the Cretaceous is determined from the fossil pollen, and this is likely to have changed if the pollination mode of the plants had changed. The first occurrences of each reasonably well identified family in the fossil record is summarised by Muller (1981). The results show that both elongate and circular aperturate anemophilous lineages date back to the Cretaceous (Fig. 3). If the lineages are grouped into Tertiary and Cretaceous lineages, they are not significantly different, thus indicating that lineages with pollen with elongate apertures can persist as long as circular aperture pollen lineages. The inability to find statistical significance may be a result of the very small sample size. Sulcate pollen has a much older pollen record than either porate or colpate pollen, but sulcate pollen cannot be placed to either wind or biotically pollinated lineages.

The second possible explanation for this pattern is that circular apertures somehow allow for more speciation and evolution to happen, and that they might act as some kind of key innovation that results in radiation (for example, Hodges and Arnold, 1995). The hypothesis that circular aperturate pollen lineages have more species than elongate aperturate lineages is equally easy to test. The pattern of circular aperturate pollen associated with lineages including several families may indicate a greater tendency to speciation or species persistence, rather than lineage persistence. This was tested by summing the number of species in elongate and circular aperturate lineages. Number of species per family was taken from Cronquist (1981), and the results indicate that lineages with less than 100 species are mostly colpate, while lineages with more than 100 species are mostly circular aperturate (Fig. 4). Grouping this into two categories, one with less than 100 species, and one with more than 101 species, shows that the two are significantly different, with p = 0.0769, by Fishers Exact Test.

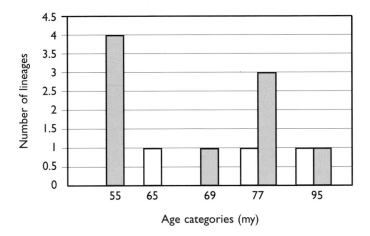

FIG. 3. Bar-graph showing the number of circular aperturate (open) and elongate aperturate (solid) wind-pollinated lineages in age-classes of 0-55, 56-65, 66-69, 70-77 and 78-95 million years. The age of the lineages is based on data in Muller (1981). If the lineages are grouped into Tertiary and Cretaceous, lineages then there is no statistically significant difference between them, by the Fisher Exact Test.

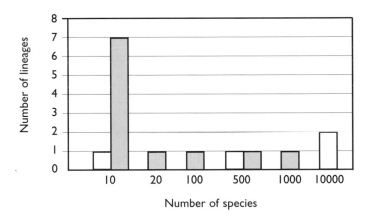

FIG. 4. Bar-graph showing the number of species in wind-pollinated lineages, separated into circular aperturate (open) and elongate aperturate (solid) lineages. If the lineages are grouped into those with less than 100 species, and those with more than 100 species, then they are significantly different, by the Fisher Exact Test (p = 0.0769).

Ideally this should be evaluated using sister-lineages, which would factor out the possibility that these lineages have existed for different times (De Queiroz, 1998), and would therefore have had the same time during which to generate species. Of the three occurrences of circular aperturate pollen in wind-pollinated lineages, we can say the following:

- In the monocots, the Poales – Cyperales clade does not have a colpate sister-lineage. This could be interpreted as an extinct sister-clade, consequently with 0 extant species.
- Urticales (1850 spp.) have Barbeyaceae (1 sp.) as sister-lineage.
- Juglandaceae (60 spp.) has Rhoipteleaceae (1 sp.) as sister. This interpretation is not dependent on the optimisation assumptions tested below for the Betulaceae – Juglandaceae clade.
- If we use DELTRAN optimisation, then circular aperturate pollen is interpreted as originating on the node below the Betulaceae – Juglandaceae clade, and the sister to this large clade would then be Fagaceae. The Betulaceae – Juglandaceae clade has ca. 250 spp. against Fagaceae with 800 spp. With ACCTRAN optimisation, the sister clade to the Betulaceae clade would be Fagaceae.

These are not significantly different, neither by the sign test, nor by the Wilcoxon test. It makes no difference whether the Fagaceae are included in the analysis or not. The basic statistical problem is that the sample size is too small, and clearly for this to work better it would be necessary to increase the sample size, which is not readily possible. Another explanation might be that the aperture shape is a consequence of the grain shape, with circular pores resulting from selection for spherical grains, while elongate pores are associated with prolate grains. Consequently, searching for explanations for the pore shape in the pollination system might be misguided, and it might be necessary to find explanations for the variation in pollen grain shape. I did not attempt to establish a correlation between aperture and grain shape. But even if there were a close correlation, it would still not be clear whether the grain shape is a result of the aperture shape, or *vice versa*. Clearly, there could be numerous other correlations to account for the variation in aperture shape – in this paper only the possible correlations with the pollination mode were examined.

Conclusions

Circular (i.e., porate, ulcerate and foraminate) pollen apertures are more common in anemophilous families than in biotically-pollinated families, and this is amply illustrated by a correlation analysis. Mapping pollen aperture type over pollination type on cladograms of monocots and dicots indicates that in many anemophilous families the plesiomorphic, elongate pollen aperture type persists, thus supporting the contention of Ferguson and Pearce (1986) that this can happen. It is striking that in the three most diverse anemophilous lineages circular pollen apertures are typical. Furthermore, using the statistical tests developed by Síllen-Tullberg, it can be shown that circular pollen apertures evolve much more commonly in wind-pollinated families than in biotically-pollinated families.

Anemophilous lineages with circular pollen apertures are not more ancient than elongate aperturate lineages, indicating that the circular apertures do not confer greater persistence on their owners. Though curiously, these lineages are more species-rich. It is not evident whether the pores generate speciation, or whether it is some other, co-incident, mechanism.

The patterns suggest that there is some selective advantage to circular pollen apertures in anemophilous lineages, an advantage which results in a greater species and family diversity in the possessors of this feature. However, what this advantage could be remains obscure, and therefore within the realm of speculation and discussion.

Acknowledgements

I would like to thank the Foundation for Research Development (Pretoria) for funding this research, the organisers of the meeting for contributing to my travel costs, and the University of Cape Town for leave to attend the pollen meeting. Mike Crisp, Tony Verboom, Sioban Munro and two referees are thanked for reading the manuscript, and for numerous constructive comments.

References

Baum, D.A. and Larson, A. (1991). Adaptation reviewed: a phylogenetic methodology for studying character macroevolution. *Systematic Zoology* 40: 1–18.

Bolick, M.R. (1990). The pollen surface in wind-pollination with emphasis on the Compositae. *Plant Systematics and Evolution, Supplementary volume* 5: 39–51.

Buchmann, S.L., O'Rourke, M.K. and Niklas, K.J. (1989). Aerodynamics of *Ephedra trifurca*. III. Selective pollen capture by pollination droplets. *Botanical Gazette* 150: 122–131.

Camazine, S. and Niklas, K.J. (1984). Aerobiology of *Symplocarpus foetidus*: interactions between the spathe and spadix. *American Journal of Botany* 71: 843–850.

Carpenter, J.M. (1989). Testing scenarios: wasp social behaviour. *Cladistics* 5: 131–144.

Chase, M.W., Soltis, D.E., Olmstead, R.G., Morgan, D., Les, D.H., Mishler, B.D., Duvall, M.R., Price, R.A., Hills, H.G., Qiu, Y.-L., Kron, K.A., H., R.J., Conti, E., Palmer, J.D., Manhart, J.R., Sytsma, K.J., Michaels, H.J., Kress, W.J., Karol, K.G., Clark, W.D., Hedren, M., Gaut, B.S., Jansen, R.K., Kim, K.-J., Wimpee, C.F., Smith, J.F., Furnier, G.R., Strauss, S.H., Xiang, Q.-Y., Plunkett, G.M., Soltis, P.S., Swensen, S.M., Williams, S.E., Gadek, P.A., Quinn, C.J., Eguiarte, L.E., Golenberg, E., Learn, G.H., Graham, S.W., Barrett, S.C.H., Dayanandan, S. and Albert, V.A. (1993). Phylogenetics of seed plants: an analysis of nucleotide sequences from the plastid gene *rbcL*. *Annals of the Missouri Botanical Garden* 80: 528–580.

Chase, M.W., Stevenson, D.W., Wilkin, P. and Rudall, P.J. (1995). Monocot systematics: a combined analysis. In: P.J. Rudall, P.J. Cribb, D.F. Cutler and C.J. Humphries (editors). Monocotyledons: systematics and evolution, pp. 685–730. Royal Botanic Gardens, Kew.

Coddington, J.A. (1988). Cladistic tests of adaptational hypotheses. *Cladistics* 4: 3–22.

Coddington, J.A. (1994). The roles of homology and convergence in studies of adaptation. In: P. Eggleton and R.I. Vane-Wright (editors). Phylogenetics and Ecology, pp. 53–78. Academic Press, London.

Crane, P.R. (1986). Form and function in wind dispersed pollen. In: S. Blackmore and I.K. Ferguson (editors). Pollen and spores: form and function, pp. 179–202. Academic Press, London.

Cronquist, A. (1981). An integrated system of classification of flowering plants. Columbia University Press, New York.

Dahlgren, R.M.T., Clifford, H.T. and Yeo, P.F. (1985). The families of the monocotyledons: structure, function and evolution. Springer-Verlag, Berlin.

Dallwitz, M.J. (1980). A general system for coding taxonomic descriptions. *Taxon* 29: 41–46.

De Queiroz, A. (1998). Interpreting sister-group tests of key innovation hypotheses. *Systematic Biology* 47: 710–718.

Donoghue, M.J. (1989). Phylogenies and the analysis of evolutionary sequences, with examples from seed plants. *Evolution* 43: 1137–1156.

Duvall, M.R., Clegg, M.T., Chase, M.W., Clark, W.D., Kress, W.J., Hills, H.G., Equiarte, L.E., Smith, J.F., Gaut, B.S., Zimmer, E.A. and Learn, G.H. (1993). Phylogenetic hypotheses for the monocotyledons constructed from *rbc*L sequence data. *Annals of the Missouri Botanical Garden* 80: 607–619.

Endress, P.K. (1989). The systematic position of the Myrothamnaceae. In: P.R. Crane and S. Blackmore (editors). Evolution, systematics, and fossil history of the Hamamelidae, Volume 1: Introduction and lower Hamamelidae, pp. 193–200. Clarendon Press, Oxford.

Faegri, K. and Van der Pijl, L. (1979). The principles of pollination biology. Pergamon Press, Oxford.

Ferguson, I.K. and Pearce, K.J. (1986). Observations on the pollen morphology of the genus *Bauhinia* L. (Leguminosae: Caesalpinioideae) in the neotropics. In: S. Blackmore and I.K. Ferguson (editors). Pollen and spores: form and function, pp. 283–296. Academic Press, London.

Ferguson, I.K. and Skvarla, J.J. (1982). Pollen morphology in relation to pollinators in Papilionoideae (Leguminosae). *Botanical Journal of the Linnean Society* 84: 183–193.

French, J.C., Chung, M.G. and Hur, Y.K. (1995). Chloroplast DNA phylogeny of the Ariflorae. In: P.J. Rudall, P.J. Cribb, D.F. Cutler and C.J. Humphries (editors). Monocotyledons: systematics and evolution, pp. 255–275. Royal Botanic Gardens, Kew.

Gould, S.J. and Lewontin, R.C. (1979). The spandrels of San Marco and the Panglossian paradigm: a critique of the adaptationist programme. In: J. Maynard-Smith and R. Holliday (editors). The evolution of adaptation by natural selection, pp. 147–164. Royal Society, London.

Gould, S.J. and Vrba, E.S. (1982). Exaptation - a missing term in the science of form. *Paleobiology* 8: 4–15.

Harvey, P.H. and Pagel, M.D. (1991). The comparative method in evolutionary biology. Oxford University Press, Oxford.

Harvey, P.H., Read, A.F. and Nee, S. (1995). Why ecologists need to be phylogenetically challenged. *Journal of Ecology* 83: 535–536.

Haynes, R.R., Les, D.H. and Holm-Nielsen, L.B. (1998a). Alismataceae. In: K. Kubitzki (editor). The families and genera of vascular plants, pp. 11–18. Springer Verlag, Berlin.

Haynes, R.R., Les, D.H. and Holm-Nielsen, L.B. (1998b). Limnocharitaceae. In: K. Kubitzki (editor). The families and genera of vascular plants, pp. 271–275. Springer Verlag, Berlin.

Hemsley, A.J. and Ferguson, I.K. (1985). Pollen morphology of the genus *Erythrina* (Leguminosae: Papilionoideae) in relation to floral structure and pollinators. *Annals of the Missouri Botanical Garden* 72: 570–590.

Hesse, M. (1978). Entwicklungsgeschichte und Ultrastruktur von Pollenkitt und Exine bei nahe verwandten entomophilen und anemophilen Angiospermensippen: *Ranunculaceae, Hamamelidaceae, Platanaceae* und *Fagaceae. Plant Systematics and Evolution* 130: 13–42.

Hesse, M. (1979a). Entwicklungsgeschichte und Ultrastruktur von Pollenkitt und Exine bei nahe verwandten entomo- und anemophilen Angiospermen: *Polygonaceae. Flora* 168: 558–577.

Hesse, M. (1979b). Entwicklungsgeschichte und Ultrastruktur von Pollenkitt und Exine bei nahe verwandten entomo- und anemophilen Angiospermen: *Salicaceae, Tiliaceae* und *Ericaceae. Flora* 168: 540–557.

Hesse, M. (1979c). Entwicklungsgeschichte und Ultrastruktur von Pollenkitt und Exine bei nahe verwandten entomophilen und anemophilen Sippen der *Oleaceae, Scrophulariaceae, Plantaginaceae* und *Asteraceae. Plant Systematics and Evolution* 132: 107–139.

Hodges, S.A. and Arnold, M.L. (1995). Spurring plant diversification: are floral nectar spurs a key innovation? *Proceedings of the Royal Society of London B* 262: 343–348.

Kohn, J.R., Graham, S.W., Morton, B., Doyle, J.J. and Barrett, S.C.H. (1996). Reconstruction of the evolution of reproductive characters in Pontederiaceae using phylogenetic evidence from chloroplast DNA restriction-site variation. *Evolution* 50: 1454–1469.

Kremp, G.O.W. (1965). Morphological Encyclopedia of Palynology. The University of Arizona Press, Tucson.

Les, D.H., Cleland, M.A. and Waycott, M. (1997). Phylogenetic studies in Alismatidae, II: evolution of marine angiosperms (seagrasses) and hydrophily. *Systematic Botany* 22: 443–463.

Linder, H.P. and Kellogg, E.A. (1995). Phylogenetic patterns in the commelinid clade. In: P.J. Rudall, P.J. Cribb, D.F. Cutler and C.J. Humphries (editors). Monocotyledons: systematics and evolution, pp. 473–496. Royal Botanic Gardens, Kew.

Lloyd, D.G. and Webb, C.J. (1992). The evolution of heterostyly. In: S.C.H. Barrett (editor). Evolution and function of heterostyly, pp. 151–178. Springer-Verlag, Berlin.

Maddison, W.P. (1990). A method for testing the correlated evolution of two binary characters: are gains or losses concentrated on certain branches of a phylogenetic tree? *Evolution* 44: 539–557.

Maddison, W.P. and Maddison, D.R. (1993). MacClade. Analysis of phylogeny and character evolution. Sinauer Associates, Sunderland, Massachusetts.

Manos, P.S., Nixon, K.C. and Doyle, J.J. (1993). Cladistic analysis of restriction site variation within the chloroplast DNA inverted repeat region of selected Hamamelididae. *Systematic Botany* 18: 551–562.

Manos, P.S. and Steele, K.P. (1997). Phylogenetic analyses of "higher" Hamamelididae based on plastid sequence data. *American Journal of Botany* 84: 1407–1419.

Mayo, S.J., Bogner, J. and Boyce, P. (1995). The Arales. In: P.J. Rudall, P.J. Cribb, D.F. Cutler and C.J. Humphries (editors). Monocotyledons: systematics and evolution, pp. 277–286. Royal Botanic Gardens, Kew.

Mayo, S.J., Bogner, J. and Boyce, P.C. (1997). The genera of Araceae. Royal Botanic Gardens, Kew.

Muller, J. (1981). Fossil pollen records of extant angiosperms. *Botanical Review* 47: 1–142.

Niklas, K.J. (1981). Airflow patterns around some early seed plant ovules and cupules: implications concerning efficiency in wind pollination. *American Journal of Botany* 68: 635–650.

Niklas, K.J. (1985). The aerodynamics of wind pollination. *Botanical Review* 51: 328–386.

Niklas, K.J. (1987). Pollen capture and wind-induced movement of compact and diffuse grass panicles: implications for pollination efficiency. *American Journal of Botany* 74: 74–89.

Niklas, K.J. and Buchmann, S.L. (1985). Aerodynamics of wind pollination in *Simmondsii chinensis. American Journal of Botany* 72:

Niklas, K.J. and Buchmann, S.L. (1987). The aerodynamics of pollen capture in two sympatric *Ephedra* species. *Evolution* 41: 104–123.

Nixon, K.C. (1989). Origins of Fagaceae. In: P.S. Crane and S. Blackmore (editors). Evolution, systematics, and fossil history of the Hamamelidae, vol. 2, Higher Hamamelidae, pp. 23–44. Clarendon Press, Oxford.

Proctor, M., Yeo, P. and Lack, A. (1996). The natural history of pollination. Harper Collins, London.

Reeves, P.A. and Olmstead, R.G. (1998). Evolution of novel morphological and reproductive traits in a clade containing *Antirrhinum majus* (Scrophulariaceae). *American Journal of Botany* 85: 1047–1056.

Rice, K.A., Donoghue, M.J. and Olmstead, R.G. (1997). Analysing large data sets: *rbc*L 500 revisited. *Systematic Biology* 46: 554–563.

Síllen-Tullberg, B. (1993). The effect of biased inclusion of taxa on the correlation between discrete characters in phylogenetic trees. *Evolution* 47: 1182–1191.

Stevenson, D.W. and Loconte, H. (1995). Cladistic analysis of monocot families. In: P.J. Rudall, P.J. Cribb, D.F. Cutler and C.J. Humphries (editors). Monocotyledons: systematics and evolution, pp. 543–578. Royal Botanic Gardens, Kew.

Wanntorp, H. (1983). Historical constraints in adaptation theory: traits and non-traits. *Oikos* 41: 157–160.

Watson, L. and Dallwitz, M.J. (1991). The families of Angiosperms: automated descriptions, with interactive identification and information retrieval. *Australian Journal of Systematic Botany* 4: 681–695.

Whitehead, D.R. (1969). Wind pollination in the angiosperms: evolutionary and environmental considerations. *Evolution* 23: 28–35.

Whitehead, D.R. (1983). Wind pollination: some ecological and evolutionary perspectives. In: L. Real (editor). Pollination Biology, pp. 97–107. Academic Press, Inc., Ontario.

Zavada, M.S., Anderson, G.J., and Taylor, T.N. (2000). The role of apertures in pollen germination: a case study from *Solanum appendiculatum*. In: M.M. Harley, C.M. Morton and S. Blackmore (Editors). Pollen and Spores: Morphology and Biology, pp. 89–97. Royal Botanic Gardens, Kew.

THE ROLE OF APERTURES IN POLLEN GERMINATION: A CASE STUDY FROM *SOLANUM APPENDICULATUM*

MICHAEL S. ZAVADA[1], GREGORY J. ANDERSON[2] AND THOMAS N. TAYLOR[3]

[1]Department of Biology, Providence College, Providence, RI 02918
[2]Department of Ecology and Evolutionary Biology U-43, The University of Connecticut, Storrs, CT 06269
[3]Department of Botany, University of Kansas, Lawrence, KS 66045

The apertures of angiosperm pollen may play many roles. We examined two of the most fundamental: that the apertures represent thin areas in the exine for egress of pollen tubes, and that they serve as avenues of communication with the external environment. To do so, we studied the distribution of hydrolytic enzymes during pollen tube initiation and growth in an unusual *Solanum*. *Solanum appendiculatum* is functionally dioecious, with males and females both bearing pollen, but of different types. Males of this species typically bear viable, tricolporate pollen that produces a normal pollen tube as is typical of *Solanum*. In the tricolporate pollen type, after incubation to stimulate *in vitro* pollen tube growth, there is increased enzymatic activity in the cytoplasm and the intine, and subsequent association of enzymatic activity with the growing tip of the pollen tube. The females of this species bear inaperturate pollen that, while viable and serving as a pollinator reward, is incapable of producing a pollen tube. This study shows that the inaperturate exine sequesters the pollen cytoplasm from the environmental stimuli that initiate pollen tube growth. This supports the assertion that the apertures in the pollen can play a fundamental role in communication between the pollen cytoplasm and the external (stigmatic) environment.

Introduction

Solanum appendiculatum Humb. & Bonpl. ex Dunal. is one of a few dioecious species of *Solanum* (Anderson 1979; Anderson and Symon 1989; Knapp *et al.* 1998). This species is unusual among dioecious angiosperms because it is cryptically dioecious in the sense that both males and females are morphologically hermaphroditic, though functionally unisexual. The ovaries of the males bear ovules, but these never develop into fruits or seeds respectively. The anthers of both sexes bear pollen, but the pollen is dimorphic: that of the male flowers is typical for the genus (for example, Anderson and Gensel 1979) in being tricolporate, while that of the female flowers is inaperturate. Both the tricolporate and the inaperturate bears the usual constituents, and both types are viable (stainable with several different stains, including tetrazolium) (Anderson and Levine 1982; Levine and Anderson 1986). Thus, both types function as pollinator rewards, in fact the only reward to pollinators in the nectarless *Solanum* flowers (Anderson and Symon 1989). However, only the tricolporate pollen is capable of germinating and producing pollen tubes (Anderson and Levine 1982; Levine and

Anderson 1986). The tricolporate pollen has a tectate columellate exine subtended by an intine that thickens in the apertural region (Zavada and Anderson, 1997). A notable feature of the inaperturate pollen is the occurrence of thickenings in the inner exine and intine that are reminiscent of the intinous thickenings associated with the apertures in the tricolporate pollen type (Zavada and Anderson, 1997). The outer exine of the inaperturate pollen is tectate columellate and is continuous.

The pollen aperture is defined as a modification of the sporopolleninous exine from which the pollen tube emerges (Kremp, 1965 and references therein; Traverse, 1988). How the morphological modifications of the exine in the apertural region affect the developing, incipient pollen tube is a matter of speculation. It can be hypothesised that under favourable hydrologic conditions the force exerted on the pollen wall in the apertural region by the developing pollen tube physically breaks open the wall and permits pollen tube exit. Inherent in this hypothesis is the assumption that the pollen wall in the apertural region is a frangible structure. In *S. appendiculatum* the inability of the inaperturate pollen to produce a pollen tube may be due to the continuous exine that does not provide an avenue of exit for the developing pollen tube: that is, it acts as a structural barrier preventing the rupture of the pollen wall and exit of the initiated pollen tube. However, apertures are known to be multifunctional structures. Apertures may also act as avenues for the passage of substances that may provide a stimulus for the initiation of the pollen tube between the pollen cytoplasm and the external environment (Heslop-Harrison, 1976; Kress and Stone 1983). In contrast, the continuous exine in the inaperturate pollen of *S. appendiculatum* may act as a barrier preventing the environmental signals from the stigma that initiate pollen tube development from reaching the cytoplasm of the pollen. Thus, the pollen tube is never initiated (Zavada and Anderson, 1997).

Dioecy is an effective way of promoting outbreeding. *Solanum appendiculatum* is unusual among dioecious plants because this was the first known angiosperm to incorporate this kind of pollen morphology with this type of sex expression.

In this study we examine the changes in the distribution and frequency of hydrolytic enzymes (particularly cellulase) at the onset of pollen tube initiation in the tricolporate and inaperturate pollen of this *Solanum* species. It might be expected that if the continuous exine in the inaperturate pollen is acting as a physical barrier to pollen tube egress, the distribution and frequency of the hydrolytic enzymes would be similar in the tricolporate and inaperturate pollen during pollen tube initiation. However, if the distribution and frequency of these enzymes is different in the two pollen types, that is, high activity in the tricolporate pollen type, and low or no appreciable change in activity in the inaperturate pollen when compared to the endogenous levels of the enzymes prior to pollen tube initiation, this would suggest that the continuous exine may effectively sequester the pollen cytoplasm from the external stimuli that initiate pollen tube germination.

Materials and Methods

Pollen Preparation for Immuno-staining

The pollen of *Solanum appendiculatum* was collected from living plants grown in the University of Connecticut greenhouses. The seeds of *S. appendiculatum* came from Mexico and Guatemala, collected by G. Anderson. Pollen was collected from the plants at anthesis and in vitro pollen tubes were initiated by incubating the pollen in a 0.01% w/v solution of boric acid (Rosen, 1968). Untreated desiccated pollen (time 0), and pollen treated for 1 hours, 2 hours and 5 hours was collected and fixed in a freshly prepared solution of 3% paraformaldehyde and 1.25% gluteraldehyde in 0.05M phosphate buffer, pH 7.0. Pollen was fixed at room temperature for 8 hours with gentle agitation. Dehydration was carried out for 1 hour in the following percentages of EtOH 10, 20, 30, 50, 75, 95, 100 and 100. Following dehydration the pollen was infiltrated and embedded in L.R. White resin. The pollen was sectioned on an LKB -1 ultramicrotome and the sections mounted on Ni grids. For SEM study, pollen was prepared as above and mounted on SEM stubs with the high vacuum wax Apiezon W-100 (Wachtel, 1980). Pollen for SEM was viewed on a Zeiss 982 Gemini.

Polyclonal Antibodies

Polyclonal antibodies from rabbits immunised with cellulase, and small amounts of hemicellulase, protease and amylase from *Trichoderma virdae* (ICN #152337) were received from the antibody service of the Department of Microbiology, The Ohio State University, Columbus OH. To determine the presence of the appropriate polyclonal antibodies in the rabbit serum and to estimate the titre of the antibodies, a series of precipitation reactions were done where the dilution of the antibody and the dilution of the antigen were varied with the appropriate controls (after Garvey *et al.* 1977).

Western Blot

Protein extraction from the pollen was carried out by excising anthers from both male and female flowers. The anthers and pollen were stored at 4°C overnight. The following day 20 anthers from female flowers and 20 anthers from male flowers were chopped into quarters, placed in micro-centrifuge tubes, agitated to release the pollen from the anthers, and the anthers from the male and female flowers were separately incubated in 1 ml of 15% sucrose and 0.01% boric acid (w/v in dH_2O) for 1 hour. The tubes were then centrifuged at 4000 rpm for 5 minutes. The anther debris and pollen (the pellet) was then transferred to small volume glass piston-like tissue homogeniser with a minimum volume of cold 0.05M phosphate buffer, pH 7.5, and the tissue was homogenised. The homogenate was transferred to 1.5 ml and centrifuged at 12,000 rpm for 10 minutes. The supernatant was drawn off and mixed 4:1 with 4× Sample Buffer (Sample Buffer = 0.25 M Tris, 40% glycerol, 12% SDS with Bromophenol Blue and beta-mercaptoethanol) and approximately 40-50 μL of the sample was applied to the PAGE gel sample wells.

Between 40-50 μL of various concentrations (10.0, 1.0 and 0.25 mg/ml) of the enzymes derived from *Trichoderma virdae* (ICN # 152337) were mixed 4:1 with 4× Sample Buffer and applied to the PAGE gel's sample wells.

The standard we used is the GibcoBRL/Life Technologies brand 'BenchMark' prestained protein ladder (#10748-010), which provided molecular weight standards from 10 to 200 kD. Gels were run on a Hoefer SE 400 at 30 mA, constant current for approximately 2.5 hours. Gels were stained with Coomassie Blue or a Pharmacia Biotech Silver Staining Kit (#17-1150-01). For the transfer of the proteins from the PAGE gels to the nitrocellulose (Bio-Rad # 162-0145) the gels were first trimmed and then run on the Bio-Rad Mini-Trans-Blot Electrophoretic Transfer Cell (#170-3930 and 170-3935). The Bio-Rad Immun.-Blot Assay Kit (#170-6460) was used for the Western Blot detection of proteins.

Immuno-staining

Thin sections were first incubated in 10% sheep serum in PBS for 1 hour at room temperature. The thin sections were then incubated at room temperature in about 20-30 μL drops of rabbit serum with the polyclonal antibodies at a 1:50 dilution in 10%

sheep serum for approximately 1 hour. Although various dilutions were tried, the 1:50 and 1:60 dilutions gave the best results. The thin sections were then washed 4-5 times for 10 m with PBS. The thin sections were then incubated with 10 nm gold-conjugated anti-rabbit Ig at 1:150 dilution in 10% sheep serum for 1 hour at room temperature. Subsequently, the sections were washed 4-5 times for 10 min in PBS, then overnight in a large volume of PBS (about 20 ml). The following morning the sections were washed three times for 10 m each in PBS and double stained for 2 minutes each in uranyl acetate and lead citrate. Each staining was followed by five washes in distilled water. In all preparations, a control was run by deleting the incubation in the primary antibody (rabbit serum) to determine any diffuse or non-specific staining. All of the stained sections were viewed with a Philips EM-300.

In order to get an estimate of the frequency of 10 nm gold particles associated with various cellular structures, a grid with a known area was placed over the TEM photographs. Although the magnifications of the photographs varied, the grid was adjusted to represent 0.324 sq μm. The number of 10 nm gold particles within the grid were then counted to determine the frequency with which the labelled antigen was associated with various cellular structures, for example, cytoplasm, intine and exine. For each wall layer, 15 or more counts were performed. Significant differences among the distribution of gold particles in the various wall layers was determined by running an ANOVA with alpha = 0.05.

Results

Western Blot

The Western Blot demonstrated that the polyclonal antibodies raised to the cellulase (with trace amounts of hemicellulase, protease and amylase from *Trichoderma virdae*) are specific to enzymes of similar molecular weight found in the pollen. Silver stained and Coomassie stained PAGE gels of the purified cellulase (with small amounts of hemicellulase, protease and amylase) showed four bands between 40-64 kD. The Western Blot showed that antibodies were present to the four bands between 40-64 kD, however, antibodies were also present to unknown components of the purified antigens at about 30 kD and 87 kD. These are presumed to be some component of the carrier that the proteins were delivered in (Text-Fig. 1). Silver stained and Coomassie stained PAGE gels of the pollen protein extract showed that a wide array of proteins were present in the pollen. The Western Blot of the pollen proteins showed three bands between 52-64 kD that were of similar molecular weight to the cellulase extract of *T. virdae*, which also contained small amounts of hemicellulase, protease and amylase (Text-Fig. 1). The pollen proteins are presumed to include the cellulase.

Immuno-staining

All of the controls where the primary antibody was omitted exhibited no diffuse staining. Thus, it is assumed that the distribution of 10 nm gold in the various treatments localises the polyclonal antibodies of cellulase, hemicellulase, protease and amylase. In the untreated dessicated tricolporate (Fig. 1) and inaperturate pollen of *Solanum appendiculatum* (time 0), the localisation of low levels of the enzymes are observed in the cytoplasm, intine and exine (Text-Fig. 2; Fig. 1). There is a significant increase in the number of 10 nm gold particles per unit area throughout the intine in the functional tricolporate pollen after 2 hours of treatment

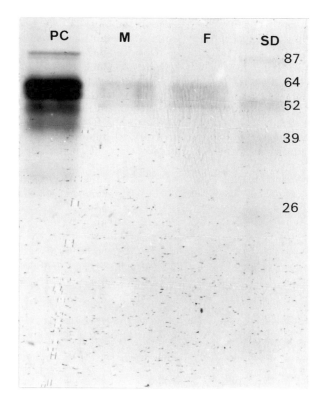

TEXT-FIG. 1. Western Blot showing that the lane with purified cellulase (**PC**) has four bands between 40 - 64 kD (0.0025 mg/ml) that are of the similar molecular weight to three proteins detectable by this method extracted from pollen of the male flowers (**M**, tricolporate pollen) and pollen from the female flowers (**F**, inaperturate pollen). **SD** = Standards, the approximate molecular weight is given to the right of the lane. Note that the additional bands observed in the **PC** lane are presumed to be components of the carrier that the purified cellulase was delivered in.

with 0.01% boric acid to induce pollen tube germination (Text-Fig. 2; Fig. 2). The exine of the tricolporate pollen exhibits no increase or only a slight increase in the number of 10 nm gold particles per unit area (Text-Fig. 2; Fig.2). Concomitant with the increase of activity in the intine, the number of particles in the cytoplasm also increases (Text-Fig. 2). Cytoplasmic activity is primarily associated with the Golgi and Golgi derived vesicles that often appear to be released extracellularly. In the viable, but non-functional inaperturate pollen of *S. appendiculatum*, the number of particles per unit area in the cytoplasm, intine and exine remain at levels similar to the control, unincubated pollen at time 0 (Text-Fig. 2; Figs. 8-10). As the pollen tube emerges from the tricolporate pollen of *S. appendiculatum* and elongates, the highest density of 10 nm gold particles occurs at and near the growing tip of the pollen tube (Figs. 3-7), falling off rapidly in the more proximal portion of the cellulose wall of the pollen tube (Figs. 3, 6-7). Activity remains high in the cytoplasm and cell wall near the growing tip of the pollen tube.

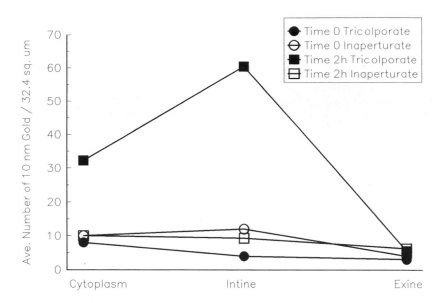

TEXT-FIG. 2. The distribution of 10 nm gold particles in the cytoplasm, intine and exine at time 0 (dessicated control pollen at anthesis) and after 2 hours incubation in 0.01% boric acid in *Solanum appendiculatum.* The distribution of gold particles is similar in the inaperturate and tricolporate pollen types at time 0. After 2 hours of incubation there is a marked increase in the distribution of gold particles in the cytoplasm and intine of the tricolporate pollen type; however, the inaperturate pollen type shows little or no response.

Discussion

The inaperturate pollen of *S. appendiculatum* shows no change in enzymatic activity after incubation in 0.01% boric acid (Text-Fig. 2; Figs. 8-10). It was expected that the appropriate conditions for pollen tube germination existed, conditions that would increase enzymatic activity in the inaperturate pollen. Apparently, the *in vitro* conditions that promote the synthesis of the enzymes in the tricolporate pollen of *Solanum appendiculatum* were not initiated in the inaperturate pollen. This clearly suggests that the signals in the external pollen environment that promote pollen tube growth never reached the cytoplasm of the inaperturate pollen to initiate pollen tube growth. That is, the pollen exine acts as a barrier preventing the passage of substances between the pollen cytoplasm and the pollen environment in the inaperturate pollen from the male flowers of *S. appendiculatum.* This suggests that the apertures, in addition to providing an avenue for pollen tube egress, act as a passage into the pollen cytoplasm for external substances (Heslop-Harrison, 1976; Kress and Stone, 1983).

One of the advantages of mediating sex expression through the pollen morphology is that the pollinators perceive the male and female flowers as identical (i.e. hermaphroditic), thus presumably attracting the same number of pollinator visits to the male and female flowers. Furthermore, floral visitors are not deceived, in that the pollen of both floral types is viable and provides identical nutritional rewards to the pollinators (Levine and Anderson, 1986). This syndrome, while maintaining the outbreeding benefits of dioecy, prevents pollinators from developing strategies to assess the nutritional value of the reward and possibly reducing visitation to pollen-less flowers, or even shifting foraging to alternative reward (i.e. nectar, given that all *Solanum* are nectarless).

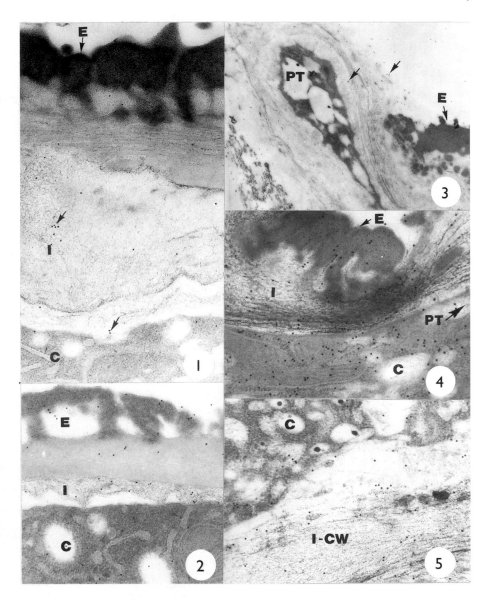

FIGS. 1–5. **Pollen of *Solanum appendiculatum*.** Fig. 1. TEM of the pollen wall of the control (time 0) tricolporate pollen, showing the cytoplasm (**C**), intine (**I**), and exine (**E**). Note the occurrence of a majority of the 10 nm gold particles in the cytoplasm and intine, and only the sparse occurrence of the particles in the exine (arrows), × 48,300. Fig. 2. TEM of the pollen wall of the 2 hour incubated tricolporate pollen after 2 hours growth, showing the cytoplasm (**C**), intine (**I**), and (**E**). Note the increased number of 10 nm gold particles in comparison to figure 1, × 51,000. Fig. 3. TEM of an emergent pollen tube (**PT**) in the tricolporate pollen. Note the abundant association of 10 nm gold particles with the fibrillar wall around the pollen tube (arrows) and low activity in the exine (**E**), × 16,100. Fig. 4. TEM of the juncture between the exine (**E**) and an emergent pollen tube (**PT**) in the tricolporate pollen. Note the distribution of numerous 10 nm gold particles in the cytoplasm (**C**) and intine (**I**), and the relatively few particles associated with the exine (**E**), × 53,900. Fig. 5. TEM of an emergent pollen tube in the tricolporate pollen showing the cytoplasm (**C**) of the pollen tube and the associated fibrillar intine - cell wall (**I-CW**) showing the distribution of 10 nm gold particles, × 65,000.

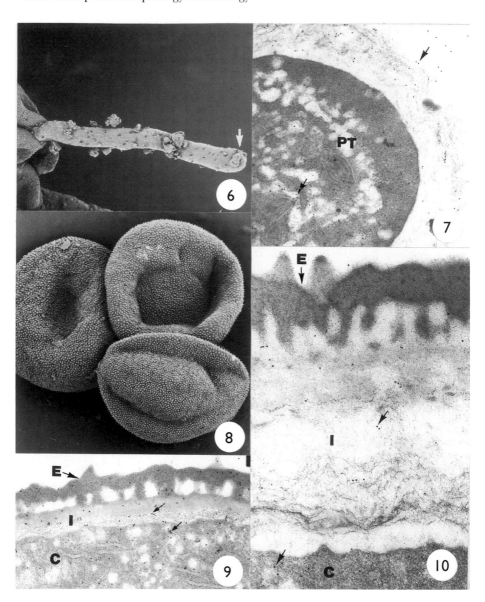

FIGS 6–10. **Pollen of Solanum appendiculatum.** Fig. 6. SEM of the emergent pollen tube from the tricolporate pollen. The arrow indicates the portion of the pollen tube where the high amount of activity is observed and also the portion of the pollen tube in figure 7, × 2,000. Fig. 7. TEM of the growing tip of an emergent pollen tube (**PT**) from the tricolporate pollen showing the abundant distribution of 10 nm gold particles associated with the fibrillar cell wall and the cytoplasm (arrows), × 34,300. Fig. 8. SEM of the inaperturate pollen, × 2,100. Fig. 9. TEM of the pollen wall of the inaperturate pollen after 1 hour of incubation in 0.01% boric acid. The number and distribution of 10 nm gold particles (arrows) in the cytoplasm (**C**), intine (**I**) and exine (E) is similar to the number and distribution of particles in the control tricolporate pollen (Text-Fig. 1), × 34,900. Fig. 10. TEM of the pollen wall of the inaperturate pollen after 2 hours of incubation in 0.01% boric acid. The number and distribution of 10 nm gold particles (arrows) in the cytoplasm (**C**), the thickened intine (**I**), and exine (**E**) is similar to the number and distribution of particles in the control tricolporate pollen (Fig. 1), × 48,300.

Although this type of pollen-mediated dioecy appears to be rare, palynologists frequently take pollen samples that are believed to be representative of a species from only a few individuals in the population. Thus, the frequency of pollen dimorphism may be underestimated, and the reasons for pollen dimorphism under-explored in angiosperms. But, most significantly, the phenomena documented for *Solanum appendiculatum* strengthen the conclusions regarding one of the functions of the apertures: that they are significant passageways for physiological communication with the external environment in mediating the most elemental of pollen functions — the initiation of pollen tube production.

Acknowledgments

The authors would like to thank Andrea Mazzacua and Douglas Mathews of Providence College for their assistance in sample preparation. Maryke Schlehofer for help with growing the plants and reading the manuscript, and Michael Lynes for comments on the manuscript. This project was funded by a grant to MSZ from Providence College, and to GJA by the National Science Foundation and the Research Foundation of the University of Connecticut.

Literature Cited

Anderson, G.J. (1979). Dioecious *Solanum* of hermaphroditic origin is an example of broad convergence. *Nature* 282: 836–838.

Anderson, G.J. and P.G. Gensel. 1976. Pollen morphology and the systematics of *Solanum* section Basarthrum. *Pollen et Spores* 18: 533–552.

Anderson G.J. and D.A. Levine. (1982). Three taxa constitute the sexes of a single dioecious species of *Solanum*. *Taxon* 31: 667–672.

Anderson, G.J. and D.E. Symon. (1989). Functional dioecy and andromonoecy in *Solanum*. *Evolution* 43: 204-219.

Garvey, J.S., N.E. Cremer and D.H. Sussdorf. (1977). Methods in immunology (3rd Edition). W.A. Benjamin, Inc., Reading, MA.

Heslop-Harrison, J. (1976). The adaptive significance of the exine, pp. 27–37. In: The evolutionary significance of the exine (I.K. Ferguson and J. Muller (eds.)), Academic Press, London.

Knapp, S., Persson, V. and S. Blackmore. (1998). Pollen morphology and functional dioecy in *Solanum* (Solanaceae). *Plant Systematics and Evolution* 210: 113–139.

Kremp, G.O.W. (1965). Morphological encyclopedia of palynology. Univ. of Arizona Press, Tucson, Arizona.

Kress, W.J. and D.E. Stone. (1983). Pollen intine structure, cytochemistry and function in monocots, pp. 159–163. In: Pollen: biology and implications for plant breeding (D.L. Mulcahy and E. Ottaviano (eds.)), Elsevier Biomedical, NY.

Levine, D.A. and G.J. Anderson. (1986). Evolution of dioecy in American *Solanum*, pp. 264–273. In: Solanaceae: biology and systematics (W.G. D'Arcy (ed.)), Columbia University Press, New York.

Rosen, W.G. (1968). Ultrastructure and physiology of pollen. *Annual Review Plant Physiology* 19: 435–462.

Traverse, A. (1988). Paleopalynology. Unwin Hyman, Boston, Massachusetts.

Wachtel, A.W. (1980). Thermoplastic wax for mounting SEM specimens. *Scanning*, 3: 302.

Zavada, M.S. and G.J. Anderson. (1997). The wall and aperture development from the dioecious *Solanum appendiculatum*: what is inaperturate pollen? *Grana*, 36: 129–134.

Lush, W.M., Grieser, F. and Spurck, T. (2000). Does water direct the initial growth of pollen tubes towards the stigma of solanaceous plants? In: M.M. Harley, C.M. Morton and S. Blackmore (Editors). Pollen and Spores: Morphology and Biology, pp. 99–108. Royal Botanic Gardens, Kew.

DOES WATER DIRECT THE INITIAL GROWTH OF POLLEN TUBES TOWARDS THE STIGMA OF SOLANACEOUS PLANTS?

W. Mary Lush[1], Franz Grieser[2] and Timothy Spurck[1]

[1]School of Botany and [2]School of Chemistry, University of Melbourne, Parkville, Victoria 3052, Australia

Abstract

Pollen tubes navigate the route from the stigma to the ovules of higher plants with great accuracy, but the mechanisms of guidance are not known. Pollen on the stigma of solanaceous plants is surrounded by an oily exudate in which it hydrates and germinates; pollen tubes grow into the stigma. The environment on the stigma of solanaceous plants was reproduced in culture by isolating the oily exudate from *Nicotiana alata* and immersing pollen in it (or a functional substitute, olive oil). Pollen in exudate or olive oil did not hydrate. When an aqueous medium was introduced to the culture system as a discrete phase, pollen surrounded by exudate or olive oil hydrated, provided it was close to the interface with the aqueous medium. Hydrated pollen grains germinated to produce tubes that grew directionally towards the aqueous medium. The most likely candidate for the pollen tube guidance cue in our *in vitro* cultures is the gradient of water within the oil phase. We propose that the hydrotropic response of pollen tubes is also responsible for their *in vivo* growth towards the stigma.

Introduction

Pollen of solanaceous plants arrives at the receptive surface of the female, the stigma, in a desiccated state (Knox, 1984). Mature stigmas of the Solanaceae are classified as 'wet', but the fluid covering them is an oil and does not contain sufficient water in solution to support pollen hydration and germination when isolated from the rest of the stigma (Konar and Linskens, 1966b). On germination, pollen tubes grow directly into the stigma via intercellular spaces at the surface (Konar and Linskens, 1966a: Herrero and Dickinson, 1979; Cresti *et al.*, 1986; Kandasamy and Kristen, 1987). Some form of guidance, usually suggested to be chemical (Heslop-Harrison and Heslop-Harrison, 1986), presumably operates to direct pollen tubes into the stigma. The pathway subsequently followed by pollen tubes through the style is, for much of its length, a structurally well-defined route. Although chemical guidance cues could operate in the defined transmitting tract, such cues have proved difficult to demonstrate and guidance may be principally by physical means (Heslop-Harrison and Heslop-Harrison, 1986).

Nearly all *in vitro* studies on chemotropism have been carried out in aqueous environments and are thus unlikely to be useful for understanding pollen behaviour in the lipid environment on the stigma of solanaceous plants. We have developed *in vitro* systems in which pollen is immersed in exudate, or a substitute oil, close to an interface with an aqueous medium. In one version of this assay, *Nicotiana tabacum* L. pollen germinated and appeared to grow towards the source of water (Wolters-Arts *et al.* 1998). However, interpretation of this assay with *N. tabacum* is difficult because water could move directly between grains, and because components of the aqueous medium and some of the oils used may have been toxic to pollen. By using a modified assay developed for *Nicotiana alata* Link et Otto, we show here that pollen tubes grow towards the aqueous phase of cultures, probably in response to a gradient in the concentration of water in the exudate or a substitute oil. We propose that the hydrotropic response of pollen tubes could also be responsible for guidance on the stigma.

Materials and Methods

Pollen culture

Plant materials and culture methods were as described in Lush *et al.* (1998) and are outlined in brief below. *N. alata* plants, self-incompatibility genotypes S_2S_2 and S_6S_6, were used as sources of pistils, stigma exudate and pollen. Incompatible reactions did not occur *in vitro*, consistent with the finding that the exudate has no role in the rejection of self-pollen in *N. alata* (Pandey, 1963), which occurs within the style (Lush and Clarke, 1997). Exudate was collected using a micromanipulator. Refined olive oil was used as a substitute for exudate (Meadowlea Foods Ltd., Mascot, NSW). The exudate of solanaceous plants is mostly triglycerides with traces of carbohydrates and proteins (Konar and Linskens, 1966b; Cresti *et al.*, 1986). The mixture of saturated and unsaturated fatty acids in the triglycerides is similar to that of olive oil. Cultures were established by injecting a drop of a pollen growth medium (12.5% PEG 6000, 0.15 M sucrose, 1.0 mM $CaCl_2$, 1.0 mM KCl, 0.8 mM $MgSO_4$, 1.6 mM H_3BO_3, 0.03% casein acid hydrolysate, 25 mM MES pH 5.9) into a pool of exudate or oil containing pollen, and covering the cultures with a coverslip.

Routine monitoring of hydration and germination was on a Zeiss IM35 inverted microscope. For time-lapse video sequences, pollen was observed using a Panasonic F250 CCD colour video camera, mounted to either a Zeiss Universal or a Leica DMRB microscope. Illumination was minimised by using a shutter (Uniblitz Model SD-10) set for an interval of 10 s. A signal controller synchronised opening of the shutter and recording of the video image on an optical memory disk recorder (Sony, LVR 6000-LVS 6000P). Images used here were digitised by capturing single frames through a Targa 2000 video capture board, and assembled into plates using Adobe Photoshop.

Observations are based on a minimum of 29 pollen grains/tubes in each treatment (usually >35), and all experiments were repeated at least once. There were at least 4 replicate cultures within each experiment with exudate, and 8 replicate cultures in experiments with olive oil. Cultures established with exudate were controls for the initial experiments with olive oil, and cultures with complete aqueous medium were controls for experiments with media modified by the omission of components. Data were pooled for statistical analysis (chi squared).

Pollen hydration and growth on stigmas

Growth of compatible pollen on stigmas held at relative humidities ranging from 76–100% was examined by light microscopy (Wild Photomakroscop M400) as described in Lush *et al.* (1998). Relative humidity in closed containers holding pistils was controlled using saturated solutions of salts.

Results

The exudate of *N. alata*

As *N. alata* flowers mature, exudate accumulates between the papilla cells of the outer surface of the stigma. When flowers open, much of the stigma is completely covered with exudate, and the amount of exudate continues to increase over the next 2–3 days. The surface of the exudate is covered by a 'skin', which is most clearly visible when the exudate is 'drawn-down' by suction. Dry pollen delivered to the stigma is hydrophobic and most grains immediately enter the exudate. Grains that enter the exudate hydrate and germinate, but these processes are difficult to observe because of light refraction by the exudate and lack of contrast between the cells of the stigma and pollen grains. *N. alata* flowers are pendulous, suggesting that most pollen grains will be suspended in exudate, and this interpretation is supported by the timing of germination (Lush *et al.*, 1998). Some pollen is deposited on papillae that project above the exudate, but these grains do not hydrate.

Exudate collected from stigmas using a micromanipulator was examined by light microscopy. The exudate has some bizarre properties. For example, air blown into the exudate from a syringe tended to maintain the cylindrical shape of the bore of the syringe instead of adopting the usual spherical shape of air bubbles. Most of the particles visible within the exudate soon after it was placed on slides disappeared from view over the following two to three hours (Figs. 1–4). The liquid phase of the exudate was continuous, i.e. it was not an emulsion (Fig. 1), and there was no loss of volume during prolonged (1 to 2 days) exposure to air in the laboratory.

Hydration and germination of pollen grains in oil

The hydration of pollen grains was monitored by observing the change in shape from ellipsoidal in dehydrated grains to spherical in hydrated grains. Pollen placed in exudate isolated from stigmas did not hydrate, consistent with previous findings that the exudate on solanaceous stigmas is not a reservoir of water for pollen germination (Konar and Linskens, 1966b). We found that olive oil could serve as a functional substitute for exudate with respect to pollen hydration, germination and the directional growth of pollen tubes (Figs. 1–4 c.f. Figs. 5–8). The following description applies to pollen in both culture systems (i.e. based on exudate or olive oil). We shall use 'oil' as a generic term to refer to either exudate or olive oil.

When water was injected into pools of oil containing pollen, pollen close to water or in direct contact with water, burst. The bursting of grains within the oil phase created an emulsion that obscured the sharp boundary between the oil and water phases, and resulted in the supply of water to pollen grains that remained viable becoming non-directional.

Pollen grains remained intact when an aqueous medium developed for growth of pollen tubes was injected into the oil. Grains in direct contact with the aqueous medium (i.e. within the aqueous medium or spanning the interface between the medium and the oil) hydrated within 3 minutes and started to germinate after about 30 minutes. Pollen in oil hydrated more slowly than pollen in contact with the aqueous medium (Fig. 5), and the closer grains were to the interface with the aqueous medium, the more rapidly they hydrated (Figs. 1–8). The time required for germination also increased with the distance of pollen from the interface (Figs. 1–8). Grains >100μm from the interface became partially hydrated, but did not germinate (experiments lasting 24 h).

Pollen tubes grew towards the aqueous medium (Figs. 2–4, 6–8, 10–11, 13–15). *N. alata* pollen has three apertures located around the short circumference (with reference to dehydrated grains) of the grain. The direction of tube growth was set by

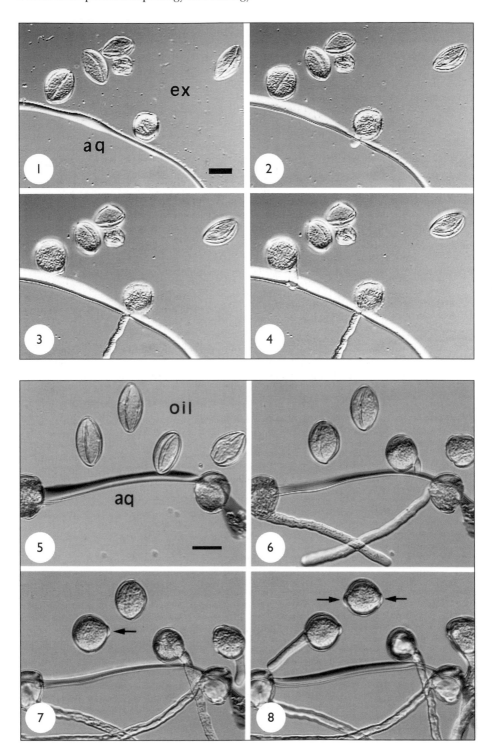

the aperture of emergence, which was usually the one closest to the interface (e.g. Figs. 9, 10), although swellings often occurred at more than one aperture (Figs. 7, 8). The angle of emergence of the tube from the grain also contributed to growth towards the interface, as did curvature of the extending tube towards the interface (Figs. 4, 6–8, 14, 15). Tubes passed into the aqueous medium. Approximately 97% of the 300 grains observed germinating within oil produced tubes that grew towards the interface. All tubes from grains spanning the interface of the aqueous medium and the oil emerged into the aqueous medium (total of 1000 tubes observed).

What is the guidance cue *in vitro*?

Our results show that in the presence of an immiscible reservoir of an aqueous medium, pollen surrounded by oil hydrates, germinates and grows towards the aqueous medium. The relative rates of hydration and germination of pollen at varying distances from the interface show that water is more available to pollen close to the interface. When a reservoir of water is introduced as an immiscible phase in oil, water molecules will enter the oil phase but at low rates and in small numbers because of the low solubility of water in oil. At equilibrium, the random motion of molecules or groups of molecules will ensure that water is uniformly distributed within the oil (Fig. 16). However, if pollen is present in the oil phase it will capture and hold water that diffuses to it. The water captured can only be replaced from the reservoir, resulting in a concentration gradient of water being established within the oil (Fig. 16). A concentration gradient of water in oil is thus one candidate for the *in vitro* guidance cue. However, it is conceivable that gradients in other molecules, for which the aqueous medium is the source, and pollen grains are sinks, could also exist within the oil phase of our cultures. We investigated the potential role of solutes in the aqueous medium by observing the directionality of pollen tube growth in the absence of individual components of the medium. These results are summarised in Table 1.

In our culture system, growth towards the aqueous medium cannot be accounted for by gravity, by the presence of directionally oriented surfaces, or by light. Both O_2 and CO_2 are more soluble in oil than in water and thus will not form directional gradients (with respect to the position of the interface). All of the components of the growth medium could be individually eliminated without affecting the direction of pollen tube growth, with the exception of PEG and water. In the absence of PEG, pollen grains and tubes tended to burst. Nevertheless, observations of the remaining tubes (in

ABBREVIATIONS FOR FIGS. 1–15. aq, aqueous medium; ex, exudate; oil, olive oil; → swellings at apertures. All scale bars represent 25 μm.

FIGS. 1–4. Progressive hydration and germination of pollen in stigma exudate. Frames were taken at the following times after the addition of the aqueous medium. Fig. 1. 45 min. Fig. 2. 100 min. Fig. 3. 140 min. Fig. 4. 155 min. Pollen closest to the aqueous phase hydrated and germinated first, and pollen tubes grew towards and into the aqueous phase.

FIGS. 5–8. Progressive hydration and germination of pollen in olive oil. Frames were taken Fig. 5. 25 min. Fig. 6. 110 min. Fig. 7. 235 min. Fig. 8. 344 min after the addition of the aqueous medium. Pollen spanning the interface between oil and the aqueous medium germinated first, and tubes emerged into the aqueous phase. Other grains germinated in order of their distance from the aqueous phase and grew towards the interface. Swellings could occur at more than one aperture prior to germination.

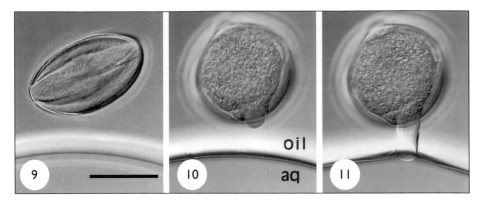

FIGS. 9–11. Pollen grain hydrating and germinating in olive oil. Fig. 9. 10 min. Fig. 10. 125 min. Fig. 11. 135 min. after addition of the aqueous medium. The pollen tube emerged from the aperture adjacent to the interface.

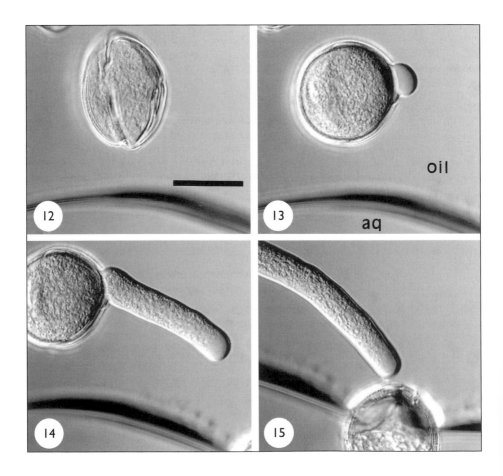

TABLE 1. Potential guidance cues for pollen tubes in *in vitro* cultures

Nature of cue	Present as a vector in in vitro cultures	Role in guidance of pollen tubes
Gravity	No	No
Touch	No	No
Light	No	No
Chemical		
O$_2$ and CO$_2$	No	No
Buffer	No	No
Borate	Possible	No
Ca^{2+}	Possible	No
Salts	Possible	No
Sucrose	Possible	No
Amino acids	Possible	No
PEG	Possible	Very unlikely
Water	Yes	Yes

regions outside regions of emulsion created by bursting tubes) indicated that growth still occurred towards the interface, although with reduced fidelity (79% grew towards the interface compared to 97% of tubes growing in the presence of PEG-containing medium, difference statistically significant).

What is the guidance cue *in vivo*?

Water could be the cue that guides pollen tubes on the stigma. To investigate the role of water gradients at the stigma, pollinated stigmas were enclosed in relative humidities ranging from 76 to 100%. At relative humidities of ≤ 88%, most pollen tubes grew directly into the stigma. At relative humidities above 88%, the incidence of tubes projecting through the exudate and into the surrounding air increased with increasing humidity (Fig. 17).

Discussion

Our culture system was designed to reproduce the environment on the stigma of solanaceous plants by supplying pollen grains in oil with a source of water from a physically separate reservoir of an aqueous medium. Pollen in our system hydrated, germinated and tubes grew towards the aqueous medium with high, but not absolute, fidelity.

No style specific molecules were involved in guidance *in vitro*, because pollen tubes in olive oil grew towards a chemically-defined aqueous medium. All components of the aqueous medium for which pollen is a sink are potential guidance cues because they

FIGS. 12–15. Pollen tube emerging from a grain in olive oil oriented such that all germinal apertures were similar distances from the interface. Fig. 12. 15 min. Fig. 13. 160 min. Fig. 14. 195 min. and Fig. 15. 220 min. after the addition of the aqueous medium. The angle of emergence of the tube resulted in growth towards the interface, and the elongating tube curved towards the interface.

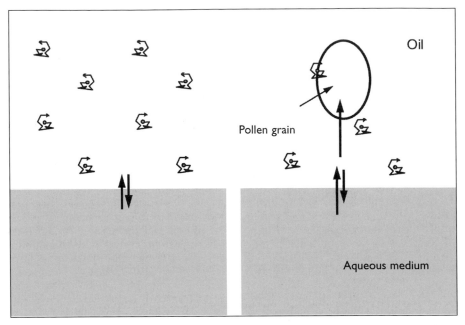

FIG. 16. Models showing the distribution of water in oil at equilibrium with the aqueous medium, and in the presence of pollen in the oil phase. See text for elaboration.

FIG. 17. Random growth of pollen tubes on a stigma held at 99% relative humidity. Scale bar represents 1 mm.

could form gradients within the oil phase. However, all components of the aqueous medium except PEG and water could be omitted without altering the direction of tube growth. Ideally we would define the roles of water and PEG by varying their concentrations independently and studying the response of pollen tubes. Pollen grains and tubes however, rarely exhibit ideal behaviour. Pollen is notoriously sensitive to the composition of *in vitro* growth media, and the response of tubes to changes in composition includes the adoption of abnormal tip morphologies and sharp changes in the direction of tip growth (Lush, unpublished). It is thus to be expected that in non-ideal growth conditions, such as in the absence of PEG, there will be a reduction in the fidelity of growth to the aqueous medium because of reduction in the competence of pollen to respond to the guidance cue.

Despite a long history of experimentation on pollen tube guidance *in vivo*, no generic signals or species-specific guidance molecules have been clearly identified (Heslop-Harrison and Heslop-Harrison, 1986; Mascarenhas, 1993; Cheung *et al.*, 1995; Sommer-Knudsen *et al.*, 1998). In solanaceous species, the presence of a water gradient within the exudate, originating from the aqueous fluid in the intercellular spaces that are open to the exudate or from the cells of the stigma themselves, could be the guidance cue. A gradient of water could also be responsible for the initial direction of growth of pollen from species with dry stigmas, which are surrounded by a lipidic pollen coat. The probability that water is the cue in solanaceous species is increased by our finding that the directionality of *N. alata* pollen tube growth is disrupted in stigmas held at high humidity. The directional growth of pollen on the dry stigmas of several species is also disrupted at high humidity (Knox, 1984; Preuss *et al.*, 1993; Dickinson and Elleman, 1994).

Directional growth of pollen tubes along a water gradient could be a consequence of some fundamental aspect of tip growth. The direction of tip growth depends upon the position of expansion of the cell wall. If tip growth is driven by turgor pressure, then growth towards the source of water will occur if the cell wall is weaker at higher water contents. Alternatively, if tip growth is amoeboid, cell wall material may be preferentially deposited at the more highly hydrated parts of the cell membrane, also resulting in growth towards the source of water. In the turgor-pressure model, the direction of tip growth could be reinforced by the opening of stretch-activated Ca^{2+} channels in the membrane underlying highly hydrated parts of the cell wall, and in the amoeboid model, by the delivery of additional Ca^{2+} channels to highly hydrated parts of the cell membrane.

Guidance of pollen tubes and other tip growing cells by a water gradient is likely to be limited to unusual environments in which substantial differences in the concentration of water can occur over the dimensions of pollen grains and tubes. For this reason we envisage that, although a water gradient is the initial cue for the guidance of pollen tubes to the ovules, water gradients are unlikely to guide pollen tubes through other parts of the style.

References

Cheung, A.Y., Wang, H. and Wu, H-M. (1995). A floral transmitting tissue-specific glycoprotein attracts pollen tubes and stimulates their growth. *Cell* **82**: 383–393.

Cresti, M., Keijzer, C.J., Tiezzi, A., Ciampolini, F. and Focardi, S. (1986). Stigma of *Nicotiana*: ultrastructural and biochemical studies. *American Journal of Botany* **73**: 1713–1722.

Dickinson, H.G. and Elleman, C.J. (1994). Pollen hydrodynamics and self-incompatibility in *Brassica oleracea*. In: A.G. Stephenson and T-H. Kao (editors). Pollen-pistil interactions and pollen tube growth, pp 45–61. American Society of Plant Physiologists, Rockville.

Herrero, M. and Dickinson, H. (1979). Pollen-pistil incompatibility in *Petunia hybrida*: changes in the pistil following compatible and incompatible intraspecific crosses. *Journal of Cell Science* **36**: 1–18.

Heslop-Harrison, J. and Heslop-Harrison, Y. (1986). Pollen-tube chemotropism: fact or delusion? In: M. Cresti and D. Romano (editors). Biology of Reproduction and Cell Motility in Plants and Animals, pp 169–174. University of Siena, Siena.

Jahnen, W., Lush, W.M. and Clarke, A.E. (1989). Inhibition of in vitro pollen tube growth by isolated *S*-glycoproteins of *Nicotiana alata*. *The Plant Cell* **1**: 501–510.

Kandasamy, M.K. and Kirsten, U. (1987). Developmental aspects of ultrastructure, histochemistry and receptivity of the stigma of *Nicotiana sylvestris*. *Annals of Botany* **60**: 427–437.

Knox, R.B. (1984). Pollen-pistil interactions. Encyclopedia of Plant Physiology New Series **17**: 508–608.

Konar, R.N. and Linskens, H.F. (1966a). The morphology and anatomy of the stigma of *Petunia hybrida*. *Planta* **71**: 356–371.

Konar, R.N. and Linskens, H.F. (1966b). Physiology and biochemistry of the stigmatic fluid of *Petunia hybrida*. *Planta* **71**: 372–387.

Lush, W.M. and Clarke, A.E. (1997). Observations of pollen tube growth in *Nicotiana alata* and their implications for the mechanism of self-incompatibility. *Sexual Plant Reproduction* **10**: 27–35.

Lush, W.M., Grieser, F. and Wolters-Arts, M. (1998). Directional guidance of *Nicotiana alata* pollen tubes in vitro and on the stigma. *Plant Physiology* **118**: 733–741.

Mascarenhas, J.P. (1993). Molecular mechanisms of pollen tube growth and differentiation. *The Plant Cell* **5**: 1303–1314.

Pandey, K.K. (1963). Stigmatic secretion and bud-pollinations in self- and cross-incompatible plants. *Naturwissenschraften* **50**: 408–409.

Preuss, D., Lemieux, B., Yen, G. and Davis, R.W. (1993). A conditional mutation eliminates surface components from *Arabidopsis* pollen and disrupts cell signalling during fertilization. *Genes and Development* **7**: 974–985.

Sommer-Knudsen, J., Lush, W.M., Clarke, A.E. and Bacic, A. (1998). Re-evaluation of the role of a transmitting tract-specific glycoprotein on pollen tube growth. *The Plant Journal* **13**: 529–535.

Wolters-Arts, M., Lush, W.M. and Mariani, C. (1998). Lipids are required for directional pollen tube growth. *Nature* **392**: 819–821.

Blackmore, S., Takahashi, M. and Uehara, K. (2000). A preliminary phylogenetic analysis of sporogenesis in pteridophytes. In: M.M. Harley, C.M. Morton and S. Blackmore (Editors). Pollen and Spores: Morphology and Biology, pp. 109–124. Royal Botanic Gardens, Kew.

A PRELIMINARY PHYLOGENETIC ANALYSIS OF SPOROGENESIS IN PTERIDOPHYTES

STEPHEN BLACKMORE[1], MASAMICHI TAKAHASHI[2] AND KOICHI UEHARA[3]

[1]The Natural History Museum, Cromwell Road, London, SW7 5BD, UK
[2]Department of Botany, Faculty of Education, Kagawa University, Takamatsu 760, Japan and
[3]Biological Laboratory, College of Arts and Sciences, Chiba University, 1-33 Yayoi-cho, Inage-ku, Chiba 263, Japan

Abstract

Ultrastructural investigations of spore wall formation have now been carried out by a number of authors so that information is available for at least 18 genera of pteridophytes representing most of the major extant lineages. This paper presents a phylogenetic analysis of sporogenesis in 18 taxa for which ultrastructural information is available. The pattern of relationships derived from this subset of information is less informative than the wider molecular and morphological analyses with which it is compared, but demonstrates the potential for new characters to be derived from comparative studies of spore development.

Introduction

Phylogenetic investigations on pteridophytes initially lagged behind studies of seed plants or bryophytes (Pryer *et al.*, 1995) but new impetus has been provided by molecular sequence data (Raubeson and Jansen, 1992; Hasebe *et al.*, 1993, 1994), the widespread application of cladistic methods (Crane, 1990; Pryer *et al.*, 1995; Stevenson and Loconte, 1996) and inclusion of evidence from the rich fossil record (Bateman and DiMichele, 1994; Rothwell, 1996; Kenrick and Crane, 1997). Despite this renewed attention (see for example contributions in Camus *et al.*, 1996) relationships among pteridophytes remain the most poorly understood within tracheophytes (Kenrick and Crane, 1997; Pryer *et al.*, 1995).

Pteridophytes are paraphyletic with respect to seed plants and the extant taxa represent five monophyletic clades: Lycopodiophyta, Psilotophyta, Equisetophyta, and two groups within the Polypodiophyta: subclasses Polypodiidae and a group comprising Marattiidae plus Ophioglossidae. The basal group is the Lycopodiophyta, comprising the fossil zosterophyllophytes and *Leclerqia* together with the three extant families Isoëtaceae, Selaginellaceae and Lycopodiaceae (Crane, 1990). This clade is sister group to all other vascular plants (Crane, 1990; DiMichele and Skog, 1992). Strong support for this relationship is provided by a chloroplast DNA inversion found in all land plants except bryophytes and Lycopodiophyta (Raubeson and Jansen, 1992). In contrast, ultrastructural characters of spermatogenesis placed *Selaginella* and bryophytes together as a monophyletic group and *Lycopodium* as sister to all other

109

vascular plants (Garbary *et al.*, 1993), but these relationships are not supported by more inclusive analyses. Psilotophyta, comprising the two genera *Psilotum* and *Tmesipteris*, have traditionally been regarded as the most primitive living vascular plants. However, Bierhorst (1977) challenged this view, arguing that *Psilotum* is a morphologically reduced fern, with affinities to *Stromatopteris* (Gleicheniaceae). Most recent analyses continue to place Psilotophyta as the second branch of tracheophytes, above Lycopodiophyta and sister group to all other vascular plants (see for example, Stevenson and Loconte, 1996). Stevenson and Loconte show that Bierhorst's placement of *Psilotum* with *Stromatopteris* is considerably less parsimonious and they highlight the need for molecular studies of *Psilotum*. Equisetophyta are represented by a single extant genus, *Equisetum*, with numerous autapomorphies. Relationships within Polypodiophyta, are resolved in various ways in recent treatments. Smith (1995) referred to a consensus view that eusporangiate ferns, Ophioglossaceae and Marattiaceae, are ancient and isolated lineages. The two families were placed in a basal monophyletic group in the molecular phylogeny of Hasebe *et al.* (1994) although not in the combined molecular and morphological analysis of Pryer *et al.* (1995). On the basis of a cladistic analysis of morphological characters, Stevenson and Loconte (1996) considered these eusporangiate ferns to be monophyletic and regarded them as fern-allies, rather than true ferns. Within the Polypodiidae, or leptosporangiate ferns, Pryer *et al.* (1995) recognised four monophyletic clades.

This study contributes to the phylogenetics of pteridophytes by extending the number of morphological characters available. It builds upon earlier contributions (Blackmore and Barnes, 1987; Blackmore and Crane, 1988; Blackmore *et al.* 1988; Blackmore, 1990) which proposed primary homologies but did not test them by means of a cladistic analysis. Ultrastructural evidence is now available from sufficient pteridophytes to merit a preliminary phylogenetic analysis of sporogenesis. The pioneering work of Bernard Lugardon (1966, 1969a,b, 1972a,b, 1973a,b, 1974, 1976, 1978, 1979, 1986, 1990) established the major details of pteridophyte sporogenesis and a number of other authors have extended the range of taxa investigated (see for example, Buchen and Sievers, 1978a,b, 1981; Parkinson, 1987; Parkinson and Pacini, 1995; Uehara and Kurita, 1989a,b, 1991; Uehara *et al.* 1991; Uffelen, 1990, 1991, 1992, 1993). Although ultrastructural studies of tapetal activity during sporogenesis are scarcer, valuable reviews are available (see for example, Pacini *et al.*, 1985; Pacini, 1990; Parkinson and Pacini, 1995).

In sporogenesis a limited number of ontogenetic processes generate a wide variety of mature forms. This is largely because the developing sporoderm results from an interplay between processes of the haploid microspores of the gametophyte generation and the diploid tapetum and sporangium of the sporophyte. Furthermore, later stages of development in spores are not always contingent upon the successful completion of earlier ones, so that deletion of an entire process may not prove fatal (Blackmore and Crane, 1988). Finally, some aspects of sporogenesis are not under direct genetic control but reflect physical factors and processes of self-assembly (Wodehouse, 1935; Uffelen, 1991; Hemsley *et al.*, 1994). This complexity makes it difficult to define developmental characters and to establish a valid frame of reference for comparison (Blackmore *et al*, 1987; Blackmore and Crane, 1988; DiMichele *et al.*, 1989; Uffelen, 1990). In heterosporous plants different developmental pathways lead to microsporangia and megasporangia. Heterospory has been the focus of many studies because it is widely regarded as a critical step in the evolution of the seed habit (see for example, Chaloner and Pettitt, 1987; Duckett and Pang, 1984; DiMichele *et al.*, 1989; Chaloner and Hemsley, 1991). Crane (1990) showed that heterospory had arisen in at least four major groups of plants and more recently, Bateman and DiMichele (1994) suggested that heterospory has arisen at least eleven times.

In this study the terminal taxa are genera. In some cases the data do derive from a single species but often they are assembled from studies of several species. Where this would result in conflicting data for a particular character that taxon is coded as unknown. In the case of *Lycopodium*, only *L. clavatum* has so far been studied at the ultrastructural level (Lugardon, 1976; Rowley, 1995). Spore development in the other segregate genera has yet to be investigated. The highly selective analysis presented here focuses only on the phylogenetic signal that may be obtained from sporogenesis. It is therefore to be seen as a preliminary analysis that sets out to extend the morphological characters that are available. The next step needs to be the integration of these characters of sporogenesis into a wider morphological and molecular data set. This has not been attempted here because, at present, a combined data set of ultrastructural, morphological and molecular characters can only be assembled for even fewer taxa than are considered here. This lack of comparative data highlights the need for continuing effort on the phylogenetics of pteridophytes, working towards the goal of a much more comprehensive analysis.

Material and Methods

The following 18 taxa were included in this study simply because they have been the subject of relatively complete ultrastructural studies of sporogenesis. The publications cited below provided the sources of information about character states. A total of 25 characters are considered (see Appendix 1 for details of the character states and Fig. 1 for the data matrix). The cladistic analyses were performed using the computer programmes Hennig86 (Farris, 1988), NONA (Goloboff, 1996a) and PIWE (Goloboff, 1996b). Details of the analyses performed are given in Results and Discussion.

Lycopodiophyta

Lycopodium   Lugardon (1976); Uehara and Kurita (1991); Rowley (1995).

Selaginella   Horner and Beltz (1970); Pettitt (1971); Robert (1971a,b,c); Lugardon (1972a); Buchen and Sievers (1978a,b, 1981); Brown and Lemmon (1985); Taylor (1991); Morbelli and Rowley (1993); Hemsley *et al.* (1994); Rowley and Morbelli (1995).

Isoëtes   Uehara *et al.* (1991); Lugardon (1973a, 1976); Robert *et al.* (1973); Takahashi and Blackmore (in prep).

Psilotophyta

Psilotum   Gabarayeva (1984); Parkinson (1987); Lugardon (1973a, 1979).

Equisetophyta

Equisetum   Lugardon (1969b); Lehmann *et al.* (1984); Uehara and Kurita (1989b).

Ophioglossidae

Ophioglossum   Sharma and Singh (1986); Uehara and Kurita (1989a).

Polypodiidae

Anemia   Schraudolf (1984).

Azolla   Konar and Kapoor (1974); Herd *et al.* (1985, 1986).

Athyrium   Schneller (1989).

Belvisia	Uffelen (1992).
Blechnum	Lugardon (1966).
Bolbitis	Hennipman (1970).
Drynaria	Van Uffelen (1990).
Marsilea	Bell (1981, 1985).
Microgramma	Van Uffelen (1992).
Osmunda	Lugardon (1969a, 1972b).
Pteridium	Sheffield and Bell (1979).
Schizaea	Parkinson and Pacini (1995); Parkinson (1995).

	1	2	3	4	5	6	7	8	9	10	11	12	13	14	15	16	17	18	19	20	21	22	23	24	25
Anemia	1	0	0	1	0	0	3	0	0	1	0	0	0	1	0	0	1	0	0	1	1	1	1	0	0
Athyrium	1	0	0	1	0	0	2	0	0	1	0	0	0	1	0	0	2	0	0	1	1	2	1	0	0
Azolla	1	0	0	1	0	0	3	0	0	1	0	0	1	0	0	1	1	1	0	1	1	2	1	0	1
Belvisia	1	0	0	1	0	0	2	0	0	1	0	0	1	1	0	0	2	0	0	1	1	2	1	0	0
Blechnum	1	0	0	1	0	0	2	0	0	1	0	0	0	1	0	0	2	0	0	1	1	2	1	0	0
Bolbitis	1	0	0	1	0	0	2	0	0	1	0	0	0	1	0	0	2	0	0	1	1	2	1	0	0
Drynaria	1	0	0	1	0	0	2	0	0	1	0	0	0	1	0	0	2	0	0	1	1	2	1	0	0
Equisetum	0	0	0	1	0	1	1	0	0	1	0	0	0	1	0	1	1	0	0	1	0	1	2	1	0
Isoëtes	0	1	1	1	0/1	0	2/3	0	1	0	1	1	1	1	1	0	1	0	1	0	0	1	0	0	1
Lycopodium	0	1	0	1	0	0	3	0	1	0	0	2	0	0	0	0	0	0	0	0	0	0	0	0	0
Marsilea	1	0	0	1	0	0	3	0	0	1	0	0	1	1	0	0	2	0	0	0	1	2	1	?	1
Microgramma	1	0	0	1	0	0	2	0	0	1	0	0	0	1	0	0	2	0	0	1	1	2	1	0	1
Ophioglossum	0	0	0	1	0	0	3	0	0	1	0	1	0	1	0	0	0	0	0	0	1	2	1	0	0
Osmunda	1	0	0	1	0	0	3	0	0	1	0	1	0	1	0	0	1	0	1	0	0	2	?	2	0
Psilotum	0	1	0	1	0	0	2	0	0	1	0	0	0	1	0	0	0	0	0	1	1	1	1	0	0
Pteridium	1	0	0	1	0	0	3	0	0	1	0	0	0	1	0	0	2	0	0	1	1	2	1	0	0
Schizaea	1	0	0	1	0	0	2	0	0	1	0	0	0	1	0	0	1	0	0	1	1	1	?	0	0
Selaginella	0	1	1	1	0	0	3	0	0	1	1	1	0	1	0	1	0	1	0	0	0	0	?	0	1

FIG. 1. Data matrix of sporogenesis characters for selected pteridophytes. See Appendix 1 for discussion of characters. Note that for *Isoëtes*, where megaspores differ from microspores, the character state of the former is given first and was used in the analyses. Thus figures 2 and 3 are based on the characters of microspores in the case of *Isoëtes*. Missing data or inapplicable characters are marked "?" in the data matrix.

Results and Discussion

Although information is available on spore development for 18 genera, these provide an inadequate sampling of the full diversity of extant pteridophytes. In particular, the segregate genera based on *Lycopodium sensu lato*, the eusporangiate ferns of the Marattiidae and Ophioglossidae, and the major clades of leptosporangiate ferns have not yet been adequately studied with regard to spore development.

The characters coded here (Fig. 1) are insufficient to produce a fully resolved, or even a well-resolved, cladogram. Characters 4 and 8 are uninformative because, all the taxa share a single character state. They have not simply been abandoned because in a wider analytical context, they would be informative, for example, one that included other land plants. Characters, 5, 6, 15 and 24, are uninformative autapomorphies confined to a single taxon. Both characters 6 and 15, for example, are autapomorphies of *Equisetum*, a genus that has many such unique characters. Given the degree of variation between the taxa it is apparent from ultrastructural studies of spore development, that it ought to be possible to recognise a greater number of new morphological characters. However, because of the relatively limited sampling of taxa and frequent lack of comparative information, we have taken a conservative approach in proposing characters. Many seemingly unique structures and events could potentially provide additional characters, but at present these would appear as autapomorphies and not contribute to better resolution.

Using the option "mhennig*" with the characters coded as non-additive, Hennig86 finds 203 equally parsimonious trees of length 51 steps, consistency index 52 and retention index 61. In the analysis an additional hypothetical taxon with character state zero for each character, was included in order to root the tree. If trees with branches that have zero-length, that is they are not supported by any data, are removed, this reduces to 44 trees. A consensus tree drawn from these 44 trees is shown in Fig. 2, and this also represents the consensus of all 203 trees generated by the initial analysis, since zero-length branches disappear in strict consensus trees. The consensus tree is poorly resolved but places the lycopodiophytes as basal to the other taxa, as in all other analyses. The heterosporous water ferns appear as sister taxa within a largely unresolved group comprising *Ophioglossum*, *Equisetum* and the leptosporangiate ferns. Within this same group, containing all taxa except the lycopodiophytes, *Psilotum* is grouped in a trichotomy with *Anemia* and *Schizaea*.

When the same data (Fig. 1) is analysed using the PIWE programme only four trees are recovered, and after the ambiguous character optimisations are removed a single tree remains (Fig. 3). The PIWE programme implements a weighting protocol that favours the characters with least homoplasy. In this analysis the characters that exhibit least homoplasy are characters 3, 4, 5, 6, 11, 15, 22, 23 and 24. The evidence of these characters gives the final tree which shows better resolution than the consensus tree. However, it is far from fully resolved and still contains polychotomies, for example, within the leptosporangiate ferns. The lycopodiophytes are basal as in the consensus tree but *Osmunda* is placed as basal within the non-lycopodiophyte taxa. This differs from traditional treatments and from the results of most other recent analyses, in which the eusporangiate ferns of the Ophioglossaceae and Marattiaceae are basal (Pryer *et al.*, 1995; Stevenson and Loconte, 1996) to leptosporangiate ferns. In this tree (Fig. 3) the heterosporous water ferns are nested within the Polypodiidae. Other recent treatments also include them within the leptosporangiate ferns, but in a variety of arrangements (see for example, Pryer *et al.*, 1995; Stevenson and Loconte, 1996). The single tree also includes the group found in the consensus tree and comprises *Anemia*, *Schizaea* and *Psilotum*, but here *Psilotum* is placed as sister to the two genera from the Schizaeaceae. This is somewhat reminiscent of Bierhorst's (1977) suggestion of a relationship between *Psilotum* and *Stromatopteris* of the Gleicheniaceae and the characters of sporogenesis do indeed lend support to Bierhorst's concept of *Psilotum* as a morphologically reduced fern. However, neither of the recent, more inclusive phylogenetic analyses support this idea. For instance, in their combined morphological and *rbcL* tree, Pryer *et al.* (1995) have *Psilotum* as sister to the eusporangiate fern *Botrychium* of the Ophioglossales and a similar result was obtained by Stevenson and Loconte (1996). The latter authors pointed out that by changing the coding for *Psilotum* from microphyllous and exannulate to macrophyllous and annulate, *Psilotum* could be placed as the second branch within the true ferns (Filicophyta). However, they demonstrated that in the context of their analysis, to place *Psilotum* as sister to *Stromatopteris* was much less parsimonious and added eleven additional steps to their most parsimonious tree.

Although the characters studied here provide only limited systematic resolution it is possible to discuss the character state changes (shown in Fig. 3), implied by the topology of the most parsimonious tree, in relation to concepts of primary homology proposed by Blackmore and Barnes (1987). Blackmore and Barnes argued, for example, that the plesiomorphic mode of sporoderm formation in land plants involved white line centred lamellae. This is expressed here by character 8, and in all the taxa sampled here, the initial spore wall layer is indeed formed on white line centred lamellae. As Blackmore and Barnes (1987) pointed out, this character only changes in taxa that either have a distinct primexine or lack a sporopollenin wall, both of which are essentially restricted characters of some seed plants. Similarly, Blackmore and Barnes (1987) suggested that a tapetal contribution to the sporoderm, represented here by character 10, was absent in basal land plants, but acquired within both mosses and vascular plants. Among the taxa treated here, a tapetal contribution to the sporoderm occurs in all taxa except *Lycopodium*. However, Blackmore and Barnes (1987) discussed this tapetal contribution primarily in terms of perispore formation which is present both in mosses and pteridophytes except *Lycopodium*. Clearly a variety of structures can result from tapetal wall synthesis, including the addition of tapetal material to layers previously formed from white line centred lamellae (character 10 state 1) which characterises the non-lycopodiophytes and *Selaginella* (Fig. 3) or the formation of an epispore (character 13). This suggests that the discussion in Blackmore and Barnes (1987) provided an incomplete dissection and analysis of the potentially informative characters of sporopollenin deposition. Given the currently limited state of knowledge, the present paper also cannot be seen as definitive.

The phylogenetic pattern of sporoderm synthesis is not simply a sequence of gains of additional wall components. There also appear to have been losses of developmental processes. For example, centripetal deposition of sporoderm layers below the first-formed layer (character 12) occurs in the basal, eusporangiate taxa but not in the relatively derived leptosporangiate taxa, including *Equisetum*. Of the taxa considered here, only *Lycopodium* exhibits character 12 in state 2, involving deposition of an inner wall of amorphous or granular sporopollenin. In many seed plants the endexine constitutes a sporoderm layer deposited on white line centred lamellae, and is formed after the outer layers of ectexine. However, as Blackmore and Barnes (1987) suggested, the formation of ectexine, derived from primexine deposited in the tetrad stage, is a new developmental process representing a non-terminal addition to sporoderm ontogeny.

This analysis therefore confirms the view of Blackmore and Crane (1988) that sporoderm ontogeny is complex, but includes non-terminal additions or deletions of developmental steps, without concomitant loss of reproductive viability. In our view, this reflects the particular aspect of spore ontogeny that is unique among all plant cell walls, namely development through an interplay of haploid, gametophytic and diploid, sporophytic processes.

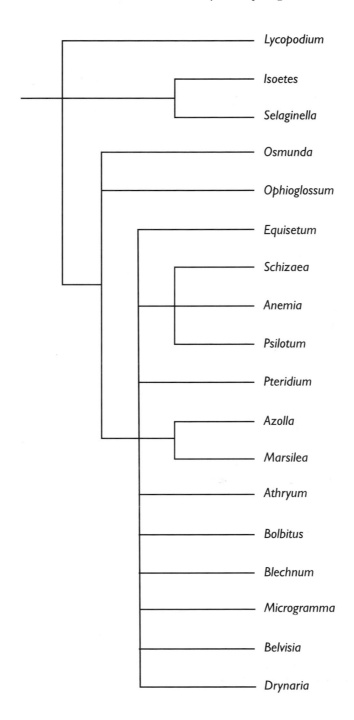

FIG. 2. Consensus of 44 trees

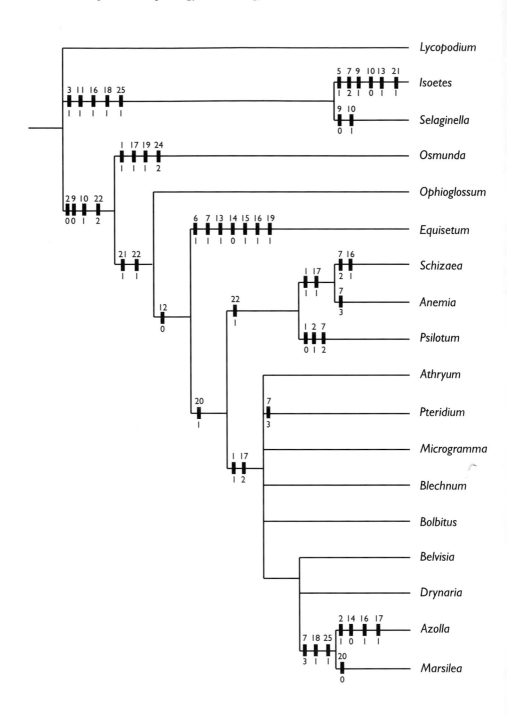

FIG. 3. Single tree obtained using PIWE.

As must be expected from any analysis of a restricted selection set of characters, the results of the analyses presented here are limited. To fully exploit the developmental characters of sporogenesis, the most pressing need is for more taxa to be investigated at the ultrastructural level. However, even at the morphological level it is clear that developmental characters of, for example, the sporangium (Edwards, 1996) would repay greater investigation. Simply to distinguish eusporangiate from leptosporangiate forms is an inadequate analysis but neither is the recognition of additional categories of sporangium a simple task. We consider that, improved resolution of relationships within pteridophytes will require not only the availability of a wider range of molecular sequences, both in terms of taxa and genes, but also a greatly improved sampling of morphological characters. The obvious sequel to the present study would be to extend the analysis of sporogenesis characters and pool this with the data available from the wider morphological and molecular datasets. Unfortunately, at present this cannot be done because there are simply too few taxa in common between the various kinds of study.

Acknowledgements

We thank David Williams for help with the phylogenetic analyses and the Japan Society for the Promotion of Science (JSPS) for a travel grant to SB in support of this study.

References

Bateman, R.M. and DiMichele, W.A. (1994). Heterospory: the most iterative key innovation in the evolutionary history of the plant kingdom. *Biological Reviews* **69**: 345–417.

Bell, P.R. (1981). Megasporogenesis in a heterosporous fern: features of the organelles in meiotic cells and young megaspores. *Journal of Cell Science* **51**: 109–119.

Bell, P.R. (1985). Maturation of the megaspore in *Marsilea vestita*. *Proceedings of the Royal Society of London* **B223**: 485–494.

Blackmore, S. (1990). Sporoderm homologies and morphogenesis in land plants, with a discussion of *Echinops sphaerocephala* (Compositae). *Plant Systematics and Evolution* (Suppl. **5**): 112.

Blackmore, S. and Barnes, S.H. (1987). Embryophyte spore walls: origin, development and homologies. *Cladistics* **3**: 185–195.

Blackmore, S., and Crane, P.R. (1988). Systematic implications of pollen and spore ontogeny. In: C.J. Humphries. (editor). Ontogeny and systematics, pp. 83–115. Columbia University Press. New York.

Blackmore, S., McConchie, C.A. and Knox, R.B. (1988). Phylogenetic analysis of the male ontogenetic programme in aquatic and terrestrial monocotyledons. *Cladistics* **3**: 333–347.

Bierhorst, D.W. (1977). The systematic position of *Psilotum* and *Tmesipteris*. *Brittonia* **29**: 3–13.

Brown, R.C. and Lemmon, B.E. (1985). A cytoskeletal system predicts division plane in *Selaginella*. *Protoplasma* **127**: 101–109.

Brown, R.C. and Lemmon, B.E. (1990). Sporogenesis in bryophytes. In, S. Blackmore and R.B. Knox (editors). Microspores: evolution and ontogeny, pp. 1–10. Academic Press, London.

Buchen, B. and Sievers, A. (1978a). Megasporogenese von *Selaginella*. I. Ultra-strukturelle und cytochemische Untersuchungen zur Sekretion von Polysacchariden. *Protoplasma* **96**: 293–317.

Buchen, B. and Sievers, A. (1978b). Megasporogenese von *Selaginella*. II. Ultrastrukturelle und cytochemische Untersuchungen zur Sekretion von Lipiden. *Protoplasma* **96**: 319–328.

Buchen, B. and Sievers, A. (1981). Sporogenesis and pollen grain formation. In, O. Kiermayer (editor). Cytomorphogenesis in plants, pp. 349–376. Springer-Verlag, Wien.

Camus, J., Gibby, M., Johns, R.J. (1996). Pteridology in perspective. Proceedings of the Holttum Memorial Pteridophyte Symposium, Kew, 1995. Royal Botanic Gardens, Kew.

Chaloner, W.G. and Pettitt, J.M. (1987). The inevitable seed. *Bulletin de la Société Botanique de France, Actualités Botaniques* **134**: 39–49.

Chaloner, W.G. and Hemsley, A.R. (1991). Heterospory: cul-de-sac or pathway to the seed? In: S. Blackmore and S.H. Barnes (editors). Pollen and spores, patterns of diversification, pp. 151–167. Clarendon Press, Oxford.

Crane, P.R. (1990). The phylogenetic context of microsporogenesis. In, S. Blackmore and R.B. Knox (editors). Microspores: evolution and ontogeny, pp. 11–41. Academic Press, London.

DiMichele, W.A., Davis, J.I. and Olmstead, R.G. (1989). Origins of heterospory and the seed habit: the role of heterochrony. *Taxon* **38**: 1–11.

DiMichele, W.A. and Skog, J.E. (1992). The lycopsida: a symposium. *Annals of the Missouri Botanical Garden* **79**: 447–449.

Duckett, J.G. and Pang, W.C. (1984). The origins of heterospory: A comparative study of sexual behaviour in the fern *Platyzoma microphyllum* R. Br. and the horsetail *Equisetum giganteum* L. *Botanical Journal of the Linnean Society* **88**: 11–34.

Edwards, P.J. (1996). The leptosporangium: developing a new awareness of, and a reliable database of these neglected organs. In: J.M. Camus, M. Gibby and R.J. Johns (editors). Pteridology in perspective, pp. 517–521, Royal Botanic Gardens, Kew.

Farris, S. (1988). Hennig86. Version 1.5. Published by the author, Port Jefferson Station, New York.

Gabarayeva, N.I. (1984). The development of spores in *Psilotum nudum* (Psilotaceae). The changes of cytoplasm and organelles of spore mother cells in the premeiotic interphase to leptotene prophase of the first meiosis. *Botaniska Zhurnal* **69**: 1441–1450.

Garbary, D.J., Renzaglia, K.S. and Duckett, J.G. (1993). The phylogeny of land plants: a cladistic analysis based on male plant gametogenesis. *Plant Systematics and Evolution* **188**: 237–269.

Goloboff, P. (1996a). NONA version 1.50. Published by the author, San Miguel de Tucumán, Argentina.

Goloboff, P. (1996b). PIWE version 2.51. Published by the author, San Miguel de Tucumán, Argentina.

Hasebe, M.T., Ito, M., Kofuji, R., Ueda, K. and Iwatsuki, I. (1993). Phylogenetic relationships of ferns deduced from *rbcL* gene sequences. *Journal of Molecular Evolution* **37**: 476–482.

Hasebe, M.T., Omori, T., Nakazawa, M., Sano, T, Kato, M. and Iwatsuki, K. (1994). RbcL gene sequences provide evidence for the evolutionary lineages of leptosporangiate ferns. *Proceedings of the National Academy of Sciences, USA.* **91**: 5730–5734.

Hemsley, A.R., Collinson, M.E., Kovach, W.L., Vincent, B. and Williams, T. (1994). The role of self-assembly in biological systems: evidence from iridescent colloidal sporopollenin in *Selaginella* megaspore walls. *Philosophical Transactions of the Royal Society, London* B **345**: 163–173.

Hennipman, E. (1970). Electron and light microscopical observations on the perine of the spores of some *Bolbitis* species (Filices). *Acta Botanica Neerlandica* **19**: 671–680.

Herd, Y.R., Cutter, E.G. and Watanabe, I. (1985). A light and electron microscopic study of microsporogenesis in *Azolla microphylla*. *Proceedings of the Royal Society, Edinburgh* **86B**: 53–58.

Herd, Y.R., Cutter, E.G. and Watanabe, I. (1986). An ultrastructural study of post-meiotic development in the megasporocarp of *Azolla microphylla*. *Canadian Journal of Botany* **64**: 822–833.

Horner, H.T. and Beltz, C.K. (1970). Cellular differentiation of heterospory in *Selaginella*. *Flora (Jena)* **78**: 335–341.

Kenrick, P. and Crane, P.R. (1997). The origin and early diversification of land plants: a cladistic study. Smithsonian Series in Comparative Evolutionary Biology. Smithsonian Institution Press, Washington. 441pp.

Konar, R.N. and Kapoor, R.K. (1974). Embryology of *Azolla pinnata*. *Phytomorphology* **24**: 228–261.

Lehmann, H., Neidhart, H.V. and Schlenkermann, G. (1984). Ultrastructural investigations on sporogenesis in *Equisetum fluviale*. *Protoplasma* **123**: 38–47.

Lugardon, B. (1966). Formation de l'exospore chez *Blechnum spicant* (L.) Roth. *Comptes Rendus de l'Academie des Sciences de Paris* **262**: 2029–2031.

Lugardon, B. (1969a). Sur la formation de l'exospore chez *Osmunda regalis* L. *Comptes Rendus de l'Academie des Sciences de Paris* **268**: 2879–2882.

Lugardon, B. (1969b). Sur la structure fine des parois sporales d'*Equisetum maximum* Lamk. *Pollen et Spores* **11**: 449–474.

Lugardon, B. (1972a). Sur la structure fine et la nomenclature des parois microsporales chez *Selaginella denticulata* (L.) Link et *S. selaginoides* (L.) Link. *Comptes Rendus de l'Academie des Sciences de Paris* **274**: 1256–1259.

Lugardon, B. (1972b). La structure fine de l'exospore et de la périspore des Filicinées isosporées: I, Généralités, Eusporangiées et Osmundales. *Pollen et Spores* **14**: 227–261.

Lugardon, B. (1973a). Sur les parois sporales de *Psilotum triquetrum* Sw. et leur structure fine. *Comptes Rendus de l'Academie des Sciences de Paris* **276**: 1277–1280.

Lugardon, B. (1973b). Nomenclature et structure fine des parois acéto-résistantes des microspores d'*Isoëtes*. *Comptes Rendus de l'Academie des Sciences de Paris* **276**: 3017–3020.

Lugardon, B. (1974). La structure fine de l'exospore et de la périspore des Filicinées isosporées: II. Filicales. Commentaires. *Pollen et Spores* **16**: 161–226.

Lugardon, B. (1976). Sur la structure fine de l'exospore dans les divers groupes de ptéridophytes actuelles (microspores et isospores). In: I.K. Ferguson and J. Muller (editors). The Evolutionary significance of the exine, pp. 231–250. Academic Press, London.

Lugardon, B. (1978). Comparison between pollen and pteridophyte spore walls. *IV International Palynological Conference, Lucknow (1976-77)* **1**: 199–206.

Lugardon, B. (1979). Sur la formation du sporoderme chez *Psilotum triquetrum* Sw. *Grana* **18**: 145–165.

Lugardon, B. (1986). Donnés ultrastructurales sur la fonction de l'exospore chez les Ptéridophytes. In: S. Blackmore and I.K. Ferguson (editors). Pollen and spores: form and function. pp.251–264. Academic Press, London.

Lugardon, B. (1990). Pteridophyte sporogenesis: a survey of spore wall ontogeny and fine structure in a polyphyletic plant group. In: S. Blackmore and R.B. Knox (editors). Microspores: evolution and ontogeny, pp. 95–120. Academic Press, London.

Lugardon, B. and Husson, P. (1982). Ultrastructure exosporale et caractères généraux du sporoderm dans le microspores et les mégaspores des Hydroptéridées. *Comptes Rendus de l'Academie des Sciences de Paris* **294**: 789–794.

Morbelli, M.A. and Rowley, J.R. (1993). Megaspore development in *Selaginella*. I. "Wicks", their presence, ultrastructure, and presumed function. *Sexual Plant Reproduction* **6**: 98–107.

Pacini, E., Franchi, G.G. and Hesse, M. (1985). The tapetum: its form, function and possible phylogeny in Embryophyta. *Plant Systematics and Evolution* **149**: 155–185.

Pacini, E. (1990). Tapetum and microspore function. In: S. Blackmore and R.B. Knox (editors) Microspores: evolution and ontogeny. pp. 213–237. Academic Press, London.

Parkinson, B.M. (1987). Tapetal organization during sporogenesis in *Psilotum nudum*. *Annals of Botany* **60**: 353–357.

Parkinson, B.M. (1995). Spore wall development in *Schizaea pectinata* (Schizaeaceae: pteridophyta). *Grana* **34**: 217–228.

Parkinson, B.M. and Pacini, E. (1995). A comparison of tapetal structure and function in pteridophytes and angiosperms. *Plant Systematics and Evolution* **198**: 55–88.

Pettitt, J.M. (1971). Developmental mechanisms in heterospory. I. Megasporocyte degeneration in *Selaginella*. *Botanical Journal of the Linnean Society* **64**: 237–246.

Pryer, K.M., Smith, A.R. and Skog, J.E. (1995). Phylogenetic relationships of extant ferns based on evidence from morphology and rbcL sequences. *American Fern Journal* **85**: 205–282.

Raubeson, L.A. and Jansen, R.K. (1992). Chlorophyll DNA evidence on the ancient evolutionary split in vascular plants. *Science* **255**: 1697–1699.

Robert, D. (1971a). Nouvelle contribution à l'étude de l'origine des parois microsporales chez le *Selaginella krausiana* A. Br. *Comptes Rendus de l'Academie des Sciences de Paris* **272**: 385–388.

Robert, D. (1971b). Etude, en microscopie électronique, des modalités d'édification des parois microsporales chez le *Selaginella selaginoides* (L.). Mise en place du feuillet externe. *Comptes Rendus de l'Academie des Sciences de Paris* **273**: 332–335.

Robert, D. (1971c). Etude, en microscopie électronique, des modalités d'édification des parois microsporales chez le *Selaginella selaginoides* (L.). Mise en place du feuillet interne. *Comptes Rendus de l'Academie des Sciences de Paris* **273**: 1933–1936.

Robert, D., Roland-Heydacker, F., Denizot, J., Laroche, J., Fougeroux, P. and Davignon, L. (1973). La paroi mégasporale de l'*Isoëtes setacea* Bosc ex Delile. *Adansonia* **13**: 313–332.

Rowley, J.R. (1995). Are the endexines of pteridophytes, gymnosperms and angiosperms structurally equivalent? *Review of Palaeobotany and Palynology* **85**: 13–34.

Rowley, J.R., and Morbelli, M.A. (1995). Megaspore wall growth in *Selaginella* (Lycopodiatae). *Plant Systematics and Evolution* **194**: 133–162.

Rothwell, G. (1996). Phylogenetic relationships of ferns; a palaeobotanical perspective. In: J.M. Camus, M. Gibby and R.J. Johns (editors). Pteridology in perspective, pp. 395–404, Royal Botanic Gardens, Kew.

Schneller, J.J. (1989). Remarks on hereditary regulation of spore pattern in *Athyrium*. *Botanical Journal of the Linnean Society* **99**: 115–123.

Schraudolf, H. (1984). Ultrastructural events during sporogenesis of *Anemia phyllitidis* (L.) Sw. II. Spore wall formation. *Beiträge Biologische Pflanzen* **59**: 237–260.

Sharma, B.D. and Singh, R. (1986). Sporogenesis in *Ophioglossum gramineum*. *Phytomorphology* **36**: 107–109.

Sheffield, E. and Bell, P.R. (1979). Ultrastructural aspects of sporogenesis in a fern, *Pteridium aquilinum* (L.) Kuhn. *Annals of Botany* **44**: 393–405.

Singh, J. and Devi, S. (1989). Ultrastructural aspects of spore development in *Pteris vittata* L. *Beiträge Biologische Pflanzen* **64**: 29–38.

Smith, A.R. (1995). Non-molecular phylogenetic hypotheses for ferns. *American Fern Journal* **85**: 104 –122.

Stevenson, D.W. and Loconte, H. (1996). Ordinal and familial relationships of pteridophytes. In: J.M. Camus, M. Gibby and R.J. Johns (editors). Pteridology in perspective, pp. 435–467, Royal Botanic Gardens, Kew.

Taylor, W.A. (1991). Ultrastructural aspects of wall ultrastructure in *Selaginella galeotii* (Lycophyta). *Plant Systematics and Evolution* **174**: 171–182.

Uehara, K. and Kurita, S. (1989a). Ultrastructural study of spore wall morphogenesis in *Ophioglossum thermale* Kom. Var. *nipponicum* (Miyabe et Kudo) Nishida. *Botanical Magazine Tokyo* **102**: 413–427.

Uehara, K. and Kurita, S. (1989b). An ultrastructural study of spore wall morphogenesis in *Equisetum arvense*. *American Journal of Botany* **76**: 939–951

Uehara, K. and Kurita, S. (1991). Ultrastructural study on spore wall morphogenesis in *Lycopodium clavatum* (Lycopodiaceae). *American Journal of Botany* **78**: 24–36.

Uehara, K., Kurita, S., Sahashi, N. and Ohmoto, T. (1991). Ultrastructural study on microspore wall morphogenesis in *Isoëtes japonica* (Isoëtaceae). *American Journal of Botany* **78**: 1182–1190.

Uffelen, G.A. van (1990). Sporogenesis in Polypodiaceae (Filicales). I, *Drynaria sparsisora* (Desv.)Moore. *Blumea* **35**: 177–215.

Uffelen, G.A. van (1991). The control of spore wall formation. In: S. Blackmore and S.H. Barnes (editors). Pollen and spores, patterns of diversification, pp. 89–102. Clarendon Press, Oxford.

Uffelen, G.A. van (1992). Sporogenesis in Polypodiaceae (Filicales). II. The genera *Microgramma* Presl. and *Belvisia* Mirbl. *Blumea* **36**: 515–540.

Uffelen, G.A. van (1993). Sporogenesis in Polypodiaceae (Filicales). III. Several Species. Spore characters and their value in phylogenetic analysis. *Blumea* **37**: 529–561.

Wodehouse. R.P. (1935). Pollen grains, their structure, identification and significance in science and medicine. McGraw-Hill, New York.

Appendix 1. Characters Used in the Analysis

1. Sporangium ontogeny. Eusporangiate (0), leptosporangiate (1).

    Sporangium ontogeny has long been used to distinguish between major groups of pteridophytes. However, although this has not been explored here, it is clear that there is sufficient variation within the two main classes to merit a more thorough analysis and the differentiation of a greater number of characters.

2. Tetrad wall. A temporary tetrad wall not formed beneath the primary wall of the sporocyte (0), a distinct tetrad wall develops beneath the primary wall of the sporocyte (1).

    The term special cell wall is avoided here because it has often been linked to specific interpretations of chemical composition. According to Lugardon (1976, 1986) most pteridophytes lack a distinct tetrad wall.

3. Monoplastidic meiosis. Meiosis monoplastidic (0), not monoplastidic (1).

    A detailed discussion of monoplastidic meiosis is given by Brown and Lemmon (1990). The relationship between organelle behaviour and the cytoskeleton during meiosis may ultimately be a source of further characters. There are clearly variations, but they have been insufficiently explored for a thorough comparison (see for example, Sheffield and Bell, 1979).

4. Mode of cytokinesis in meiosis. Centripetal furrowing contributes to partitioning of daughter cells (0), partitioning of daughter cells is by centrifugal cell plates, without any centripetal furrowing (1).

    Studies of meiosis have generally concentrated only on temporal distinctions, between simultaneous and successive cytokinesis, but there are clearly variations in the mode of partitioning that can be systematically informative at the level of the land plants (for discussion see, Blackmore and Crane, in press).

5. Timing of cytokinesis in meiosis. No cell wall formed after first division of meiosis (0), a dyad wall formed after the first division of meiosis (1).

    The differences in relative timing between simultaneous and successive cytokinesis are relatively minor and a few taxa show both kinds of division within a sporangium.

6. Aperture definition. Aperture defined by cytoplasmic constriction and folding of plasma membrane (0), defined by an oncus like plug (1), defined by an apertural shield without cytoplasmic folding (2).

    For a discussion of aperture formation in land plants see Blackmore and Crane (in press).

7. Apertures. Apertures absent (0), hilate (1), proximal and monolete (2) proximal and trilete (3).

    This corresponds to character 71 of Stevenson and Loconte (1995) but is coded differently.

8. Exine deposition. Initial exine deposition involves white line centred lamellae formed on the plasma membrane (0), white line centred lamellae not involved in the initial deposition of the exine, primexine absent (1), white line centred lamellae not involved in the initial deposition of the exine, primexine present (2).

    Deposition of sporoderm on white line centred lamellae corresponds to the first of four modes of spore wall formation described by Blackmore and Barnes (1987, p. 188).

9. Exine patterning. Plasma membrane of early free microspores smooth in non-apertural regions (0), plasma membrane strongly undulating in sectional view, giving rise to a reticulate pattern during exospore deposition (1).

In *Lycopodium* (Uehara and Kurita, 1991) the plasma membrane becomes strongly patterned soon after the completion of meiosis. This is unusual for a pteridophyte but somewhat resembles events in some bryophytes (see for example, Brown and Lemmon, 1990).

10. Tapetal contribution to exine. Exine formation does not involve layer(s) formed by deposition of material contributed by the tapetum onto the surface of the developing spore (0), exine formation involves a tapetum-derived component incorporated onto surfaces formed by the microspore and usually based on white line centred lamellae (1).

Deposition of sporoderm material from the tapetum onto the surface of the microspores was the second mode of sporoderm formation described by Blackmore and Barnes (1987, p. 188). In this analysis, the formation of a distinct layer of tapetal origin, the epispore, is treated as a different character, character 13.

11. Paraexospore. Outer exine layer deposited continuously (0), outer exine layer with a large discontinuity separating structurally similar layers formed by white line centred lamellae (1).

Lugardon (1976) introduced the term paraexospore for an outer layer of the exospore that is separated from an inner component by a space.

12. Inner exine. Exine formation does not involve additional layers (differing in staining properties and or mode of deposition) underlying the first deposited exine layer (0), additional exine layer(s) deposited below the initial layer by accumulation of white line centred lamellae (1), additional exine layer(s) formed by deposition of granular or amorphous sporopollenin (2).

Sporoderm formation below an earlier formed layer was the fourth mode of sporoderm deposition described by Blackmore and Barnes (1987, p.188).

13. Epispore. Absent (0), present (1).

The term epispore was introduced by Lugardon to refer to a layer within the sporoderm of heterosporous pteridophytes which is formed during the latest stages of exospore development, is composed of sporopollenin and is largely, if not exclusively of tapetal origin (for a discussion see Lugardon, 1990).

14. Perispore. Perispore absent (0), perispore present (1).

Perispore consists of material derived from the degenerating tapetum which condenses in one or more continuous layers over the surface of the spores. The middle layer of *Equisetum* sporoderm is treated here as a thin perispore layer, although Lugardon (1976) and Uehara and Kurita (1989b) regarded it as potentially a unique component of sphenophyte spore walls. In *Azolla*, the layer that is formed is, strictly, an epispore, rather than perispore (Lugardon and Husson, 1982).

15. Elaters. Elaters absent (0), elaters present as part of the sporoderm (1).

The details of development of elaters in *Equisetum* have been elucidated by Lugardon (1969b) and Uehara and Kurita (1989b). The cellular elaters of hepatics are not included here.

16. Silica. Silica absent from spore wall (0), silica present (1).

    This character has not been adequately surveyed and tends only to have been noted in cases where abundant silica is incorporated into the spore wall.

17. Spore output. More than 250 microsporocytes give rise to more than 1000 spores (0), small determinate number of sporocytes give rise to <100 spores (often 64) (1), 25 - 250 microsporocytes giving rise to 100-1000 spores (2).

    Spore output is correlated with the number of microsporocytes and has been coded in several different ways in recent studies. Our treatment follows that for character 46 of Pryer *et al.* (1995), and differs from Stevenson and Loconte's (1996) character 67.

18. Homospory/heterospory. Spores homosporous (0), heterosporous (1).

    This corresponds to character 68 of Stevenson and Loconte (1996).

19. Dormancy. Spores with dormant cytoplasm at maturity (0), spores chlorophyllous at maturity (1).

    This corresponds to character 69 of Stevenson and Loconte (1996).

20. Endospore/intine. Pecto-cellulosic cell wall (endospore or intine) develops prior to maturation of spores (0), endospore develops at start of germination (1).

21. Tapetal thickness. Tapetum of one layer of cells (0), more than one layer of cells (1).

22. Tapetal activity. Tapetal cells remaining parietal throughout development (0), tapetum comprising a parietal layer and a plasmodial layer (1), tapetal cells invading the locule and losing their cell walls to become fully plasmodial (2).

    Parkinson and Pacini (1995) described a "combination type" tapetum with a parietal component and an invasive, plasmodial component in certain pteridophytes.

23. Orbicules. Tapetum not producing orbicules (0), orbicules produced by the tapetum (1).

    Lugardon (1978) discussed the homology of orbicules in pteridophytes and seed plants.

24. Germination. Germination at proximal pole (0), equatorially (1), or at distal pole (2) This character is equivalent to character 85 of Stevenson and Loconte (1996), but is coded differently.

25. Gametophyte growth. Gametophyte grows outside spore (0), gametophyte within spore wall (1).

Van Uffelen, G.A. (2000). Studying spores of the Polypodiaceae: a comparison of SEM with other microscope techniques. In: M.M. Harley, C.M. Morton and S. Blackmore (Editors). Pollen and Spores: Morphology and Biology, pp. 125–131. Royal Botanic Gardens, Kew.

STUDYING SPORES OF THE POLYPODIACEAE: A COMPARISON OF SEM WITH OTHER MICROSCOPE TECHNIQUES

GERDA A. VAN UFFELEN

Hortus Botanicus, P.O. Box 9516, 2300 RA Leiden,
The Netherlands

Abstract

Different methods can be applied to the study of fern spores. The specific advantages and disadvantages of each method are discussed in relation to Polypodiaceae, together with some practical hints on making spore study as easy and informative as possible.

Introduction

In the Netherlands, over the last fifteen years the palaeotropical representatives of the Polypodiaceae have been studied, resulting in their treatment in Flora Malesiana (Hovenkamp *et al.*, 1998). This project has generated revisions of the palaeotropical genera, as well as studies on different character complexes such as the spores (Van Uffelen and Hennipman, 1985). The family Polypodiaceae in its entirety now accommodates about 30 genera and 800 species (Hennipman *et al.*, 1990).

Spores of Polypodiaceae are monolete and bean-shaped, and show an astonishing variety of spore wall sculpture and structure. Polypodiaceous spores usually measure between 40 and 100 μm, so they are fairly large compared to many other spores and pollen. If the study of spores is in conjunction with a taxonomic or systematic revision of a genus or family, it is essential to study reliably identified material, especially if spore surface and wall ultrastructure are to be used as systematic characters.

The basis for the detailed study of fern spores has been laid by the observations of Lugardon, who uses transmission electron microscopy (for example, 1971, 1972, 1974, 1976, 1978a, 1978b) and, among others, by the use of scanning electron microscopy by Tryon and Tryon (1982). These observations resulted in a standard work on fern spores, integrating scanning and transmission electron microscope images (Tryon and Lugardon, 1991). However, Tryon and Lugardon only included micrographs of a selection of spores from each included genus, and study of many more species is still needed for the elucidation of relationships of both species and genera.

Materials and Methods

Light microscopy

When I started my studies of spore wall morphology at the species level, I enjoyed unlimited access to a scanning electron microscope (SEM), which was an essential part of the research equipment. Thus, it was only later that I realised the great potential of a more universally available instrument, the light microscope. A number of aspects of spore morphology, in particular, overall dimensions of spores and spore walls, the external morphology of spores, and the general structure of spore walls are routinely described from light microscopy, and in fern taxonomy LM descriptions of spores are often included in keys. LM is also used to assess the fertility of spores.

The preparation of thin sections for routine or fluorescence LM allows spore wall structure to be studied in some detail. However, the study of sections for light microscopy has the same drawback as the study of ultrathin sections for transmission electron microscopy (TEM). It is very difficult, if not impossible, to reconstruct surface pattern from a series of sections.

Scanning electron microscopy

As more data about the intricate structures on the surface of fern spores are resolved with SEM, it has become apparent that light microscopy alone cannot provide sufficient detail. Ultra fine structures may yield interesting characters for use in the assessment of relationship between species or groups of species (Hennipman, 1990), and this is where SEM should be applied.

Preparation of fern spores for study by SEM is very easy (Van Uffelen and Hennipman, 1995), as critical point drying is not necessary. Material from herbarium specimens may be used, provided it is not too old (there is a danger that in old herbarium specimens all the spores may have fallen out of the sporangia, or the spores may have been infected by fungi). Spores can be lifted from the specimen using easily available tools. Lifting whole sporangia is a sure method of obtaining spores from a certain specimen, although there is a cautionary note that whole sporangia may yield unripe spores. After mounting on electron microscope stubs pressure can be applied to the spores with the flat surface of another clean stub; a gentle grinding action often fractures spores to expose the spore wall layers for ultrastructural study.

A thin spreading of adhesive on an electron microscope stub, for example, Scotch Tape adhesive dissolved in undiluted acetone, then left to air dry for a few minutes, is usually sufficient to fix the dry spores firmly. The specimen can then be sputter-coated with an ultra thin layer of gold or other fine-grained heavy metal.

Occasionally the coating is too thin, resulting in horizontally banded SEM micrographs. This is the effect of secondary electrons being absorbed, rather than bounced back off the specimen (spores). It is usually caused, either by insufficient contact of the spores with the stub, or by inadequate coating. To avoid banded images it is important to take particular care during mounting spores onto the stub, while at the same time not "burying" or sinking the spores into the adhesive. It is also important to take care when setting up the parameters for sputter coating, for example the specimen distance from the metal target, versus sputtering time. Recommendations can usually be found in the sputtering device manual. Bear in mind that overcoating specimens obscures detail at high magnification.

Access to a scanning electron microscope provides the opportunity to study details, at resolutions far in excess of the 1,500x available to light microscopists. At the same time, the whole stub can be searched very carefully, methodically using X and Y co-ordinates, just as with LM. Searches may reveal some abnormal spores but always give a good insight into the range of variation possible between spores within a specimen sample. The sample will usually include some broken spores with ultrastructural details exposed, for example the SEM pictures (Figs. 1-4) show one spore of *Aglaomorpha meyeniana* Schott, a drynarioid polypodiaceous fern (Roos, 1986). The perispore covering the exospore has broken spontaneously. The whole spore (Fig. 1) shows the outer surfaces of both exospore and perispore. A detail of this image (Fig. 2) shows broken perisporal echinae lying around on the exospore surface. The perispore is quite smooth on the inside; note also the details of the echinae scattered over the basic warty pattern of the perispore (Figs. 1- 4). Details of the outer exospore surface on the proximal side, with the openings of microchannels, and of the perispore in cross section are shown in Fig. 3. The broken perispore (Fig. 4), shows details of the outer perispore surface, the thinnest part of the perispore is indicated by an arrow, a detail of the outer exospore surface on the distal side is also apparent.

It is important that any newly initiated project involving SEM, should build in exploratory time at the outset of the study. It is during this period that the potential range of morphological data available can be assessed and a *modus operandi* established. Furthermore, it cannot be over-emphasised that the specimens should be studied, and the spore images for SEM micrographs should be taken by the researcher concerned. Technical staff, no matter how skilled, should not be entrusted with this work before researcher and technical assistant have established a close working relationship over an extended period of time.

Detailed study of spore wall surface and structure has also yielded ideas about spore function (Van Uffelen, 1985) and the mechanisms behind spore wall formation (Van Uffelen, 1991), such as self-assembly in a colloidal system (Gabarayeva, 2000; Hemsley, 2000; Hemsley *et al.*, 1998; see also Adams *et al.*, 1998).

The great advantage of SEM, in detailing surface structure, can be taken one step further by applying freeze fracturing, as developed by Blackmore and Barnes (1985), which has now become an accepted technique. It has been applied to several species of Polypodiaceae in various stages of sporangium development in order to study spore wall formation (Van Uffelen, 1990, 1992, 1993). This is, however, a far more time-consuming method of preparation, comparable in time investment to making preparations for TEM. The technique (Van Uffelen, 1990, 1992, 1993) involves fixation in paraformaldehyde and/or OsO_4, freeze fracturing in dimethylsulfoxide, a long period of post-fixation in OsO_4 and tannic acid, followed by dehydration and critical point drying. A note should be made on the use of paraformaldehyde as a long-term fixative, without refrigeration. Lugardon (pers. comm.), uses it quite often as a first fixative for TEM procedures, in order to build up a stock of material for future use. Although the use of paraformaldehyde has been generally abandoned as it may damage ultra fine structure, it does not seem to damage the spore wall material unduly. The advantage of using paraformaldehyde is that it allows for the gathering of material if and when it becomes available, even during field trips.

Freeze fracturing has yielded a wealth of new and unexpected information on the formation of the pteridophyte spore wall, showing a succession of different surface patterns superimposed upon each other. This may even serve to reconstruct phylogenetic relationships, as attempted by Blackmore *et al.* (2000).

Transmission Electron Microscopy
Since the late sixties TEM has become the standard technique for ultrastructural study of fern spores (for example, Lugardon, 1966). Lugardon's studies form the basis of our present knowledge of fern spore ultrastructure and sporoderm development, and two advantages of the technique are obvious:
1) it is possible to observe fine detail at high magnifications and, 2) it is possible to obtain information on the chemical composition of structural elements. Apart from ultrastructural (Lugardon 1972, 1974, 1976 etc.) and developmental studies (for example, Lugardon 1990), Lugardon also applied TEM for the description of spore function (1986), and for the comparison of spore and pollen walls (1978b, 1995).

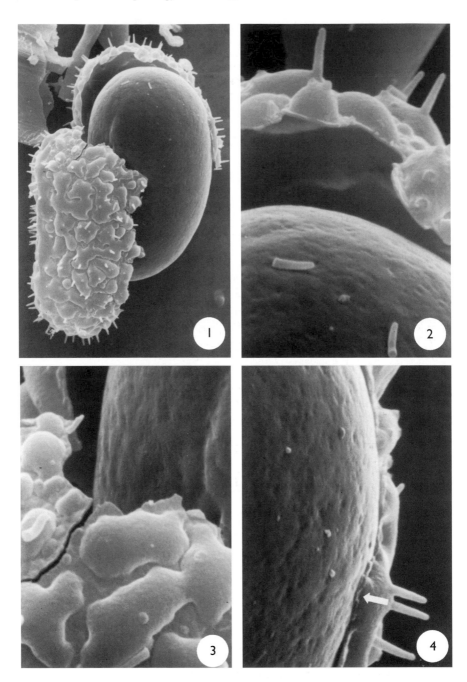

FIGS. 1–4. Spore of *Aglaomorpha meyeniana* Schott (M. Ramos 13649). Fig. 1. Spore with fractured perispore, ×1500. Fig. 2. Detail of the exospore (bottom), note detached perispore echinae lying on exospore surface. Inner and outer details of fractured perispore (above) are also apparent, ×7500. Fig. 3. Detail of the exospore surface, showing the openings of microchannels, and of the perispore in cross section, ×7500. Fig. 4. Detail of the broken perispore (arrow indicates thinner parts of perispore wall), and of the exospore surface, ×7500.

There are, however, some drawbacks associated with TEM studies, these are:

• Highly specialised techniques require great care to yield good results
• Techniques are very time-consuming
• All information is two-dimensional. Three dimensional reconstructions require serial thin sectioning, which is a highly skilled technique

Conclusions

In conclusion, each microscope technique has its own purpose:
- The light microscope is a low resolution instrument, excellent for study at the specimen level to ascertain basic characteristics of spore morphology. It may also be used in combination with more advanced techniques such as fluorescence microscopy.
- The scanning electron microscope is the preferred instrument for studying spore wall surfaces (usually perispore but also exospore), and wall fractures at high resolution. It can yield characters of value in the assessment of relationships between species, and also at higher taxonomic levels.

The transmission electron microscope can also yield characters to assist in defining relationships between taxa. Furthermore, it is an excellent tool for developmental and chemical studies.
- Paraformaldehyde is a particularly flexible alternative to other fixatives, since it not only allows for fixation under field conditions, but also permits for a stock of material to accumulate without unacceptable damage to the spores.

Finally, it pays to experiment in order to discover new, easier or more informative procedures. An apparently disproportionate amount of time spent in careful examination and assessment of samples, in the initial stages of a new study, will reap dividends later in terms of the amount of potentially valuable morphological data extracted in the long term.

Acknowledgements

I want to thank Ms Bertie Joan van Heuven for technical assistance with the preparation of the micrographs made for this manuscript.

References

Adams, M., Dogic, Z., Keller, S.L., and Fraden, S. (1998). Entropically driven microphase transitions in mixtures of colloidal rods and spheres. *Nature* 393: 349–352.

Blackmore, S. and Barnes, S.H. (1985). *Cosmos* pollen ontogeny: a scanning electron microscope study. *Protoplasma*, 126: 91–99.

Blackmore, S., Takahashi, M., and Uehara, K. (2000). A preliminary phylogenetic analysis of sporogenesis in pteridophytes. In: M.M. Harley, C.M. Morton and S. Blackmore (editors). Pollen and spores: morphology and biology, pp. 109–124. Royal Botanic Gardens, Kew.

Gabarayeva, N.I. (2000). Principles and recurrent themes in sporoderm development. In: M.M. Harley, C.M. Morton and S. Blackmore (editors). Pollen and spores: morphology and biology, pp. 1–16. Royal Botanic Gardens, Kew.

Hemsley, A.R., Vincent, B., Collinson, M.E., and Griffiths, P.C. (1998). Simulated self-assembly of spore exines. *Annals of Botany* 82: 105–109.

Hemsley, A.R. (2000). Self-assembly of colloidal units in exine development. In: M.M. Harley, C.M. Morton and S. Blackmore (editors). Pollen and spores: morphology and biology, pp. 31–44. Royal Botanic Gardens, Kew.

Hennipman, E. (1990). The significance of the SEM for character analysis of spores of Polypodiaceae (Filicales). In: D. Claugher (editor). Scanning electron microscopy in taxonomy and functional morphology, pp. 23–44. Clarendon Press, Oxford.

Hennipman, E., P. Veldhoen, K.U. Kramer, and Price, M.G. (1990). Polypodiaceae. In: K. Kubitzki (editor). The families and genera of vascular plants. I. pteridophytes and gymnosperms, pp. 203–230. Springer Verlag, New York.

Hovenkamp, P.H., Roos, M.C., Franken, N.A.P., Hennipman, E., Hetterscheid, W.L.A., Rödl-Linder, G., Nooteboom, H.P. and Bosman, M.T.M. (1998). Polypodiaceae. In: C. Kalkman and H.P. Nooteboom (editors). *Flora Malesiana*, Series 2 (3), pp. 1–234. Foundation Flora Malesiana, Leiden, The Netherlands.

Lugardon, B. (1966). Formation de l'exospore chez *Blechnum spicant* (L.) Roth. C. R. *Acad. Sc. Paris ser.* D 262: 2029–2031.

Lugardon, B. (1971). Contribution à la connaissance de la morphogenèse et de la structure des parois sporales chez les Filicinées isosporées. Thesis, Univ. P. Sabatier, Toulouse, 257 pp.

Lugardon, B. (1972). La structure fine de l'exospore et de la périspore des filicinées isosporées. I - Généralités. Eusporangiées et Osmundales. *Pollen et Spores* 14: 227–261.

Lugardon, B. (1974). La structure fine de l'exospore et de la périspore des filicinées isosporées. II - Filicales. Commentaires. *Pollen et Spores* 16: 161–226.

Lugardon, B. (1976). Sur la structure fine de l'exospore dans les divers groupes de Pteridophytes actuelles (microspores et isospores). In: I.K. Ferguson, and J. Muller (editors). The evolutionary significance of the exine, pp.231–250. Academic Press, London.

Lugardon, B. (1978a). Isospore and microspore walls of living pteridophytes: identification possibilities with different observation instruments. In: D.C. Bharadwaj, K.M. Lele, R.K. Kar, H.P. Singh, R.S. Tiwari, Vishnu-Mittre and H.K. Maheshwari (editors). Proceedings of the Fourth International Palynological Conference, volume I, pp.152–163. Birbal Sahni Institute of Palaeobotany: Lucknow, India.

Lugardon, B. (1978b). Comparison between pollen and pteridophyte spore walls. In: D.C. Bharadwaj, K.M. Lele, R.K. Kar, H.P. Singh, R.S. Tiwari, Vishnu-Mittre and H.K. Maheshwari (editors). Proceedings of the Fourth International Palynological Conference, volume I, pp.199–206. Birbal Sahni Institute of Palaeobotany: Lucknow, India.

Lugardon, B. (1986). Données ultrastructurales sur la fonction de l'exospore chez les Ptéridophytes. In: S. Blackmore and I.K. Ferguson (editors). Pollen and spores: form and function, pp. 251–264. Academic Press, London

Lugardon, B. (1990). Pteridophyte sporogenesis: a survey of spore wall ontogeny. In: S. Blackmore and R.B. Knox (editors). Microspores: evolution and ontogeny, pp. 95–120. Academic Press, London..

Lugardon, B. (1995). Exine formation in *Chamaecyparis lawsoniana* (Cupressaceae) and a discussion on pteridophyte exospore and gymnosperm exine ontogeny. *Review of Palaeobotany and Palynology* 85: 35–51.

Roos, M.C. (1986). Phylogenetic systematics of the Drynarioidae (Polypodiaceae). Thesis, Utrecht, 203pp.

Tryon, A.F. and Lugardon, B. (1991). Spores of the Pteridophyta. Surface, wall structure, and diversity based on electron microscope studies. Springer Verlag, New York, 648 pp.

Tryon, R.M. and Tryon, A.F. (1982). Ferns and allied plants. With special reference to tropical America. Springer Verlag, New York, 857 pp.

Uffelen, G.A. van (1985). Synaptospory in the fern genus *Pyrrosia* (Polypodiaceae). *Blumea* 31: 57–64.

Uffelen, G.A. van (1990). Sporogenesis in Polypodiaceae (Filicales). I. *Drynaria sparsisora* (Desv.) Moore. *Blumea* 35: 177–215.

Uffelen, G.A. van (1991). The control of spore wall formation. In: S. Blackmore and S.H. Barnes (editors), Pollen and spores: patterns of diversification, pp. 89–102. Clarendon Press, Oxford.

Uffelen, G.A. van (1992). Sporogenesis in Polypodiaceae (Filicales). II. The genera *Microgramma* Presl and *Belvisia* Mirbel. *Blumea* 36: 515–40.

Uffelen, G.A. van (1993). Sporogenesis in Polypodiaceae (Filicales). III. Species of several genera. Spore characters and their value in phylogenetic analysis. *Blumea* 37: 529–561.

Uffelen, G.A. van, and Hennipman, E. (1985). The spores of *Pyrrosia* Mirbel (Polypodiaceae), a SEM study. *Pollen et Spores* 27: 155–197.

Johns, R.J. (2000). Spore ornamentation and the species of simple-fronded *Asplenium* (Aspleniaceae) in West Africa. In: M.M. Harley, C.M. Morton and S. Blackmore (Editors). Pollen and Spores: Morphology and Biology, pp. 133–146. Royal Botanic Gardens, Kew.

SPORE ORNAMENTATION AND THE SPECIES OF SIMPLE-FRONDED *ASPLENIUM* (ASPLENIACEAE) IN WEST AFRICA

R. J. JOHNS

Royal Botanic Gardens, Kew, Richmond, Surrey TW9 3AE, UK

Abstract

Studies of perispore ornamentation have made a major contribution to our understanding of the taxonomy of the pteridophytes. In the present paper a re-evaluation is made of the taxonomy of the simple-fronded species of *Asplenium* from The Cameroon based on a detailed study of perispore ornamentation and the frond morphology of available collections. Preliminary results of this study are presented and it appears that several 'new species' need to be described from The Cameroon. The contribution of perispore ornamentation is reinforced as a taxonomic tool to gain a better understanding of species complexes within difficult taxonomic groups such as *Asplenium*.

Introduction

While preparing the account of the Aspleniaceae for the Flora of Tropical East Africa it became apparent that there are problems in the delimitation of the simple-fronded species of *Asplenium*. *Asplenium* is the largest genus of ferns in continental Africa, including around 180 species, in excess of 20 percent of the total diversity of pteridophytes in Africa. Two 'groups' of simple-fronded *Asplenium* occur in Tropical Africa. The species with a distinct intramarginal vein (*Asplenium* section *Thamnopteris*) are restricted in their African distribution to the coastal ranges of Kenya, Tanzania and their offshore islands. Of greater complexity are the species of simple-fronded *Asplenium* which have free, divided or, rarely, simple veins.

Three simple-fronded species have traditionally been accepted from The Cameroon in the floristic accounts published on West Tropical Africa (Alston, 1959) and The Cameroon (Tardieu-Blot, 1964). In his account of *Asplenium* for the Flora of Tropical West Africa, Alston (1959) lists these species: *A. africanum* Desv., *A. currori* Hook. and *A. subintegrum* C.Chr. a *nomen novum* for *Asplenium coriaceum* Baker (1868) *non* Bory (1833). Alston (*loco citato*) listed *A. venosum* Hook., *A. guineense* Schum and *A. sinuatum* P. Beauv. as synonyms of *A. africanum*. *Asplenium currori* was treated as a variety of *A. africanum* by Tardieu-Blot (1953). Like *A. holstii* Hieron. in tropical East Africa, *A. africanum* has been accepted by regional workers as a widely distributed species reported from West Africa to the mountains of Uganda where it is interpreted as a West African element in the flora. Studies of perispore ornamentation in *A. africanum* specimens from Uganda do not support the presence of this species, which probably represents an undescribed taxon.

Previous studies which use perispore ornamentation

Perispore ornamentation has been used extensively in the study and delimitation of the species of *Asplenium*. Tryon and Lugardon (1990) concluded that the spores in the family are exceptionally varied, particularly in their ornamentation. Such studies have been used to enhance our understanding for local and regional revisions of *Asplenium* in Australia (Puttock and Quinn, 1980) and New Zealand (Brownsey, 1977). More detailed studies aimed at the taxonomic understanding of particular groups within *Asplenium* have been made, particularly with reference to section *Hymenasplenium* (Smith, 1976) and section *Thamnopteris* in Irian Jaya (Johns, 1996). Viane and Van Cotham (1977) studied the perispores of several species of *Asplenium* and compared their groupings with groups based on stomatal characters. Several groups of *Asplenium* spores are recognised from the SEM spore studies. Viane and Van Cotham (1977) recognised three groups: costate-alate, echinate and reticulate in their spore studies from East Africa. Tryon and Lugardon (1990) recognised four common groups in *Asplenium*: alate, fenestrate, echinate and reticulate. Spores with echinate or reticulate surfaces are considered by Tryon (1990) to be 'more derived' than species with folds and fenestrate surfaces. Epilithic and terrestrial species of *Asplenium* have a majority of species with alate and fenestrate spores (Tryon, 1990). Similar conclusions were drawn from studies of the spores of *Pyrrosia,* showing a correlation between ecology and surface ornamentation (van Uffelen,1985). The groups recognised by Tryon (1990) and Tryon and Lugardon (1990) include alate spores (folded) - these are the most common and have prominent or low folds, prominent wings or cristae. Cristate spores can have thin irregular folds to coarse (nearly tuberculate) folds, fenestrate spores which are characterised by the prominent perforate areas between the wings, cristae or folds (alate forms with fenestrate areolae). Echinate spores have projections from the folds (wings or fenestrate surfaces). Reticulate spores have a different meshwork which often covers the entire surface. Because of the large variation within each 'group' this classification system is difficult to apply to the spores of *Asplenium.*

Four basic 'rules' are evident from the above studies of spore ornamentation in *Asplenium*:

1. All species appear to have a uniform ornamentation throughout their known geographic range. This is well shown by unpublished studies of *Asplenium monanthes* from East Africa and South East Asia.
2. Each species appears to show a distinct ornamentation, which is unique for that species. Where species are described, based on changes in ploidy, the spores generally have similar surface features. This is well illustrated in the studies of *A. adiantum-nigrum* L. (Shivas, 1969) and its diploid *Adiantum onopteris* L. (Roberts, 1979).
3. Related species often show some common features in their patterns of ornamentation. This is illustrated in the studies of section *Hymenasplenium* (Iwatsuki, 1975) and section *Thamnopteris* (Johns, 1996).
4. Changes in spore size are often observed, and possibly reflect changes in ploidy level (Brownsey, 1977). Aborted or shrivelled spores are indicative of hybrids. Large spores in populations of *Asplenium bipinnatifidum* Baker from the western Pacific are indicative of an apomyctic condition (Johns, in prep.).

   In the present study variation in the perispore ornamentation is compared with studies of frond morphology. A key, based on frond morphology, is included for the species from Mount Cameroon.

Materials and Methods

Detailed morphological studies were made of the collections from Mokoko Forest Reserve (Fig. 1). Measurements of frond length and breadth, stipe length and the presence/absence of wings on the stipe, the position of sori along the frond and the detailed size and arrangement of the sori. Three measurements were made of the sorus in the middle section of the frond: the distance from the mid-vein to the lower end of the sorus, the length of the sorus and the distance from the outer end of the sorus to the margin. These were averaged for analysis. The degree of branching of the veins is recorded. Some taxa have simple veins, in others there are 2–3 branches. A total of 16 measurements were taken from each frond and all fertile fronds in each collection were measured. Analysis of the data was carried out using UNISTAT. The results of this analysis showed that several distinct species could be recognised (Fig. 2). The groups resulting from this analysis were then compared with the perispore ornamentation from each of the collections. Spore samples were removed from selected specimens, coated in platinum using a Balzer's Union SCD 050 coater, and the surface ornamentation studied using an Hitachi S 2400 scanning electron microscope (SEM). All specimens are deposited in the Kew Herbarium.

Results

Detailed examination of the perispore ornamentation of specimens collected from Mount Cameroon and the Mokoko Forest Reserve (Fig. 1) show that this group of simple-fronded *Asplenium* consists of a large number of distinct, unrelated species.

I. Notes on perispore ornamentation of the species in the Mokoko Forest Reserve

The Mokoko Forest Reserve is an undulating plateau at 300–500m altitude (Acwood pers. comm). Recently some 20 collections were received from Mokoko Forest Reserve and a series of measurements made from herbarium collections. Five groups of specimens can be identified using this data (Fig. 2). Representatives of all groups were studied using the SEM (Figs. 3–12). Although the spore ornamentation is similar in the three taxa, A, D and E, they are separated by frond measurements into three groups as shown in Fig. 2. Taxa B (Figs. 5–6) and C (Figs. 7–8) are distinguished by the presence of spines but these differ between the two taxa. The specimens were all forest epiphytes in closed forest.

Taxon A

Perispores with low, narrow, alate wings, and fenestrate areolae.

Specimens examined: Ackworth 271; Sonké 1051; Ekema 1035, 1061 (Figs. 3–4); Nere 2136.

Taxon B

The spines are coarse and irregular. This contrasts with the spines in Taxon C which are narrower and longer.

Specimens examined: Sonké 1143; Ekema 1043; Ndam 1043, 1077, 1323 (Fig. 5), 1332.

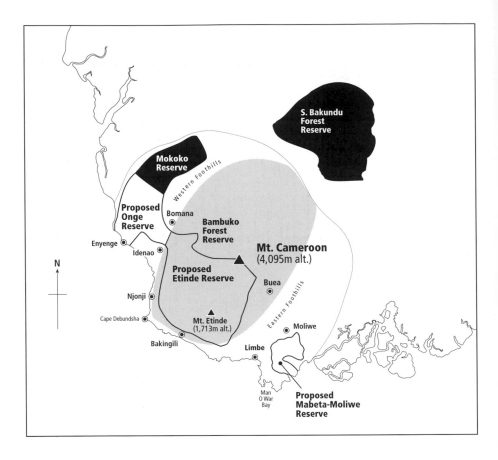

FIG.1. Map showing the location of Mokoko Reserve and the Mount Cameroon area. Reproduced from Cable and Cheek (1998).

Taxon C

The long and narrow spines on the perispore are quite distinctive from Taxon B.

Specimens examined: Cervani 1910; Chervet 183; Ekema 930; Mabani 440A; Mbani 444 (Fig. 6), 451, 457; Mere 747.

Taxon D

Perispore with low, narrow, alate wings and fenestrate areolae.

Specimens examined: Ekema 911 (Figs. 7–8); Ndam 1148; Schliebien 3002.

Taxon E

Perispore with low, narrow, alate wings and fenestrate areolae.

Specimens examined: Ackworth 208 (Figs. 9–10); Dany 42; Desvaux 1; Fleurg 12; Fraser 370; Halle 2234; Letouzey 3139; Watts 1133.

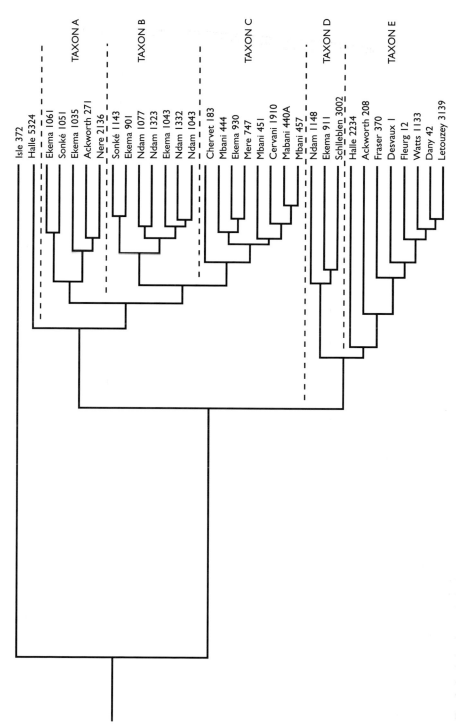

FIG. 2. Analysis of morphological data from the collections from Mokoko Forest Reserve. Taxa A – E.

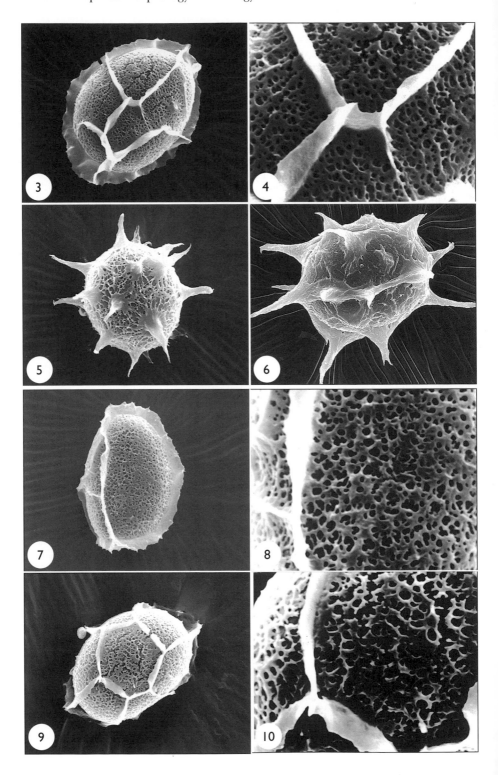

II. Notes on perispore ornamentation of Mount Cameroon and Mount Kupe specimens
Over 20 collections have been made by Earthwatch teams from Mount Cameroon and Mount Kupe since 1987 and these have provided the basis for the study of frond morphology and variation of perispore morphology of the simple-fronded species from Mount Cameroon. Most specimens were provisionally identified under one of the three epithets listed by Tardieu-Blot *loco citato* however, the detailed studies of the perispore ornamentation indicated that it is a complex of ten species. The field key was prepared to illustrate that the variation in the morphological characteristics of the fronds, of the different species, reflect the differences in perispore ornamentation as discussed in the notes below.

Asplenium variabile Hook. var. *variabile*
Both collections from Mount Cameroon were recorded as epiphytes at 100–300m altitude. The perispore has a delicate, echinate ornamentation (Fig. 11) which is only found in this species. Herbarium collections are immediately recognisable because of the fine wiry stipes and the dentate frond margin. Tardieu-Blot *loco citato*, keys out this species based on the presumed terrestrial habit, as this variety has only been collected as a terrestrial taxon from The Cameroon, Guinea, Nigeria, Liberia, Ghana, Congo and Fernando Po. Both collections from Mount Cameroon are epiphytes, so further studies are required to check on the identification.
Specimens examined: known from five collections from The Cameroon: Cable 544 (14/12/1993) from Diculu (transect C). Mabeta-Moliwe District; Cheek 5577 (25/11/1993) from Njonji, Etinde; Mbani 443 from Mokoko River; Wheatley 121 and Wheatley 239 both from Mabeta-Moliwe.

Asplenium Species 1
In forest and farm bush at 800–1000m altitude. The spores have broad, coarsely serrate ridges, the sides of the ridges lack fine sculpturing, and alate, reticulate areolae (Fig. 12).
Specimen examined: Groves 126. Eastern slopes, Likombe Village.

Asplenium Species 2
An epiphyte growing to 3m up the bole of a tree in secondary forest at 850m altitude. The ridges of the perispore are broad with a distinct surface ornamentation, the outer surface of the ridges is almost smooth (Figs. 13–14). This is probably *Asplenium subintegrum* C.Chr. This species is recorded in the literature from Cameroon, but reports from Tanzania and Uganda are not supported by perispore studies.
Specimen examined: Cable 861 (29/1/1995) collected in the Mount Kupe forest along a trail near Nyasoso High School.

Figs. 3–10. Ornamentation of spores from Mokoko Reserve. Figs. 3–4. Taxon A. Ekema 1061. Fig. 3. Whole spore × 1500. Fig. 4. Surface details × 5000. Fig. 5. Taxon B. Ndam 1323. Whole spore × 1500. Fig. 6. Taxon C. Mbani 444. Whole spore × 1500. Figs. 7–8. Taxon D. Ekema 911. Fig. 7. Whole spore × 1500. Fig. 8. Surface details × 5000. Figs. 9–10. Taxon E. Ackworth 208. Fig. 9. Whole spore × 1500. Fig. 10. Surface details × 5000.

Asplenium Species 3

The spores are very distinct because of their reduced size (Fig. 15), being approximately half the size of all other species of *Asplenium* studied from Mount Cameroon.

Specimen examined: Start 457.

Asplenium Species 4

A lithophyte in secondary forest with an open canopy at 1050m. The perispores have broadly based ridges with fenestrate areolae (Figs. 16–17). This species is probably *Asplenium currori* Hook. (*Asplenium africanum* var. *currori* (Hook.) Tardieu). It is reported in the literature from The Cameroon to Uganda and Angola.

Specimen examined: Cable 69 (29/8/1992) from Mount Kupe, Nyasoso District.

Asplenium Species 5

Grows in open secondary forest as an epiphyte on the bole of *Aningeria* sp. as 1050m altitude. The perispores have narrow ridges with fenestrate areolae (Fig. 18). The perispores are identical to a Ghana collection (Adams 057–56–05701).

Specimen examined: Lane 31. (30/7/1992) Mount Kupe, Max's Trail, Nyasoso District.

Asplenium Species 6

Epiphytes in tropical moist forest on volcanic soils, or in secondary forest dominated by *Musanga*. The perispore consists of narrow, alate wings with fenestrate areolae (Figs. 19–20). The surface structure of the fenestrate areolae is distinct from species 7 where it forms a pattern of low secondary ridges. These specimens possibly represent *Asplenium africanum* which has been reported Ghana, The Ivory Coast, Cameroon and Angola.

Specimens examined: Groves 13 (19/1/1995), Kupe Village, Mount Kupe; Tchouto 893, Onge; Watts 131 (13/1/ 1992), Mabeta, 1.5 km north of Bimbia.

Asplenium Species 7

Epiphyte with fronds to 60cm long in disturbed evergreen forest on old volcanic soils, along watercourses at 30–200m altitude. The perispore consists of narrow, alate wings with fenestrate areolae (Figs. 21–22).

Specimens examined: Watts 290 (19/5/1992), Chop Farm, Limbe District; Sunderland 1239 (22/4/1992) Mabeta-Moliwe, Limbe District.

Asplenium Species 8

Epiphyte on *Elaeis* in secondary forest along the seashore at 5m altitude. The perispores are irregularly echinate with fenestrate areolae (Figs. 23–24).

FIGS. 11–18. Ornamentation of spores from Mount Cameroon and Mount Kupe. Fig. 11. *Asplenium variabile*. Cheek 5577. Fig. 11. Whole spore × 1500. Figs. 12. *Asplenium* sp. 1. Groves 126. Fig. 12. Whole spore × 1500. Figs. 13–14. *Asplenium* sp. 2. Cable 861. Fig. 13. Whole spore × 1500. Fig. 14. Surface details × 5000. Fig. 15. *Asplenium* sp. 3. Start 457. Fig. 15. Whole spore × 1500. Figs. 16–17. *Asplenium* sp. 4. Cable 69. Fig. 16. Whole spore × 1500. Fig. 17. Whole spore × 5000. Fig. 20. *Asplenium* sp. 5. Lane 31. Fig. 18. Whole spore × 1500.

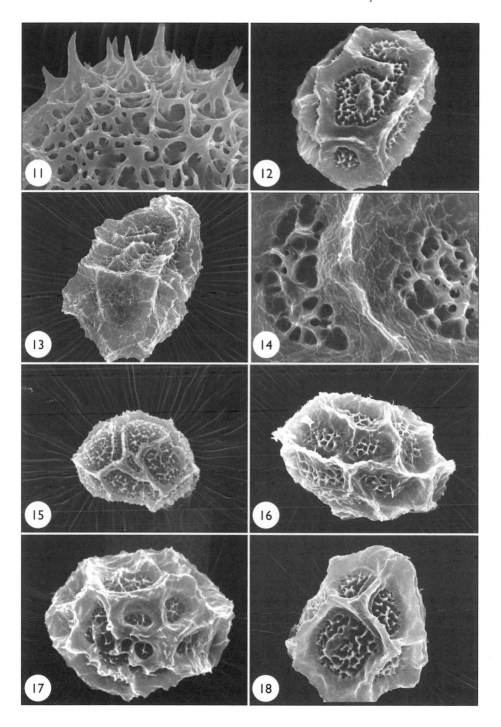

Specimen examined: Groves 230 (7/3/1995) collected from along the road to the north of Dikulu Village. Mabeta-Moliwe.

Asplenium Species 9

Recorded as an epiphyte, up to 1.5m above ground, in low altitude forest above an oil palm plantation at 300m altitude. The plant forms a funnel-shaped rosette of 8 fronds. Perispore with long, narrow spines (Figs. 25–26). *Asplenium repandum* Mett. ex Kuhn (Lectotype Kew) has similar long spines to the spores from Cheek 5576 but the spores from the type collection are much smaller. Cheek 5576 may be a tetraploid.

Specimen examined: Cheek 5576 (25/11/1993), along path to Lake Njonji, Etinde.

Provisional key to the simple-fronded species from Mount Cameroon and Mount Kupe

1. Plants with simple or compound fronds, usually mixed. Stipe to 7–8cm long, thin and wiry, not winged . . . . *Asplenium variabile*

1. Plants only with simple fronds. Stipe variable in length but thick, not wiry; if upper portion of stipe is winged this is not included in measurement of stipe length but as an extension of frond 2

2. Stipes (excluding winged sections) very short, less than 2–2.5cm long 3

2. Stipes (excluding winged sections) long, over 6–8cm long 6

3. Fronds over 5cm broad, and (30–)70+cm in length 4

3. Fronds narrow (less than 3cm broad), and usually less than 20cm in length . . . 6

4. Fronds to 6cm in width, length variable 5

4. Fronds broad (over 10cm in width), over 70cm in length. Sori 3–4cm long with a marginal infertile zone on frond to 1.5–2cm broad . . . . *Asplenium* sp. 9

5. Stipes short to 2–3cm. Frond to 34 × 4.8cm. Sori 1–1.5cm long with a marginal infertile zone on frond to 1cm broad . . . . *Asplenium* sp. 6

5. Stipes lacking, or to 0.5cm long. Fronds 55–72cm × (4–)5.5–6cm broad. Gradually narrowed to base over lower 15–25cm of frond. Sori 1.5–2cm long. Infertile marginal zone to 1cm broad . . . . *Asplenium* sp. 7

6. Stipes short, to 1.2cm long. Fronds to 20×2cm. Base of frond narrow, forming a wing to 8cm long. Sori short (0.6–1cm long), arising near costa. Narrow infertile zone to 0.2–0.3cm along margins of frond . . . . *Asplenium* sp. 2

FIGS. 19–26. Ornamentation of spores from Mount Cameroon and Mount Kupe. Figs. 19–20. *Asplenium* sp. 6. Groves 13. Fig. 19. Whole spore × 1500. Fig. 20. Surface details × 5000. Figs. 21–22. *Asplenium* sp. 7. Watts 290. Fig. 21. Whole spore × 1500. Fig. 22. Surface details × 5000. Figs. 23–24. *Asplenium* sp. 8. Groves 230. Fig. 23. Whole spore × 1500. Fig. 24. Surface details × 5000. Figs. 25–26. *Asplenium* sp. 9. Cheek 5576. Fig. 25. Whole spore × 1500. Fig. 26. Surface details × 5000.

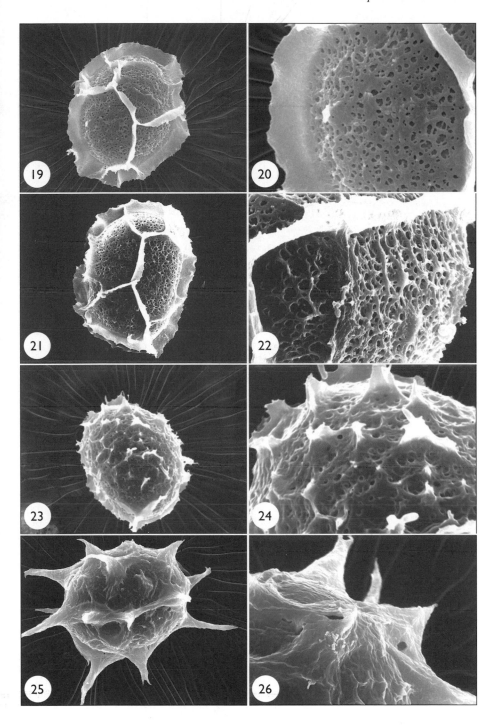

6. Stipes short, to 1.5–2cm long. Fronds to 17 × 2.8cm. Base of frond rapidly narrowed to a short section forming a wing along an extension stipe. Sori short (0.5–0.6cm long), scattered along frond with only (5–)10–12 sori on either side of costa in fertile fronds. Sori arise away from costa leaving a distinct infertile zone between costa and soriferous zone. Narrow infertile zone to 1cm long along margins of frond . . . . *Asplenium* **sp. 4**

7. Fronds over 60cm long **8**

7. Fronds less than 45cm long **9**

8. Stipe 8–10cm long. Fronds to 44 × 4cm narrowed towards base in lower third of the lamina. Sori 1–2cm long, rising near costa. Infertile marginal region of lamina to 0.5cm . . . . *Asplenium* **sp. 1**

8. Stipe 8–9cm long. Fronds to 35 × 4cm, gradually narrowed towards base of lamina. Sori 1–1.5cm long, rising near costa. Infertile marginal region of lamina to 0.5cm wide. Widest portion of frond in top third of the blade . . .
. . . . *Asplenium* **sp. 5**

9. Stipe to 6–8cm long. Fronds to 60 × 5cm. Sori 1.0–1.8cm long. Lamina gradually narrowed towards the base. Infertile marginal region to 1cm broad. Fronds growing in shade can be significantly longer . . . . *Asplenium* **sp. 6**

9. Stipe to 6–7cm long. Fronds to 60 × 5cm, widest in the upper section of lamina. Lamina gradually decreasing towards base, forming a winged stipe. Sori 1.5–2cm long. Infertile marginal region to 2cm broad . . . . . *Asplenium* **sp. 8**

Species 3, 8 and 9 are apparently endemic and known from only single collections in The Cameroon. Some species are more widespread in Tropical Africa but all species are under-collected and additional material is necessary to understand the African species.

Discussion and Conclusions

Problems in the identification of the simple-fronded species of *Asplenium* have been compounded by the incorrect application of the two widely used names in regional floras throughout Tropical Africa. *Asplenium africanum* has been used to name most collections from West and Central Africa. *A. holstii* is misapplied for the identification of most collections from Tropical East Africa. Similar problems also occur in the application of the name *Asplenium nidus* L. which is incorrectly recorded from Africa. It is possible to draw the following conclusions from the study of perispore morphology:

1. All the species in The Cameroon can be recognised by their distinctive perispore morphology.
2. Where species are widespread in Africa they have a consistent perispore morphology throughout their range.
3. Some monophyletic groups, such as section *Thamnopteris*, show an overall uniformity in perispore ornamentation and these characteristics could be used to define this section.
4. The simple-fronded species from The Cameroon are highly diverse in perispore ornamentation. This suggests that they do not form a natural group within the genus.

The study of perispore ornamentation provides a powerful tool for the recognition of species within *Asplenium* (Aspleniaceae). However, much work remains to be done in order to gain an understanding of the 'natural groups' within the genus. Preliminary anatomical studies of material from The Cameroon, where excellent collections have been made as part of an Earthwatch programme, indicate that considerable variation exists between the species. DNA studies may further help to elucidate these relationships and provide a better understanding of the evolutionary trends within the Aspleniaceae. Recognition of distinct species within the simple-fronded 'group' of *Asplenium* will help to identify sites of special scientific interest as targets for conservation in Tropical Africa. Several Cameroon species appear to be local endemics. Additional collections will help to better understand the taxonomy of the African species as there are probably several species of simple-fronded *Asplenium* not yet collected in Tropical Africa.

Acknowledgements

This research was stimulated by the extensive collections made by members of The Cameroon Project at Kew, and also the collections made by several Earthwatch Teams associated with The Cameroon Project. I wish to particularly thank Martin Cheek for these collections. James Clarkson provided assistance with the SEM studies of the specimens from Mokoko and in the preparation of the plates for this paper. Dr David Simpson advised on the analysis of morphological data using UNISTAT. Peter Edwards has provided continual help during the project. I wish to also thank the reviewers of the paper who made several useful suggestions and particularly to Dr Madeline Harley for her continued advice and encouragement.

References

Alston, A. H. G. (1959). The ferns and fern allies of West Tropical Africa. Supplement to edition 2 of Flora of West Tropical Africa. Crown Agents, London

Belbin, L. (1989). PATN pattern analysis package. Users and technical reference manuals. CSIRO Division of Wildlife and Ecology, Canberra.

Brownsey, P. J. (1977). A taxonomic revision of the New Zealand species of *Asplenium*. *New Zealand Journal of Botany* **15**: 39–86.

Cable, S. and Cheek, M. (1998). The plants of Mount Cameroon. A conservation checklist. Royal Botanic Gardens, Kew, Mount Cameroon Project, Limbe.

Holttum, R. E. (1974). *Asplenium* L. section *Thamnopteris* C. Presl. *Gardens Bulletin Singapore* **27**: 143–154.

Iwatsuki, K. (1975). Taxonomic studies of Pteridophyta 10. *Acta Phytotaxonomica et Geobotanica* **27**: 39–55.

Johns, R. J. (1996). A field study of *Asplenium* section *Thamnopteris* in the NE Kepala Burung, Irian Jaya, Indonesia. In: J. M. Camus, M. Gibby and R. J. Johns (editors). Pteridology in perspective, pp. 337 – 340, Royal Botanic Gardens, Kew.

Puttock, C. F. and Quinn, C.J. (1980). Perispore morphology and the taxonomy of Australian Aspleniaceae. *Australian Journal of Botany* 28: 305–322.

Reichstein, T. (1981). Hybrids in European Aspleniaceae (Pteridophyta). *Botanica Helvetica* **91**: 89–139.

Roberts, R. H. (1979). Spore size in *Asplenium adiantum-nigrum* L. and *A. onopteris* L. *Watsonia* **12**: 233–238.

Shivas, M. G. (1969). A cytotaxonomic study of the *Asplenium adiantum-nigrum* complex. *British Fern Gazette* **10**: 68–80.

Smith, A. R. (1976). *Diplazium delitescens* and the neotropical species of *Asplenium* sect. *Hymenasplenium. American Fern Journal* **66**: 116–120.

Tardieu-Blot, M.-L. (1953). 28. Les Pteridophytes de l'Afrique Intertropicale Francaise. Memoires de l'Institut Francais d'Afrique Noire. Ifan-Dakar.

Tardieu-Blot, M.-L. (1964). Flore du Cameroun. 3. Pteridophytes. Museum National d'Histoire Naturelle. Paris.

Tryon, A.F. (1990). Fern spores: evolutionary levels and ecological differentiation. In: Hesse, M. and F. Ehrendorfer (editors). Morphology, development and systematic relevance of pollen and spores. *Plant Systematics and Evolution. Supplement* B 5: 71–79. Springer-Verlag Vienna.

Tryon, A.F. and Lugardon, B. (1990). Spores of the Pteridophyta. Springer-Verlag. 1–648.

Van Uffelen, G. A. (1985). Synaptospory in the fern genus *Pyrrosia* (Polypodiaceae) *Blumea* 31: 57–64.

Viane, R. and Van Cotham, W. (1977). Spore morphology and stomatal characters of some Kenyan *Asplenium* species. *Bericht der Deutschen Botanischen Gesellschaft* **90**: 219–239.

Tomlinson, P.B. (2000). Structural features of saccate pollen types in relation to their functions. In: M.M. Harley, C.M. Morton and S. Blackmore (Editors). Pollen and Spores: Morphology and Biology, pp. 147–162. Royal Botanic Gardens, Kew.

STRUCTURAL FEATURES OF SACCATE POLLEN TYPES IN RELATION TO THEIR FUNCTIONS

P.B. TOMLINSON

Harvard University, Harvard Forest, Petersham, MA 01366 USA

"They went to sea in a sieve, they did, in a sieve they went to sea"
The Jumblies. E. Lear (1906)

Abstract

Saccate pollen is one of the most distinctive of pollen types but is exclusive to gymnosperms. It may be ancestral in conifers, but with frequent loss of the condition. Several explanations have been offered to account for the adaptive benefits of saccate pollen. Here I review some of these in relation to the detailed structure of pollen sporoderm and features of the corresponding ovules that pollen serves since unexpected functional correlations seem to exist. Saccate pollen is associated with an inverted pollination drop in the corresponding ovule. Saccate pollen floats in a water surface because sacci are air-filled. Air enters via minute sporoderm pores with hydrophobic walls, but is trapped when pollen enters water. Saccate pollen thus provides a refined example of sporoderm structure being interpretable in functional terms.

Introduction

Despite the seeming simplicity of wind pollination, the Coniferales offer a diversity of pollen types that exceeds that of any other group of non-flowering seed plants. Why is this? The answer lies in the behaviour of pollen at the time of pollen capture, i.e., when the pollen reaches the exposed micropyle of the ovules in the ovulate cone. In most taxa pollen becomes incorporated into a pollen chamber, the cavity into which the micropyle leads internally. It is not generally appreciated that pollen diversity can be correlated not with pollen dispersal itself, but to pollen capture and the method of pollen germination, together with features of cone and ovule morphology at the time of pollen reception (Tomlinson and Takaso, 1998). This leads to the recognition of a "pollen capture syndrome" in which pollen structure is only one aspect of several features that work in concert (Table 1). The only limitation would appear to be one of pollen grain size that makes wind dispersal possible; nevertheless pollen diameters for all conifers range over almost an order of magnitude with saccate pollen at both extremes (Ueno, 1960c). The ratio of saccus volume to corpus volume also varies considerably (Tomlinson, 1994), but this variation remains unexplained in functional terms.

Saccate pollen represents the most distinctive kind of conifer pollen and here I attempt some analysis of its structure-function relationships. In mature saccate pollen there are one or more air-filled extensions of the exine. These sacci are often referred to as "bladders" or "wings," which is misleading in terms of both structure and function. Of particular interest is the sporoderm morphology of the saccus and its physical properties, especially in water. A comparative functional approach shows that variation in sporoderm morphology is under tight selective control and can account for the variation observed. This comparative approach was pioneered by Joseph Doyle (1945) and most recent work is an extension of some of his ideas. I have emphasised Phyllocladaceae and Podocarpaceae since these families are relatively inaccessible to most botanists, but they show considerable pollen diversity (Ueno, 1960a, b, c; Pocknall, 1981a, b).

Saccate pollen occurs in at least seven orders of seed plants, six of them extinct (Table 2). In modern gymnosperms (Figs. 2-17) it occurs only in Pinaceae and Podocarpaceae (Coniferales), but not in all genera. Phyllocladaceae could be included, but the sacci might be described as vestigial because they are small and lack an obvious hydrodynamic function since pollen does not float. The existence of saccate pollen in so many seemingly independent lineages, going back to the Callistophytales of Middle Pennsylvanian age, suggests that such pollen may have evolved independently in several groups of gymnosperms. However, in the lineage leading to modern conifers, beginning in the Paleozoic and represented by the Cordaitales and continuing with the Mesozoic transition conifers (for example, Voltziales) saccate pollen occurs in many, but not all taxa, and may be a plesiomorphy within the coniferophyte clade. If so, the existence of saccate and non-saccate pollen types in related taxa needs an explanation. This also leads to an assumption that saccate pollen is ancestral within Coniferales and the assumption is here tested.

TABLE 1. Characters and some character states in the "pollen capture" syndrome of conifers.

Character	Contrasted States[1]	
Pollination drop	Present	Absent
Drop withdrawal	Pollen stimulated	Not pollen stimulated
Cone orientation	Inverted	Neutral
Ovule orientation	Inverted or neutral	Erect
Mechanism of orientation	Epimatium (Podocarpaceae)	Others
Pollen morphology	Saccate	Non-saccate
Exine ultrastructure	Alveolate	Non-alveolate (granular)
Pollen hydrodynamics	Non-wettable, floating	Wettable, sinking
Exine shedding in water	Absent	Present
Intine thickness	Thin	Thick
Site of pollen germination	Inside pollen chamber	Outside pollen chamber
Degree of siphonogamy	Limited	Extended
Integument elaboration	Present	Absent
Integument invagination	Present (*Larix, Pseudotsuga*)	Absent
Nucellar extrusion	Present (*Saxegothaea*)	Absent

[1]Note: the features shown together in each column are not necessarily correlated. Actual correlations are shown in Tomlinson and Takaso 1998.

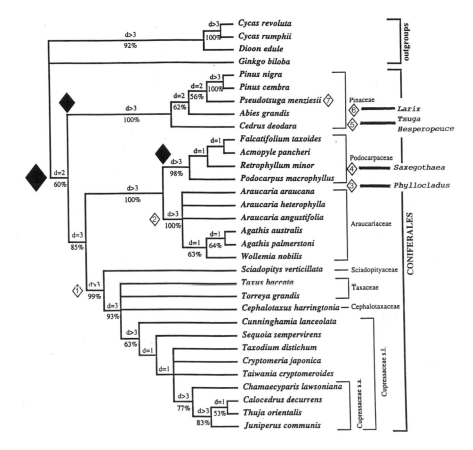

Fig. 1. Strict consensus of the six most parsimonious trees from a family-level analysis comprising 32 28S rRNA sequences from most conifer families (modified from Stevanovi *et al.*, 1998). The cladogram establishes the monophyly of conifers, using extant gymnosperms as outgroups. The modification involves the addition of 5 taxa not in the original cladogram (*Larix*, *Tsuga* and *Hesperopeuce*, *Saxegothaea* and *Phyllocladus*), but with their likely position suggested. If one assumes that saccate pollen is ancestral in conifers (large black diamond), open diamonds at seven positions indicate where the saccate pollen has been lost from certain major clades or genera (the numbers do not necessarily indicate an order of appearance). Small black diamonds indicate where the ancestral saccate condition was initially retained. *Phyllocladus* retains non-functional (vestigial) sacci.

d = decay values above branches; % = bootstrap values below branches.

Evolution of pollen in Coniferales

Recent phylogenies proposed for the conifers allow one to speculate about the changes in pollen morphology during their evolutionary history. The recent molecular analysis (using 32 28S rRNA sequences) by Stevanovi *et al.* (1998), although based on a limited number of taxa, allows this to be done (Fig. 1). Its original purpose was to establish the monophyly of conifers, which is now well accepted. If one assumes that in the ancestral condition conifers possessed saccate (non-wettable) pollen, then it has been retained in Pinaceae and Podocarpaceae, but lost from these modern taxa with

149

non-saccate (wettable) pollen. The latter taxa themselves fall into distinct categories. In the group of five families ("taxads") with very uniform pollen morphology and a consistently present pollination drop (Cephalotaxaceae, Cupressaceae, Sciadopityaceae, Taxaceae, Taxodiaceae), pollen is wettable and sheds the exine in water by enlargement of the intine. These families form a discrete and well supported clade in Fig. 1. Araucariaceae is also an independent lineage (Fig. 1) with pollen that does not respond to water, but with extended siphonogamy. Genera in this family uniformly lack a pollination drop (information is not available for *Wollemia*). This leaves a highly derived group of taxa within Pinaceae and Podocarpaceae that have more recently lost sacci from their pollen, i.e., *Larix, Pseudotsuga* and *Tsuga* (including *Hesperopeuce*) in Pinaceae, *Saxegothaea* from Podocarpaceae and, more distinctly, *Phyllocladus* from Podocarpaceae. In *Tsuga* and *Hesperopeuce* transitional stages may be observed (Ueno, 1957). This implies that the non-saccate condition has been derived independently on at least seven occasions (numbered open triangles in Fig. 1).

An alternative interpretation is that in the ancestral condition conifer pollen was non-saccate. This would imply that the non-alveolate ("granular") exine of the taxads was ancestral and has persisted in six families. One then assumes that the condition in the Araucariaceae is also derived, but now from a non-saccate ancestor. However, this scenario also requires, according to Fig. 1, that saccate pollen is derived independently in Pinaceae and Podocarpaceae, with the secondary loss of the condition in about five genera in the two families. Otherwise, if Pinaceae and Podocarpaceae do constitute a single lineage characterised by the ancestral presence of saccate pollen, then a major restructuring of the base of the cladogram is required, which is problematical. The point to emphasise is that evolution of pollen types in conifers is closely dependent on changes in pollen capture mechanisms (for example, Tomlinson and Takaso, 1998). Research on the functions of these pollen types will be useful in unravelling phyletic processes.

Sporoderm morphology in saccate pollen

Previous structural and especially ultrastructural research, in combination with developmental study, has established that sacci are localised lateral extensions of the exine (for example, Kurmann, 1989). They seem to be an exaggeration of the alveolate sporoderm texture that otherwise characterises the non-saccate portion of the grain, particularly its proximal region or corpus. By differential floating of pollen on drops of water that can be inverted, contrasted orientations can easily be observed (Figs. 2, 3).

Figs. 2–9. SEM photographs of pollen structure in *Pinus* and *Phyllocladus*. Figs. 2–7. *Pinus mugo* Turra.; Figs. 8, 9. *Phyllocladus glaucus* Carr. Fig. 2. Saccate pollen dried down from a drop of water, grains all oriented with distal face uppermost; Fig. 3. the same, but drop of water inverted and orientation of most grains reversed; Fig. 4. dehydrated grain, distal views, body collapsed but sacci rigid; Fig. 5. dehydrated grain in lateral view, with contrasted sculpturing of saccus and body; Fig. 6. detail of saccus surface to show micropores; Fig. 7. detail of saccus after mechanical fracture to show internal reticulations; Fig. 8. dehydrated grain with reduced sacci ("protosaccate" or "quasisaccate"); Fig. 9. detail of surface of saccus with few micropores. Scale bar = 100 μm in Fig. 3 (same for Fig. 2); 20 μm in Fig. 5 (same for Fig. 4); 3 μm in Fig. 6; 5 μm in Fig. 7; 10 μm in Fig. 8; 3 μm in Fig. 9.

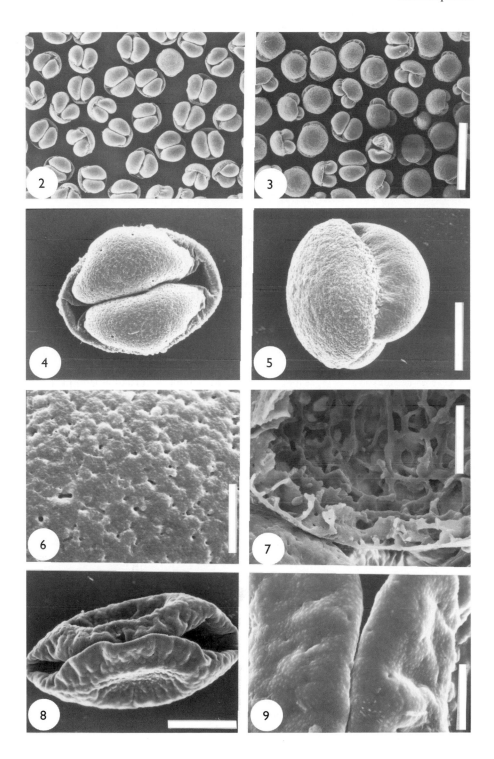

Three main morphological types exist: monosaccate (only developed in extinct orders); bisaccate (the most common condition); and trisaccate (in modern genera only occurring in *Dacrycarpus* (Figs. 12, 13), *Microcachrys* and *Microstrobos*, all Podocarpaceae. These three types may be seen as simple variations on the saccus as an encircling region that may or may not be continuous. Monosaccate and trisaccate aberrant grains in taxa that are normally either bisaccate or trisaccate (Cranwell, 1961) may be used to support this interpretation. Sacci are usually relatively smooth on their external surface (Figs. 4, 5), but show a pronounced internal alveolate elaboration of the outer part of the sexine that is readily visible under the light microscope and in fractured grains in the SEM (Fig. 7). The floor of the saccus consists of the innermost layer of the sexine and is typically lamellate. An important feature that has received little comment is that the outer wall of the saccus is usually provided with minute pores so that there is a continuous passage between the saccus lumen and the outside. However, the presence of pores in all saccate pollen types is still incompletely documented (the recent paper by Runions *et al.* (1999) provides a conspicuous exception). These pores are sometimes visible in sections of the saccus viewed by transmission electron microscopy (for example, Ueno, 1960a) and may be seen under the scanning electron microscope (Fig. 6), although this technique does not necessarily demonstrate that they are continuously open pores. Their diameter is of the order of 100-300 nm (but often with a considerable range (Figs. 14, 15)). At the time of release from the microsporangium sacci are air-filled; this is evident in grains examined microscopically in water, since the refractive properties of the air renders the sacci dark. Later I will explain how they become air-filled.

The size of saccate grains, measured as the length of the body, encompasses a wider range than any other kind of pollen in conifers, but Podocarpaceae have consistently smaller grains (range 20-50 μm) than Pinaceae (range 50-100 μm) with scarcely any overlap (Tomlinson, 1994). *Dacrycarpus dacrydioides* (Rich.) De Laub. (Fig. 12, 13) represents the upper limit for podocarps. Saccus diameter varies considerably, but surprisingly Podocarpaceae tend to have proportionately larger sacci than Pinaceae (Tomlinson, 1994; data from Ueno, 1960c). We will later explore in detail diversity in Podocarpaceae.

Adaptive significance of saccate pollen

Five major hypotheses have been proposed to account for the existence of saccate pollen; three refer to the function of sacci in air (aerodynamic) and two to their function in water (hydrodynamic).

A. Aerodynamic

1. Sacci reduce grain density; this slows their terminal velocity (according to Stokes' Law), which in turn promotes wider dispersibility in moving air, and is presumed to favour outcrossing (text-book explanation).

2. Sacci close together in dry air such that they close the distal sulcus ("germ furrow"), thus minimising water loss, since the distal region always has a thin sporoderm. This harmomegathic function was proposed by Wodehouse (1935). The "closed" appearance of sacci is easily seen in unprocessed grains dried down for scanning microscopy (for example, Figs. 4, 12, 16).

3. Sacci facilitate discharge of the pollen from microsporangia, i.e., they "entrain" pollen into the air (Niklas, 1985).

B. Hydrodynamic

4. Sacci orientate the grain in the pollination drop such that they float upwards in the micropyle, so that the pollen lands on the nucellus with the germ furrow appropriately applied to the surface that will immediately receive the pollen tube (Doyle, 1945).

5. Saccate pollen is selectively captured by the process of pollen scavenging in an inverted pollination drop such that all other particles are rejected by the drop. Apart from non-saccate pollen, rejected particles would include pathogen spores or insect eggs (Tomlinson *et al.*, 1991).

It should be noted that these mechanisms are not mutually exclusive, and all have the sanction of physics, so the problem is to establish the evolutionary adaptive process involved, rejecting other explanations as exaptations. The first three may be rejected because they do not explain why saccate pollen should have a restricted distribution within modern conifers. The harmomegathic explanation of Wodehouse (1935) is particularly difficult to accept, since the pollen grain must reach its minimum volume (i.e., is maximally dehydrated) before the mechanism can begin to operate. The hydrodynamic explanation is more acceptable because it is based on a very precise correlation between the presence of saccate pollen and ovular features at the time of pollination, as first pointed out by Doyle (1945). These include the presence of a pollination drop, and an ovule that is topographically inverted. Topographic inversion is the result of morphological inversion in an erect cone or by some other means. The reduced density of saccate grains is such that they always float upwards in water. They behave like bubbles in that they aggregate at an air-water interphase ("non-wettable" in Tomlinson, 1994), and will float upwards if they become free within a column of water. Doyle's explanation is rational in this physical context, but there are difficulties in accepting his interpretation if the process of pollen scavenging is either observed directly or simulated by adding saccate pollen to an existing drop. Pollen is then seen to be largely drawn into the micropyle by the meniscus of the receding drop. Furthermore, the micropyle is not necessarily straight and, in addition, direct observation of naturally captured pollen suggests that the arrangement of pollen grains on the nucellus becomes random by mutual crowding (Figs. 10, 11) but this does not seem to inhibit the directionality of pollen tube growth. Pollen does not have to be seated on the nucellus in order to germinate (Figs. 10, 11).

Even though I favour the pollen selection mechanism (Tomlinson *et al.*, 1991), I realise that it is not very specific, because it does not discriminate between saccate pollen from different species. For example, in mixed populations of two genera, trisaccate grains of *Dacrycarpus* can be found in the pollen chambers of the ovules of *Podocarpus*, which has bisaccate pollen. Nevertheless, in the absence of known mechanisms for pollen screening in gymno-ovulate plants, this mechanism can be interpreted as the beginnings of a more efficient system that could work well in species-poor forests that might have characterised Mesozoic vegetation, the process accentuated by different phenological strategies. With these disputable conclusions drawn we now need to explore variations on this central theme, and especially diversity of pollen types.

Pollination in Pinaceae

Building on the pioneer work of Doyle (1945) with this family, John Owens of the University of Victoria, British Columbia, has extensively enhanced our understanding of the pollination process in its members. The observations of his co-workers have been summarised in Tomlinson and Takaso (1998, their Table 3). There is a strong correlation between the presence of sacci and a pollination drop, providing evidence to substantiate Doyle's ideas, but the subject requires more detailed analysis than can be provided here. Exceptional taxa, in so far as they have been studied, show variation on the central theme. In some taxa there is the possibility of pollen scavenging, since in *Pinus* and most species of *Picea*, air-borne pollen is first received by integumentary extensions, adhesion to these structures being facilitated by the initial secretion of minute water droplets. Subsequently, pollen initially captured in this way can become incorporated into a later secreted pollination drop.

Most recently Runions *et al.* (1999) have demonstrated a distinctive correlation in *Picea orientalis* (L.) Link between saccate pollen structure and the mechanism of pollen capture. Here the pollination drop is erect, not inverted, and the sacci only function temporarily as floats, since air is rapidly displaced in water and pollen rapidly sinks and enters the drop. These authors emphasise that sporoderm pores facilitate loss of air from the sacci.

TABLE 2. Seed plants with saccate pollen.

1. Cycadophyte line[1]

A. PALEOZOIC	
Callistophytales	monosaccate ("Vesicaspora-type")
	Middle Pennsylvanian
Glossopteridales[2]	bisaccate
	Pennsylvanian to Jurassic
B. MESOZOIC	
Caytoniales	monosaccate ("Vitreisporites-type")
	Triassic to Cretaceous
Corystospermales	bisaccate
	Triassic

2a. Coniferophyte line (extinct taxa)

A. PALEOZOIC	
Cordaitales	monosaccate ("Florinite &
	Felixipollenites-types")
	Pennsylvanian to Permian
B. MESOZOIC	
Voltziales	mono- & bisaccate
	Pennsylvanian to Jurassic

2b. Coniferophyte line (extant taxa)

Coniferales	
(1) Pinaceae	most genera bisaccate[3]
	Jurassic to present
(2) Podocarpaceae	bisaccate or trisaccate
	(except *Saxegothaea*)
	Jurassic to present
(3) Phyllocladaceae	vestigially bisaccate?
	Cretaceous to present

[1] *Lasiostrobus* (Pennsylvanian) is described as "polysaccate," but the sacci are more like the colpae of angiosperms.
[2] Last occurrence is late Triassic
[3] *Abies, Cathaya, Cedrus, Keteleeria, Picea, Pinus, Pseudolarix.*

TABLE 3. Pollen and ovule diversity in Podocarpaceae and Phyllocladaceae.

1. Pollen non-saccate, spherical, with uniform surface sculpturing . . . . *Saxegothaea*
    [Pollination drop absent, epimatium reduced, nucellus protruding through micropyle, pollen germinating outside pollen chamber, gametes delivered by extended siphonogamy.]

1A. Pollen saccate, not spherical . . . (Phyllocladaceae and other Podocarpaceae) 2
    [Pollination drop probably always present, nucellus not protruding through micropyle, pollen germinating in pollen chamber, siphonogamy not extended.]

2. Pollen trisaccate 3

2A. Pollen bisaccate 5

3. Body of pollen grain > 40 μm diameter; endoreticulations of "*Podocarpus*" type . . . .
    . . . . *Dacrycarpus*
    [Ovules usually one per cone, epimatium enclosing ovule except for micropyle, pollen scavenging possible.]

3A. Body of pollen grain < 35 μm diameter 4

4. Sacci with distinct endoreticulations . . . . *Microcachrys*
    [Epimatium present; ovule perpendicular to cone axis, pollen scavenging minimal.]

4A. Sacci without distinct endoreticulations . . . . *Microstrobos*
    [Epimatium absent, ovule inverted by recurvature of cone axis.]

5. Grains small, body 35 μm long; sacci reduced or vestigial with obscure endoreticulations; pollen sinking in water . . . . *Phyllocladus* (Phyllocladaceae)
    [Epimatium absent, retraction of pollination drop stimulated by addition of pollen.]

5A. Grains 15-50 μm long; sacci usually well-developed, pollen probably always floating on water; withdrawal of pollination drop probably not stimulated by addition of pollen 6

6. Sacci small (< 1/10 volume of body of grain), body of grain > 30μm . . . . *Lagarostrobos*
    [Epimatium present; cone, but not ovule, inverted.]

6A. Sacci large (normally 1/3 to 1/2 volume of body of grain) 7
    [Ovule morphologically and topographically inverted, cones erect.]

7. Endoreticulations of saccus radially columellate, i.e., extending the full radius of the saccus
    . . . ."*Dacrydium*-type"
    i.e. includes *Dacrydium*, *Falcatifolium*, *Halocarpus* and *Manoao*.
    [Pollination drop present; epimatium producing inverted ovule but not enclosing mature seed.]

7A. Endoreticulations alveolate, restricted to inner surface of saccus . . . . "*Podocarpus*-type"
    i.e., includes *Acmopyle*, *Afrocarpus*, *Lepidothamnus*[1], *Nageia*, *Parasitaxus*, *Podocarpus*, *Prumnopitys*, *Retrophyllum*, *Sundacarpus*.
    [Pollination drop probably always present, pollen scavenging probably well developed; epimatium usually enclosing ovule or mature seed.]

---

*According to Ueno, 1960b, pollen of *Podocarpus lambertii* is about the smallest of that of all podocarps, the body of the grain < 20 μm long.
[1]*Lepidothamnus* is somewhat anomalous, since it has "*Podocarpus*-type" pollen, but resembles the "*Dacrydium*-type" in the seed not becoming enclosed by epimatium at maturity; it has a distinctive hooked micropyle.

Pollen structure in Podocarpaceae and Phyllocladaceae

Pollen morphology and saccus structure in Podocarpaceae is particularly diverse (Figs. 12-17). Information for the New Zealand species has been provided by Pocknall (1981a) and aspects of microsporogenesis most recently by Del Fueyo (1996). We must emphasise that grains are uniformly small compared with those of most Pinaceae. The smallest reported is *Podocarpus lambertii* Klotzsch. ex Endl. (length 15 μm) and with very small sacci (Ueno, 1960b). Those of *Afrocarpus*, *Microcachrys*, *Microstrobos* and *Lagarostrobos* (s.l.), all with a body length less than 30 μm, are among the smallest of all conifer pollen grains. Table 3 sets out an appropriate division of pollen types for the family, in conjunction with some comments on the distinctive features of ovule morphology at the time of pollination.

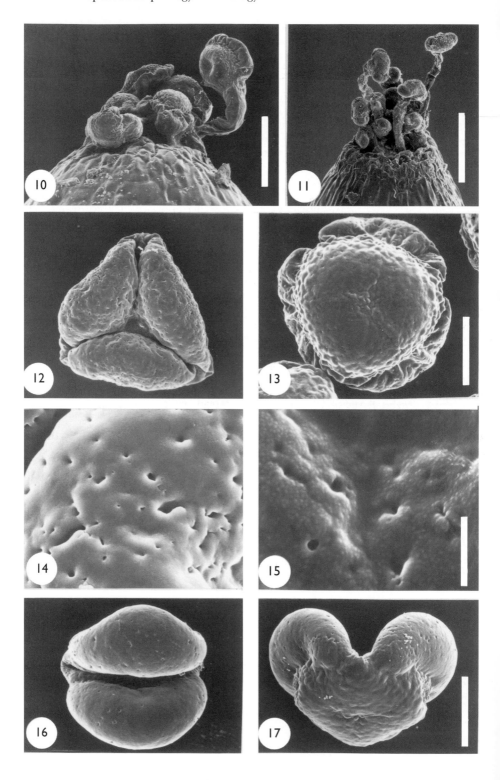

The most common type ("*Podocarpus*-type" of Ueno, 1960b) includes species in *Acmopyle, Afrocarpus, Dacrycarpus, Lepidothamnus, Parasitaxus, Podocarpus, Prumnopitys, Retrophyllum* and *Sundacarpus*. The sacci are of the alveolate type comparable to that in Pinaceae, with some variation in the coarseness and depth of the endoreticulations. The "*Dacrycarpus*-type" (of Ueno, 1960b) includes *Dacrydium, Falcatifolium, Halocarpus* and *Manoao* and is represented by grains in which the endoreticulations extend more or less completely and radially across the saccus. Also the outer surface of the saccus is frequently rugulate rather than smooth, a feature which is most pronounced in *Dacrydium cupressinum* Sol. ex Forst. Noteworthy is that the pollen of the parasitic *Parasitaxus* and its host, *Falcatifolium* are of contrasted types. Pollen apparently of the *Podocarpus* type, but with reduced sacci includes that of *Lagarostrobos franklinii* (Hook. f.) Quinn whose method of pollen capture is not known. Trisaccate pollen occurs in three distinctive genera. Pollen is large in *Dacrycarpus* while the smaller grains of *Microstrobos* and *Microcachrys* may be distinguished because sacci in the former lack endoreticulations.   •

Phyllocladus illustrates a significant departure in sporoderm morphology and function from the saccate condition, even though it is often described as having saccate pollen. It is assumed that the genus shares a common ancestor with Podocarpaceae, in which it may be included, or regarded as a distinct family (Page, 1990). Sacci are present (Fig. 8) but reduced and regarded as vestigial ("protosaccate" or "quasisaccate"). It is best to describe the sacci as "non-functional" because although they become air-filled, seemingly by virtue of their microporous state (Fig. 9) the pollen sinks in water and are of the "wettable" type. Sacci are relatively thick-walled so that enclosed air is insufficient to make them float. Their persistence may be related to the distinctive pollen capture mechanism of *Phyllocladus*, in which pollen stimulates withdrawal of the pollination drop (Tomlinson *et al.*, 1997).

This diversity of pollen types in Podocarpaceae suggests that there is diversity of pollen capture mechanisms, although the topic remains little explored. For example, the peculiar cone morphology of *Dacrydium cupressinum*, in which the micropyle is almost sealed by an opposing bract, is associated with an elusive pollination drop, the pollen itself floating upward only sluggishly. In other *Dacrydium* species access to the micropyle seems very limited. The genus is in need of extended field-oriented study.

Saccate pollen mechanisms and hydrodynamics

Two aspects of saccate sporoderm morphology need emphasising in order that hydrodynamic mechanisms can be correctly interpreted. These relate to the mechanical function of endoreticulations and the accessibility of the saccus lumen to air.

FIGS. 10–17. Pollen orientation and structure in Podocarpaceae (*Podocarpus, Dacrycarpus* and *Prumnopitys*). Figs. 10–11, 15. *Podocarpus totara* D. Don ex Lamb.; Figs. 12–14. *Dacrycarpus dacrydioides* (Rich.) D. Laub.; Figs. 16, 17. *Prumnopitys taxifolia* (Sol. ex D. Don) De Laub. Fig. 10. germinating pollen grains on the nucellus, grain to right demonstrates lateral development of pollen tube; Fig. 11. cluster of grains with varying orientation, many germinating remote from the nucellus; Fig. 12. trisaccate grain in dehydrated condition, distal view; Fig. 13. same in proximal view with triradiate ridge; Fig. 14. detail of saccus surface to show micropores; Fig. 15. detail of saccus surface to show micropores; Figs. 16–17. dehydrated grain in distal and end view, micropores readily visible.

Scale bars: 60 μm in Fig. 10; 100 μm in Fig. 11; 20 μm in Fig. 13 (same for Fig. 12); 3 μm in Fig. 15 (same for Fig. 14); 20 μm in in Fig. 17 (same for Fig. 16).

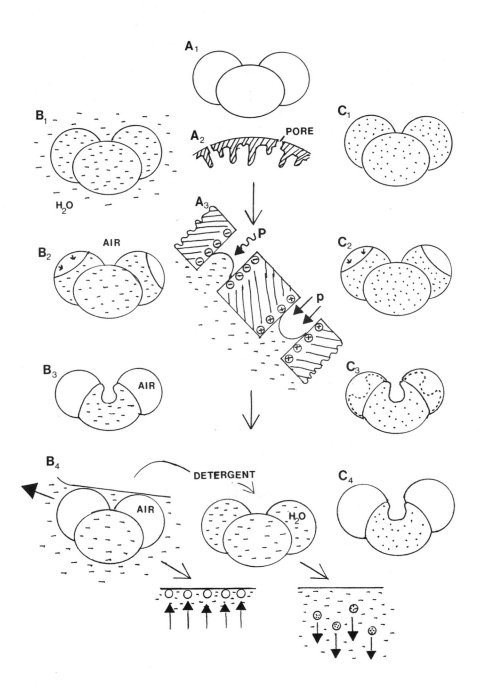

A. Mechanics

From the internal sculpturing of the saccus, with its network of wall ingrowths (Fig. 7), it is likely that sculpturing serves a mechanical function, maintaining the saccus in an inflated position, as is essential for the flotation mechanism, i.e., the saccus "balloon" should not collapse in water. A similar requirement is necessary for the harmomegathic function for sacci suggested by Wodehouse (1935). Nevertheless, the sacci sometimes collapse in the vacuum of the scanning electron microscope without prior treatment. Collapse and re-inflating of sacci can be observed, using a technique initially described by Hess *et al.* (1973) to remove lipids. Saccate pollen may be treated with a lipid solvent (for example, chloroform:methanol - 1:1) for several minutes and mounted under the same solvent for microscopic examination. As the solvent evaporates the sacci collapse as the last liquid disappears, but almost immediately re-inflate as air enters the sacci. This demonstrates considerable flexibility and extensibility of the saccus sporoderm, albeit in a lipid-free condition. Sacci are therefore not necessarily very rigid structures and seemingly maintain their shape by means of internal gas pressure. Sacci are thus miniature dirigibles as well as geodesic domes!

B. Hydrodynamic

A particular question about sacci is how they become gas-filled in the first instance, presumably as they dry out in the microsporangium at the time of pollen release. Sacci possess micropores that provide an air-passage between exterior and interior of the saccus, but such pores, although documented by a number of workers, both by SEM (for example, Pocknall, 1981a) and particularly by TEM (Ueno, 1960a) have received insufficient attention. They are illustrated in Figs. 6, 14, 15 at high magnification, but can usually be seen in SEM photographs at standard magnification (Figs. 4, 5, 12, 16, 17). Since pollen develops essentially in an aqueous medium, the initial displacement of any enclosed water through the pores seems impossible under natural conditions. Equally, the displacement of enclosed air from floating pollen should be almost impossible.

Superficially it would seem that saccate pollen, like the Jumblies "goes to sea in a sieve", because the sacci functioning as water wings are porous. A number of simple experiments reveal the efficiency of the several processes that are involved, as presented in the cartoon diagram (Fig. 18) in which Figs. $18A_{1-3}$ represent structure features at successively higher magnifications.

FIG. 18. Diagram showing behaviour of saccate pollen in natural and experimental conditions.

A_1 Outline of pollen exoderm, which is the functionally significant feature. A_2. Detail of saccus exine, which is internally reticulate and porous. A_3. Physical nature of an air-water interphase when pore walls are either hydrophobic (above, - negative symbols), or hydrophilic (below, + positive symbols). When hydrophobic, meniscus cannot be forced through the pore even by appreciable external pressure (P), when hydrophilic the meniscus passes through pore easily with slight pressure (p).

B. Behaviour of pollen under natural conditions. B_1. Pollen in aqueous medium of microsporangium, sacci are water filled. B_2. On exposure to air, water is withdrawn through the body of the grain, the meniscus always retreating in the arrowed direction. B_3. Pollen in air during dispersal, body is collapsed and wings more or less appressed (harmomegathy). B_4 (left-hand). Pollen in water, floats under surface meniscus, always (below) rising to the highest point of the meniscus (arrows). This is the trajectory in an inverted pollination drop. Even in water air cannot be displaced by vacuum from the pores because of the hydrophobic nature of the pore surface (A_3 upper). B_4 (right-hand) Sacci become air-filled and sink when under vacuum and a detergent added since this destroys the meniscus properties shown in A_3.

C. Saccate pollen in organic solvents that reduce surface tension and/or dissolve lipids. C_1. Sacci become solvent filled. C_2. In air solvent evaporates rapidly and the meniscus withdraws through the pollen body. C_3. Body collapses, sacci may initially collapse, but are immediately re-inflated. C_4. Saccus remains in inflated condition and air-filled, such grains float readily.

First, observation of pollen grains dissected under water from wet, immature sporangia show that the sacci are initially water-filled, because the pollen sinks (Fig. 18B$_1$). When allowed to dry out under the microscope, the fluid in the saccus always recedes through the body of the grain. As the meniscus within the saccus retracts, always starting very irregularly in the alveolate outer portion of the saccus, air is drawn into the saccus through the pores (Fig. 18B$_2$), and the saccus becomes air-filled (Fig. 18B$_3$) and the body collapses. It should be emphasised that in nature this process always occurs within the microsporangium, before it dehisces. This meets the requirement for pollen entrainment in the Niklas (1985) theory. Consequently, when one ruptures mature but undehisced sporangia under water, the pollen immediately floats because the sacci are air-filled. In nature sacci remain air-filled through the subsequent cycle of wind-dispersal, initially, in air when the grains have minimal volume because of the harmomegathic process (Fig. 18B$_3$), subsequently, in the pollination drop when the hydrated grains have maximum volume but minimal density (Fig. 18B$_4$). The medium for the volume changes is presumably the cellulosic, hydrophilic intine and possibly the lamellated inner layer of the exine. When captured (or scavenged) by the pollination drop, saccate pollen rapidly rehydrates and expands (Fig. 18B$_4$), but the sacci remain permanently air-filled because the air-water meniscus within the pores remains intact (upper part of Fig. 18A$_3$). Pollen floats in contact with the ambient meniscus, but rises up any sloping meniscus surface (Fig. 18B$_4$). This observation could support Doyle's (1945) hypothesis concerning saccus function. Certainly pollen can become incorporated into a drop of water, but only if the meniscus becomes inverted. This is the basis for the method used to show pollen orientation in Fig. 3.

Second, one must address the question of how porous sacci remain permanently air-filled in water, the essential requirement for pollen scavenging and pollen acquisition by the pollen chamber. This apparently is simply the effect of the pore wall/water/air interphase and surface tension. Surface tension is amply sufficient to exclude water from the saccus because its enclosed air cannot be pulled out under moderate vacuum (Fig. 18A$_3$, above). This experiment further suggests that the pore wall is hydrophobic (i.e. presumably lipidised). Addition of a mild detergent (dish-washing liquid) rapidly causes the sacci to flood, under the same vacuum (Fig. 18A$_3$, below). This destroys the surface repellant nature of the pores; pollen now sinks demonstrating that it is inherently denser than water. The function of air-filled sacci as flotation devices in mature pollen is thus verified.

Third, the physical nature of the saccus sporoderm can be demonstrated by placing the pollen in a diversity of organic solvents that reduces surface tension (Fig. 18C). In a graded series of ethyl alcohol solutions, the air in the saccus is dissolved - instantly in 100%, slowly in 70% and not at all in 50% ethyl alcohol. The process involves retraction of the air from the internal walls of the saccus to form a detached bubble, which shrinks to nothing as it dissolves. Lipid solvents (for example, methyl alcohol:chloroform, 1:1 - cf. Hess et al. (1973)) instantly displace the air (Fig. 18C$_1$). In these volatile solvents the subsequent rapid disappearance of the liquid evaporating from the sacci can be observed under the microscope and the process of refilling of the saccus with air as the meniscus recedes can also be observed, the meniscus retreating across the saccus (arrows in Fig. 18C$_2$) repeating the natural process of Fig. B$_2$, but with a different liquid. Under these conditions the saccus may collapse and immediately re-inflate, demonstrating its inherent flexibility (Fig. 18C$_3$). However, this property does not come into play, seemingly, in natural conditions. In the final circumstance, saccate pollen is now again air-filled (Fig. 18C$_4$).

The question of the anomalous behaviour of *Picea orientalis* (Runions et al., 1999) in which sacci are naturally "leaky" remains unanswered. Possibly the numerous sporoderm pores are hydrophilic, rather than hydrophobic, or the sporoderm includes a natural surfactant, released on contact with water.

These simple experiments are very revealing of the physical nature of the saccus sporoderm and need to be repeated on a range of taxa that encompass the total diversity of saccate pollen types. They illustrate how beautifully adapted is sporoderm structure to the physical processes involved in pollen capture in those conifers that possess this type of pollen.

Conclusions

Despite our seemingly intimate knowledge of sporoderm structure and variation in saccate pollen and its association with structural attributes of ovules at the time of pollination, we are still a long way from wholly understanding functional attributes. There is still a requirement for direct observation of pollen capture, especially in taxa with distinct cone morphologies. These should be attempted in conditions as natural as possible. In another direction we need to know more about the physics of saccate pollen, a subject that lends itself to experimentation. Our present limited understanding does not permit us to generalise about this pollen type and it would be unwise to equate it with a single adaptive mechanism. Observational and simple experimental approaches are needed before the diversity and taxonomic distribution of saccate pollen is fully explained and are a necessary precursor to the use of pollen characters in phylogenetic analysis.

Acknowledgements

Support for field work on Podocarpaceae and Phyllocladaceae in the South Pacific has been provided by grants from the Committee for Research and Exploration of the National Geographic Society. I much appreciate Dr. N. M. Holbrook's clarification of many features of the physics of saccate pollen. Dr. M. H. Kurmann has been extremely helpful with photographic documentation of sporoderm features in saccate pollen. SEM photography that produced Figs. 2-17 was done at the microscope facility of the University of Waikato, Hamilton, New Zealand, under the auspices of Professor W.B. Silvester and with the assistance of Mr. Alfred Harris.

Bibliography

Cranwell, L.M. (1961). Coniferous pollen types of the southern hemisphere. I. Aberration in *Acmopyle* and *Podocarpus dacrydioides*. *Journal of the Arnold Arboretum* 42: 416–423.

Del Fueyo, G.M. (1996). Microsporogenesis and microgametogenesis of the Argentinian species of *Podocarpus* (Podocarpaceae). *Botanical Journal of the Linnean Society of London* 122: 171–182.

Doyle, J. (1945). Developmental lines in pollination mechanisms in the Coniferales. *Scientific Proceedings of the Royal Dublin Society* 24: 43–62.

Hess, W.M., Weber, D.J., Allen, J.V. and Laseter, J.L. (1973). Ultrastructural changes caused by lipid extraction of pollen of *Pinus echinata*. *Canadian Journal of Botany* 51: 1685–1688.

Kurmann, M.H. (1989). Pollen wall formation in *Abies concolor* and a discussion of wall layer homologies. *Canadian Journal of Botany* 67: 2489–2504.

Lear, E. (1906). The Jumblies. In: Nonsense songs. Chatto and Windus, London.

Niklas, K.J. (1985). The aerodynamics of wind pollination. *Botanical Review* 51: 328–386.

Page, C.N. (1990). Pinatae. In: K. Kubitzki and P.S. Green (editors), The families and genera of vascular plants. Vol. 1. Pteridophytes and Gymnosperms. Springer Verlag, Berlin.

Pocknall, D.T. (1981a). Pollen morphology of the New Zealand species of *Dacrydium* Solander, *Podocarpus* L'Heritier, and *Dacrycarpus* Endlicher (Podocarpaceae). *New Zealand Journal of Botany* 19: 67–95.

Pocknall, D.T. (1981b). Pollen morphology of *Phyllocladus* L. C. et A. Rich. *New Zealand Journal of Botany* 19: 259–266.

Runions, C.J., Rensing, K.H., Takaso, T. and Owens, J.N. (1999). Pollination of *Picea orientalis* (Pinaceae): saccus morphology governs pollen buoyancy. *American Journal of Botany* 86: 190–197.

Stefanovi, S., Jager, M., Deutsch, J. Broutin, J. and Masselot, M. (1998). Phylogenetic relationships of conifers inferred from partial 28S rRNA gene sequences. *American Journal of Botany* 85: 688–697.

Tomlinson, P.B. (1994). Functional morphology of saccate pollen in conifers, with special reference to Podocarpaceae. *International Journal of Plant Science* 155: 699–715.

Tomlinson, P.B., Braggins, J.E. and Rattenbury, J.A. (1991). Pollination drop in relation to cone morphology in Podocarpaceae (Coniferales): a novel reproductive mechanism. *American Journal Botany* 78: 1289–1303.

Tomlinson, P.B., Braggins, J.E. and Rattenbury, J.A. (1997). Contrasted pollen capture mechanisms in Phyllocladaceae and certain Podocarpaceae (Coniferales). *American Journal of Botany* 84: 214–223.

Tomlinson, P.B. and Takaso, T. (1998). Hydrodynamics of pollen capture in conifers. In: S.J. Owens and P.J. Rudall (editors), Reproductive biology, pp. 265–275. Royal Botanic Gardens, Kew.

Ueno, J. (1957). Relationships of genus *Tsuga* from pollen morphology. *Journal of the Institute of Polytechnics of Osaka City University*, ser. D, 8: 191–202.

Ueno, J. (1960a). On the fine structure of the cell walls of some gymnosperm pollen. *Biological Journal of Nara Women's University* 10: 19–25.

Ueno, J. (1960b). Palynological notes of Podocarpaceae. *Acta Phytotaxonomica and Geobotany* 18: 198–207.

Ueno, J. (1960c). Studies on pollen grains of Gymnospermae. Concluding remarks to the relationships between Coniferae. *Journal of the Institute of Polytechnics of Osaka City University*, ser. D, 11: 109–136.

Wodehouse, R.P. (1935). Pollen grains. McGraw Hill, New York

Osborn, J.M. (2000). Pollen morphology and ultrastructure of gymnospermous anthophytes. In: M.M. Harley, C.M. Morton and S. Blackmore (Editors). Pollen and Spores: Morphology and Biology, pp. 163–185. Royal Botanic Gardens, Kew.

POLLEN MORPHOLOGY AND ULTRASTRUCTURE OF GYMNOSPERMOUS ANTHOPHYTES

JEFFREY M. OSBORN

Division of Science, Truman State University, Kirksville, Missouri 63501, USA

Abstract

In recent years, phylogenetic analyses of seed plants have suggested a close relationship among angiosperms and three orders of gymnosperms, including Gnetales, Bennettitales, and Pentoxylales. On the basis of several vegetative and reproductive features, these studies have linked the three gymnosperm groups along with angiosperms into a single 'anthophyte' clade. Furthermore, reports of Cretaceous fossils with *in situ* *Eucommiidites*-type pollen indicate that plants that produced these types of pollen grains (Erdtmanithecales) may also have their affinities nested within the same clade of highly derived seed plants. Two unifying palynological characters are shared among all anthophyte groups; these include a non-saccate condition and a granular sexine infrastructure. In the present paper, pollen morphology and ultrastructure are reviewed for the extant and fossil genera of the four anthophyte orders of gymnosperms that have been studied with transmission electron microscopy. The taxa reviewed include Gnetales (*Ephedra*, *Welwitschia*, *Gnetum*, *Ephedripites*, *Equisetosporites*), Bennettitales (*Leguminanthus*, *Cycadeoidea*, *Monosulcites*), Pentoxylales (*Sahnia*), and Erdtmanithecales (*Eucommiidites*, *Erdtmanitheca*, *Eucommiitheca*). Pollen of these taxa is compared with regard to size, shape, aperture type and location, surface ornamentation, and exine infrastructure. These characters are also discussed with regard to their systematic and phylogenetic significance.

Introduction

Many botanists generally consider the term 'anthophyte' as a reference to flowering plants, or members of the Anthophyta (for example, Raven *et al.*, 1999). However, in 1986 the term was also used as an informal designation for a clade encompassing several major groups of highly derived seed plants (Doyle and Donoghue, 1986a). Since the term's initial use in this cladistic context it has been increasingly incorporated into the literature, particularly in other phylogenetic analyses (for example, Donoghue and Doyle, 1989; Crane, 1990; Rothwell and Serbet, 1994).

The anthophyte clade includes angiosperms as well as Gnetales, Bennettitales, and Pentoxylales. The clade, whether specifically referred to as 'anthophyte' or not, is held together by a number of vegetative and reproductive features (for example, Doyle and Donoghue, 1986a, 1992; Crane, 1985, 1990; Rothwell and Serbet, 1994). These characters include syndetocheilic stomata, complex, 'flower-like' reproductive structures (lacking in Pentoxylales according to Rothwell and Serbet, 1994) with aggregated microsporophylls, non-saccate pollen with granular exine infrastructure, and bitegmic ovules. Moreover, reports of Cretaceous fossils with *in situ* *Eucommiidites*-type pollen suggest that plants that produced these types of grains may also have their affinities nested within the anthophyte clade (Pedersen *et al.*, 1989a; Friis and Pedersen, 1996). A new order and a new family (Erdtmanithecales, Erdtmani-thecaceae) have been established to accommodate dispersed *Eucommiidites* pollen, as well as *Eucommiidites*-producing plants (Friis and Pedersen, 1996).

In the present paper, pollen morphology and ultrastructure are reviewed for the extant and fossil genera of the four anthophyte orders of gymnosperms that have been studied with transmission electron microscopy (Table 1).

Pollen morphology and ultrastructure

Gnetales

The Gnetales are the only group of gymnospermous anthophytes with extant representatives, comprising three living genera: *Ephedra*, *Welwitschia*, and *Gnetum*. Despite marked dissimilarity in habit and habitat among these taxa, as well as one study suggesting that the order is paraphyletic (Nixon *et al.*, 1994), most phylogenetic analyses indicate that the Gnetales are monophyletic. This phylogenetic assessment is based on a suite of shared morphological features (Crane, 1985, 1988; Doyle and Donoghue, 1986a, b; Doyle, 1996), as well as evidence of molecular similarity (*rbc*L data; Hasabe *et al.*, 1992; Price, 1996). *Ephedra* segregates independently, whereas *Welwitschia* and *Gnetum* appear to be more closely related (for example, Doyle, 1996).

Limited information is available about the geologic history of the Gnetales. *Drewria potomacensis* Crane & Upchurch, from the Early Cretaceous Potomac Group of Virginia, U.S.A., is the only described megafossil species with unequivocal gnetalean affinities (Crane and Upchurch, 1987). Additionally, several other Triassic and Cretaceous megafossils have been suggested to have affiliation within the Gnetales (see Crane, 1988, 1996). The majority of data about gnetalean evolution through geologic time comes from palynological evidence. The record of dispersed, 'polyplicate' pollen resembling that of *Ephedra* and *Welwitschia*, at the light microscopical level, extends from Lower Permian to Recent sediments (see Osborn *et al.*, 1993, and references therein). Polyplicate pollen is also produced by several angiosperm groups (for example, Arales, Laurales, Zingiberales). Although some dispersed polyplicate grains are thought to be angiospermous (see below, and Osborn *et al.*, 1993), Hesse *et al.* (2000) have shown that the pollen walls of these pollen types are structurally and chemically different from those of the Gnetales.

EXTANT TAXA
Ephedra

Ephedra consists of approximately 35-45 species (Kubitzki, 1990). No comprehensive palynological survey has been conducted with electron microscopy; micrographs of pollen wall ultrastructure have been published for only eight species (Table 1). However, Steeves and Barghoorn (1959) used transmitted light to systematically examine the pollen of 44 species primarily to generate a database with which to compare fossil pollen.

Pollen of *Ephedra* is elliptic to elongate and ranges from 20-80 μm in length and 16-50 μm in width. Grains are characterised by a series of longitudinal ribs, or plicae, that are typically psilate (Figs. 1-4). Plicae number ranges from 4-19 (Steeves and Barghoorn, 1959). The polyplicate (=striate in many phylogenetic studies) pollen of the genus is typically considered inaperturate, although the exine is considerably thinner between the plicae, or within the furrows (Figs. 5, 7). El-Ghazaly *et al.* (1998) have experimentally shown that the entire exine is completely discarded prior to pollen tube germination. Palynologically, species are principally distinguished from one another by the number of plicae, plica height and degree of slope, and furrow morphology. The furrow may either be straight (Figs. 1-2) or undulated/branched (Figs. 3-4).

TABLE 1. Gymnospermous anthophytes for which pollen ultrastructure has been investigated with transmission electron microscopy.

Taxon	Reference(s)
GNETALES	
EXTANT TAXA	
Ephedra americana	Hesse, 1984; El-Ghazaly *et al.*, 1998
E. californica	Zavada, 1984
E. campylopoda	Hesse, 1984
E. distachya	Van Campo and Lugardon, 1973; Kurmann, 1992; Kurmann and Zavada, 1994; El-Ghazaly *et al.*, 1998
E. foliata	Rowley, 1995; El-Ghazaly and Rowley, 1997
E. intermedia	Bernard and Meyer, 1972
E. monosperma	Afzelius, 1956; Gullvåg, 1966
E. sinica	Ueno, 1960
Gnetum africanum	Oryol *et al.*, 1986
G. gnemon	Gullvåg, 1966; Hesse, 1980; Kurmann, 1992
G. indicum	Bernard and Meyer, 1972
G. montanum	Gullvåg, 1966
G. ula	Gullvåg, 1966
G. sp.	Zavada, 1984
Welwitschia mirabilis	Ueno, 1960; Gullvåg, 1966; Bernard and Meyer, 1972; Hesse, 1984; Kedves, 1987; Zavada and Gabarayeva, 1991; Hesse *et al.*, 2000
FOSSIL TAXA	
Ephedripites sp. 1	Trevisan, 1980
Equisetosporites spp.	Osborn *et al.*, 1993
BENNETTITALES	
Cycadeoidea dacotensis	Taylor, 1973; Osborn and Taylor, 1995
Leguminanthus siliquosus	Ward *et al.*, 1989
Monosulcites sp. 1	Trevisan, 1980
PENTOXYLALES	
Sahnia laxiphora	Osborn *et al.*, 1991
ERDTMANITHECALES (*EUCOMMIIDITES* PLANTS)	
Eucommiidites sp.	Doyle *et al.*, 1975
E. sp. 1	Trevisan, 1980
E. sp. 2	Trevisan, 1980
E. sp. from *Erdtmanispermum balticum*	Pedersen *et al.*, 1989a
E. troedssonii	Scheuring, 1970, 1978; Batten and Dutta, 1997
Erdtmanitheca texensis	Pedersen *et al.*, 1989a
Eucommiitheca hirsuta	Friis and Pedersen, 1996

Overall exine thickness, as well as thicknesses of individual exine strata vary depending on where the pollen wall is measured (i.e., at the crest of a plica vs. grading into a furrow). The sexine consists of a homogeneous tectum and a granular infratectum (Figs. 5-6). Within the plicae, the tectum is typically thinnest at the crest and gradually thickens down the sides (Fig. 5). The tectum typically becomes thinner as it grades into the furrow. Within the furrow, the tectum may either be present as a thin layer (Fig. 7), or may be almost completely absent (Hesse, 1984). The granular infratectum is also absent within the furrow (Figs. 5, 7). The nexine is electron-dense, uniform in thickness in both the plica and furrow regions, and consists of well-defined, sometimes anastomosed lamellae (Figs. 5-7). A thin, electron-lucent "footlayer" has been reported to overlay the lamellated layer in some taxa (for example, Van Campo and Lugardon, 1973; Kurmann, 1992), whereas it is absent in others (Fig. 6).

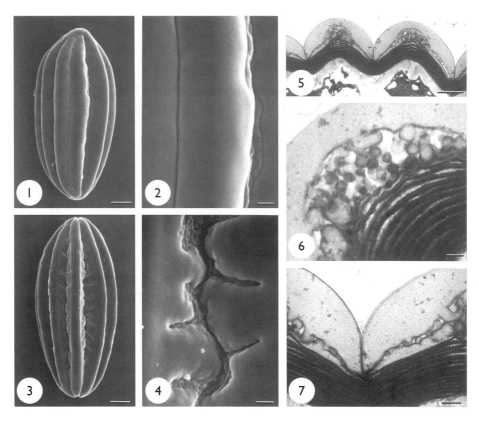

FIGS. 1–7. Gnetales – *Ephedra.* Fig. 1. Pollen of *Ephedra californica.* Bar = 5 μm. Fig. 2. Detail of the pollen surface of *E. californica* showing unbranched furrows. Bar = 1 μm. Fig. 3. Pollen of *Ephedra aspera.* Bar = 5 μm. Fig. 4. Detail of the pollen surface of *E. aspera* showing branched, scabrate furrows. Bar = 1 μm. Fig. 5. Transverse section through the exine of *Ephedra trifurca* showing two complete plicae and three furrows. Note the electron-lucent sexine and the electron-dense nexine. Bar = 1 μm. Fig. 6. Detail of the exine from *E. trifurca* at the crest region of a single plica. Note the homogeneous tectum, large infratectal granules, absence of an electron-lucent 'foot layer' and the white-line-centered, electron-dense lamellae. Bar = 0.1 μm. Fig. 7. Detail of the exine from *E. trifurca* at a furrow region showing thin tectum and lack of infratectal granules. Bar = 0.2 μm.

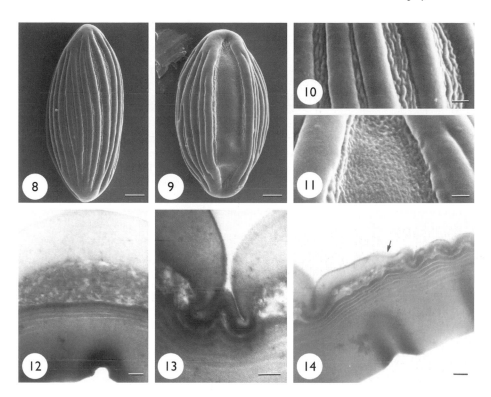

FIGS. 8–14. Gnetales – *Welwitschia mirabilis.* Fig. 8. Proximal view of a pollen grain. Bar = 5 μm. Fig. 9. Distal view of a pollen grain showing a single, broad sulcus. Bar = 5 μm. Fig. 10. Detail of the proximal wall showing five plicae and four furrows. Bar = 1 μm. Fig. 11. Detail of the distal sulcus showing scabrate surface. Bar = 1 μm. Fig. 12. Transverse section through the crest region of a single plica. Note the thick, homogeneous tectum, finely granular infratectum, absence of a well-defined, electron-lucent 'foot layer,' and electron-dense nexine. Lamellae are only detectable in the uppermost region of the nexine. Bar = 0.1 μm. Fig. 13. Detail of a furrow region of the proximal exine showing thin tectum and absence of infratectal granules. Bar = 0.1 μm. Fig. 14. Detail of the distal wall showing the margin of the apertural membrane. Note that the tectum (arrow) gradually thins toward the apertural membrane (at right). Bar = 0.1 μm.

Welwitschia

Pollen of the monotypic *Welwitschia mirabilis* J.D. Hook. (Kubitzki, 1990) is like that of *Ephedra* in being elongate to elliptic in shape and polyplicate (Figs. 8-9). Grains average 51 μm in length and 31 μm in width and have 15-20 psilate plicae (for example, Sahashi *et al.*, 1976; Kedves, 1987). In contrast to *Ephedra*, pollen of *Welwitschia* typically has more rounded plicae and is monoaperturate (Figs. 8-11). The sulcus is relatively broad, extending nearly the entire length of each grain, and the apertural membrane is psilate to slightly scabrate (Figs. 9, 11).

The tectum is generally thicker than that of *Ephedra*, and it is more uniform in thickness within the plicae (i.e., both at the crests and laterally toward the furrows; Figs. 12-13). However, the tectum becomes thinner and is present over both the non-apertural furrows and the apertural membrane (Figs. 13-14). A distinct infratectum is present within the plicae, consisting of relatively small, tightly packed granules (Fig. 12). Zavada and Gabarayeva (1991) have shown that tectum thickness within the plicae of *Welwitschia* is in part due to ontogenetic fusion of infratectal granules.

167

Distinguishable granules are absent at the non-apertural furrows (Fig. 13) and the apertural membrane (Fig. 14). A well-defined, electron-lucent foot layer is also lacking (Figs. 12-13). The nexine is typically more electron-dense than the sexine, and it is uniform in thickness in both non-apertural and apertural regions (Figs. 12-14). The nexine is lamellate (Figs. 12-14); however, the lamellae are not as easily detectable as they are in the pollen of *Ephedra* and *Gnetum*.

Gnetum

Gnetum consists of approximately 30 species (Kubitzki, 1990), yet pollen wall ultrastructure has been illustrated for only six taxa (Table 1). However, a palynological survey of the genus has been undertaken, and a preliminary summary of the systematic significance of surface morphology has been reported (Gillespie and Nowicke, 1994).

Pollen of *Gnetum* is spheroidal, inaperturate, and relatively small, averaging 12-20 μm in diameter (Figs. 15-16). *Gnetum* is unique among gymnosperms in being the only taxon with a spinose to spinulose ornament, although surface sculpture varies among species (Gillespie and Nowicke, 1994). Asian species are characterised by conical, blunt-tipped spines that are supported by a uniform tectum (Fig. 15). By contrast, the pollen of Neotropical and African taxa have smaller, more rounded, and more numerous spinules. The spinules of these species appear to rest on a discontinuous, irregularly thickened tectum that forms plate-like areas (Fig. 16).

Seen in thin section, it is evident that the surface spines and spinules are not supratectal. Rather, these 'sculptural elements' are sexinous regions in which the tectum folds over conical areas of densely packed infratectal granules (Figs. 17-18). Given this ultrastructure, the spines and spinules appear to be homologous with the longitudinally oriented plicae of *Ephedra* and *Welwitschia* pollen. Interestingly, in *Gnetum*, pollen thickness of the tectum is uniform in these regions, as it is in non-spinose regions (Figs. 17, 19). A distinguishable electron-lucent foot layer is absent. The nexine is more electron-dense than the sexine, and it is evenly thickened in all wall regions. Well-defined and relatively thick nexine lamellae are present (Figs. 17, 19).

FOSSIL TAXA

Equisetosporites

Several dispersed palynomorphs assignable to *Equisetosporites* spp. from the Lower Cretaceous Santana Formation of Brazil have been studied (Osborn *et al.*, 1993). The polyplicate grains are ellipsoidal, average 41 μm in length and 13 μm in width, and have a variable number of plicae (ranging from 5 to 11; Figs. 20-21). Plicae have psilate to slightly scabrate surface sculpture; these grade into straight furrows that do not reach the grain ends (Fig. 20).

The overall exine averages 1.0 μm in thickness in plicae regions (Figs. 21-22). The tectum is uniform in thickness within the plicae (i.e., at the crest and laterally toward the furrows). Infratectal granules range in size from 0.09-0.26 μm in diameter. Small granules are present in the lower portion of the infratectum and are directly adjacent to the underlying, electron-dense nexine; these gradually grade into larger granules in the upper portion and appear to fuse with the tectum (Fig. 22). Within the furrow regions, the tectum thins abruptly at the margins, and it is the only layer of sexine present over the furrow (Fig. 23). The nexine is relatively thick throughout, but thins slightly under the sexine furrows (Figs. 21, 23). Although the nexine is lamellate throughout, this was only detectable in grains that had undergone significant preservational folding (Fig. 24). Most grains studied only exhibited a few lamellae near the nexine/sexine interface or in slightly folded regions near the furrows (Osborn *et al.*, 1993).

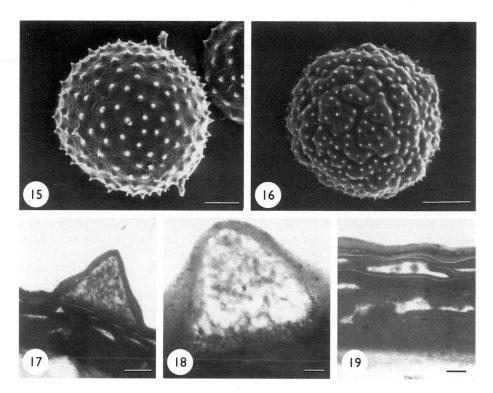

FIGS. 15–19. Gnetales – *Gnetum.* Fig. 15. Pollen of *Gnetum gnemon* showing continuous tectum and surface spines. Bar = 3 μm. Fig. 16. Pollen of *Gnetum nodiflorum* showing plate-like tectum and surface spinules. Bar = 3 μm. Fig. 17. Transverse section of the exine of *G. montanum.* Note the thin, homogeneous tectum, densely packed infratectal granules under the 'spine,' and thick, electron-dense lamellae of the nexine. Bar = 0.3 μm. Fig. 18. Detail of the exine in a spine region of *G. montanum* showing the infratectal granules in direct contact with the electron-dense nexine. Bar = 0.1 μm. Fig. 19. Detail of the exine in a non-spinous region of *G. montanum* showing the tectum in direct contact with the electron-dense nexine. Note also the thick, white-line-centered lamellae of the nexine. Bar = 0.1 μm.

Another species of *Equisetosporites* (*E. chinleana* Daugherty; Zavada, 1984, 1990; Pocock and Vasanthy, 1988), as well as *Cornetipollis reticulata* Pocock & Vasanthy (separated from *E. chinleana*; Pocock and Vasanthy, 1988) have also been studied at the ultrastructural level. Based on surface morphology of the plicae, as well as the presence of infratectal columellae, these two Upper Triassic grains are now thought to be angiospermous (see Osborn *et al.*, 1993) and will not be considered here.

Ephedripites

Ultrastructural details for one dispersed taxon of *Ephedripites* have been reported. *Ephedripites* sp. 1, from the Lower Cretaceous of Italy (Trevisan, 1980), is ellipsoidal, averages 44 μm in length and 19 μm in width, and is polyplicate. The palynomorph is also distinctly monoaperturate. The sulcus is relatively broad and extends the entire length of the grain, but does not reach the grain ends (Fig. 25A).

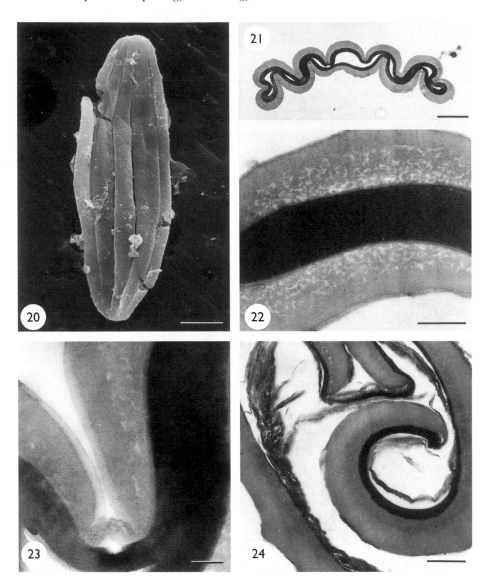

FIGS. 20–24. Gnetales – *Equisetosporites* spp. Fig. 20. Polar view of a pollen grain showing five, broad plicae. Bar = 5 μm. Fig. 21. Transverse section through a grain showing electron-lucent sexine and uniformly thick, electron-dense nexine. Bar = 2 μm. Fig. 22. Detail of two opposing plicae. The pollen lumen is not detectable as the opposing electron-dense nexines are compressed against each other. Note the homogeneous tectum, well-defined infratectal granules, as well as the absence of both a detectable foot layer and nexine lamellae. Bar = 0.5 μm. Fig. 23. Detail of the exine at a furrow showing the absence of a granular infratectal layer and a folded tectum in this region. Bar = 0.2 μm. Fig. 24. Section through a preservationally folded grain showing separated lamellae within regions of the electron-dense nexine. Bar = 0.5 μm.

The exine of *Ephedripites* sp. 1 within the plicae regions consists of an inner lamellated nexine ("layer A" of Trevisan, 1980) and an outer, electron-dense, five-layered sexine ("layers B_1, B_2, B_3, C, D" of Trevisan, 1980). The most prominent sexine layers are the inner three (Fig. 25B). These consist of a thin, homogeneous band (=foot layer?; B_1) appressed to the lamellate nexine, a layer of "anastomosing units" (=granules; B_2), and a homogeneous layer (=tectum; B_3). Layer C is a thin, discontinuous zone just outside the tectum and is intermixed to overlaid by layer D. Layer D consists of variably sized "globulets" (Trevisan, 1980) that have subsequently been suggested to represent debris (Osborn *et al.*, 1993) or tapetal remains (Zavada, 1984). The tectum appears to be uniform in thickness under the plicae and abruptly thins at the margins of the furrows, where it is present as a thin layer (Fig. 25B). The apertural membrane is described as having a similar ultrastructure to that of the non-apertural furrows (Trevisan, 1980).

Bennettitales

The Bennettitales are represented by fossils of Triassic to Cretaceous age. Although the group is now recognised to be somewhat diverse taxonomically, it is best known from the permineralised, cone-bearing trunks of *Cycadeoidea* (see Crepet, 1974, and references therein), as well as a number of compression-impression leaf taxa (for example, Thomas and Bancroft, 1913).

In situ pollen is known from many taxa (Osborn and Taylor, 1995); however, pollen of only two genera has been described at the ultrastructural level, *Leguminanthus* and *Cycadeoidea*. In addition to these taxa, Ward *et al.* (1989) suggested that the dispersed palynomorph *Monosulcites*, for which fine structural data have been published, may have affinities within the Bennettitales. Furthermore, monosulcate pollen identified within micropylar tubes of the Triassic bennettitalean ovules *Vardekloeftia* are reported to be similar to *Monosulcites minimus* Couper (Pedersen *et al.*, 1989b). Given these descriptions, *Monosulcites* will be addressed here, despite previous classification as *Incertae sedis* (Osborn *et al.*, 1991; Osborn and Taylor, 1994).

Leguminanthus

Leguminanthus siliquosus (Leuthardt) Kräusel & Schaarschmidt is a compressed microsporophyll known from the Upper Triassic of Luntz, Austria and Neuewelt, Switzerland (Crane, 1988, and references therein). Ward *et al.* (1989) examined pollen of *L. siliquosus* in a comparative context as part of a study of *Lethosmasites fossulatus* Ward, Doyle, & Hotton, a putative angiosperm monosulcate grain with a granular exine from the Early Cretaceous Potomac Group, U.S.A.

Pollen of *Leguminanthus siliquosus* is elliptic, relatively small, measuring 22 μm in length and 13 μm in width, and monosulcate (Fig. 25C). The exine is psilate in external ornament, 0.70 μm in overall thickness, and two-parted. The electron-lucent sexine consists of a thick tectum and an infratectal layer of relatively large, tightly packed granules (Fig. 25D). The granular infratectum is reported to be "fused basally into a foot layer" (Ward *et al.*, 1989). The underlying nexine is electron-dense, faintly lamellate (Fig. 25D), and consistent in thickness in both non-apertural and apertural regions. The aperture spans almost the entire length of each grain (Fig. 25C). The apertural membrane also appears to have a thin layer of sexine present.

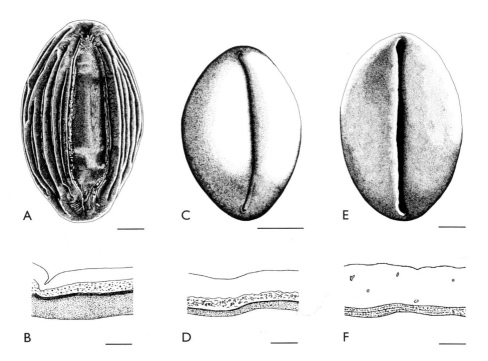

FIG. 25. Gnetales and Bennettitales. A-B. Gnetales - *Ephedripites* sp. 1 (Modified from Trevisan, 1980). A. Distal view of a pollen grain showing elongate sulcus. Bar = 5 μm. B. Section through the non-apertural wall showing thick tectum, granular infratectum, and partially lamellate nexine (stippled). Bar = 0.5 μm. C-D. Bennettitales - *Leguminanthus siliquosus* (Modified from Ward *et al.*, 1989). C. Distal view of a pollen grain showing thin sulcus. Bar = 5 μm. D. Section through the non-apertural wall showing thick tectum, compressed granular infratectum, and partially lamellate nexine (stippled). Bar = 0.5 μm. E-F. Bennettitales - *Monosulcites* sp. 1 (Modified from Trevisan, 1980). E. Distal view of a pollen grain showing broad, elongate sulcus. Bar = 5 μm. F. Section through the non-apertural wall showing relatively homogeneous sexine and lamellate nexine (stippled). Bar = 0.5 μm.

Cycadeoidea

Specimens of *Cycadeoidea* have been recovered from a wide range of Northern Hemisphere localities (Crepet, 1974). However, most information about these plants comes from studies of fossils collected at several Lower Cretaceous sites in the Black Hills of South Dakota and Wyoming, U.S.A. Many species have been described (Wieland, 1906, 1916), but those bearing cones preserved with intact pollen organs and *in situ* pollen grains, or in a bisporangiate condition, are relatively rare (Crepet, 1974). Pollen fine structure has been investigated from a single species, *C. dacotensis* (McBride) Ward (Taylor, 1973; Osborn and Taylor, 1995).

Pollen of *Cycadeoidea dacotensis* is monosulcate and typically elliptic in shape (Fig. 26), although occasional spheroidal grains also occur. Grains average 25 μm in length and 12 μm in width and exhibit punctate to psilate surface ornamentation (Fig. 26). The length and surface sculpture of the sulcus has proven difficult to ascertain, because the apertural region of most grains is highly folded. The exine averages 0.73 µm in overall thickness. The sexine is two-zoned, consisting of a relatively thin, homogeneous tectum and a granular infratectum (Figs. 27-29). In some grains, the tectum is easily delimited from the underlying infratectum because granule size and spacing is fairly uniform (Fig. 27). However, in many grains granule packing is so dense that the sexine superficially appears homogeneous throughout (Figs. 28-29). The nexine lacks detectable lamellae (Figs. 27-29) and is uniform in thickness all around the grain (Fig. 28). The apertural membrane consists of a thin, homogeneous sexine layer (=tectum), resulting from lateral thinning of both the tectum and infratectum, and an electron-dense nexine (Fig. 29).

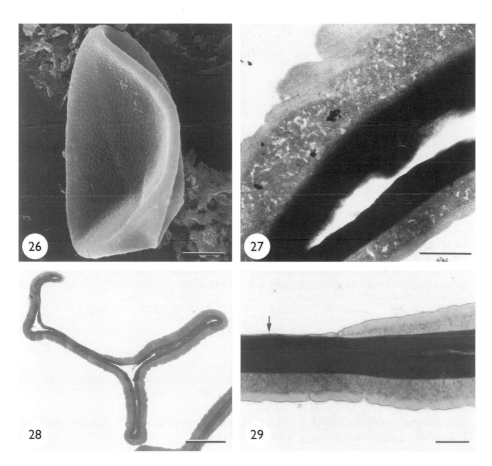

FIGS. 26–29. Bennettitales – *Cycadeoidea dacotensis.* Fig. 26. Partially folded pollen grain showing elongate shape and punctate surface. Bar = 5 µm. Fig. 27. Oblique section through the non-apertural wall showing homogeneous tectum, well-defined granular infratectum, absence of a foot layer, and thick, electron-dense nexine. Note also the absence of nexine lamellae. Bar = 0.5 µm. Fig. 28. Transverse section of a folded grain showing electron-lucent sexine and uniformly thick, electron-dense nexine. Bar = 3 µm. Fig. 29. Transverse section of a compacted grain in the apertural (arrow) and opposing non-apertural regions. The apertural membrane consists of a thin, electron-lucent tectum and thick nexine. Note also that in non-apertural regions the infratectal granules are highly compacted and not individually detectable. Bar = 0.5 µm.

Monosulcites

Two investigations have reported fine structural data for the dispersed taxon *Monosulcites* (Trevisan, 1980; Zavada and Dilcher, 1988). *Monosulcites* sp. 1, from the Lower Cretaceous of Italy (Trevisan, 1980), is ovoid with somewhat pointed ends and averages 47 µm in length and 25 µm in width. The palynomorph is psilate and has a single, slightly folded aperture extending almost the entire length of the grain (Fig. 25E). *Monosulcites* sp. 1 has a thick exine (1.5 - 2.0 µm), consisting of two principal layers (Fig. 25F). The inner layer (nexine) is electron-dense, lamellate, and uniform in thickness in both apertural and non-apertural regions. The outer, electron-lucent layer (sexine) is five to six times thicker than the nexine and appears to consist of a thin tectum and a thick, granular infratectum (Fig. 25F). However, individual granules are not distinct. Granules appear large and highly fused throughout the infratectum, especially in the lower region where they form a basal, homogeneous layer.

The second *Monosulcites* taxon described, *M.* sp. from the Upper Cretaceous of Minnesota, U.S.A. (Zavada and Dilcher, 1988), will not be considered here. Osborn and Taylor (1995) have suggested that this palynomorph more than likely has its affinities outside of the anthophyte clade, within either Cycadales or Ginkgoales. This suggestion was based on the large grain size and apparent alveolar infratectal elements.

Pentoxylales

The Pentoxylales range from Jurassic to Cretaceous in age, and are known only from Southern Hemisphere localities. The taxon was originally established to classify a number of enigmatic permineralised plants (Sahni, 1948); however, a variety of compression fossils have subsequently been described and assigned to the order (see Osborn *et al.*, 1991, and references therein). Four genera are typically recognised for woody stems (*Pentoxylon*), leaves (*Nipaniophyllum*), ovulate organs (*Carnoconites*), and pollen organs (*Sahnia*). Despite the fact that these organs are not necessarily recovered in organic attachment, many studies, including phylogenetic analyses, often collectively refer to them as the *Pentoxylon* plant (for example, Bose *et al.*, 1985; Crane, 1985; Doyle and Donoghue, 1986a).

Sahnia

Two species of *Sahnia* have been described. *Sahnia nipaniensis* Vishnu-Mittre is based on silicified material of Jurassic age from the Rajmahal Hills, northeastern India (Vishnu-Mittre, 1953), whereas *S. laxiphora* Drinnan & Chambers is known from Lower Cretaceous compression specimens of the Strzelecki Group, southeastern Victoria, Australia (Drinnan and Chambers, 1985, 1986). Although both species contain *in situ* pollen, data on pollen fine structure are only known from *S. laxiphora* (Osborn *et al.*, 1991).

Pollen grains of *Sahnia laxiphora* are ovoid and small, averaging 26 µm in length and 23 µm in width (Fig. 30). Grains are monosulcate, with the relatively broad sulcus extending nearly the entire length of each grain (Fig. 30). The exine is psilate and typically highly folded in the apertural region (Fig. 30). The exine averages 0.95 µm in overall thickness and is two-parted. The sexine consists of a relatively thick, homogeneous tectum and a granular infractectum (Fig. 31). The boundary between the tectum and infractectum is not well-defined, with the infratectal granules grading into the tectum. Granules are most pronounced at the base of the sexine where they are directly contiguous with the underlying nexine (Fig. 31). The nexine is electron-dense and relatively uniform in thickness in both non-apertural and apertural regions (Fig. 31). Faint nexine lamellae are occasionally present and are most prominent at the nexine/sexine interface.

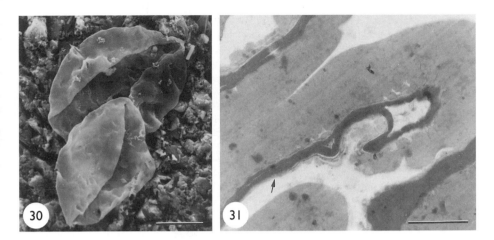

FIGS. 30–31. Pentoxylales – *Sahnia laxiphora.* Fig. 30. Distal view of two compressed pollen grains. Note that both grains are folded in the sulcus regions. Bar = 10 μm. Fig. 31. Transverse section of a grain in the apertural (arrow) and opposing non-apertural region showing electron lucent sexine and electron-dense nexine. Note that the infratectal granules are compacted and appear to be fused with the thick tectum, as well as that the nexine is uniform in thickness in apertural and non-apertural regions. Bar = 1 μm.

Erdtmanithecales (*Eucommiidites* plants)

The *sporae dispersae* genus *Eucommiidites* was established by Erdtman (1948) to describe 3 'colpate' pollen grains from the Lower Jurassic of Sweden that resembled pollen of the extant angiosperm *Eucommia.* Subsequent studies have shown that *Eucommiidites* is gymnospermous and a variety of dispersed species have been established (see Friis and Pedersen, 1996, and references therein). Furthermore, several intact pollen organs and seeds with *in situ Eucommiidites* pollen have been described. On the basis of these descriptions, Friis and Pedersen (1996) have established Erdtmanithecaceae and Erdtmanithecales to accommodate both dispersed *Eucommiidites* species and *Eucommiidites*-producing plants. The latter includes two pollen organs (*Erdtmanitheca texensis* Pedersen, Crane, & Friis and *Eucommiitheca hirsuta* Friis & Pedersen) and four seeds (*Erdtmanispermum balticum* Pedersen, Crane, & Friis, *Spermatites pettensis* Hughes, *S. patuxensis* Brenner, and *Allicospermum retemirum* Harris). The family is based on a suite of reproductive characters, including three pollen features: elliptic shape having a "distal main colpus" with expanded ends and "flanked by two colpi or a single ring colpus in a proximal to almost equatorial position," granular infratectum, and "thick laminated endexine" (Friis and Pedersen, 1996).

Eucommiidites

Ultrastructural details have been studied from four dispersed taxa of *Eucommiidites*, as well as from *Eucommiidites*-type pollen found within the micropyle of *Erdtmanispermum balticum* seeds (Table 1). All of these taxa are Lower Cretaceous in age, except for *E. troedssonii* Erdtman, which is Lower Jurassic. Most dispersed taxa studied are ovoid to slightly elliptic in outline, have a psilate to slightly scabrate ornament, and range from 22-29 μm in length to 15-26 μm in width (Fig. 32A-C). Furthermore, most palynomorphs have a well-defined sulcus with slightly expanded, round ends and two lateral furrows with pointed ends (Fig. 32A-C). Variation in surface sculpture is seen in the *Eucommiidites*-type grains recovered within *Erdtmanispermum balticum* micropyles (Pedersen *et al.*, 1989a); these *in situ* grains have a distinctly foveolate surface (Fig. 32D-F).

Exine ultrastructure varies widely among the *Eucommiidites* grains sectioned. Most grains have an electron-lucent, outer layer (sexine) and an electron-dense, inner layer (nexine). However, the entire exine is uniformly electron-dense in *E.* sp. (Doyle *et al.*, 1975), whereas it is uniformly electron-lucent in *E. troedssonii* (Batten and Dutta, 1997). Furthermore, infratectal granule structure and 'foot layer' structure vary. Granules range from being well-defined and easily detectable in section view (Fig. 32G, M) to variably fused and slightly pillar-like (Fig. 32J). The presence of an electron-lucent layer that separates the granules from the electron-dense nexine also varies. Such a 'foot layer' may be absent (Fig. 32L-M), present but thin (Fig. 32H), to present and very thick (Fig. 32G). The presence of distinguishable nexine lamellae is also a variable character (Fig. 32G-H, J, L-M).

Another dispersed species of *Eucommiidites* has also been examined at the fine structural level (*E.* sp.; Zavada, 1984), but is not being considered in the present paper. This Jurassic palynomorph is approximately twice as large, measuring 45 μm in length and 43 μm in width, and does not have as distinctive an aperture and lateral furrows in comparison with other *Eucommiidites* taxa. Because the lateral furrows are not well-illustrated in the light micrograph of *E.* sp. (Zavada, 1984), the identification of this palynomorph to *Eucommiidites* has been questioned (Pedersen *et al.*, 1989a; see also Batten and Dutta, 1997).

Erdtmanitheca

Pedersen *et al.* (1989a) described *Erdtmanitheca texensis* from the Upper Cretaceous of Arthur City, Texas, U.S.A., the first pollen organ unequivocally known to produce *Eucommiidites*-type pollen. *In situ* pollen of *E. texensis* is ellipsoidal to ovoid and averages 24 μm in length and 18 μm in width. The pollen surface is psilate to slightly scabrate. Pollen grains have a well-defined sulcus with somewhat rounded ends and two, more slit-like lateral furrows with pointed ends. All three are approximately the same length and do not reach the grain ends. The exine ranges from 0.7-1.3 μm in overall thickness. The electron-lucent sexine consists of a thick, homogeneous tectum and thinner granular infratectum (Fig. 32K). Granules are relatively small, tightly packed, and appear to be directly adjacent to the underlying, electron-dense nexine. Although the

FIG. 32. Erdtmanithecales. A-C. Views of the psilate surface of most *Eucommiidites*-type pollen grains. Bars = 5 μm. A. Distal view showing the sulcus with rounded ends. B. Equatorial view showing sulcus (top) and two, lateral furrows. C. Proximal view showing two lateral furrows with pointed ends. D-F. Views of the foveolate surface of *Eucommiidites*-type pollen found within the micropyles of *Erdtmanispermum balticum* seeds. Bars = 5 μm. D. Distal view. E. Equatorial view. F. Proximal view. G-M. Sections of *Eucommiidites*-type grains shown at the same scale. Bars = 0.5 μm. G-I. Sections of grains with a detectable, sexinous foot layer present above the nexine (stippled). G. Non-apertural exine of *E.* sp. from *Erdtmanispermum balticum* seeds. (Modified from Pedersen *et al.*, 1989a). H. Non-apertural exine of *Eucommiidites* sp. 2. (Modified from Trevisan, 1980). I. Non-apertural exine of *Eucommiitheca hirsuta.* (Modified from Friis and Pedersen, 1996). J. Non-apertural exine of *Eucommiidites* sp. 1 showing infratecum with ill-defined granules fused into pillar-like units. (Modified from Trevisan, 1980). K-M. Sections of grains with infratectal granules in direct contact with the underlying nexine (stippled). K. Non-apertural exine of *Erdtmanitheca texensis*. (Modified from Pedersen *et al.*, 1989a). L. Non-apertural exine of *Eucommiidites troedssonii.* (Modified from Batten and Dutta, 1997). M. Non-apertural exine of *Eucommiidites* sp. (Modified from Doyle *et al.*, 1975).

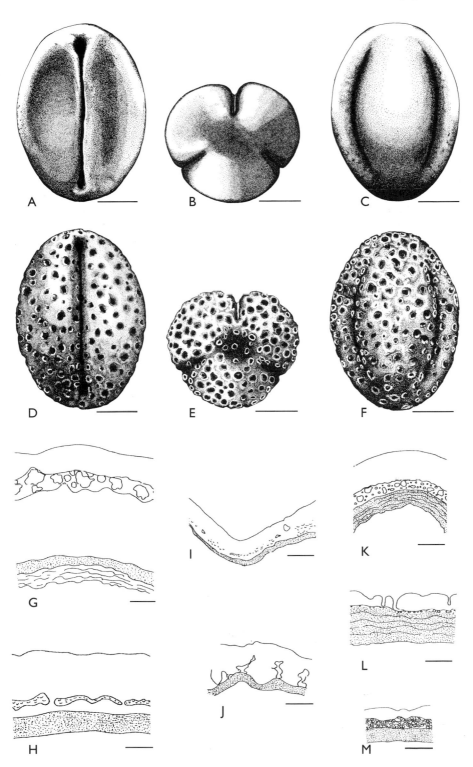

nexine is not particularly well-preserved, lamellae are clearly detectable (Fig. 32κ). The nexine appears to be uniform in thickness in all grain regions, thickening slightly below the apertural membrane and the two furrows.

Eucommiitheca

In situ pollen of *Eucommiitheca hirsuta,* from the Early Cretaceous of Portugal, has been described ultrastructurally (Friis and Pedersen, 1996). Grains are elliptic with pointed to truncate ends and range from 15-20 μm in length and 10-12 μm in width. Grains have a psilate to slightly foveolate surface and are "3 colpate" (Friis and Pedersen, 1996). The colpi are parallel, approximately the same length, and extend nearly to the grain ends. The exine is distinctly two-parted, consisting of an electron-lucent "ektexine" and an electron-dense "endexine" (Friis and Pedersen, 1996). In non-apertural regions the ektexine averages 0.35 μm in thickness, and it gradually thins towards the margins of the colpi. Although the grains are not well-preserved at the ultrastructural level, a thick, homogeneous tectum and a granular infratectum have been described (Fig. 32I). Furthermore, although an electron-dense basal layer is present, preservation has affected characterisation of this layer in apertural regions. Lamellae are also not detectable in the basal layer (Fig. 32I).

Discussion

Pollen of the four groups of gymnospermous anthophytes exhibits several common character trends. These include the absence of sacci, a relatively small size (12-50 μm), an elliptic shape (except for *Gnetum*), a monosulcate aperture type (except for the inaperturate gnetalean taxa: *Ephedra, Gnetum,* and *Equisetosporites*), a granular infratectum, and a thick basal layer (nexine) that is uniform in thickness in non-apertural and apertural regions. In addition to reviewing structural data for the gymnosperm taxa addressed above, an initial goal of this paper was to provide a comparative table that included a variety of ultrastructural and morphological characters (for example, sexine and nexine thicknesses, tectum architecture, granule size and ultrastructure, lamellae thickness, etc.). It was hoped that such a table would be useful in standardising the descriptions of salient characters and thereby provide for more accurate and consistent character scoring in phylogenetic analyses. However, the summary table was omitted because direct comparisons of many palynological characters are not possible, primarily due to the significant variation that was identified among the taxa.

Character Variation

Clearly, natural variation exists among pollen types, and this variability is critically important in a phylogenetic context. However, palynological characters can also reflect a wide range of other biological and physical processes that are often phylogenetically uninformative and ambiguous. Osborn and Taylor (1994) have discussed the phylogenetic utility of several ultrastructural characters from an array of fossil gymnosperms. Four palynological features are relevant to the present paper with regard to gymnospermous anthophytes. These include staining properties of ultrathin sections, nexine structure, infratectal granules, and foot layer.

Overall exine structure, as well as particular layers, may exhibit a range of electron densities after staining for transmission electron microscopy. For example, different sections of the same pollen grain of *Equisetosporites* have been shown to exhibit both typical staining (i.e., electron-lucent sexine and electron-dense nexine), as well as an opposite staining pattern (i.e., electron-dense sexine and electron-lucent nexine) (Osborn *et al.*, 1993). The latter pattern is also evident in sections of *Ephedripites* (Trevisan, 1980). Similar examples of stain variation have been reported in *Ephedra foliata* Boiss. ex K.C.A. Mey, in which El-Ghazaly and Rowley (1997) have suggested that differential leaching of pollen wall materials may contribute to the electron-lucent endexine in this species. The relative staining pattern in turn plays an important role in perceived ultrastructure, especially the ability to resolve whether or not particular characters are present (for example, nexine lamellations and foot layer).

Nexine structure is a particularly variable character among the pollen of gymnospermous anthophytes, especially with regard to the presence of lamellations. Preservational aspects clearly influence the ability to detect nexine lamellae. For example, lamellations were only identifiable in *Equisetosporites* pollen after ultrathin-sectioning numerous palynomorphs; this sample size provided the opportunity to identify grains that had undergone significant preservational folding thereby separating the highly compressed, individual lamellae (Osborn *et al.*, 1993). In other cases, although lamellae-like structures may be distinguished in ultrathin sections, these may represent poorly preserved intine remains that become tightly appressed against the nexine as observed in *Cycadeoidea* pollen (Osborn and Taylor, 1994, 1995). Furthermore, a significant portion of the basal layer may be absent altogether in fossil pollen, thereby precluding the identification of lamellae that may have in fact been present before fossilisation. Under analogous conditions, the endexine of many extant taxa may be completely lost following acetolysis. This occurs because the endexine is often structurally supported by the underlying intine (Blackmore and Crane, 1988), which is not acetolysis-resistant (Erdtman, 1960).

Although a granular infratectum is a shared character among gymnospermous anthophytes, considerable variation occurs in granule size, shape, and packing. The wide range of granular infratecta observed in the pollen of both *Sahnia* (Osborn *et al.*, 1991) and *Cycadeoidea* (Osborn and Taylor, 1995) is due primarily to preservational influences. In fact, the sexine appears entirely homogeneous in most grains of *Cycadeoidea* that have been ultrathin-sectioned (Taylor, 1973; Osborn and Taylor, 1995); infratectal granules were only detected in this fossil after sectioning multiple grains (Osborn and Taylor, 1995). Exine development also plays a role in the ultrastructural detection of infratectal granules. For example, during sexine ontogeny in *Welwitschia*, the outermost granules fuse with, and contribute to, the developing tectum (Zavada and Gabarayeva, 1991). Infratectal granules also undergo ontogenetic aggregation and fusion to form columellae-like elements in *Ephedra foliata* (El-Ghazaly and Rowley, 1997).

The pollen of many gymnospermous anthophytes lacks an ultrastructurally detectable foot layer. In most taxa, the electron-lucent infratectal granules are directly contiguous with the underlying, electron-dense basal layer. It is possible that an ectexinous foot layer is present in some pollen grains, but is not identifiable because of variation in staining. Foot layer and endexine commonly have similar electron densities following conventional staining in both gymnosperms and angiosperms (for example, Osborn and Taylor, 1994; Weber, 1998; Kreunen and Osborn, 1999).

Future Research: Identifying Phylogenetically Relevant Characters

Determining which palynological characters are phylogenetically valuable while taking into account the external factors (for example, preservation, specimen preparation, ontogeny, pollination) that can introduce significant variation in pollen ultrastructure and morphology is particularly challenging. Osborn and Taylor (1994) have emphasised a conservative approach with regard to 'homologising' characters when a limited number of ultrathin sections have been prepared from a single or a small number of pollen grains. Clearly, it is evident from the review above that important structural data are available for the pollen of gymnospermous anthophytes; however, greater phylogenetic resolution will come from additional research that focuses on a greater diversity of taxa, larger sample sizes, pollen ontogeny, and megafossils with *in situ* pollen.

Pollen morphology and ultrastructure from relatively few taxa of gymnospermous anthophytes have been studied. New insight will come from investigation of additional species clearly nested within anthophyte groups, as well as other putatively allied taxa. For instance, phylogenetic analyses by Doyle (1996) have linked Gnetales with *Piroconites*, an enigmatic microsporophyll with *in situ Ephedripites*-type pollen from the Lower Jurassic of Germany (van Konijnenburg-van Cittert, 1992). Although the pollen of *Piroconites kuespertii* Gothan appears polyplicate in surface view, when ultrathin-sectioned the plicae appear to be major folds in the whole exine (Osborn and van Konijnenburg-van Cittert, unpublished data) rather than the alternating thick and thin regions that characterise gnetalean pollen. Furthermore, three additional dispersed polyplicate taxa from the Lower Cretaceous of Brazil and Ecuador have recently been described at the fine structural level (Dino *et al.*, 1999). Although ultrastructural data on these unusual 'elater-bearing' palynomorphs were not available at the time the current paper was prepared, *Elateroplicites africaensis* Herngreen, *Elaterosporites klaszii* (Jardiné & Magloire) Jardiné, and *Sofrepites legouxiae* Jardiné appear to have gnetalean affinities (Dino *et al.*, 1999).

In addition to the need for new species of gymnospermous anthophytes to be studied, larger sample sizes of pollen grains from individual taxa need to be investigated. This is particularly important for fossils. As discussed above, preservational influences can alter phylogenetically informative data and introduce dubious characters. The degree of preservational alteration can be determined and overcome in many cases by studying multiple pollen grains. Increased sample sizes can also be achieved by examining entire reproductive organs with *in situ* pollen.

Ultrastructural details about pollen development in gymnospermous anthophytes, or related taxa, are also not well-understood. As discussed above, infratectal granules undergo various degrees of fusion during pollen ontogeny in *Welwitschia* (Zavada and Gabarayeva, 1991) and *Ephedra foliata* (El-Ghazaly and Rowley, 1997). Therefore, the developmental stage at which a pollen grain is prepared for study will have significant bearing on the observed ultrastructure and potential character scoring. Ontogenetic data may also help resolve questions regarding nexine structure. Mature gymnosperm pollen typically has a uniformly thick and lamellated basal layer, whereas mature angiosperm pollen has a thinner basal layer with fewer lamellae that are typically restricted to apertural regions (see Osborn and Taylor, 1994). Developmental data may indicate that the thinner basal layer in angiosperms is due to loss of the innermost endexine lamellae. In some basal angiosperms a layer of electron-dense granules develops below a layer of well-defined, endexine lamellae during ontogeny (for example, *Nymphaea*, Nymphaeaceae, Gabarayeva and El-Ghazaly, 1997; and *Nelumbo*, Nelumbonaceae, Kreunen and Osborn, 1999). It is possible that these granules represent flocculents of later-formed sporopollenin of the endexine that were unable to aggregate onto phylogenetically lost sub-structural, white-line units. Hemsley *et al.* (2000) have modelled flocculation patterns using polystyrene latex and suggested that white-line units may be sites for sporopollenin deposition. Furthermore, the endexines of gymnosperm and angiosperm pollen are considered by Rowley (1995) to be structurally and functionally equivalent. If correct, this provides support for the white-line loss hypothesis presented in the current paper.

Ultrastructural details about ancient developmental patterns can be ascertained by studying whole pollen organs with *in situ* grains (see Taylor *et al.*, 1996). Limited information is known about pollen ontogeny in fossils (Taylor, 1990), but investigations of *in situ* pollen have the potential to yield phylogenetically valuable data. For example, although mature pollen of the Jurassic fossil *Caytonanthus arberi* (Thomas) Harris is saccate and has an alveolar infratectum (Pedersen and Friis, 1986; Zavada and Crepet, 1986; Osborn, 1994), identification of early ontogenetic stages indicates that granular sexine units are present and later aggregate to form the plate-like alveolae observed in mature pollen (Osborn, 1994). Although the Caytoniales has been viewed as more distantly related to anthophytes (for example, Crane, 1990; Rothwell and Serbet, 1994), Doyle (1996) has suggested that this group of 'Mesozoic seed ferns' may be linked to angiosperms.

It is likely that new data on pollen development from both fossil and extant taxa will require a re-evaluation of how traditional ultrastructural characters such as 'granular infratectum' are interpreted in a phylogenetic context. Although pollen characters alone will not definitively answer all of the systematic questions about anthophyte relationships and angiosperm origins, when enough palynological information becomes available these data can add considerable resolution to phylogenetic analyses.

Acknowledgements

The author thanks Bradford L. Day for rendering the whole-grain illustrations depicted in Figs. 25 and 32; Ranessa L. Cooper for assistance with TEM preparation of extant gnetalean pollen; the Botanical Society of America (*American Journal of Botany*) for permission to reproduce two micrographs from Osborn *et al.* (1991) and three micrographs from Osborn and Taylor (1995); and Elsevier Science (*Review of Palaeobotany and Palynology*) for permission to reproduce five micrographs from Osborn *et al.* (1993). This study was supported in part by a Faculty Research Grant (J.M. Osborn) and an Undergraduate Research Stipend (B.L. Day) from Truman State University. Acknowledgement is also made to the donors of The Petroleum Research Fund, administered by the ACS, for partial support of this research (ACS-PRF 29004-GB8).

Literature cited

Afzelius, B.M. (1956). Electron-microscope investigations into exine stratification. *Grana Palynologica* 1: 22–37.

Batten, D.J. and Dutta, R.J. (1997). Ultrastructure of exine of gymnospermous pollen grains from Jurassic and basal Cretaceous deposits in Northwest Europe and implications for botanical relationships. *Review of Palaeobotany and Palynology* 99: 25–54.

Bernard, V.V. and Meyer, N.R. (1972). Pollen grains of *Ephedra*, *Welwitschia* and *Gnetum*. *Vestnik Moskovskogo universiteta. Serija 6. Biologija, pocvovedenie* 27: 86–88.

Blackmore, S. and Crane, P.R. (1988). The systematic implications of pollen and spore ontogeny. In: C.J. Humphries (editor), Ontogeny and systematics, pp. 83–115. Columbia University Press, New York, NY.

Bose, M.N., Pal, P.K., and Harris, T.M. (1985). The *Pentoxylon* plant. *Philosophical Transactions of the Royal Society of London* 310B: 77–108.

Crane, P.R. (1985). Phylogenetic analysis of seed plants and the origin of angiosperms. *Annals of the Missouri Botanical Garden* 72: 716–793.

Crane, P.R. (1988). Major clades and relationships in the "higher" gymnosperms. In: C.B. Beck (editor), Origin and evolution of gymnosperms, pp. 218–272. Columbia University Press, New York, NY.

Crane, P.R. (1990). The phylogenetic context of microsporogenesis. In: S. Blackmore and R.B. Knox (editors), Microspores: evolution and ontogeny, pp. 11–41. Academic Press, London.

Crane, P.R. (1996). The fossil history of the Gnetales. *International Journal of Plant Sciences* 157 (6, supplement): S50–S57.

Crane, P.R. and Upchurch, Jr., G.R. (1987). *Drewria potomacensis* gen. et sp. nov., an Early Cretaceous member of Gnetales from the Potomac Group of Virginia. *American Journal of Botany* 74: 1722–1736.

Crepet, W.L. (1974). Investigations of North American cycadeoids: the reproductive biology of *Cycadeoidea. Palaeontographica Abt. B* 148: 144–169.

Dino, R., Pocknall, D.T., and Dettmann, M.E. (1999). Morphology and ultrastructure of elater-bearing pollen from the Albian to Cenomanian of Brazil and Ecuador: implications for botanical affinity. *Review of Palaeobotany and Palynology* 105: 201–235.

Donoghue, M.J. and Doyle, J.A. (1989). Phylogenetic analysis of angiosperms and the relationships of Hamamelidae. In: P.R. Crane and S. Blackmore (editors), Evolution, systematics and fossil history of the Hamamelidae, pp. 17–45. Academic Press, London.

Doyle, J.A. (1996). Seed plant phylogeny and the relationships of Gnetales. *International Journal of Plant Sciences* 157 (6, supplement): S3–S39.

Doyle, J.A. and Donoghue, M.J. (1986a). Seed plant phylogeny and the origin of angiosperms: An experimental cladistic approach. *The Botanical Review* 52: 321–431.

Doyle, J.A. and Donoghue, M.J. (1986b). Relationships of angiosperms and Gnetales: a numerical cladistic analysis. In: R.A. Spicer and B.A. Thomas (editors), *Systematic and taxonomic approaches in palaeobotany*, pp. 177–198. Clarendon Press, Oxford.

Doyle, J.A. and Donoghue, M.J. (1992). Fossils and seed plant phylogeny reanalyzed. *Brittonia* 44: 89–106.

Doyle, J.A., Van Campo, M. and Lugardon, B. (1975). Observations on exine structure of *Eucommiidites* and Lower Cretaceous angiosperm pollen. *Pollen et Spores* 17: 429–486.

Drinnan, A.N. and Chambers, T.C. (1985). A reassessment of *Taeniopteris daintreei* from the Victorian Early Cretaceous: A member of the Pentoxylales and a significant Gondwanaland plant. *Australian Journal of Botany* 33: 89–100.

Drinnan, A.N. and Chambers, T.C. (1986). Flora of the Lower Cretaceous Koonwarra Fossil Bed (Korumburra Group), South Gippsland, Victoria. *Memoirs of the Association of Australasian Palaeontologists* 3: 1–77.

El-Ghazaly, G. and Rowley, J.R. (1997). Pollen wall of *Ephedra foliata. Palynology* 21: 7–18.

El-Ghazaly, G., Rowley, J.R., and Hesse, M. (1998). Polarity, aperture condition and germination in pollen grains of *Ephedra* (Gnetales). *Plant Systematics and Evolution* 213: 217–231.

Erdtman, G. (1948). Did dicotyledonous plants exist in Early Jurassic time? *Förhandlingar, Geologiska Föreningen i Stockholm* 70: 265–271.

Erdtman, G. (1960). The acetolysis technique: a revised description. *Svensk Botanisk Tidskrift Utgifven af Svenska Botaniska Föreningen, Stockholm* 54: 561–564.

Friis, E.M. and Pedersen, K.R. (1996). *Eucommiitheca hirsuta*, a new pollen organ with *Eucommiidites* pollen from the Early Cretaceous of Portugal. *Grana* 35: 104–112.

Gabarayeva, N.I. and El-Ghazaly, G. (1997). Sporoderm development in *Nymphaea mexicana* (Nymphaeaceae). *Plant Systematics and Evolution* 204: 1–19.

Gillespie, L.J. and Nowicke, J.W. (1994). Systematic implications of pollen morphology in *Gnetum. Acta Botanica Gallica* 141: 131–139.

Gullvåg, B. (1966). The fine structure of some gymnosperm pollen walls. *Grana Palynologica* 6: 435–475.

Hasabe, M., Ito, M., Kofuji, R., Iwatsuki, K., and Ueda, K. (1992). Phylogenetic relationships in Gnetophyta deduced from *rbcL* gene sequences. *Botanical magazine of Tokyo* 105: 385–392.

Hemsley, A.R., Collinson, M.E., Vincent, B., Griffiths, P.C., and Jenkins, P. (2000). Self-assemby of colloidal units in exine development. In: M.M. Harley, C.M. Morton, and S. Blackmore (editors), Pollen and spores: morphology and biology, pp. 31–44. Royal Botanic Gardens, Kew.

Hesse, M. (1980). Pollenkitt is lacking in *Gnetum gnemon* (Gnetaceae). *Plant Systematics and Evolution* 136: 41–46.

Hesse, M. (1984). Pollenkitt is lacking in Gnetatae: *Ephedra* and *Welwitschia*; further proof for its restriction to the angiosperms. *Plant Systematics and Evolution* 144: 9–16.

Hesse, M., Weber, M., and Halbritter, H. (2000). A comparative study of the polyplicate pollen types in Arales, Laurales, Zingiberales and Gnetales. In: M.M. Harley, C.M. Morton, and S. Blackmore (editors), Pollen and spores: morphology and biology, pp. 227–239. Royal Botanic Gardens, Kew.

Kedves, M. (1987). LM and EM studies on pollen grains of recent *Welwitschia mirabilis* Hook. and *Ephedra* species. *Acta Botanica Hungarica* 33: 81–103.

Kreunen, S.S. and Osborn, J.M. (1999). Pollen and anther development in *Nelumbo*. (Nelumbonaceae). *American Journal of Botany* 86: 1662–1676.

Kubitzki, K. (editor). (1990). The families and genera of vascular plants, vol. 1. pteridophytes and gymnosperms (K.U. Kramer and P.S. Green, vol. editors). Springer-Verlag, Berlin.

Kurmann, M.H. (1992). Exine stratification in extant gymnosperms: A review of published transmission electron micrographs. *Kew Bulletin* 47: 25–39.

Kurmann, M.H. and Zavada, M.S. (1994). Pollen morphological diversity in extant and fossil and gymnosperms. In: M.H. Kurmann and J.A. Doyle (editors), Ultrastructure of fossil spores and pollen, pp. 123–137. Royal Botanic Gardens, Kew.

Oryol, L.I., Kuprijanova, L.A. and Golubcva, E.A. (1986). Ultrastructure of acetolysis-resistant wall of tapetal cells and pollen grains in *Gnetum africanum* (Gnetaceae). *Botanicheskij Zhurnal* (Moscow & Lenningrad) 71: 750–754.

Osborn, J.M. (1994). The morphology and ultrastructure of *Caytonanthus. Canadian Journal of Botany* 72: 1519–1527.

Osborn, J.M. and Taylor, T.N. (1994). Comparative ultrastructure of fossil gymnosperm pollen and its phylogenetic implications. In: M.H. Kurmann and J.A. Doyle (editors), Ultrastructure of fossil spores and pollen, pp. 99–121. Royal Botanic Gardens, Kew.

Osborn, J.M. and Taylor, T.N. (1995). Pollen morphology and ultrastructure of the Bennettitales: *In situ* pollen of *Cycadeoidea. American Journal of Botany* 82: 1074–1081.

Osborn, J.M., Taylor, T.N. and Crane, P.R. (1991). The ultrastructure of *Sahnia* pollen (Pentoxylales). *American Journal of Botany* 78: 1560–1569.

Osborn, J.M., Taylor, T.N. and de Lima, M.R. (1993). The ultrastructure of fossil ephedroid pollen with gnetalean affinities from the Lower Cretaceous of Brazil. *Review of Palaeobotany and Palynology* 77: 171–184.

Pedersen, K.R. and Friis, E.M. (1986). *Caytonanthus* pollen from the Lower and Middle Jurassic. In: J.T. Møller (editor), 25 years of geology in Aarhus, pp. 255–267. Geologisk Institut Aarhus Universitet (Geoskrifter, No. 24).

Pedersen, K.R., Crane, P.R., and Friis, E.M. (1989a). Pollen organs and seeds with *Eucommiidites* pollen. *Grana* 28: 279–294.

Pedersen, K.R., Crane, P.R., and Friis, E.M. (1989b). The morphology and phylogenetic significance of *Vardekloeftia* Harris (Bennettitales). *Review of Palaeobotany and Palynology* 60: 7–24.

Pocock, S.A.J. and Vasanthy, G. (1988). *Cornetipollis reticulata*, a new pollen with angiospermid features from Upper Triassic (Carnian) sediments of Arizona (U.S.A.), with notes on *Equisetosporites*. *Review of Palaeobotany and Palynology* 55: 337–356.

Price, R.A. (1996). Systematics of the Gnetales: a review of morphological and molecular evidence. *International Journal of Plant Sciences* 157 (6, supplement): S40–S49.

Raven, P.H., Evert, R.F., and Eichhorn, S.E. (1999). Biology of Plants, 6th edition. W.H. Freeman/Worth, New York, NY.

Rothwell, G.W. and Serbet, R. (1994). Lignophyte phylogeny and the evolution of spermatophytes: a numerical cladistic analysis. *Systematic Botany* 19: 443–482.

Rowley, J.R. (1995). Are the endexines of pteridophytes, gymnosperms and angiosperms structurally equivalent? *Review of Palaeobotany and Palynology* 85: 13–34.

Sahashi, N., Takeda, T., and Ikuse, M. (1976). Pollen morphology of *Welwitschia mirabilis* Hook. f. *Journal of Japanese Botany* 51: 27–32.

Sahni, B. (1948). The Pentoxyleae: a new group of Jurassic gymnosperms from the Rajmahal Hills of India. *The Botanical Gazette* 110: 47–80.

Scheuring, B.W. (1970). Palynologische und palynostratigraphische Untersuchungen des Keupers im Bölchentunnel (Solothurner Jura). *Schweizerische Paläontologische Abhandlungen* 88: 1–119.

Scheuring, B.W. (1978). Mikrofloren aus den Meridekalken des Mte. San Giorgio (Kanton Terrsin). *Schweizerische Paläontologische Abhandlungen* 100: 1–205.

Steeves, M.W. and Barghoorn, E.S. (1959). The pollen of *Ephedra*. *Journal of the Arnold Arboretum* 40: 221–255.

Taylor, T.N. (1973). A consideration of the morphology, ultrastructure, and multicellular microgametophyte of *Cycadeoidea dacotensis* pollen. *Review of Palaeobotany and Palynology* 16: 157–164.

Taylor, T.N. (1990). Microsporogenesis in fossil plants. In: S. Blackmore and R.B. Knox (editors), Microspores: evolution and ontogeny, pp. 121–145. Academic Press, London.

Taylor, T.N., Osborn, J.M., and Taylor, E.L. (1996). The importance of *in situ* pollen and spores in understanding the biology and evolution of fossil plants. In: J. Jansonius and D.C. McGregor (editors), Palynology: principles and applications. Volume 1, pp. 427–441. American Association of Stratigraphic Palynologists Foundation, Salt Lake City, UT.

Thomas, H.H. and Bancroft, N. (1913). On the cuticles of some recent and fossil cycadean fronds. *Transactions of the Linnean Society, London. Botany* 8: 155–204.

Trevisan, L. (1980). Ultrastructural notes and considerations on *Ephedripites*, *Eucommiidites* and *Monosulcites* pollen grains from Lower Cretaceous sediments of southern Tuscany (Italy). *Pollen et Spores* 22: 85–132.

Ueno, J. (1960). On the fine structure of the cell walls of some gymnosperm pollen. *Biological Journal of Nara Women's University* 10: 19–25.

Van Campo, M. and Lugardon, B. (1973). Structure grenue infratectale de l'ectexine des pollens de quelques gymnospermes et angiospermes. *Pollen et Spores* 15: 171–187.

van Konijnenburg-van Cittert, J.H.A. (1992). An enigmatic Liassic microsporophyll, yielding *Ephedripites* pollen. *Review of Palaeobotany and Palynology* 71: 239–254.

Vishnu-Mittre. (1953). A male flower of the Pentoxyleae with remarks on the structure of the female cones of the group. *The Palaeobotanist* 2: 75–84.

Ward, J.V., Doyle, J.A. and Hotton, C.L. (1989). Probable granular magnoliid angiosperm pollen from the Early Cretaceous. *Pollen et Spores* 31: 113–132.

Weber, M. (1998). The detection of pollen endexines - with special reference to Araceae pollen. Pollen and spores: morphology and biology, Abstracts: 48. Royal Botanic Gardens, Kew. (Abstract).

Wieland, G.R. (1906). American fossil cycads. Carnegie Institution of Washington (Publication no. 34), Washington, D.C.

Wicland, G.R. (1916). American fossil cycads, volume II. Taxonomy. Carnegie Institution of Washington (Publication no. 34, vol. II), Washington, D.C.

Zavada, M.S. (1984). Angiosperm origins and evolution based on dispersed fossil gymnosperm pollen ultrastructure. *Annals of the Missouri Botanical Garden* 71: 444–463.

Zavada, M.S. (1990). The ultrastructure of three monosulcate pollen grains from the Triassic Chinle Formation, western United States. *Palynology* 14: 41–51.

Zavada, M.S. and Crepet, W.L. (1986). Pollen wall ultrastructure of *Caytonanthus arberi*. *Plant Systematics and Evolution* 153: 259–264.

Zavada, M.S. and Dilcher, D.L. (1988). Pollen wall ultrastructure of selected dispersed monosulcate pollen from the Cenomanian, Dakota Formation, of central USA. *American Journal of Botany* 75: 669–679.

Zavada, M.S. and Gabarayeva, N. (1991). Comparative pollen wall development of *Welwitschia mirabilis* and selected primitive angiosperms. *Bulletin of the Torrey Botanical Club* 118: 292–302.

Dettmann, M.E. and Jarzen, D.M. (2000). Pollen of extant *Wollemia* (Wollemi pine) and comparisons with pollen of other extant and fossil Araucariaceae. In: M.M. Harley, C.M. Morton and S. Blackmore (Editors). Pollen and Spores: Morphology and Biology, pp. 187–203. Royal Botanic Gardens, Kew.

POLLEN OF EXTANT *WOLLEMIA* (WOLLEMI PINE) AND COMPARISONS WITH POLLEN OF OTHER EXTANT AND FOSSIL ARAUCARIACEAE

MARY E. DETTMANN AND DAVID M. JARZEN

Department of Botany, University of Queensland, Q. 4072, Australia and Florida Museum of Natural History, University of Florida, Gainesville, Fl 32611, USA

Abstract

Comparative analyses of pollen of the recently discovered araucarian conifer *Wollemia nobilis* W.G. Jones, K.D. Hill and J.M. Allen (Wollemi pine) and fossil pollen included in *Dilwynites granulatus* Harris confirm their morphological and ultrastructural similarity. Grains of both taxa are asaccate, spheroidal, inaperturate with an exine composed of granular ectexine and lamellated endexine; sculpture comprises irregular grana and pila, but sculptural elements of *Wollemia* pollen are larger than those of *Dilwynites granulatus*. Exine stratification/ultrastructure is consistent with that of pollen of extant Araucariaceae (*Agathis* Salisb., *Araucaria* Henkl. & Hochst.) and with fossil pollen included within *Dilwynites tuberculatus* Harris and *Araucariacites* Cookson ex Couper. Type material of *Araucariacites* (*A. australis* Cookson ex Couper) from the Miocene of Kerguelen has also been reinvestigated. Whereas *Araucariacites* has been reported worldwide in the Late Mesozoic, *Dilwynites* is known only from Turonian and younger sediments of Antarctica, New Zealand, and Australia. The Late Cretaceous-early Tertiary distribution range of *Dilwynites* is similar to that recorded for several Gondwanan relicts in the present day vegetation of Australasia. Among these elements are diverse Podocarpaceae (*Podocarpus, Dacrydium, Lagarostrobos, Microcachrys, Dacrycarpus*), *Nothofagus,* Proteaceae, and Myrtaceae.

Introduction

The recent discovery of *Wollemia*, a monotypic genus within the Araucariaceae (Jones *et al.,* 1995), growing as two isolated small stands west of Sydney, Australia (Map 1), has prompted comparative studies of its reproductive and vegetative organs with araucarian-like fossils preserved in Cretaceous and Tertiary sediments of austral areas. Fossil shoots, cone scales, winged seeds, and pollen have been shown to bear a striking resemblance to those of *Wollemia nobilis* (Macphail *et al.,* 1995; Chambers *et al.,* 1998).

Pollen of *Wollemia* shares morphological characters in common with pollen of the two other extant genera (*Agathis, Araucaria*) of the Araucariaceae. All possess inaperturate, spheroidal pollen that are globally sculpted by grana. Pollen of similar generalised form occurs in several other conifer families (Taxodiaceae, Cupressaceae,

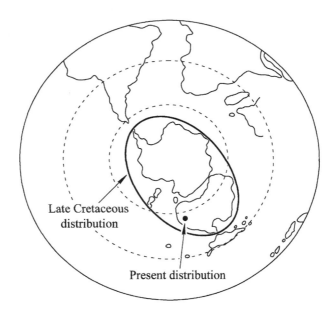

Map 1. Present distribution of *Wollemia nobilis* and distribution range (oval area) of *Dilwynites*. Map is South Polar Lambert equal-area map for the Santonian (after Smith *et al.*, 1981, map 24).

Pinaceae, Cephalotaxaceae, Taxaceae), but characters that distinguish pollen of *Agathis* and *Araucaria* include their larger size and their exine stratification/ultrastructure. With respect to *Wollemia* pollen, Macphail *et al.* (1995) using light and scanning electron microscopy, demonstrated its morphological similarity to fossil pollen included in *Dilwynites granulatus*. Thus, the present investigation was directed at detailing exine stratification and ultrastructure of pollen of *Wollemia* and the similar fossil pollen included in *Dilwynites granulatus* based on transmission electron microscopy. For additional comparisons, morphology of pollen of other extant Araucariaceae (*Agathis robusta, Araucaria laubenfelsii*) and of other fossil araucarian pollen taxa (*Dilwynites tuberculatus*, type material of *Araucariacites australis*) has been investigated.

Materials and Methods

Pollen of extant Araucariaceae

Pollen cones of *Wollemia nobilis* W.G.Jones, K.D.Hill and J.M.Allen used in this study were kindly provided by Dr Ken Hill, Royal Botanic Gardens, Sydney (NSW). Those of *Agathis robusta* C. Moore were collected by MED from trees growing on the grounds of the University of Queensland. Pollen studied of *Araucaria laubenfelsii* Corbeson is on slide 12218, of the pollen collection of the Canadian Museum of Nature.

Extant pollen material was prepared by the acetolysis method of Erdtman (1960) prior to mounting in glycerine jelly for light microscope (LM) study. Specimens for scanning electron microscope (SEM) study were transferred to a small coverslip which had previously been smeared with an adhesive. After attaching the coverslip to a stub with aluminium paint, the pollen was gold plated and examined using a Jeol 8400 instrument. Preparation technique for transmission electron microscope (TEM) analysis follows that utilised by Dettmann and Jarzen (1998), and based on a method outlined by Milne (1998). Both acetolysed and non-acetolysed specimens were prepared. The specimens were fixed in 5% glutaraldehyde, followed by 2% osmium tetroxide for 2 hrs and uranyl acetate for 1 hr prior to surface embedding in Spurr's medium. Sections were cut with a diamond knife, stained with lead citrate and viewed with a Hitachi 300 instrument.

The terms "sexine" and "nexine" are used *sensu* Erdtman (1952) for the outer and inner layers of the exine as observed using LM. "Ectexine" and "endexine" (*sensu* Faegri, 1956) are employed for the outer, more electron dense, and inner, less electron dense, wall layers revealed by TEM. Relationships between the two nomenclatures are outlined by Punt *et al.* (1994).

Fossil *Araucariacites* Cookson ex Couper and *Dilwynites* Harris

Material examined of *Araucariacites australis* Cookson includes the type material described by Cookson (1947) and other specimens from new preparations of the type sample from Kerguelen. In addition, specimens from Lower Tertiary sediments from Victoria and Upper Cretaceous exposures on the Antarctic Peninsula were examined. *Dilwynites granulatus* Harris and *D. tuberculatus* Harris were also obtained from the Victorian and Antarctic Peninsula samples.

1. Kerguelen, southern Indian Ocean. Waterfall Gorge, near Port Jeanne d'Arc, sample 85a, British Australian New Zealand Antarctic Research Expedition, 1929-31 (BANZARE) Collection (see Cookson 1947). Age: Miocene (Nougier, 1970).
2. Victoria, Australia. Dilwyn Formation, type section in Dilwyn Bay, E of mouth of Gellibrand River (Princetown 1:25 000 topographic map 7520-4-2, zone 54 co-ords 57123N 6885E), collected Dr. J. Douglas and MED January, 1993. Age: late Paleocene-Eocene (Harris, 1965).
3. James Ross Island, Antarctic Peninsula. Hidden Lake Formation, S side of Brandy Bay (British Antarctic Survey sample D.8632.10). Age: Turonian.
4. Vega Island, Antarctic Peninsula. Marambio Group, Cape Lamb (British Antarctic Survey sample D.31223). Age: Maastrichtian (Dettmann and Thomson, 1987).

Sample processing involved mineral digestion with 50% hydrofluoric acid followed by 2-3 minutes oxidation with nitric acid. Soluble humic acids were removed with weak (1-2%) potassium or ammonium hydroxide. Residues were mounted in glycerine jelly, and specimens for SEM and TEM analyses were prepared as described above.

The Kerguelen type and other illustrated specimens as well as those from the Dilwyn Formation, Victoria are housed in the Museum of Victoria and bear catalogue numbers prefixed by "P". Specimens illustrated from Antarctica will be housed in the collection of the British Antarctic Survey, Cambridge.

Pollen of *Wollemia* W.G. Jones, K.D. Hill & J.M. Allen

Morphology (Figs. 1–5, 10, 11)

Pollen inaperturate, asaccate; spheroidal with a circular outline (35–55 μm in diameter), but in dehydrated and overmacerated specimens with broad depressions and an elliptical outline (long axis up to 75 μm) (Figs. 1–3, 10). Exine 1.5–3 μm thick; comprising nexine 0.5–1 μm thick and thicker (1–2 μm thick) sexine that has granulate/pilate surface sculpture (Figs. 1–5, 10, 11). Sculptural elements up to 1.2 μm

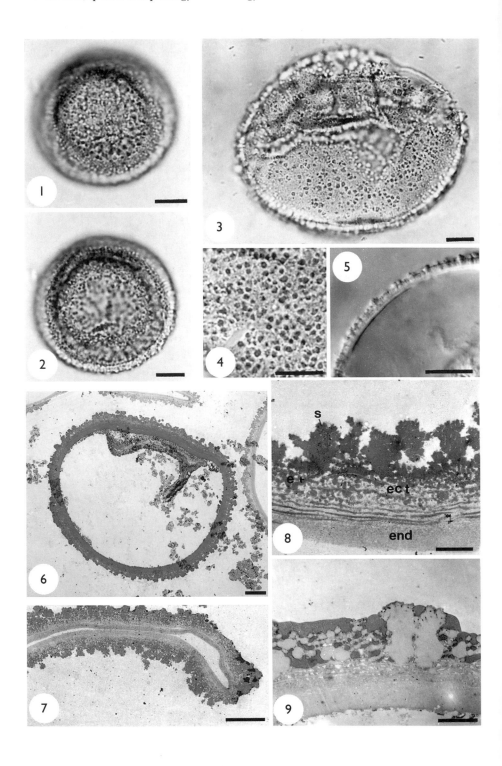

in height and 1.2 μm in basal diameter, irregularly spaced 1–3 μm apart; bases circular to polygonal in outline; crests rounded or planar or more often irregularly tapering (Figs. 4, 5, 10, 11). Orbicules associated with some specimens.

Ultrastructure (Figs. 6–9)

Exine atectate, 1.5–3 μm thick, with ectexine thinner (0.5–1 μm) than endexine (1–1.5 μm). Endexine lamellated, the lamellae tightly compacted at inner surface and separated from each other by increasingly wide cavities towards junction with ectexine (Fig. 8). Ectexine with granulate a structure, with larger grana that form surface sculpture embedded in a matrix of small grana. Surface grana/pila irregular in longitudinal section, sometimes comprising aggregates of small grana, and often tapering and capped by clusters of small grana (Figs. 7, 8). More electron dense material (possibly tapetal lipids) occurs on both the outer surface and within the interstices of the outer part of the ectexine (Fig. 9) in unacetolysed specimens stained with methanolic uranyl acetate and lead citrate.

Comparisons with fossil araucarian-like pollen

Dilwynites Harris (Figs. 12, 13, 16–28)

Pollen of *Wollemia nobilis* display a striking resemblance to fossil *sporae dispersae* referred to *Dilwynites* Harris (1965, p.88), and specifically to those included in the type species, *D. granulatus* Harris. The extant and fossil pollen share the same morphological attributes with respect to size, shape, and granulate/pilate form of exine sculpture (Macphail *et al.*, 1995; Chambers *et al.*, 1998; Figs. 12, 16–18). However, sculptural elements in *D. granulatus* (Figs. 12, 13,19, 20) are smaller in basal diameter than those of pollen of extant *Wollemia nobilis* (Figs. 3, 4, 10, 11). *Dilwynites tuberculatus* Harris is also similar, but has a sculpture of small verrucae (Figs. 23–26).

In possessing an exine composed of lamellated endexine overlain by granular ectexine, *Dilwynites* is comparable to pollen of extant Araucariaceae, including that of *W. nobilis*. Exine thickness of *D. granulatus* is sometimes less (1.5–2.5 μm) than that of *Wollemia* pollen, but in fine structural attributes the fossil and extant pollen are near-identical (*cf.* Figs. 6–9, and 21, 22). The exine of *D. tuberculatus* is broadly similar but differs in having thick endexine in which the lamellae are loosely arranged, thinner ectexine, and ectexine sculpture of verrucae (Figs. 27, 28).

Araucariacites Cookson ex Couper (Figs. 36–45)

Fossil pollen consistent with *Araucariacites* is inaperturate, with a circular or near circular amb, and an intectate exine with scabrate/granulate sculpture. Type material of the genus (see description below; Figs. 39–45) is similar to that of extant *Araucaria* (Figs. 29–31) and *Agathis* (Figs. 14, 15, 32–35). *Wollemia* pollen differs in possessing coarser sculpture (Macphail *et al.*, 1995; Figs.1–5, 10, 11).

FIGS. 1–9. Pollen of *Wollemia nobilis*. Figs. 1–5. Light micrographs of whole specimens (1–3) and sculptural detail (4–5); scale bars = 10 μm. Figs. 6–9. Thin sections of exine (6–8 acetolysed pollen, 9 unacetolysed pollen) showing lamellated endexine (end) comprising inner zone of tightly compacted lamellae and outer zone of loosely arranged lamellae, granular ectexine (ect), and sculptural elements (s); note more electron-dense material (possibly tapetal material) on and within the ectexine of unacetolysed pollen (9); scale bars (6, 7) = 5 μm, (8, 9) = 1 μm.

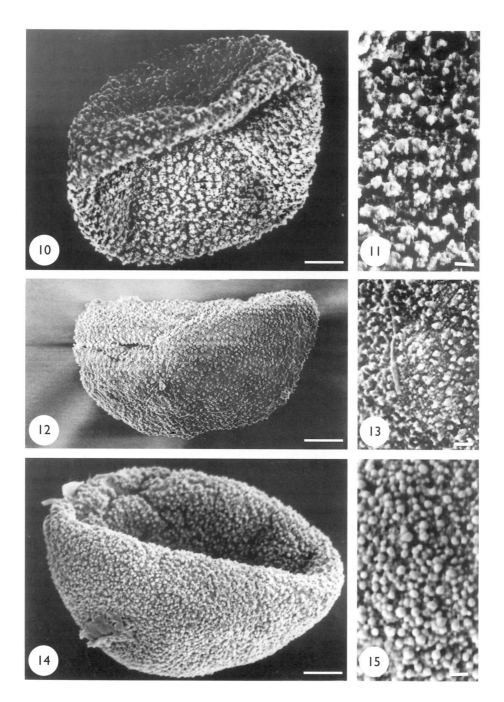

FIGS. 10–15. Scanning electron micrographs of extant and fossil araucarian pollen. Figs. 10, 11. Pollen of *Wollemia nobilis*. Fig. 12, 13. Pollen of *Dilwynites granulatus*. Figs. 14, 15. Pollen of *Agathis robusta*. Scale bars (10, 12, 14) = 5 μm, (11, 13, 15) = 1 μm.

The araucarian affinity of some *Araucariacites* is confirmed from its *in situ* occurrence in male cones of *Nothopehuen brevis* del Fueyo from the Lower Cretaceous Baqueró Formation, Argentina (del Fueyo, 1991). Exine of the pollen has scabrate sculpture and comprises lamellated endexine overlain by granular ectexine on which orbicules are randomly distributed. Dispersed *Araucariacites* from the Baqueró Formation are similar in morphological/ ultrastructural attributes (Archangelsky, 1994) and are more similar to pollen of *Araucaria* and *Agathis* than to that of *Wollemia*. Archangelsky (1994) considered ultrastructure of *Araucariacites* from the Upper Cretaceous of northern Africa (Kedves and Párdutz, 1974) to be distinct from that of the Baqueró specimens. Nonetheless, the northern African specimens possess an exine with lamellated endexine and granular ectexine and are consistent with araucarian pollen.

Comparisons with pollen of extant Araucariaceae

Agathis Salisb. and *Araucaria* Henkl & Hochst. (Figs.14, 15, 29–35)

Pollen of extant *Agathis* and *Araucaria* has gross morphology and exine ultrastructure similar to that of *Wollemia* pollen. Pollen of *Agathis* (Figs. 14, 15, 32–35) and *Araucaria* (Figs. 29–31) is inaperturate spheroidal and the exine is intectate, comprising lamellated endexine and granular ectexine (Ueno, 1959; Erdtman, 1965; Van Campo and Lugardon, 1973; Pocknall, 1981b; Kurmann, 1992). However, pollen of *Araucaria* is generally larger (50–100 μm) and its sculptural elements are shorter and more uniform in size and shape than those of *Wollemia* (Cookson and Duigan, 1957; Erdtman, 1965; Macphail *et al.*, 1995; Figs. 29–31). *Agathis* pollen is similar in size (30–55 μm) to that of *Wollemia*, but it has finer, more densely packed sculptural elements (Pocknall, 1981b; Macphail *et al.*, 1995; Figs. 14, 15, 32–35).

Comparisons with pollen of other extant gymnosperms

Asaccate, spheroidal pollen with granulate sculpture occur in several other conifer families including the Taxodiaceae, Cupressaceae, Pinaceae, Cephalotaxaceae and Taxaceae. The exine of asaccate pollen in the Taxodiaceae and Pinaceae has ectexine that is subtended by a foot layer (Kurmann, 1992), a layer that is absent from araucarian pollen. Exine ultrastructure of the Cupressaceae, Cephalotaxaceae and Taxaceae is broadly similar to that of the Araucariaceae. However, araucarian pollen differs in its larger size and lack of a leptoma, which in the Cupressaceae is often accompanied by a pouting or splitting of the exine (Erdtman, 1965; Pocknall, 1981b).

Saccate pollen of the Podocarpaceae and Pinaceae has a tectate exine, the ectexine of which has alveolar structure and is thus distinct in morphology and ultrastructure (Ueno, 1958; Pocknall, 1981a, 1981c; Kurmann 1992). Pollen in the Cycadales and Ginkgoales are monosulcate and have ectexine with alveolar structure (Kurmann,1992). That of the Gnetales is ellipsoidal and the ectexine comprises homogeneous tectum, granular infratectum and foot layer (Kurmann 1992; Rowley, 1996).

Comparisons with pollen having granular ectexine among extant angiosperms

Certain inaperturate, spheroidal pollen that occur in several angiosperm families possess granular ectexine and superficially resemble pollen of *Wollemia nobilis*. In

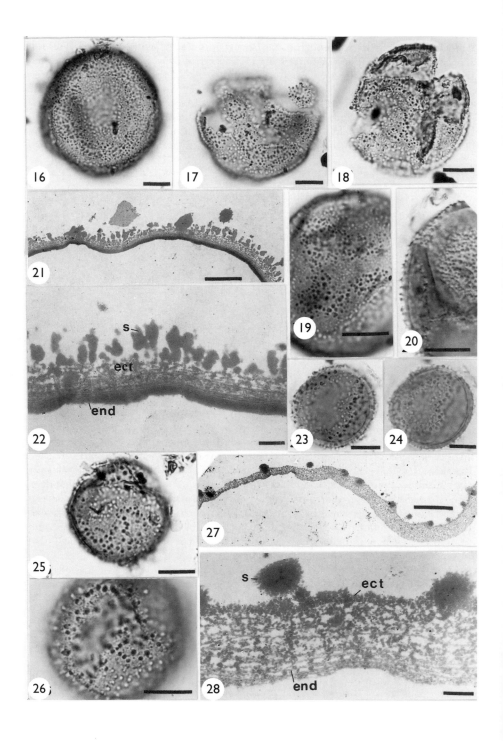

Xymalos Baillon ex Ward (Monimiaceae), the spheroidal pollen have granulate/spinulate sculpture, but the pollen lack lamellated endexine (Walker, 1976). Pollen of the Lauraceae also resemble that of *W. nobilis*, but lauraceous pollen (e.g., *Litsea* Lam., *Beilschmiedia* Nees.) have an exine that is largely endexinous and surface sculpture of larger spines and gemmae (Moar, 1993).

Distribution of Dilwynites

Recorded occurrences of *Dilwynites* express a southern, high latitudinal distribution of the parent plants during the Late Cretaceous and Tertiary (Map 1) as has been demonstrated for a range of other conifer and angiosperm plant taxa that are now relict in austral regions. The association of *Dilwynites* with pollen of diverse Podocarpaceae (*Podocarpus, Dacrydium, Dacrycarpus, Lagarostrobos, Microcachrys*) together with that of *Nothofagus*, Proteaceae, and Myrtaceae throughout the region emphasises the close relationships that existed between the vegetation of Australia, New Zealand and Antarctica during the Late Cretaceous and Early Tertiary. Moreover, the Late Cretaceous distribution range of *Dilwynites* approximates the contemporaneous ranges of *Dacrydium, Dacrycarpus, Lagarostrobos,* and *Nothofagus* and it is tempting to infer that, as for *Nothofagus*, Antarctica may have been a radiation centre of the conifers. However, inferences must be treated with caution as pollen of each taxon has considerable potential for long distance dispersal.

First reported from the Paleocene-Early Eocene in southwestern Victoria (Harris, 1965), *Dilwynites* has a Turonian-late Pliocene range in Australia (Macphail *et al.*, 1995). For New Zealand the reported range is Paleocene-Miocene (Raine, 1984; Pocknall and Tremain, 1988; Mildenhall and Pocknall, 1989), but there are suspected older (Maastrichtian-Danian) occurrences under the name *Monosulcites granulatus* Couper (Macphail *et al.*, 1995). The Antarctic record includes Turonian-Paleocene in the Peninsula region (Askin, 1990; Baldoni, 1992; present study), and recycled into recent marine sediments in the Ross Sea (Truswell, 1983).

If *Dilwynites* was shed by *Wollemia*, it would be expected that other organs of *Wollemia* may be associated in sediments of contemporaneous age within the distributional range of *Dilwynites*. Foliage and cuticles of *Araucarioides* Bigwood and R. Hill from the Eocene Regatta Point Locality in Tasmania are considered to accord well with *Wollemia* (Chambers *et al.*, 1998), and *Dilwynites* is represented in the sediments (Macphail *et al.*, 1994). Araucarian cone scales and winged seeds reported from older, Cretaceous (Barremian-Cenomanian), sediments in Australia were "favorably compared with *Wollemia*" (Chambers *et al.*, 1998; p.160), but *Dilwynites* remains unreported from sediments older than the Turonian in Australia or elsewhere.

FIGS. 16–28. Pollen of *Dilwynites granulatus* (16–22) and *Dilwynites tuberculatus* (23–28). Figs. 16–20. *Dilwynites granulatus*, light micrographs of specimens from the Dilwyn Formation, Victoria (16, 17, 19, 20) and from D.8632. 10, James Ross Island, Antarctic Peninsula (18); scale bars = 10 μm. Figs. 21, 22. *Dilwynites granulatus*, thin sections of specimen from the Dilwyn Formation, Victoria showing lamellated endexine (end) comprising inner zone of tightly compacted lamellae and outer zone of loosely arranged lamellae, granular ectexine (ect), and irregularly tapering sculptural elements (s); scale bars (21) = 2.5 μm, (22) = 0.5 μm. Figs. 23–26. *Dilwynites tuberculatus*, specimens from the Dilwyn Formation, Victoria (23–25) and sculptural detail of specimen from D.31223, Vega Island, Antarctic Peninsula (26); scale bars =10 μm. Figs. 27, 28. *Dilwynites tuberculatus*, thin sections of specimen from the Dilwyn Formation, Victoria showing lamellated endexine (end) comprising thin, inner zone of tightly compacted lamellae and thicker outer zone of lamellae arranged in a lattice-like structure, granular ectexine (ect), and rounded sculptural elements (s); scale bars (27) = 5 μm, (28) = 0.5 μm.

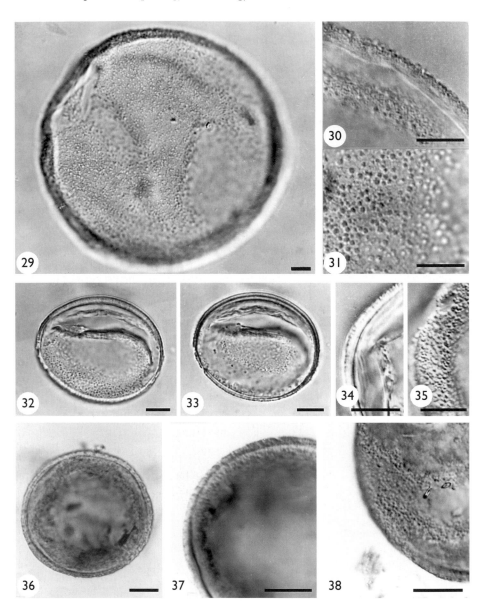

Figs. 29–38. Pollen of *Araucaria laubenfelsii* (29–31), *Agathis robusta* (32–35) and *Araucariacites australis* (36–38). Figs. 29–31. *Araucaria laubenfelsii*, surface focus (29), optical section (30) and sculptural detail (31). Figs. 32–35. *Agathis robusta* high and median foci (32, 33), optical section (34), and sculptural detail (35). Figs. 36–38. *Araucariacites australis*, specimens from the Dilwyn Formation, Victoria showing optical sections of exine and surface sculpture. Scale bars = 10 μm.

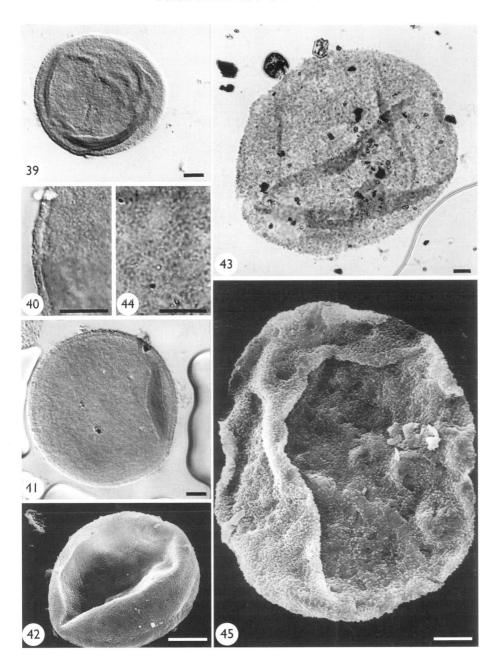

FIGS. 39–45. Pollen of *Araucariacites australis*, specimens from type sample, Waterfall Gorge, Kerguelen. Fig. 39. Lectotype (Cookson,1947; Pl. 13, fig. 3); smaller morphotype. Figs. 40, 41. Topotype and detail of exine (Cookson, 1947; Pl. 13, fig. 2); smaller morphotype. Fig. 42. SEM of smaller morphotype. Figs. 43, 44. Topotype and detail of exine (Cookson, 1947; Pl. 13, fig. 1); larger morphotype. Fig. 45. SEM of larger morphotype. Scale bars = 10 μm.

By contrast *Araucariacites* has a greater stratigraphic range and a wider distribution, being known worldwide in Jurassic and Cretaceous sediments. Tertiary occurrences appear to be concentrated in southern hemispheric regions.

Systematic Palynology

Genus *Araucariacites* Cookson ex Couper, 1953.
1947, *Granulonapites* (*Araucariacites*) Cookson, p. 132.
1953, *Araucariacites* Cookson ex Couper, p. 39.
Diagnosis: Grains inaperturate, spheroidal, but sometimes folded and ellipsoidal; amb circular to elliptical. Exine intectate comprising nexine and sexine; sexine with granular structure and sculpted by grana and/or spinulae.
Type species: *Araucariacites australis* Cookson ex Couper, 1953; by designation of Couper, 1953.
Remarks: As discussed by Jansonius and Hills (1976, card 151; 1990, card 4598), the generic name was validated by Couper (1953). Courtinat's (1987) emendation of the diagnosis was influenced by his study of pollen of extant *Araucaria excelsa* R.Br. that had been subjected to aggressive chemical treatment. However, we believe it more appropriate to base the diagnosis on the type material described by Cookson.
Comparison: *Balmeiopsis* Archangelsky has a differentially thickened exine, with thickest exine at the equator and thinnest exine on one surface of the grain. *Dilwynites* has coarser sculpture.

Araucariacites australis **Cookson ex Couper, 1953 (Figs. 36–45)**
1947, *Granulonapites* (*Araucariacites*) *australis* Cookson, p. 132; pl. 13, figs. 1–4
1953, *Araucariacites australis* Cookson ex Couper, p. 39
1958, *Araucariacites australis* Cookson ex Couper; Potonié, p.81 (lectotype designated)
Diagnosis: Pollen inaperturate, originally biconvex or spheroidal but flattened and folded; amb circular, subcircular or elliptical. Exine 1.5–2 μm thick, the nexine equal to or thicker than sexine. Sexine with surface sculpture of closely spaced, fine grana and spinulae/setae.
Dimensions: Equatorial diameter 40 (72) 100 μm.
Type specimens: One of the four specimens illustrated by Cookson (1947, pl. 13, fig. 3) was selected (Potonié, 1958, p. 81) as lectotype (Fig. 39). This and other topotype examples (Figs. 40, 41, 43, 44) in the original preparations have equatorial diameters in the range 72–120 μm; they are swollen as are other palynomorphs in the slides; specimens in new preparations are some 20–30 percent smaller and fall within the size range specified by Cookson (Figs. 42, 45).
Lectotype: Specimen P31031, Fig. 39. Pollen inaperturate, flattened with a circular amb 80 μm in diameter. Exine 2 μm thick with nexine as thick as sexine. Surface with fine grana and spinae.
Remarks: The description is based on specimens in Cookson's (1947) original slides and in slides of new preparations of the type sample. Two morphotypes are recognised in Cookson's material. Larger grains (Cookson 1947, pl.13, fig.1; Figs. 43–45) have a thinner (1.5 μm) exine and coarser sculpture than the smaller lectotype and comparable grains (Cookson 1947; pl. 13, figs.2–4; Figs. 39–42) in which the exine is 2 μm thick and the sculpture very fine. It is probable that the type material includes pollen from more than one araucarian taxon and, furthermore, that pollen attributed to *Araucariacites australis* from Mesozoic and Tertiary sediments elsewhere are derived from a disparate array of araucarian conifers.

Genus *Dilwynites* Harris 1965
Diagnosis: Grains inaperturate, spheroidal, but often folded; amb circular to ellipsoidal. Exine intectate comprising nexine and sexine; sexine with granular structure and sculptural elements of grana, verrucae or spinulae that may exceed 1 μm in basal diameter and height.
Type species: *Dilwynites granulatus* Harris, 1953; by original designation.
Comparison: Harris (1965) discriminated the genus from *Araucariacites* on the basis of its thicker exine and coarser sculpture. However, some *Araucariacites* studied by us have an exine as thick as that of *Dilwynites* (cf. Figs. 20 and 37, 40). The main distinguishing character is that the sculptural elements of *Dilwynites* are larger and less crowded than those of *Araucariacites*. *Hoegisporis* Cookson has differentially thickened exine and with discrete lenticular thickenings at the equator.

Dilwynites granulatus **Harris, 1965 (Figs. 12, 13, 16–22)**
Dimensions: 34 (41) 48 μm
Remarks: Thin sections reveal lamellated endexine with tightly compacted lamellae at inner surface and more loosely arranged lamellae at junction with the thin, granular ectexine in which the granular/pilate sculptural elements are embedded (Figs. 21, 22).

Dilwynites tuberculatus **Harris, 1965 (Figs. 23–28)**
Dimensions: 30 (40) 46 μm
Remarks: The exine comprises thick, lamellated, lattice-like endexine overlain by thinner granular ectexine in which are embedded the surface verrucae (Figs. 27, 28).

Discussion

Fossil pollen included in *Dilwynites* have morphological and structural attributes that underscore congruence with pollen of the Araucariaceae. Furthermore, *D. granulatus* is near identical in morphological and ultrastructural attributes to pollen of *Wollemia nobilis*. The only difference between the fossil and extant pollen is that the former have finer sculpture than that of *Wollemia*. A close relationship between the fossil pollen and *Wollemia* seems likely, as was recognised also on morphological characters (Macphail *et al.*, 1995; Chambers *et al.*, 1998). Exine stratification/ultrastructure does not support a close relationship between *Dilwynites* and the Lauraceae as was suggested by Harris (1965) and Askin (1990). Pollen of Lauraceae are spheroidal, inaperturate but differ from *Dilwynites* and that of *Wollemia nobilis* in possessing thick endexine overlain by very thin ectexine with surface sculpture of spinae and gemmae. In terms of architecture, surface sculpturing and exine stratification/ultrastructure, pollen of *Wollemia nobilis* is consistent with pollen of other extant Araucariaceae. It may be differentiated from *Agathis* pollen, which is smaller and has finer sculpture of rounded grana, and pollen of *Araucaria*, which is larger and with sculpture of shorter, uniform, and more densely distributed grana. Asaccate pollen of certain other extant conifers within the Cupressaceae, Taxodiaceae, Pinaceae, Cephalotaxaceae, and Taxaceae have similar ultrastructural and sculptural attributes, but are mostly smaller and have a leptoma, a feature not represented in the Araucariaceae. Moreover, in asaccate pollen of the Pinaceae and Taxodiaceae the ectexine is subtended by a foot layer, which appears to be absent in the Araucariaceae.

Acknowledgements

Sincere thanks to Dr. K. Hill, Royal Botanic Gardens, Sydney for provision of pollen cones of *Wollemia nobilis*; Dr J. Douglas, Warrnambool for guidance in collecting from the Dilwyn Formation, Victoria; Dr M.R. Thomson, British Antarctic Survey for provision of samples from James Ross and Vega Islands, Antarctic Peninsula; and to Wendy Armstrong, Department of Botany, University of Queensland for expertise in ultrasectioning and assistance with transmission electron microscopy. Helpful comments were provided for improvement of the manuscript from two anonymous reviewers. MED acknowledges the support of ARC during the tenure of an ARC Senior Research Fellowship.

References

Archangelsky, S. (1994). Comparative ultrastructure of three Early Cretaceous gymnosperm pollen grains: *Araucariacites, Balmeiopsis* and *Callialasporites. Review of Palaeobotany and Palynology* 83: 185–198.

Askin, R.A. (1990). Campanian to Paleocene spore and pollen assemblages of Seymour Island, Antarctica. *Review of Palaeobotany and Palynology*: 65: 105–113.

Baldoni, A. (1992). Palinologia de la formacion Santa Marta, Cretacico superior de la Isla James Ross, Antartida. In: C.A. Rinaldi (ed.) Geologia de la Isla James Ross, Antartida, pp. 359–374. Instituto Antarctico Argentino, Buenos Aires.

Chambers, T.C., Drinnan, A.N. and McLoughlin, S. (1998). Some morphological features of Wollemi Pine (*Wollemia nobilis*: Araucariaceae) and their comparison to Cretaceous plant fossils. *International Journal of Plant Science* 159: 160–171.

Cookson, I.C. (1947). Plant microfossils from the lignites of Kerguelen Archipelago. *B.A.N.Z. Antarctic Research Expedition 1929–31 Reports*, Series A2: 127–142.

Cookson, I.C. and Duigan, S.L. (1957). Tertiary Araucariaceae from south-eastern Australia, with notes on living species. *Australian Journal of Scientific Research, Series B* 4: 415–449.

Couper, R.A. (1953). Upper Mesozoic and Cainozoic spores and pollen grains from New Zealand. *New Zealand Geological Survey Paleontological Bulletin* 22: 1–77.

Courtinat, B. (1987). Interprétation des grains de pollen du groupe *Araucaricites* Cookson, 1947. Comparison avec les formes actuelles (*Araucaria excelsa* R. Brown). *Revue de Micropaléontologie* 30: 79–90.

Del Fueyo, G. (1991). Una nueva Araucariaceae cretácica de Patagonia, Argentina. *Ameghiniana* 28: 149–161.

Dettmann, M.E. and Jarzen, D.M. (1998). The early history of the Proteaceae in Australia: the pollen record. *Australian Journal of Systematic Botany* 11: 401–438.

Dettmann, M.E. and Thomson, M.R.A. (1987). Cretaceous palynomorphs from the James Ross Island area, Antarctica - a pilot study. *British Antarctic Survey Bulletin* 77: 13–59.

Erdtman, G. (1952). Pollen morphology and plant taxonomy. Angiosperms. Almqvist and Wiksell, Stockholm.

Erdtman, G. (1960). The acetolysis method. A revised description. *Svensk Botanisk Tidskrift* 54: 561–564.

Erdtman, G. (1965). Pollen and spore morphology/plant taxonomy, v. 2. Gymnospermae, Pteridophyta, Bryophyta. Almqvist and Wiksell, Stockholm.

Faegri, K. (1956). Recent trends in palynology. *Botanical Review* 22: 639–664.

Harris, W.K. (1965). Basal Tertiary microfloras from the Princetown area, Victoria, Australia. *Palaeontographica* 115B: 75–106.

Jansonius, J. and Hills, L.V. (1976, 1990). Genera file of fossil spores and pollen. Special Publication, Department of Geology, University of Calgary, Canada.

Jones, W.G., Hill, K.D. and Allen, J.M. (1995). *Wollemia nobilis*, a new living Australian genus and species in the Araucariaceae. *Telopea* 6: 173–176.

Kedves, M. and Párdutz, Á. (1974). Ultrastructural studies on Mesozoic inaperturate Gymnospermatophyta pollen grains. *Acta Biologica Szeged* 20: 81–88.

Kurmann, M.H. (1992). Exine stratification in extant gymnosperms: a review of published transmission electron micrographs. *Kew Bulletin* 47: 25–39.

Macphail, M.K., Alley, N.F., Truswell, E.M. and Sluiter, I.R.K. (1994). Early Tertiary vegetation: evidence from spores and pollen. In: R.S. Hill (editor). History of the Australian vegetation: Cretaceous to Recent. pp. 189–261. Cambridge University Press, Cambridge.

Macphail, M., Hill, K., Partridge, A., Truswell, E. and Foster, C. (1995). 'Wollemi Pine' - old pollen records for a newly discovered genus of gymnosperms. *Geology Today* (March-April): 48–50.

Mildenhall, D.C. and Pocknall, D.T. (1989). Miocene-Pleistocene spores and pollen from Central Otago, South Island, New Zealand. *New Zealand Geological Survey Paleontological Bulletin* 59: 1–128.

Milne, L.M. (1998). Surface embedding of fossil pollen for time- and cost-effective ultramicrotomy (TEM) and multiple microscopy (LM, SEM, TEM) of single grains. *American Association of Stratigraphic Palynologists, Contribution Series* 33: 95–105.

Moar, N.T. (1993). Pollen grains of New Zealand dicotyledonous plants. Manaaki Whenua Press, Lincoln, New Zealand.

Nougier, J. (1970). Contribution a L'étude geologique et geomorphologique de Îles Kerguelen verified. *Terres Australea et Antarctiques Francaises* 48: 11–21.

Pocknall, D.T. (1981a). Pollen morphology of the New Zealand species of *Dacrydium* Solander, *Podocarpus* L'Heritier, and *Dacrycarpus* Endlicher (Podocarpaceae). *New Zealand Journal of Botany* 19: 67–95.

Pocknall, D.T. (1981b). Pollen morphology of *Phyllocladus* L.C. et A. Rich. *New Zealand Journal of Botany* 19: 259–266.

Pocknall, D.T. (1981c). Pollen morphology of the New Zealand species of *Librocedrus* Endlicher (Cupressaceae) and *Agathis* Salisbury (Araucariaceae). *New Zealand Journal of Botany* 19: 267–272.

Pocknall, D.T. and Tremain, R. (1988). Tour LB1, 7th International Palynological Conference, Brisbane, Australia, August 1988. *New Zealand Geological Survey Record* 33: 1–107.

Punt, W., Blackmore, S., Nilsson, S., and Le Thomas, A. (1994). Glossary of pollen and spore terminology. Laboratory of Palaeobotany and Palynology Foundation, Utrecht.

Raine, J.I. (1984). Outline of a palynological zonation of Cretaceous to Paleogene terrestrial sediments in West Coast region, South Island, New Zealand. *New Zealand Geological Survey Report* 109: 1–82.

Rowley, J.R. (1996). Exine origin, development and structure in pteridophytes, gymnosperms and angiosperms, In: J. Jansonius and D.C. McGregor (editors). Palynology: principles and applications, vol. 1, pp.443–462. American Association of Stratigraphic Palynologists Foundation, Dallas, Texas.

Smith, A.G., Hurley, A.M. and Briden, J.C. (1981). Phanerozoic paleocontinental world maps. Cambridge University Press, Cambridge.

Truswell, E.M. (1983). Recycled Cretaceous and Tertiary pollen and spores in Antarctic marine sediments: a catalogue. *Palaeontographica B* 186: 121–174.

Ueno, J. (1958). Some palynological observations of Pinaceae. *Journal of the Institute Polytechnics Osaka City University, Series D, Biology* 9: 163–187.

Ueno, J. (1959). Some palynological observations of Taxaceae, Cupressaceae and Araucariaceae. *Journal of the Institute Polytechnics Osaka City University, Series D, Biology* 10: 75–87.

Van Campo, M. and Lugardon, B. (1973). Structure grenue infratectale de l'ectexine des pollens de quelques Gymnospermes et Angiospermes. *Pollen et Spores* 15: 171–187.

Walker, J.W. (1976). Evolutionary significance of the exine in the pollen of primitive angiosperms. In: I.K. Ferguson and J. Muller (editors). The evolutionary significance of the exine. pp. 251–308. Academic Press, London, New York.

Appendix

Register of illustrated specimens in Museum of Victoria, Melbourne (catalogue number prefixed "P") and British Antarctic Survey, Cambridge (BAS). ICC denotes original slides of Cookson (1947).

Species		Fig.	Locality	Prep./ Slide	England Finder Coordinates	Catalogue No.
Araucariacites australis; lectotype (smaller morphotype)		39	Kerguelen, sample 85A	ICC	M33/0	P31031
" topotype (smaller morphotype)	;	40, 41	"	ICC	V45/4	P3103
" topotype (smaller morphotype)	;	42	"	B1104/S1	L41/1	P208585
" topotype (larger morphotype)	;	43, 44	"	ICC	F35/2	P21029
" topotype (larger morphotype)	;	45	"	B1104/S2	N37/0	P208586
Araucariacites australis		36, 37	Victoria, Dilwyn Formation	M001/2	D24/3	P208594
"	;	38	"	M001/2	R35/4	P208595
Dilwynites granulatus		12, 13	Victoria, Dilwyn Formation	B908/S91	P36/2	P208587
"	;	16	"	M001/2	Q20/3	P208588
"	;	17	"	M001/2	P53/4	P208589
"	;	18	James Ross Is. D.8632.10	B1361/1	V31/0	BAS
"	;	19	Victoria, Dilwyn Formation	M001/2	T21/0	P208590
"	;	20	"	M001/2	B2 8/1	P208591
Dilwynites tuberculatus		23, 24	Victoria, Dilwyn Formation	M001/2	V38/2	P208592
"	;	25	"	M001/2	Q43/2	P208593
"	;	26	Vega Is. D.31223	B1264/3	N56/3	BAS

Pozhidaev, A.E. (2000). Pollen variety and aperture patterning. In: M.M. Harley, C.M. Morton and S. Blackmore (Editors). Pollen and Spores: Morphology and Biology, pp. 205–225. Royal Botanic Gardens, Kew.

POLLEN VARIETY AND APERTURE PATTERNING

ANDREW E. POZHIDAEV

Palynological Laboratory, Komarov Botanical Institute, Russian Academy of Sciences, 2 Prof. Popov Street, St. Petersburg, 197376, Russia

Abstract

It has been observed that distantly related angiosperm taxa may have identical deviant pollen aperture patterns, in addition to the common condition. Pollen forms of both types (deviant and common forms) can be arranged in a continuous branched series in such a way that the deviant form with a meridional circular colpus is gradually transformed into one of the common patterns. This series is supposed to reflect some hypothetical spatial transformations in the developing microspore cytoplasm, which are usually completed before the deposition of the exine layers, predetermining the common aperture arrangement. The formation of the exine before these processes are completed gives rise to the pollen form series. The behaviour of endoaperture in colporate deviant forms and the existence of predictable variety of polycolpate pollen isomers prove to be a direct consequence of the suggested mechanism of aperture patterning. The aperture arrangement and orientation in the tetrads of the initial zona-aperturate form was found to be similar to the zonasulcate pattern. This enables us to homologise these two conditions and to consider the zona-aperturate pattern as the basis for variety in pollen form.

Introduction

At present, there is neither a clear understanding of why the pollen grains of flowering plants have a geometrically perfect and regular aperture arrangement, nor is it known by what mechanisms this is determined in phylogenesis and ontogenesis. The most generally accepted theory of pollen aperture evolution, based on systematic and paleobotanical data, states that one of the sulcate pollen forms (monosulcate or trichotomosulcate) or the inaperturate condition was the initial form, antecedent to the 3-colpate pattern of the eudicots (Wodehouse, 1935; Straka, 1963; Meeuse, 1965; Nair, 1967; Muller, 1970; Walker, 1974; Walker and Doyle, 1975; Kuprianova, 1979; Melville, 1981). All other forms are considered to be 3-colpate derivations (Van Campo, 1967, 1976; Thanikaimoni, 1986 Takhtajan, 1959, 1991). However, no undoubted transitional forms of this aperture arrangement, which could provide evidence for any of these suggestions, have been found in fossil or recent pollen. Nevertheless, there is a general agreement that the formation of 3-aperturate grains reflects the duplication and repositioning of developmental processes originally restricted to the distal surface (Blackmore and Crane, 1998).

Much effort has been made to understand the mechanisms determining the pollen wall pattern and the aperture region during the pollen development. Wodehouse (1935) supposed that the symmetry of pollen aperture patterns is determined by the number of contact points of the grains in the tetrad. When pollen grains are placed in a tetrahedral arrangement (the most characteristic tetrad arrangement for dicots), each of them has three equally spaced points of contact. This suggests the explanation of the prevailingly 3-colpate character of dicot pollen. Pollen grains in a rhomboidal arrangement can have only two, asymmetric, points of contact. In this case, a degree of symmetry may be achieved by the development of symmetrical supplementary apertures which give rise to other aperture patterns. Thus, the number and arrangement of elements in the patterns of pollen grain symmetry are considered to result from cellular interrelation. Melville (1981) re-examined this model and supposed that the processes involved in the development of aperture symmetry depended upon the action of surface tension, surface energy and diffusion, which provide the physical basis for pattern initiation. He assumed that after meiosis the daughter cells moved under the action of surface tension until they reached the tetrahedral arrangement characterised by minimum surface energy. The pressure of the surrounding tissues may counteract this process during the sensitive period of pollen wall formation and result in a decussate, rhomboidal or linear tetrad arrangement. The areas of the pollen surface that were destined to become colpi would have different surface energies from the rest of the cell surface. The result would be that the surface tension would swing the trifurcate sulci of the initial trichotomosulcate condition into contact in the radial positions on the tetrahedral group. The conclusion was made that the 3-colpate condition would be attained simultaneously with the tetrahedral arrangement of pollen grains.

Attempts have been made to explain the variety of pollen forms by simple geometrical rules. Wodehouse (1935) suggested that the configuration of pollen apertures might have resulted from the "trischistoclasic system" governed by what he called "the law of equal triconvergent angles". Following this law, the globally distributed apertures become arranged in such a way that they tend to converge toward each other in threes with equal angles of convergence, which, consequently, tend to be 120 degrees. Due to the trischistoclasic system, the number of possible polygons and corresponding pantoaperturate patterns was suggested to be fixed. Van Campo (1967, 1976) described some series of pollen aperture configurations within angiosperm taxa, which obey a limited number of rules and produce a limited number of variations in aperture pattern. She designated a succession of 3-colpate-pericolpate-periporate pollen types, which belongs to the trischistoclasic system of Wodehouse, as "successiformy". Besides, she identified two more patterns of variation: spiralisation and breviaxy and suggested all three patterns of morphological variation as three trends in pollen evolution (Van Campo, 1976). Melville (1981) examined the successiform series in several families and devised a simple formula for a calculation of pore numbers from the grain diameter and the distance between the pores.

The arrangement of the cortical microtubule system in the young postmeiotic microspore, the microtubule-organising centres and the endoplasmic reticulum shields were considered as possible structures affecting aperture position (Heslop-Harrison, 1963, 1971; Dover, 1972; Sheldon and Dickinson, 1983; Dickinson and Sheldon, 1986; Tiwari, 1989). The action of endoplasmic reticulum shields, locally disrupting the primexine deposition in the place destined to be the aperture, is generally accepted as the basic mechanism, by which the aperture regions are defined. However, it remains unclear how the structures assume the geometrically perfect spatial aperture arrangement. The quadripolar cytoskeleton spindles that arise in telophase-II of simultaneous meiocytes in many higher plant taxa (Juniperaceae, Alexandrovsky 1971; Nymphaeaceae, Guignard, 1897; Asteraceae, Wodehouse, 1930; Eupomatiaceae, Hotchkiss, 1958; Ranunculaceae, Waterkeyn, 1962; Winteraceae, Sampson 1963; Asphodelaceae, Van Lemmeren, et al., 1985) are suggested to be the structure, which could be responsible for establishing the initial pollen cytoplasm symmetry (Blackmore and Crane, 1998). However, the nature of the processes, which could have transformed this initial polarisation to the final condition and govern the spatial co-ordination of developmental factors, is still unknown.

Materials and Methods

The pollen samples were acetolysed and mounted in glycerine jelly for study with the light microscope (LM). The untreated, acetolysed and FAA fixed pollen were coated with gold and studied with a scanning electron microscope (SEM), JEOL JSM-35.
Since a wide variety of pollen forms was encountered, from 200 to 400 grains were typically examined in each sample.

Specimens investigated
Pollen samples from more than 850 species, belonging to 130 families, have been examined. The results to be discussed here are based on this extensive survey, but this paper refers explicitly only to the following specimens, which are all illustrated. The samples were obtained from the Herbarium of the Komarov Botanical Institute, St.-Petersburg (LE). The sample of *Rhodospatha venosa* was obtained from the Herbarium of the Missouri Botanical Garden, St. Louis (MO) and *Aesculus parryi* from the Herbarium of the New York Botanical Garden, New York (NY).

Dicots

Aceraceae	*Acer tataricum* L. Ukraine, Crimea. Kuprianova s.n.	(Fig. 30)
Austrobaileyaceae	*Austrobaileya scandens* C. T. White s.n. Bot. Gard. Massachusetts University	(Fig. 55)
Basellaceae	*Basella diffusa* Ruiz et Pav. ex Mog. Herb. Fisheri s.n.	(Fig. 42)
Bonnetiaceae	Kielmeyera variabilis Mart. Brazil. K.S.Irvin 7321	(Fig. 75)
Canellaceae	*Cinnamodendron axillare* Endl. ex Walp. Brazil: Rio-Janeiro. Glaziou 14531	(Figs. 68-69)
Chenopodiaceae	*Atriplex sagittata* Borckh. Russia: St.-Petersburg region. Medvedeva s.n.	(Fig. 49)
Clusiaceae	*Hypericum davisii* N. Robson. Armenia. Bourgeau 64	(Fig. 36)
Convolvulaceae	*Merremia pterygocaulos* (Stend.) Hallier f. Zaire. Lisowski 10627	(Figs. 19-20)
Cunoniaceae	*Belangera glabra* Cambess. Riedel 1202	(Figs. 50-54; 57-60)

Cucurbitaceae	*Echinopepon cirrhopedunculatus* Rose. Mexico: State of Jalisco. Pringle s.n.	(Fig. 28)
Eupomatiaceae	*Eupomatia laurina* R.Br. Australia. A. Cronquist and B. Gray 11627; Russia: St.-Petersburg Bot. Gard. s.n.	(Fig. 7) (Figs. 76-80)
Flacourtiaceae	*Scolopia schreberi* J.F. Gmel. Sri Lanka. Coll. no. 15506	(Figs. 4-5)
Fumariaceae	*Corydalis buschii* Nakai. Russia: St.-Petersburg Bot.Gard. s.n.	(Figs. 37-38)
	C. cava (Miller) Schweig. & Kort. France. Bersier 347	(Fig. 34)
	Dicentra spectabilis Lem. Northern China. Bunge s.n.	(Fig. 31)
	Fumaria officinalis L. Bulgaria. Marnova and Ivanova 918	(Figs. 45-47)
Hippocastanaceae	*Aesculus parryi* A. Gray. USA: California. C.R. Orcutt s.n. (NY)	(Fig. 6)
Lamiaceae	*Scutellaria dubia* Taliew & Schirjaew. Russia: Choper reservation. Tzvelev s.n.	(Fig. 8)
	Sideritis gomeraea De Noë ex Bolle. Canary Islands: Gomera. Bourgeau 60	(Figs. 23-26)
Limnanthaceae	*Limnanthes alba* Hartw. ex Benth. USA: California. Bigelow s.n.	(Fig. 9)
Lentibulariaceae	*Utricularia lilacina* Griff. Australia: Port Phillip. F. Müller s.n.	(Fig. 21)
Malpighiaceae	*Banisteria oxyclada* (A. Juss.) Gates. Brazil. Glaziou 18945	(Fig. 39)
Martyniaceae	*Martynia violacea* Engelm. Mexico. Emory s.n.	(Fig. 40)
Nymphaeaceae	*Nymphaea capensis* Thunb. Russia: St.-Petersburg Bot. Gard. s.n.	(Fig. 73)
	Victoria regia Lindl. Russia: St.-Petersburg Bot. Gard. s.n.	(Fig. 74)
Papaveraceae	*Dendromecon rigidum* Benth. USA: California. Torne and Everett 34476	(Fig. 27)
	Papaver argemone L. Sweden. Fries s.n.	(Fig. 48)
	Papaver detritophilum V.V. Petrovskii. Russia: Chukchi Peninsula. Petrovskii 14	(Figs. 43-44)
Ranunculaceae	*Ficaria verna* Huds. Russia: St.-Petersburg Bot. Gard. s.n.	(Figs. 29, 33)

Salicaceae	*Salix amaniana* Willd. Switzerland, Herb. Mertens	(Fig. 1)
	S. candidula Fluegge ex Willd. Switzerland. Herb. Mertens	(Figs. 2-3)
	S. hexandra Ehrh. Austria: Central Tyrol. Hunter s.n.	(Fig. 22)
	S. geyeriana Anderson. USA: Colorado. Yones 45276	(Fig. 35)
Scrophulariaceae	*Pedicularis oederi* Vohl. Canada: Yukon. Golder and Gillett 26036	(Fig. 32)
Violaceae	*Alsodeia echinocarpa* Korth. Sumatra. Forbes 2754	(Figs. 17-18)

Monocots

Araceae	*Rhodospatha venosa* Gleason. Croat 59390	(Figs. 10, 11)
Calochortaceae	*Calochortus kennedyi* Porter. USA: Northern Arizona. Herb. A. Gray 4175	(Fig. 16)
Haemodoraceae	*Anigozanthos flavida* DC. Russia: St. Petersburg Bot. Gard. s.n.	(Figs. 14-15)
Iridaceae	*Juno nicolai* Vved. Tadzikistan: Mt Kushten. Chukivik 12766	(Fig. 41)
Phormiaceae	*Dianella ensifolia* (L.) DC. Vietnam. Averianov and Kudriavceva 178	(Figs. 66-67)
Rapateaceae	*Rapatea spruceana* Koern. Colombia. Schultes *et al.* 17988	(Figs. 12-13)
	R. longipex Spruce ex Koern. Colombia. R.E. Shultes, I. Carbera 19550	(Fig. 81)
Tecophilaeaceae	*Cyanella racemosa* Schinz. South Africa: western region. Coll. no. 11258	(Fig. 56)
	C. uniflora L. f. South Africa: Cape Province. Coll. no. 8686	(Figs. 64-65)
Uvulariaceae	*Uvularia sessilifolia* L. Brooklyn Bot. Gard. Dodd s.n.	(Figs. 61-63)

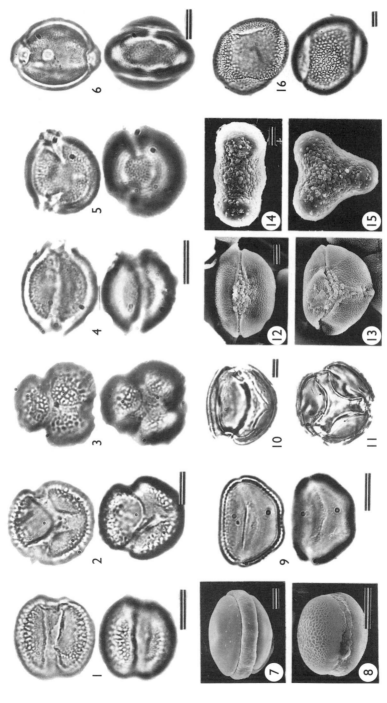

FIGS. 1–16. Deviant and common pollen forms. Fig. 1. *Salix amaniana*, form A. LM. Figs. 2–3. *S. candidula*, form B and C, polar view. LM. Figs. 4–5. *Scolopia schreberi*, forms A and B, polar view. LM. Fig. 6. *Aesculus parryi*, form B, equatorial view. LM. Fig. 7. *Eupomatia laurina*, common pollen. SEM. Fig. 8. *Scutellaria dubia*, form A. SEM. Fig. 9. *Limnanthes alba*, common pollen, polar view. LM. Figs. 10–11. *Rhodospatha venosa*, forms B and C, polar view. LM. Figs. 12–13. *Rapatea spruceana*, forms B and C, polar view. SEM. Figs. 14–15. *Anigozanthos flavida*, 2- and 3-aperturate forms, polar view. SEM. Fig. 16. *Calochortus kennedyi*, form F. LM. [The upper and lower LM foci are given for each pollen grain except figs 10–11. For letter symbols of deviant forms see Fig. 70. Scale = 10 μm].

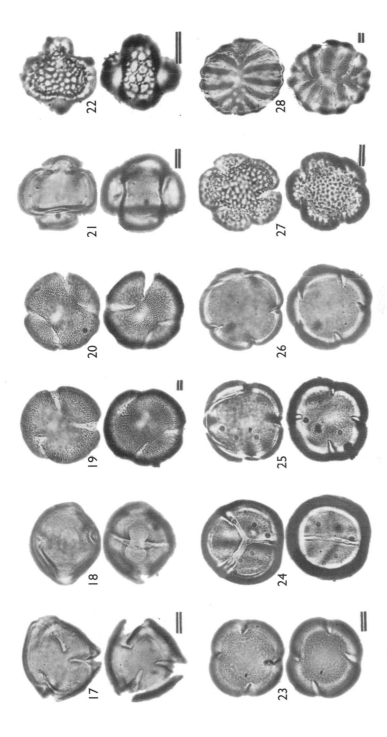

FIGS. 17–28. Deviant and common pollen forms. Figs. 17–18. *Alsodeia echinocarpa*, form D, polar and equatorial view. LM. Figs. 19–20. *Merremia pterygocaulos*, forms D and E, polar view. LM. Fig. 21. *Utricularia tilaxina*, form G, polar view. LM. Fig. 22. *Salix hexandra*, form G, polar view. LM. Figs. 23–26. *Sideritis gomeraea*. LM: Fig. 23. Form H, polar view. Figs. 24–25. Form I, equatorial and polar view. Fig. 26. Form J, polar view. Fig. 27. *Dendromecon rigidum*, form J, polar view. LM. Fig. 28. *Echinopepon cirrhopedunculatus*, polycolpate pattern, common pollen, polar view. LM. [The upper and lower LM foci are given for each pollen grain. For letter symbols of deviant forms see Fig. 70. Scale = 10 μm].

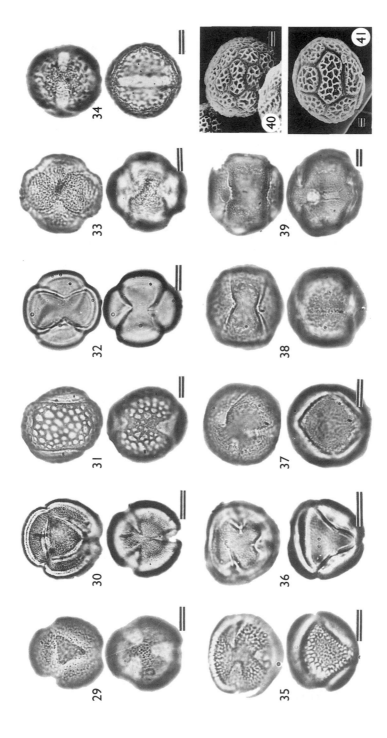

FIGS. 29–41. Deviant and common pollen forms. Fig. 29. LM; *Ficaria verna*, form L, polar view. LM. Fig. 30. *Acer tataricum*, form L, polar view. LM. Fig. 31. *Dicentra spectabilis*, form K. LM. Fig. 32. *Pedicularis oederi*, form K. LM. Fig. 33. *Ficaria verna*, form K. LM. Fig. 34. *Corydalis cava*, form K. LM. Fig. 35. *Salix geyeriana*, form M. LM. Fig. 36. *Hypericum davisii*, form M. LM. Figs. 37–38. *Corydalis buschii*, form M and N. LM. Fig. 39. *Banisteria oxiclada*, form N. LM. Fig. 40. *Martynia violacea*, polyrugate pattern with divided areola. SEM. Fig. 41. *Juno nicolai*, polyrugate pattern with divided areola. SEM. [The upper and lower LM foci are given for each pollen grain. For letter symbols of deviant forms see Fig. 70. Scale = 10 μm].

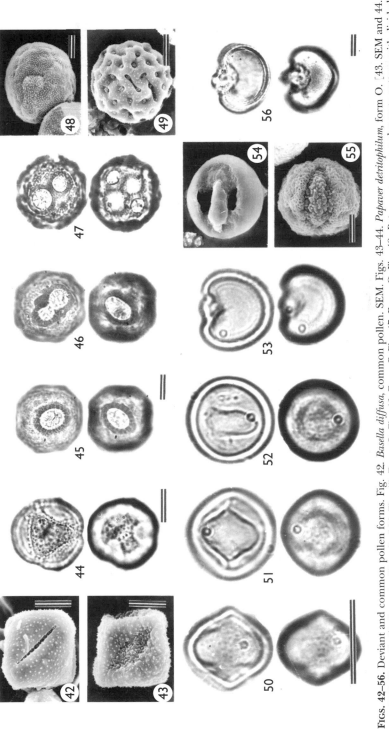

FIGS. 42–56. Deviant and common pollen forms. Fig. 42. *Basella diffusa*, common pollen. SEM. Figs. 43–44. *Papaver detritophilum*, form O. [43. SEM and 44. LM]. Figs. 45–47. *Fumaria officinalis*. LM: Fig. 45. Form O. Fig. 46. Form P. Fig. 47. Form Q. Fig. 48. *Papaver argemone*, polyporate pattern with divided pore. SEM. Fig. 49. *Atriplex sagittata*, polyporate pattern with divided pore. SEM. Figs. 50–52. *Belangera glabra*. LM: Fig. 50. Form R, equatorial view. Fig. 51. Form S, equatorial view. Figs. 52–53. Form T, equatorial and polar view. LM. Fig. 54. Form T, equatorial view. SEM. Fig. 55. *Austrobaileya scandens*, common pollen. SEM. Fig. 56. *Cyanella racemosa*, common pollen. LM. [The upper and lower LM foci are given for each pollen grain. For letter symbols of deviant forms see Fig. 70. Scale = 10 μm].

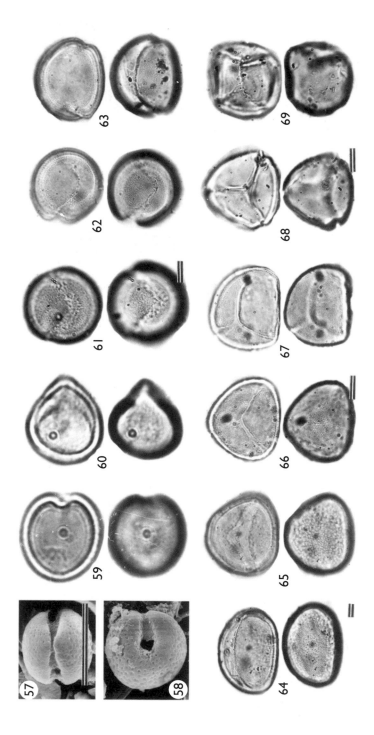

FIGS. 57-69. Deviant and common pollen forms. Figs. 57–60. *Belangera glabra*: Figs. 57–58. Forms U and V, polar and equatorial view. LM. Figs. 59–60. Form V, polar view. SEM. Figs. 59–60. Form V, polar and equatorial view. LM. Figs. 61–63. *Uvularia sessilifolia*. LM: Fig. 61. Form W, polar view. Figs. 62–63. Form X, polar and equatorial view. Figs. 64–65. *Cyanella uniflora*, forms Y and Z, polar view. LM. Figs. 66–67. *Dianella ensifolia*, forms Z and Z', polar view. LM. Figs. 68–69. *Cinnamodendron axillare*, trichotomo- and tetrachotomosulcate pollen. LM. [The upper and lower LM foci are given for each pollen grain. For letter symbols of deviant forms see Fig. 70. Scale = 10 μm].

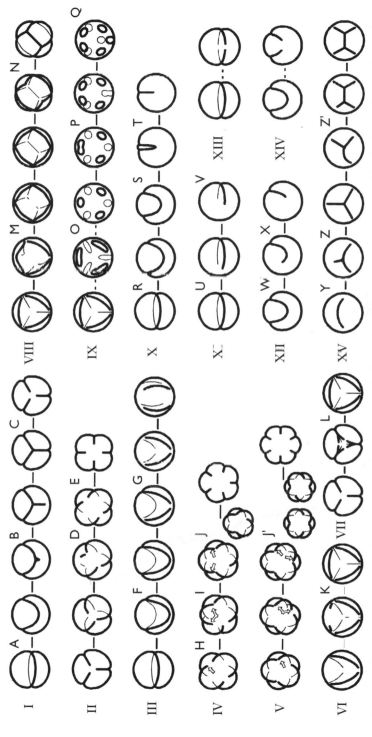

Fig. 70. The pollen form series (in series IV and V, the small images near the forms show the aperture arrangement and two 6-colpate isomers; the broad arrows indicate splitting colpi).

215

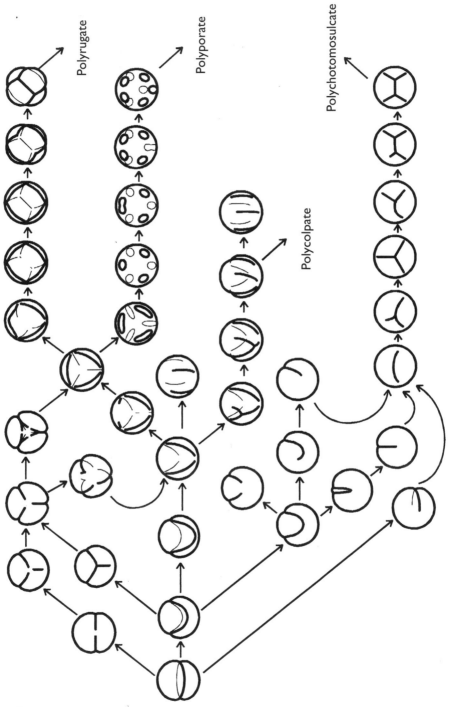

FIG. 71. To show the pattern of pollen form variants with modes of ectoaperture transformation (the curved arrows indicate that the forms are shown in different projections).

Terminology

Some terms used in the text should be explained.

Zona-aperturate – describing a pollen grain with a ring-like aperture is used here as a general term to describe dicotyledonous colpate/colporate pollen with a circular colpus, as distinct from zonasulcate/zonasulculate pollen of monocotyledons.
Colpus – a meridionally oriented elongate aperture.
Ruga – a globally distributed elongate aperture.
Polycolpate – describing a zonocolpate pollen with more than three colpi.
Isomers- a loxocolpate variant of polycolpate pollen grains with the same number of apertures, but differing by the alternating arrangement of inclined colpi.

Results

Only geometrical characteristics of the pollen grains have been studied, namely, the arrangement of pollen apertures and the positioning of a pollen grain in a tetrad. All aperture patterns, independent of their abundance, were recorded. The pollen forms were compared without taking systematic relationships into consideration. In spite of this obviously one-sided approach, the following observations were made:

1. As well as the common pollen form (i.e. the typical and most abundant aperture pattern), diverse and rare deviant patterns (i.e. differing from the common pollen form by the number and arrangement of apertures) were found within many samples. The distribution and frequency of these deviations in different specimens is quite fortuitous. The forms occurring abundantly in the pollen of some collections may be entirely absent from the pollen of other collections of the same species (Wodehouse, 1935; Santisuk, 1979). Thus, the abundance of different deviations in the samples studied varies widely, from single findings to a significant percentage. In some samples, the total number of deviant forms may exceed 20-30%.

2. Similar deviant pollen forms with the identical aperture arrangement have been found in remote taxa, even in very different groups of flowering plants.

3. It is suggested that all forms, whether deviant or common, can be arranged in a series of aperture configurations in such a way that the deviant forms provide the transitional arrangements intermediate between the common patterns. Similar form-series have been observed not only in individual stamens and flowers or single plants and populations, but in different samples of one and the same species as well as in different genera, families or distantly related taxa of a higher level.

For convenience of description, these series may be represented as the transformation of one form into another one. The following transformation modes of the ectoaperture geometry can be distinguished (Fig. 70: I-XV). The details of deviant pollen morphology are considered in a series of papers (Pozhidaev, 1998; Pozhidaev, press.). However, the variety of forms and their interpretation, are considered here.

1. Transformation of the zona-aperturate and zonasulcate forms into the 3-colpate and the 3-sulcate condition (Fig. 70: I). The ring-shaped colpus of zona-aperturate form is gradually bent by 120° and occupies the position of two colpi of the final 3-colpate

configuration. The third colpus arises via merging two incipient elements on the pollen grain equator that appeared in each polar zone (Figs. 1-5, 8-9; 70: I.B). Several modifications of this series have been identified in many groups of dicots (Pozhidaev, 1998). Similar series have been observed in some monocot species with zonasulcate pollen, whose aperture typically has the same configuration as the colpus of zona-aperturate form of the dicots (Fig. 7). In this case, the transformation results in the 3-aperturate condition (Figs. 10-15, 79-80).

2. Transformation of the 3-colpate and zona-aperturate forms into the 4-colpate condition (Fig. 70: II-III). The 4-colpate condition can be acquired in two ways (Pozhidaev, subm.): either as a result of the equatorial merging of two incipient elements that appeared in each polar zone of the 3-colpate pollen grain (Figs. 17-20; 70: II) or due to the bending of the ring-like colpus of zona-aperturate pollen which is twisted like the seam on a tennis ball (Fig. 70: III). Subsequent disruption of a twisted colpus at the two points on each pole (Figs. 21-22; 70: III.G) gives rise to the pattern with four inclined colpi with a W-shaped arrangement in a plan ('W-form' (Clarke, 1975)). It should be particularly noted that the increase in colpus number of the 3-colpate configuration (Fig. 70: II.D) initially results in the W-form, and only later do the inclined colpi acquire the meridional orientation. This probably testifies to a special significance for W-symmetry. The tennis ball arrangement may also be acquired by the sulcate apertures of monocotyledons (Fig. 16).

3. Transformation of the W-form into the 5-, 6-colpate and polycolpate conditions (Fig. 70: IV). One of the inclined W-colpi can undergo splitting into two branches like a zipper, giving rise to the 5-colpate pattern with inclined colpi (Figs. 23-27). Due to the geometry of the W-form, the splitting can occur in two mirror symmetrical ways. As a result, the left and the right 5-colpate variants appear (Fig. 70: IV.J). Then one of the inclined colpi of the 5-colpate form can split again (Fig. 70: V). In the case of 5-colpate form, the splitting can occur in four different ways and result in three isomers of the 6-colpate condition, one of which exists in left and right mirror symmetrical variants (Fig. 70: IV.J'). The isomers are distinguished by the order of alternation of the narrow and broad intervals between the apices of the inclined colpi. In the same way, the patterns of the 7-, 8- and the polycolpate isomers have been predicted and many of them have been found in the samples studied (Pozhidaev, in press). One can see that even 10-, 15-colpate isomers can be traced through the changing of the number of W-colpi and have the same initial form of symmetry (Fig. 28). Thus, the polycolpate isomer geometry is a direct consequence of the W-form symmetry, which may indicate the reality of the suggested mechanism of colpus multiplication.

4. Transformation of the W-form and 3-colpate form into the 6-rugate and polyrugate conditions (Fig. 70: VI-VIII). This transformation turns the meridional aperture arrangement into the pantoaperturate condition with a global arrangement of the apertures. The simplest pantarugate condition is the 6-rugate pattern with the six rugae arranged along the edges of a tetrahedron. This pattern may result from the W-form (Figs. 31-34; 70: VI) by the merging of the incipient elements of two additional rugae in each polar zone of the W-form (Fig. 70: VI.K). The same condition may arise from the 3-colpate condition (Figs. 29-30; 70: VII) by the splitting of the colpus ends in one of the apocolpia (Fig. 70: XV.L). All the other polyrugate patterns can arise from the initial 6-rugate form due to the subsequent division of the spaces between the rugae by the additional apertures (Figs. 35-40; 70: VIII). Monocot pollen can also have the polyrugate pattern (Fig. 41).

5. Transformation of the 6-rugate form into 6-porate and polyporate conditions (Fig. 70: IX). Some species have 6-porate pollen with oval-shaped pores in which the long axis extends along the tetrahedron edges like the rugae of the 6-rugate pattern (Figs. 42-45). This, probably, indicates a relationship between these two conditions. The pores of the initial 6-porate form can undergo a subsequent division into two, giving rise to a series of polyporate patterns (Figs. 46-49).

6. Transformation of zona-aperturate forms into the monocolpate, monosulcate and monosulculate conditions (Fig. 70: X-XII). The bending of the meridional ring-shaped colpus of the zona-aperturate form can reduce the size of one of the mesocolpia and change it to an operculum-like structure (Figs. 50-54; 70: X). An operculum-like mesocolpium may become very narrow. Although the forms of this series are very rare in the dicots (even as deviations) they demonstrate that dicot pollen can acquire an operculate sulcate condition, resembling operculate sulcate pollen, typical of some magnoliid dicot, and many monocot taxa (Figs. 55-56; 70: X.T). The operculum of this 'sulcate' form (probably homologous with one half of a zonasulcate form) can eventually be completely reduced, resulting in the monosulcate condition.

There are at least two more modes of transformation into the monoaperturate condition. Disruption to the ring-like zona-aperture on the pollen equator may give rise to the monocolpate condition with a meridional colpus orientation (Figs. 57-60; 70: XI). In addition, the zonasulculate pattern with a bent zonasulculus may undergo a similar lateral disruption (Figs. 61-63; 70: XII) resulting in the monosulculate condition with a characteristic slight shift of the sulculus from the perfectly aligned polar position (Fig. 64). It should be noted that a single rupture of the circular aperture in the polar zones has never been recorded with certainty in the samples studied.

7. Transformation of the zona-aperturate forms into the 2-colpate and disulculate condition (Fig. 70: XIII-XIV). This transformation remains doubtful because of the lack of forms with a single polar rupture of circular apertures, although the aperture arrangement (Fig. 70: XIII) and the tetrad positioning of 2-colpate pollen (Huynh, 1976) are very similar to that of the zona-aperturate form (Pozhidaev, 1998). The transformation of the zonasulculate form with a bent zonasulcus into the disulcate condition (Fig. 70: XIV) has not been clearly observed either.

8. Transformation of the monosulcate form into the trichotomosulcate and polychotomosulcate condition (Fig. 70: XV). The appearance of additional sulcus arms or the splitting of the sulcus ends of the monosulcate, or monosulculate, form (Fig. 70: XV.Y) changes this form into the trichotomosulcate and, subsequently, into the polychotomosulcate condition (Figs. 65-69).

Discussion

Pollen series from distantly related taxa have been found to overlap considerably: the final aperture configurations in the series from the same taxon may be the initial condition in the series from another taxon. This has allowed all of the series to be grouped into one branching succession (Fig. 71) which suggests that the main types of pollen apertures (colpi with meridional orientation, or the globally distributed pori and rugae of the eudicotyledons, or the polar (sulci) or latitudinal (sulculi) apertures in the monocotyledons and magnoliid dicotyledons) may result from a zona-aperturate form (zonasulcate or zonasulculate) by a gradual and geometrically regular transformation. Species with permanent tetrads allow tetrad orientation of deviant forms to be studied, and lend support to this idea. In the pollen of *Kielmeyera variabilis.* it has been observed that the circular colpus of the zona-aperturate form can have one or other of two possible meridional orientations within the tetrad (Pozhidaev, 1998, Fig. II, 16, 17). The meridional orientation of zonasulcate monocot pollen, and of the zona-aperturate condition in the dicotyledons suggests that the two conditions may be homologous. Furthermore, the fact that monocotyledonous taxa with zonasulcate pollen may also acquire the zona-aperturate derived patterns of dicotyledons testifies to the same homology, and provides a reason to consider the zona-aperturate condition, rather then the mono- or trichotomosulcate condition, as the initial pattern for pollen form variety as a whole.

There is the question of how to interpret the zonasulculate condition when the aperture has an equatorial rather than a meridional orientation (Figs. 73-74), because no series of deviant forms has been clearly recorded in the taxa with zonasulculate pollen. The flat (monoplanar) types of tetrad (tetrahedral and rhomboidal) in species with permanent tetrads, or during the tetrad period, are most convenient for the study of pollen orientation. The study of *Rapatea longipex* and *Eupomatia laurina* tetrads has shown that the zona-aperture pollen can have an intermediate type of orientation between the equatorial and the meridional one (Figs. 75, 78). In additional to equatorially oriented zonasulculate pollen, *Eupomatia laurina* has been found to have meridionally oriented zonasulcate pollen (Figs. 76-77) as well sometimes within the same tetrad. This demonstrates that orientation is not the principal distinction of the two types of zona-apertures: zonasulcate and zonasulculate apertures may be considered as two extreme modifications of one and the same aperture type. That is why the 3-sinaperture pollen of *Eupomatia* (Fig. 80), *Rhodospatha* (Fig. 11), and *Rapatea* (Fig. 13) and similar pollen series may, putatively, result both from the zonasulcate and zonasulculate or the intermediate condition in spite of their different tetrad orientation. However, the positioning in the tetrad of 3-aperture monocot pollen has not been clearly detected.

The resulting succession appears to differ from the generally accepted trends in pollen aperture evolution in a number of ways:

1. The zona-aperturate form rather than the mono- or trichotomosulcate form with a polar aperture, is suggested geometrically to be the basis of all other aperture types.
2. The derivation of meridional 3-colpate and distal mono- and trichotomosulcate is considered as two independent trends rather than successive steps.
3. The geometry of zona-aperturate derived patterns with tennis ball symmetry (4-colpate W-form and 6-rugate tetrahedron form), rather than the 3-radiate symmetry of the 3-colpate form, is considered to have more potential for further transformation into other types of eudicotyledonous aperture arrangement (zonocolpate and pantoaperturate)

It has also been found that the aperture arrangements which are deviant in the same species may be common in the other taxa, and it is very likely that any pattern of variety may be consolidated as a common one. For example, a polycolpate isomer, found to be a rare or very rare deviation in the Ranunculaceae, Sarraceniaceae, Papaveraceae and Lentibulariaceae (Pozhidaev, in prep.), is a common form in some species of *Echinopepon* (Cucurbitaceae) (Fig. 28). Another example is the tetrahedron-form which occurs as a deviation through many dicot families (Hippocastanaceae, Salicaceae) but is common in *Corydalis* (Fumariaceae), in the section *Empedoclea* of *Sideritis* (Lamiaceae) (Huynh, 1972) and in *Basella* (Basellaceae). It is interesting that in *Basella* the rare deviation is the 3-colpate pattern.

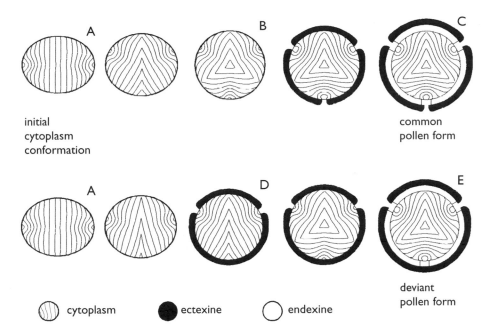

FIG. 72. Common and deviant aperture patterning. Upper row: A hypothetical spatial transformation in the developing microspore cytoplasm (A) is usually completed before the deposition of the exine (B), predetermining the common 3-colporate condition (C). Lower row: The formation of exine layers before these processes are completed (D) gives rise to the condition with a deviant aperture (E).

It is suggested that the pattern of variety of the morphological characteristics is the property which allows us to verify our ideas about the nature of evolutionary processes. The structure of the succession obtained exhibits three essential characteristics: the series is continuous, regular, and this regularity does not depend on the systematic position. These three characteristics indicate that the origin of the succession cannot be explained in terms of the theory of selective evolution because this approach is based on the typological idea of divergence and can describe only the origin of discrete and random diversity. The pattern continuity and the existence of all intermediate forms of pollen in recent plants means that no form has been eliminated by natural selection or by other causes in the course of their evolution. The regularity means that the source of pollen variability is not the process of stochastic mutation. Consequently, the series with such a pattern could not be described as a diversity (which should have a discrete and random pattern, by definition). They should be considered as a variety (which may have any pattern), whose origin can be explained without using the ideas of phylogenetic evolution like divergence, selection of adaptive advantages, specialisation, functional progress and parallelism or convergence, that is, the historical and functional approaches as a whole. The origin of the variety may be explained by the preformation of inner regularities of the morphological structure itself and the evolution of variety by changes to these regularities. The study of pollen grain geometry allows us to assume only that

221

these regularities are the realisation of pollen symmetry properties and that this is the symmetry of four conjugated spheres (for simultaneous tetrads) or two spheres (for successive ones). The high level of parallelism between common and deviant pollen forms cannot be accounted for by similar adaptations or convergence but by the fact that these regularities are common to a very high number of taxa. Following this idea, a particular pollen form may be considered as the initial one in terms of geometry only, but not historically.

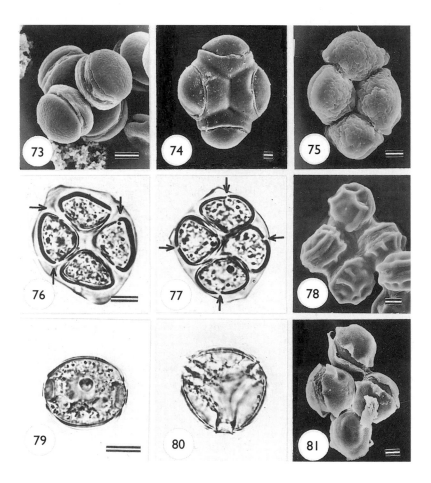

FIGS. 73–81. Tetrad orientation of zona-aperturate pollen. Fig. 73. *Nymphaea capensis*, zonasulculate pollen in a tetrahedral tetrad. SEM. Fig. 74. *Victoria regia*, zonasulculate pollen in a permanent rhomboidal tetrad. SEM. Fig. 75. *Kielmeyera variabilis*, zona-aperturate pollen with meridional aperture orientation (the circular colpi align with the frame of the photo) in a permanent rhomboidal tetrad. SEM. Figs. 76–80. *Eupomatia laurina*: Fig. 76. Zonasulculate pollen with equatorial aperture orientation (arrows) in a tetragonal tetrad, tetrad stage. LM. Fig. 77. Zonasulcate pollen with meridional aperture orientation (arrows) in a rhomboidal tetrad, tetrad stage. LM. Fig. 78. Zona-aperturate pollen with intermediate orientation in rhomboidal tetrad, SEM. Figs. 79–80. Zona-aperturate and 3-sinaperturate pollen, post-tetrad stage. LM. Fig. 81. *R. longipex*, zona-aperturate pollen with intermediate orientation in a rhomboidal tetrad. SEM. [Scale = 10 μm].

It is probable that the idea of transformation may, probably, not only be a convenient way of describing the pollen form series. For example, the transformation series for colporate pollen demonstrates some regularities in the endoaperture arrangement, which give indirect evidence that transformation may be used to explain deviant pollen form development and the morphological appearance of the series of variants (Pozhidaev, 1998). It has been found that, in the course of pollen form succession, the completion of endoaperture formation outstrips that of the colpus. As a result, in some deviant forms, although the endoaperture is in its usual (pre-destined) position, it is not under the colpus, as in a normal compound aperture, but under the un-interrupted ectexine layer (Figs. 6 & 18). Such strange endoaperture behaviour, coupled with the well-known fact that endoapertures are formed in the endexine, which is initiated after the ectexine and ectoapertures have already developed, allows us to suggest the following mechanism for aperture patterning: it is probable that the series of pollen forms depicted reflect spatial transformations in the developing microspore cytoplasm, the hypothetical process which predetermines the common aperture arrangement, usually completed before the beginning of exine deposition (Fig. 72.B). The initial stage of conformation (that cell cytoplasm acquires immediately after meiosis) may correspond to the zona-aperturate pattern (Fig. 72.A), while the final stage is the common aperture configurations (Fig. 72.C). Series of deviant pollen forms may arise if the formation of the exine starts before the process of conformation is completed. In this case the ectexine of the deviant forms, as the older layer, has the earlier stage of microspore cytoplasm conformation imprinted in the ectoaperture arrangement (Fig. 72.D). By the moment of endexine deposition, the cytoplasm reaches the next, more advanced stage of conformation, close to the final one in which the endoaperture takes its normal position (Fig. 72.E). Thus, pollen succession proves to be connected with the sequence of pollen wall developmental processes.

Conclusions

The mechanism of pollen aperture patterning which has been described, suggests that a microspore has a system for maintaining the spatial organisation and motion coordination of the cytoplasm. The possible nature of such a hypothetical system remains unknown. Furthermore, direct observation can be technically difficult due to the transience of these developmental stages and ephemerality of morphological structures.

Acknowledgments

I am particularly thankful to Dr. D.B. Archangelsky for making the materials of the Palynological Ikonotheca (Archangelsky, 1982) of the Palynological Laboratory at Komarov Botanical Institute available. My warm thanks are due to Dr. L.N. Smirnova for the reading of the English manuscript and Mrs. L.A. Kartzeva for the use of the SEM, (Laboratory of Electron Microscopy at the Komarov Botanical Institute).

References

Alexandrovsky, E.S. (1971). Development of ovules and microsporogenesis in the s species of *Juniperus*. *Botanical Journal* 2: 193–201. (in Russian, with English summary)

Archangelsky, D.B. (1982). Palynological Ikonotheka, its aims and tasks. *Botanicheski Zhurnal* **67**: 667–671 (in Russian, with English summary).

Blackmore, S. and Crane, P.R. (1998). The evolution of apertures in the spores and pollen grains of embryophytes. In: S.J. Owens and P.J. Rudall (editors), Reproductive biology, pp. 159–182. Royal Botanic Gardens, Kew.

Clarke, G.C.S. 1975. Irregular pollen grains in some *Hypericum* species. *Grana* **15**: 117–125

Dickinson, H.G. and Sheldon, J.M. (1986). The generation of patterning at the plasma membrane of young microspore of *Lilium*. In: S. Blackmore and I.K. Ferguson (editors). Pollen and spores: form and function, pp: 1–17. Royal Botanic Gardens, Kew.

Dover, G.A. (1972). The organization and polarity of pollen mother cell of *Triticum aestivum*. *Journal of Cell Science* **11**: 699–711.

Guingard, L. (1897). Les centres cinéticues chez les végétaux. *Annales des Sciences Natureles.* 8 Ser. Bot. t **6**: 177–220.

Harley, M.M. (1999). Tetrad variation: its influence on pollen form and systematics in the Palmae. In: M.H. Kurmann and A.R. Hemsley (editors). The evolution of plant architecture, pp. 289–304. Royal Botanic Gardens, Kew.

Heslop-Harrison, J. (1963). An ultrastructural study of pollen wall ontogeny in *Silene pendula*. *Grana Palynologica* **4**: 7–24.

Heslop-Harrison, J. (1971). Wall pattern formation in angiosperm microsporogenesis. In: Control Mechanisms of Growth and Differentiation. *Symposia of the Society for Experimental Biology* **25**: 277–300.

Hotchkiss, A.T. (1958). Pollen and pollination in the Eupomatiaceae. *Proceedings of the Linnean Society of New South Wales.* **83**: 86–91.

Huynh, K.-L. (1968). Etude de la morphologie du pollen du genre *Utricularia* L. *Pollen et Spores* **10**: 11–55.

Huynh, K.-L. (1970). Le pollen du genre *Anemone* et du genre *Hepatica* (Ranunculaceae) et leur taxonomie. *Pollen et Spores* **12**: 329–364.

Huynh, K.-L. (1972). Le pollen et la systématique du genre *Sideritis* L. (Labiatae).- *Bulletin du Museum National d'Histoire Naturelle.* 3ᵉ Serie, Botanique 1, **45**: 1–28.

Huynh, K.-L. (1976). Arrangement of some monosulcate, disulcate, trisulcate, dicolpate, and tricolpate pollen types in the tetrads, and some aspects of evolution in the angiosperms. In: I. K. Ferguson and J. Muller (editors). The evolutionary significance of the exine, pp. 101–124, Academic Press, New York and London.

Kuprianova, L.A. (1979). On the possibility of the development of tricolpate pollen from monosulcate. *Grana* **18**: 1–4.

Meeuse, A.D.J. (1965). The message of the pollen grains. Angiosperms - past and present. Inst. *Advancing Frontiers of Plant Science* **11**: 112–124.

Melville, R. (1981). Surface tension, diffusion and the evolution and morphogenesis of pollen aperture patterns. *Pollen et Spores* **23**: 179–203.

Mohl, H. von. (1835). Sur la structure et les formes des grains de pollen. *Annales des Sciences Natureles.* Ser. Bot. t **3**: 220–236.

Muller, J. (1970). Palynological evidence on early differentiation of angiosperms. *Biological Review* **45**: 417–450.

Nair, P.K.K. (1967). Pollen morphology with reference to the taxonomy and phylogeny of the Monochlamideae. *Review of Palaeobotany and Palynology* **3**: 81–91.

Pozhidaev, A.E. (1998). Hypothetical way of pollen aperture patterning. 1. Formation of 3-colpate patterns and endoaperture geometry. *Review of Palaeobotany and Palynology* **104**: 67–83.

Pozhidaev, A.E. (2000). Hypothetical way of pollen aperture patterning. 2. Formation of polycolpate patterns and pseudoaperture geometry. *Review of Palaeobotany and Palynology* **109**: 235–254.

Sampson, F.B. (1963). The floral morphology of *Pseudowintera*, the New Zealand member of the vesseless Winteraceae. *Phytomorphology* **13**: 403–423.

Sheldon, J.M. and Dickinson, H.G. (1983). Determination of patterning in the pollen wall of *Lilium henryi*. *Journal of Cell Science* **63**: 191–208.

Straka, H. (1963). Ueber die mögliche phylogenetische Bedeutung der Pollen-morphologie der madagassichen *Bubbia perrieri* R. Cap. (Winteraceae). *Grana Palynologica* **4**: 355–360.

Santisuk, T. (1979). A palynological study of the tribe Ranunculeae. *Opera Botanica* **48**: 1–74.

Takhtajan, A.L. (1959). Die Evolution der Angiospermen. G. Fischer, Jena.

Takhtajan, A.L. (1991). Evolutionary trends in flowering plants. Colombia University Press, New York.

Thanikaimoni, G. (1986). Pollen and spores: form and function. In: S. Blackmore and I.K. Ferguson (editors). Pollen and spores: form and function. pp. 119–136. Academic Press. New York and London.

Tiwari, S.C. (1989). Cytoskeleton during pollen development in *Tradescantia virginiana*: a study employing chemical fixation, freeze-substitution, immunofluorescence, and colchicine administration. *Canadian Journal of Botany* **67**: 1244–1253.

Van Campo, M. (1967). Pollen and classification. *Review of Palaeobotany and Palynology*. **3**: 65–71.

Van Campo, M. (1976). Patterns of pollen morphological variation within taxa. In: I.K. Ferguson and J. Muller (editors). The evolutionary significance of the exine, pp. 125–137. Academic Press, London.

Van Lemmeren, A.A.M., Keijzer, C.J., Willemse, M.T.M. and Kieft, H. (1985). Structure and function of the microtubular cytoskeleton during pollen development in *Gasteria verrucosa* (Mill.) H. Duval. *Planta* **165**: 1–11.

Walker, J.W. (1974). Aperture evolution in the pollen of primitive angiosperms. *American Journal of Botany* **61**: 1112–1137.

Walker, J.W., Doyle, J.A. (1975). The bases of angiosperm phylogeny: palynology. *Annals of the Missouri Botanical Garden* **62**: 644–723.

Waterkeyn, L. (1962). Les parois microsporocytaires de nature callosique chez *Helleborus* et *Tradescantia*. *Cellule* **62**: 224–255.

Wodehouse, R.P. (1930). The origin of the six-furrowed configuration of *Dahlia* pollen grains. *Bulletin of Torrey Botanical Club* **57**: 371–380.

Wodehouse, R.P. (1935). Pollen grains: their structure, identification and significance in science and medicine. McGraw Hill, New York.

Hesse, M., Weber, M. and Halbritter, H. (2000). A comparative study of the polyplicate pollen types in Arales, Laurales, Zingiberales and Gnetales. In: M.M. Harley, C.M. Morton and S. Blackmore (Editors). Pollen and Spores: Morphology and Biology, pp. 227–239. Royal Botanic Gardens, Kew.

A COMPARATIVE STUDY OF THE POLYPLICATE POLLEN TYPES IN ARALES, LAURALES, ZINGIBERALES AND GNETALES

MICHAEL HESSE, MARTINA WEBER, HEIDEMARIE HALBRITTER

Institute of Botany and Botanical Garden, University of Vienna, Rennweg 14, A-1030 Vienna, Austria

Polyplicate, inaperturate pollen grains of Arales (Araceae: *Spathiphyllum*, *Holochlamys*, *Amorphophallus*, *Pseudodracontium*, *Arisarum*, *Ambrosina*, *Steudnera*, *Protarum*, *Pistia*), Laurales (*Hortonia*, *Dahlgrenodendron*), Zingiberales (*Zingiber* sect. *Cryptanthium*), Gnetales (*Ephedra*), and additionally of some fossil ephedroid pollen types were analysed with regard to their pollen walls. By tabulating our findings, and the published information about fine structural details of the respective pollen wall layers, it can be clearly demonstrated that the various polyplicate types are only superficially similar. Pollen grains differ distinctly in morphology and chemical composition of the ridges as well as the underlying pollen wall stratum. The investigation concludes that pollen grains of *Hortonia*, *Dahlgrenodendron*, *Zingiber* sect. *Cryptanthium* are not resistant to decay. Pollen of some Araceae (for example *Pistia* and *Steudnera*) have a low fossilisation potential and, if at all preserved, will be found in a highly modified form. In contrast, the pollen grains of the Araceae taxa *Spathiphyllum* and *Holochlamys* have high fossilisation potential and should be present in the fossil record.

Introduction

Polyplicate, inaperturate pollen grains, often strongly resembling the "ephedroid" type, exist in several recent gymnosperm and angiosperm taxa (surprisingly the Arales - Araceae have at least 10 genera with polyplicate pollen). The polyplicate condition refers to the presence of many, at least 6 to more than 20, nearly identical prominent ridges ("ribs" or "bands"), which are parallel to the long axis of the pollen grain. A comparison of recent versus dispersed fossil polyplicate pollen grains in Spermatophyta , especially of Araceae, is of high interest, because from the position of a palynologist working stratigraphically it is understandable that all dispersed polyplicate pollen grains would be called ephedroids, and would be considered related to *Ephedra* L. (*cf.* Crane, 1996, for review and discussion.)

Dispersed polyplicate ephedroid pollen occur in sediments ranging from Lower Permian to Recent and are especially abundant in the Mid-Cretaceous. Of the numerous ephedroids only three genera have been examined using the transmission electron microscope (TEM): *Ephedripites mediolobatus* Bolkhovitina (Trevisan, 1980), *E. virginiaensis* Brenner and *E. multicostatus* Brenner (Kedves, 1994), *Equisetosporites chinleana* Daugherty (Zavada, 1984; Pocock and Vasanthy,

1988), and *Cornetipollis reticulata* Pocock and Vasanthy (Pocock and Vasanthy, 1988), all from Laurasian localities. The ephedroid pollen grains from the Lower Cretaceous of Brazil, a Gondwana locality, are most similar to *Ephedripites* (Osborn *et al.*, 1993). Sometimes ephedroid pollen are found *in situ*, for example, attached to the Early Cretaceous pre-flower organ of *Vithimantha* (Krassilov, 1997). According to Osborn *et al.* (1993) fossil ephedroid pollen grains may represent an artificial group: Osborn *et al.* (1993) stressed that ephedroid grains may share affinities with various modern taxa and the Gnetales or Araceae in particular. We have wondered if, therefore, some of the presumed Gnetalean dispersed pollen grains might be confused with angiosperm polyplicates.

The aim of this paper is twofold. First to show that all polyplicates are only superficially similar. Even in the Araceae the genera differ in morphology and chemical composition of the ridges or bands, and also the underlying wall stratum may be distinctly different. Second, apart from the Gnetales, to check the respective fossilisation potential of angiosperm polyplicates, and to draw attention to their possible attribution to Cretaceous and Tertiary dispersed ephedroids (see Hesse *et al.*, 1999). Identification of angiosperm polyplicate pollen may provide critical evidence of the geological age of these genera and would have an influence on stratigraphic diagrams.

Materials and Methods

The material was obtained from the Botanic Garden Munich, the Botanic Garden of the University of Vienna, and the Herbarium of the Institute of Botany, University of Vienna. The following Araceae taxa are illustrated: *Spathiphyllum blandum* Schott, *Spathiphyllum cannifolium* (Dryandr.) Schott, *Holochlamys beccarii* (Engl.) Engl., *Pseudodracontium siamense* Gagnepain, and *Pistia stratiotes* L., although a much larger sample of Araceae taxa was investigated (Weber *et al.*, 1999). For both TEM and SEM investigations the pollen material was prepared according to Weber *et al.* (1998), and Halbritter (1998), respectively.

Results

Descriptions of polyplicate ("ribbed" or "banded"), inaperturate pollen grains

In the following paragraphs characteristic features of polyplicate, inaperturate pollen grains from extant polyplicate taxa of Arales, Zingiberales, Laurales and Gnetales are listed. Fossil polyplicate pollen grains, which are assigned to *Ephedripites* Bolkhovitina, *Equisetosporites* Daugherty, and *Cornetipollis* Pocock and Vasanthy, are also considered.

Each genus is characterised by the respective nature of ridges and grooves, the exine stratification, and also by its presumed fossilisation potential and known fossil record. Presence or absence of representative micrographs is indicated. The interpretation of the respective exine stratification in the Araceae follows Weber *et al.* (1998, and 1999). The interpretation of *Ephedripites* stratification follows Osborn *et al.* (1993). The subfamily classification of Araceae follows Mayo *et al.* (1997). While *Pistia* is clearly an isolated taxon, the otherwise paired placement of Araceae genera is supported by French *et al.* (1995, Fig. 1). Tables 1 and 2 refer to typical features of polyplicate, inaperturate pollen grains in dry state before and after acetolysis: the respective nature of ridges and grooves, presence and form of ectexine and endexine, and the presumed fossilisation potential. The tables do not refer to features like long axis (overlapping dimensions) or "straight" versus "zig-zag-formed" ridges, which may depend on the preparation method, or the developmental stage of the specimen.

Spathiphyllum Schott (Araceae - Monsteroideae): The psilate tectate-columellate ridges are acetolysis-resistant, and represent ectexine. The ridges (striae) converge at both apices of the pollen grain forming psilate end plates (caps). Grayum (1992) refers to this type of ornamentation as a "football pattern". The grooves represent the endexine surface. The endexine is thin and without thickenings, but not compact. Long axis c. 33 µm. Tetrad arrangement mostly isobilateral (Huynh, 1972). SEMs: Grayum (1992), Weber *et al.* (1999), this paper (Fig. 1). TEMs: Lugardon *et al.* (1987/88), Trevisan (1980), Zavada (1990), Weber *et al.* (1999), this paper (Fig. 2). Fossil record: According to Muller (1981, 1984) since the Miocene (this is generally accepted, earlier reports are doubtful).

Holochlamys Engl. (Araceae - Monsteroideae): Stratification according to Lugardon *et al.* (1987/88) and to our findings in all details practically identical with *Spathiphyllum*. But the arrangement of the striae deviates significantly from that of *Spathiphyllum*. Grayum (1992), with regard to striation, describes converging and non-converging striae at the apices in the pollen of some *Holochlamys* species, and for example also in *Steudnera*. He calls this type of ornamentation the "baseball pattern" (as opposed to the "football pattern" in *Spathiphyllum*). Our micrographs of *H. beccarii* show a rather different situation: the striae (ridges) do not run towards, but encircle, the apices which consist of distinct ridges (Figs. 3, 4). From this it is clear that psilate apical plates, typical for *Spathiphyllum*, are lacking. Long axis c. 33 µm. SEMs: Lugardon *et al.* (1987/88), Grayum (1992), this paper (Figs. 3, 4). TEMs this paper (Fig. 5). No fossil record.

Protarum Engl. (Araceae - Aroideae): Stratification poorly known. Long axis c. 21 µm. SEMs: Grayum (1992). No TEMs. No fossil record. Possibly in some features similar to the closely related genus *Steudnera*.

Steudnera K. Koch (Araceae - Aroideae): An acetolysis-susceptible layer, forming the ridges and the grooves, covers the endexine. The faint ridges are not columellate and do not represent an ectexine. The endexine is thick and spongy. The grooves and the ridges are completely smooth. After acetolysis the remaining ovoid body appears totally smooth, although often collapsed. Long axis c. 44 µm. SEMs: Grayum (1992). No published TEMs. No fossil record.

Amorphophallus Blume (Araceae - Aroideae): Within this large genus many polyplicate species exist. The polyplicate/striate pattern is highly diverse. The ridges do not fuse at the apices into smooth caps. More than 60 species were investigated by SEM, and c. 11 species by TEM (van der Ham *et al.*, 1998). The grooves and ridges are always smooth and consist of an acetolysis-susceptible layer. Under this acetolysis-susceptible layer a thick, spongy, smooth endexine is found. An ectexine is lacking. After acetolysis, at least in *A. konjac* K. Koch and *A. elatus* Ridley, a psilate body remains, while, according to van der Ham *et al.* (1998), in other species the plicate pattern may not completely disappear. Long axis widely range between c. 30 - 80 µm. SEMs: van der Ham *et al.* (1998), Grayum (1992). TEMs: van der Ham *et al.* (1998). No fossil record.

Pseudodracontium N. E. Brown (Araceae - Aroideae): According to the data presented by van der Ham *et al.* (1998) and to our own results *Pseudodracontium* is quite similar to many polyplicate *Amorphophallus* species, with one exception. The many narrow ridges are arranged in the *Spathiphyllum* manner and fuse at the apices into psilate caps. The ridges are damaged by acetolysis, and the endexine is thick and spongy. Long axis c. 48 µm (c. 60 µm: van der Ham *et al.*, 1998). SEMs: van der Ham *et al.* (1998), this paper (Figs. 8, 9), Grayum (1992). TEMs: van der Ham *et al.* (1998), this paper (Fig. 10). No fossil record.

Table 1. Polyplicate, inaperturate pollen grains (in dry state) of some Arales - Araceae. While *Pistia* is clearly an isolated taxon, the otherwise paired placement of the Araceae genera follows Mayo et al. (1997); however, this pairing is supported by French et al. (1995, Fig. 1). The data on *Amorphophallus* and *Pseudodracontium* are from van der Ham et al. (1998). All other data are from our group in Vienna (for example Weber et al. 1998, and 1999).

Characters → Taxa ↓	Are the ridges resistant to acetolysis?	Nature of ridges and grooves		Ectexine (If ridges are acetolysis-susceptible: an ectexine is by definition lacking!)	Endexine	Fossilisation potential Which part of the pollen wall will become fossil?	Fossil record
		Ornamentation and structure of ridges	Form of grooves between ridges				
Spathiphyllum sp., **Holochlamys** sp. (Araceae)	Acetolysis resistant	Psilate ridges **columellate**	Grooves are foveolate-reticulate	Present (the ridges)	Thin, spongy	High (the total exine): some ridges may separate	From Miocene (perhaps even from Cretaceous), but seldom recognised?
Protarum sp., **Steudnera** sp., (Araceae)	*Protarum*: no data *Steudnera*: not acetolysis-resistant	Psilate ridges **not columellate**	Grooves psilate	*Protarum*: no data! *Steudnera*: lacking	*Protarum*: unknown *Steudnera*: spongy	*Protarum*: unknown *Steudnera*: rather low (the endexine)	No record, not to be expected
Amorphophallus sp. **Pseudodracontium** sp. (Araceae)	Not acetolysis resistant	Psilate ridges **not columellate**	Grooves psilate	Lacking	Thick, spongy	Rather low (only the endexine)	No record, not to be expected
Arisarum sp. **Ambrosina** sp. (Araceae)	*Arisarum*: (? <u>not</u>) acetolysis-resistant *Ambrosina*: not acetolysis-resistant	*Arisarum*: foveolate ridges **columellate** *Ambrosina*: psilate ridges **columellate**	Grooves not psilate	*Arisarum*: (generally?) lacking *Ambrosina*: lacking!	*Arisarum*: thick, spongy *Ambrosina*: thick, ?spongy	*Arisarum*: probably low (only the endexine in some species) *Ambrosina*: rather low (only the endexine)	No record (*Arisarum*: not recognised?) *Ambrosina*: not to be expected)
Pistia sp. (Araceae)	Not acetolysis resistant	Psilate ridges **not columellate**	Grooves psilate	Lacking	Thick, spongy	Rather low (only the endexine)	No record, not to be expected

Table 2. Polyplicate, inaperturate pollen grains (in dry state) of some Laurales, Zingiberales, Gnetales, and some fossil ephedroids. The data on *Zingiber* sect. *Cryptanthium* are from Liang (1988) and Theilade *et al.* (1993), the data on *Hortonia* from Sampson (1993), and the data on *Dahlgrenodendron* from van der Merwe *et al.* (1988), and Zavada (1990). The data on the three ephedroid taxa are from Trevisan (1980), Zavada (1984), Pocock and Vasanthy (1988), Osborn *et al.* (1993), and Kedves (1994). The data on *Ephedra* are from El-Ghazaly *et al.* (1998), El-Ghazaly *et al.* (1998) and from our group.

Characters → Taxa ↓	Nature of ridges and grooves			Ectexine	Endexine	Fossilisation potential	Fossil record
	Are the ridges resistant to acetolysis?	Ornamentation and structure of ridges	Form of grooves between ridges			Which part of the pollen wall will become fossil?	
Zingiber sect. **Cryptanthium** (*Zingiberaceae*)	Acetolysis-resistant	Psilate ridges **not columellate**	Grooves foveolate	Present (the ridges)	± Thin, spongy?	Very low (deleted by acetolysis treatment)	No record (not to be expected?)
Hortonia sp. (*Monimiaceae*)	Acetolysis-resistant	(Hollow) twisted, foveolate ridges **columellate**	Grooves psilate	Present (the hollow ridges)	Lacking	Low (the twisted ridges get separated)	No record (not to be expected?)
Dahlgrenodendron sp. (*Lauraceae*)	Acetolysis-resistant	Slightly ornamented ridges **columellate**	Grooves not psilate	Present (the ridges)	Lacking (?)	Very low (deleted by ultrasonic and acetolysis treatment)	No record (not to be expected?)
Ephedra sp. (*Gnetales*)	Acetolysis-resistant	(Straight or twisted) psilate ridges **not columellate**	No grooves, because a thin, but continuous ectexine is present	Present (the ridges and the granular infratectum)	Thin, throughout lamellate	High (the total exine)	Pleistocene (or since the Cretaceous?) to Recent
Ephedripites	Resistant to decay	(Straight or twisted) psilate ridges **not columellate**	No grooves, because a continuous, but almost disappeared ectexine is present	Present (the ridges and the granular infratectum)	Thin, faintly lamellate	High (the total exine)	Triassic to Tertiary
Equisetosporites chinleana	Resistant to decay	(Twisted) psilate ridges **columellate**	Grooves not psilate, tectum discontinuous in grooves	Present (ridges), but no granular infratectum	Distal pericline lamellae often degraded	High (the total exine or at least the ridges)	Triassic to Tertiary
Cornetipollis reticulata	Resistant to decay	Foveo-reticulate ridges **columellate**	Grooves not psilate degraded	Present (ridges), but no granular infratectum	No lamellate, sometimes degraded	High (the total exine or at least the ridges)	Upper Triassic

231

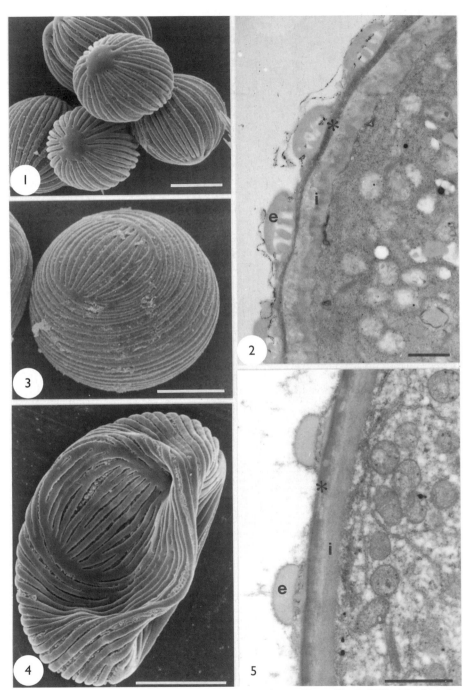

FIGS. 1–5. Polyplicate pollen grains of *Spathiphyllum* and *Holochlamys*. Fig. 1. SEM of *Spathiphyllum cannifolium* before acetolysis. Scale bar = 10 μm. Fig. 2. TEM of *Spathiphyllum blandum*, with acetolysis–resistant columellate ectexine (e), thin endexine (asterisk), and intine (i). Scale bar 1 μm. Fig. 3. SEM of *Holochlamys beccarii* before acetolysis. Scale bar = 10 μm. Fig. 4. SEM of *H. beccarii* after acetolysis. Scale bar = 10 μm. Fig. 5. TEM of *H. beccarii* with acetolysis–resistant columellate ectexine (e), generally thin endexine (asterisk), and intine (i). Scale bar = 1 μm.

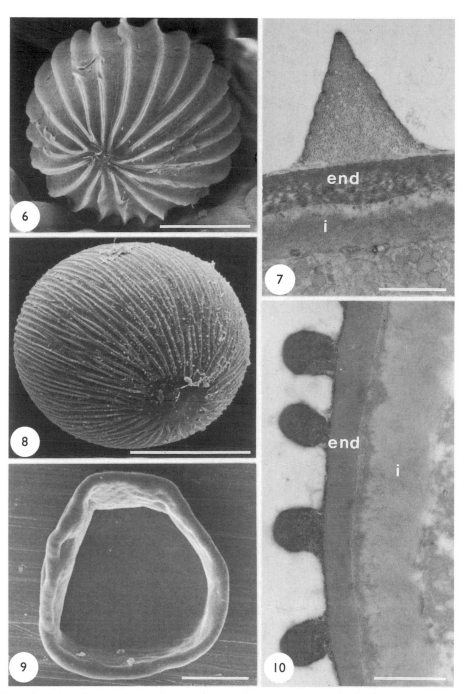

FIGS. 6–10. Polyplicate pollen grains of *Pistia* and *Pseudodracontium.* Fig. 6. SEM of *Pistia stratiotes* before acetolysis. Scale bar = 10 μm. Fig. 7. TEM of *Pistia stratiotes* with acetolysis–susceptible outer wall layer upon a spongy, thick endexine (end) and a thick intine (i). Scale bar = 1 μm. Fig. 8. SEM of *Pseudodracontium siamense* before acetolysis. Scale bar = 10 μm. Fig. 9. SEM of *Pseudodracontium siamense* after acetolysis. Scale bar = 10 μm. Fig. 10. TEM of *Pseudodracontium siamense* with acetolysis–susceptible outer wall layer upon a spongy, thick endexine (end) and a thick intine (i). Scale bar = 1 μm.

Arisarum Mill. (Araceae - Aroideae): (Note: Before acetolysis pollen of both species investigated appeared almost identical. However, after acetolysis, both species differed with respect to the ridges, this might be due to the preparation method. Acetolysed herbarium material of *A. vulgare* Targ.-Tozz. retained ridges in reduced form. In contrast, acetolysed fresh material of *A. proboscideum* (L.) Savi had lost the ridges.) The outer wall layer of *A. vulgare* is more or less resistant to acetolysis and forms the foveo-reticulate ridges and grooves, which are continuous with acetolysis-resistant smooth caps covering the apices of the pollen grains. As stated above, after acetolysis the plicate ornamentation of *A. proboscideum* pollen largely disappears. In unacetolysed material this outer wall layer consists of tectum-like structures on vertically arranged thin, columella-like structures below, while a thick, smooth, spongy endexine is present. Long axis c. 44 μm. SEMs: Grayum (1992), No published TEMs. No fossil record.

Ambrosina Bassi (Araceae - Aroideae): An outermost layer, not resistant to acetolysis, consists of smooth columellate ridges and the grooves. The ridges are arranged as in *Spathiphyllum*. Long axis c. 44 μm. TEMs are lacking, so the presence and nature of an endexine is unclear. According to Thanikaimoni (1969) the ridges get easily separated and become destroyed. After acetolysis a collapsed, but totally smooth, body remains. SEMs: Grayum (1992). No published TEMs. No fossil record.

Pistia L. (Araceae - Aroideae): This monotypic genus is placed on its own in all classifications of Araceae. An outermost layer, consisting of smooth ridges and grooves, covers a smooth, thick, spongy endexine. This outer layer is not resistant to acetolysis, and does not represent an ectexine. After acetolysis a collapsed, but totally smooth body remains. During acetolysis and even before the bands often become detached from the underlying stratum. The ridges are not columellate and are arranged in the *Spathiphyllum* pattern. Long axis c. 27 μm. SEMs: Grayum (1992), Weber *et al.* (1999), this paper (Fig. 6). TEMs: Weber *et al.* (1999), this paper (Fig. 7). No fossil record.

Zingiber sect. *Cryptanthium* Horan. (Zingiberaceae): Pollen grains were sectioned by Theilade *et al.* (1993). The thin exine appears to be bi-layered: the dark-staining outer stratum is represented by the smooth ridges, while the inner, non-homogenous stratum forms the narrow grooves. The thick inner layer below the non-homogenous stratum is interpreted by the authors as bi-layered intine. The (probably acetolysis-resistant) ridges are not columellate (Theilade *et al.*, 1993). The pollen grains do not resist acetolysis (Liang, 1988).

Hortonia floribunda Wight (Monimiaceae - Laurales): The only polyplicate taxon with hollow, acetolysis-resistant ridges, consisting of a tectum, some sort of columellae, and a thin foot layer. The hemispherical bands give rise to the typical pollen form of *Hortonia floribunda* Wright. Below the hemispherical bands a discontinuous layer, interpreted as part of the ectexine, is found. According to Sampson (1993) an endexine is lacking. Long axis 22-35 μm. SEMs and TEMs: Sampson (1993). No fossil record.

Dahlgrenodendron natalense (J. H. Ross) J. J. M. van der Merwe and Van Wyck (Lauraceae-Laurales): The slightly ornamented, columellate ridges (bands) of *D. natalense* are most probably resistant to acetolysis, but according to van der Merwe *et al.* (1988), the pollen grains get destroyed during ultrasonic treatment and acetolysis. The grooves are not psilate and represent the surface of the intine. This is consistent with the interpretation by Zavada (1990) that an exinous basal layer is lacking. Long axis 25-40 μm. SEMs: van der Merwe *et al.* (1988), Zavada (1990). TEMs: Zavada (1990). No fossil record.

Ephedra (Gnetaceae - Gnetales): The psilate ridges (with a granular infratectum) are arranged in the football pattern. Ridges and infratectal granules are acetolysis-resistant and represent the ectexine. The ridges are not columellate, but spatially fused (the tectum is continuous in the furrows according to Pocock and Vasanthy, 1988). The ridges converge at both pollen grain apices, but do not form psilate end plates. The endexine is-lamellated throughout. Tetrad arrangement mostly tetrahedral (El-Ghazaly *et al.*, 1998). Long axis c. 45 *μ*m. In *Ephedra* before pollen tube formation the exine is detached and curls into a scroll-like configuration with the ridges and grooves *perpendicular* to the long axis (El-Ghazaly *et al.*, 1998); there is no tendency to separate ectexine and endexine. SEMs and TEMs: El-Ghazaly and Rowley (1997), El-Ghazaly *et al.* (1998). Fossil record: *Ephedra* is currently viewed as a recent taxon, accepted since the Pleistocene (but see also the next paragraphs).

Ephedripites: This fossil pollen form, as pointed out by Zavada (1984), may not form a natural group. In the past, dispersed fossil polyplicate pollen was usually assigned to *Ephedripites* (Van Konijnenburg-Van Cittert, 1992). From this it is understandable that *Ephedripites* pollen grains are widespread and have been recorded from the Triassic onwards. There is a remarkable diversity in appearance, which exceeds that of living *Ephedra* (Hughes, 1994); for example, some polyplicate pollen grains with conspicuous psilate end plates (caps) were described from the Paleocene of Colombia as *Ephedripites vanegensis* by Van der Hammen and Garcia de Mutis (1966); some small, ephedroid-like pollen grains with straight ridges, reported from the Cretaceous of Japan by Takahashi *et al.* (1995), are also assigned to *Ephedripites*. According to Zavada (1984), the Triassic Chinle form *Equisetosporites chinleana* likewise falls within the morphological circumscription of *Ephedripites*. This is a complex question: see for example Pocock and Vasanthy, 1988, for circumscriptions of dispersed ellipsoidal to spheroidal palynomorphs. So far only four species have been investigated by TEM (Trevisan, 1980; Zavada, 1984; Kedves, 1994). Practically all the features seen by TEM are very similar to recent *Ephedra*, or to fossil pollen grains formerly assigned to the extant genus *Ephedra* (*cf.* Osborn *et al.*, 1993). According to Trevisan (1980), Zavada (1984), Van Konijnenburg-Van Cittert (1992) and Kedves (1994) the ectexine of the investigated *Ephedripites* pollen grains almost disappears in the furrows, but is continuous as in *Ephedra* (El-Ghazaly and Rowley, 1998; El-Ghazaly *et al.*, 1998), and so there is no difference to *Ephedra*, as claimed by Pocock and Vasanthy (1988). Another difference, the non-granular infratectum reported by Pocock and Vasanthy, is correctly questioned by Osborn *et al.* (1993): a granular infratectum is present as in *Ephedra*. The endexine is faintly lamellated, but in principal the same as *Ephedra*: the lamellae may get compressed during fossilisation. Long axis c. 20 - 50 *μ*m. SEMs: Osborn *et al.* (1993). TEMs: Trevisan (1980). Fossil record: Triassic to Tertiary.

Equisetosporites, E. chinleana Daugherty, type species: psilate, twisted ridges (bands) with distinct columellae represent the ectexine. Traverse (1988) denotes the spiral-striate forms of American south-west Triassic palynofloras as *Equisetosporites*. Some Japanese Cretaceous polyplicate pollen with spiral ridges were associated with *Equisetosporites* by Takahashi *et al.* (1995). The spiral bands are often separated or poorly connected, but also found with an inner core appearing to be intact and the bands obliquely oriented (Pocock and Vasanthy, 1988, in their key to polyplicate types describe the tectum as discontinuous in the furrows). Scott (1960) describes similar polyplicate grains from the Upper Triassic: the endexine separated easily from the ectexinous ridges, and could be detached and shrunken. According to Pocock and Vasanthy (1988) the "unique" *Equisetosporites* endexine shows some distal periclinal lamellae, but under the lamellae the endexine is compact, and the exine is columellate. The proximally compact endexine and the presence of columellae might be good ultrastructural reasons to separate *Equisetosporites* from *Ephedripites*. Long axis c. 40 - 80 *μ*m. SEMs: Pocock and Vasanthy (1988), Zavada (1990), Takahashi *et al.* (1995). TEMs: Pocock and Vasanthy (1988), Zavada (1984, 1990). Fossil record: Triassic to Tertiary.

Cornetipollis reticulata Pocock and Vasanthy: This fossil species describes the reticulately banded pollen of palynomorphs included in the species *Equisetosporites chinleana* (Pocock and Vasanthy, 1988). The foveolate-reticulate ridges (bands) of *Cornetipollis reticulata* are columellate and represent the ectexine. The endexine, which is not lamellated, is often degraded, and only the interconnected ridges are found. The grooves, if present, are not smooth and represent the surface of the endexine. Long axis c. 75 μm. SEMs and TEMs: Pocock and Vasanthy (1988). Fossil record: Upper Triassic.

Discussion

Conspectus of polyplicate pollen grains

Already Trevisan (1980) and Pocock and Vasanthy (1988) have addressed the question of convergence in the polyplicate pollen type. Our results show that polyplicate, inaperturate pollen grains of Spermatophyta are merely superficially similar, and that only a few of all polyplicate pollen types are resistant to decay. Polyplicate, inaperturate pollen is found in some unrelated taxa of the Spermatophyta including Gnetales (*Ephedra*), Laurales (Monimiaceae: *Hortonia,* Lauraceae: *Dahlgrenodendron*), Zingiberales (Zingiberaceae: *Zingiber* sect. *Cryptanthium*) and Arales (various Araceae: *Spathiphyllum, Holochlamys, Amorphophallus, Pseudodracontium, Arisarum, Ambrosina, Steudnera, Protarum, Pistia*). Dispersed fossil inaperturate polyplicate pollen is described generally as ephedroid pollen, and is usually assumed to be related to extant *Ephedra* and *Welwitschia*. Only Van der Hammen and Garcia de Mutis (1966) confer the Palaeocene *Ephedripites vanegensis* with *Spathiphyllum* because of its large smooth pollen caps, which are absent in *Ephedra*. (Note: *Welwitschia* pollen is clearly monosulcate, not inaperturate, and has parallels in the fossil record, for example in the Barremian of the UK:, *c.f.* Hughes, 1994. Satish K. Srivastava, this volume, discusses its palaeodistribution under the generic name *Jugella*).

The degree of external analogy is very variable. Some polyplicate taxa have a nearly identical appearance and may be mixed up. This is the case with some *Ephedra* species and some fossil ephedroids, when compared with *Spathiphyllum,* especially with *S. kalbreyeri* Bunting, because the *Spathiphyllum* ridges withstand acetolysis and have a high fossilisation potential. Pollen grains of *Pistia* also have a superficially similar appearance to *Ephedra,* as do some *Amorphophallus* species, but only before acetolysis; after acetolysis the ridges are gone. The other polyplicates, especially *Hortonia, Dahlgrenodendron, Zingiber* sect. *Cryptanthium,* but also *Holochlamys* (Araceae), have their own specialised characters and cannot be confused with the *Ephedra*-pollen type.

The inaperturate, polyplicate condition, the only character common to all taxa, is a superficial similarity. Profound, unique, but hidden differences concerning form, for example columellate - not columellate, and the chemistry of the ridges, and of the endexine, are masked by this superficial similarity. The ridges of Gnetales, Laurales, Zingiberales are resistant to acetolysis, while all the ephedroid forms resist decay and fossilise. This is only partly the case in Araceae because some taxa have acetolysis-resistant ridges and, simultaneously, a mostly thin endexine, while others have an acetolysis-susceptible, tapetum-borne, polysaccharidic layer and, simultaneously, a thick, spongy endexine. From this striking difference in a superficially similar feature it is concluded that in Araceae the polyplicate pollen type has evolved more than twice (Grayum, 1992): once in the Monsteroideae (tribe Spathiphylleae), and at least twice in the Aroideae (Mayo *et al.,* 1997): in the tribes Thomsonieae (*Amorphophallus* and *Pseudodracontium*), Arisareae, Ambrosineae, Colocasieae (*Protarum* and *Steudnera),* and Pistieae. The pollen features, listed in Table 1 in abbreviated form, also support the close relationships respectively of *Spathiphyllum* and *Holochlamys,* of *Protarum* and *Steudnera,* of *Amorphophallus* and *Pseudodracontium,* and even give some support to the placement of *Arisarum* and *Ambrosina* into different tribes by Mayo *et al.* (1997).

The differences in the chemical composition of the respective ridges, and in pollen wall stratification result in the fact that the different polyplicate pollen types have different fossilisation potentials (Hesse *et al.*, in press). The polyplicate Gnetales, the ephedroid genera, and few Araceae fossilise as a whole, and have the potential to be found in the fossil record. The corresponding polyplicate pollen grains of the Laurales and the Zingiberales have, for various reasons outlined above, no fossilisation potential. Similarly, the majority of Araceae, for example the polyplicate *Pistia*, have a reduced fossilisation potential. This is a major reason why fossil ephedroids in a wider sense, with the documented exception of *Spathiphyllum*, are only connected with *Ephedra*, and not with angiosperms.

There is a possible confusion concerning dispersed fossil pollen of *Ephedra* (*Ephedripites* and *Equisetosporites*) and *Spathiphyllum*. According to Huynh (1975) *Ephedra* and *Spathiphyllum* share six parallel features, including the (alleged) same tetrad arrangement. In fact there are only four parallel features: 1. the presence of acetolysis-resistant ridges, 2. the whole exine is resistant to decay, 3. the evenly thick intine, and 4. the inaperturate condition. However, the two genera are separated by five different features: 1. ridges not columellate (*Ephedra*) versus columellate ridges in *Spathiphyllum*, 2. endexine lamellated and evenly thickened (*Ephedra*) or not lamellated and thin (*Spathiphyllum*), 3. in *Ephedra* the endexine is always totally covered by the continuous ectexinous ridges, while in *Spathiphyllum* the transversally discontinuous ridges do not cover the endexinous grooves, 4. plicate pollen caps in *Ephedra* lacking, or much smaller than in *Spathiphyllum*, and 5. tetrad arrangement in *Ephedra* mostly tetrahedral (El-Ghazaly *et al.*, 1998), in *Spathiphyllum* isobilateral (Huynh 1975, Thanikaimoni 1969). So confusing *Spathiphyllum* and *Ephedra* pollen grains is unlikely, given well preserved pollen investigated by SEM at a magnification of c. 5000: at least form and nature of the ridges are clearly separating features.

Conclusion

We conclude, that the differences in exine composition and stratification, impart a variable resistance to decay, between the polyplicate grains of the taxa studied, and give rise to a very different fossilisation potential. The polyplicate pollen grains of the Laurales (*Hortonia, Dahlgrenodendron*) and also of Zingiberales (*Zingiber* sect. *Cryptanthium*) cannot be found in the fossil record. They will have been completely destroyed because there is not one stable stratum within the whole pollen wall, therefore no fossilisation potential. Pollen of *Spathiphyllum* and *Holochlamys*, of *Ephedra* and the corresponding dispersed ephedroids have excellent fossilisation potential because of an acetolysis-resistant ectexine which covers a continuous endexine. Pollen grains of *Amorphophallus, Pseudodracontium, Arisarum, Ambrosina, Steudnera* and *Pistia* will be found only in a highly modified form. After the dissolution of the ridges in *Pistia*, in *Steudnera,* in *Ambrosina,* in *Pseudodracontium,* in the many polyplicate *Amorphophallus* species, and also in *Arisarum* (in the latter two genera sometimes slow or incomplete), the smooth, thick, spongy endexine will remain as a psilate, but collapsed, indistinct ovoid body without genus or species specific characters. All these pollen types have a much lesser fossilisation potential if compared with *Spathiphyllum*. Muller (1981) pointed out that pollen from *Ambrosina* is "expected" to occur in a polyplicate condition, but in fact this pollen, if at all preserved, will not appear in a polyplicate, but in a psilate condition!

Given that the exine of *Spathiphyllum, Holochlamys* and also of some *Arisarum* species is resistant to acetolysis treatment and therefore presumably resistant to exine decay, pollen of the *Spathiphyllum-* , *Holochlamys-* and perhaps also of the *Arisarum-*type should be present in the fossil record. Identification of such fossil pollen may provide critical evidence of the age of these genera and of the Araceae.

Acknowledgements

We are very grateful to Andrea Frosch-Radivo for skilful technical assistance and to Josef Bogner (Munich) for generously providing us with material. We thank Madeline Harley, David Ferguson and an anonymous reviewer for helpful comments on the manuscript.

References

Crane, P.R. (1996). The fossil history of the Gnetales. *International Journal of Plant Science* 157 (6 Suppl.): S50-S57.

El-Ghazaly, G. and Rowley, J.R. (1997). Pollen wall of *Ephedra foliata. Palynology* 21: 7–18.

El-Ghazaly, G., Rowley, J.R. and Hesse, M. (1998). Polarity, aperture condition and germination in pollen grains of *Ephedra* (Gnetales). *Plant Systematics and Evolution* 213: 217–231.

French, J.C., Chung, M.G. and Hur, Y.K. (1995). Chloroplast DNA phylogeny of the Ariflorae. In: P. Rudall, P.J. Cribb, D.F. Cutler and C.J. Humphries (editors). Monocotyledons, Systematics & Evolution, vol. 1, pp. 255–275. Royal Botanic Gardens, Kew.

Grayum, M. (1992). Comparative external pollen ultrastructure of the Araceae and putatively related taxa. Monographs in Systematic Botany from Missouri Botanical Garden 43: 1–167.

Halbritter, H. (1998). Preparing living pollen material for scanning electron microscopy using 2,2–dimethoxypropane (DMP) and critical–point drying. *Biotechnic and Histochemistry* 73: 137–143.

Hesse, M., Weber, M. and Halbritter H. (1999). Pollen walls of Araceae, with special reference to their fossilisation potential. *Grana* 38: 19–25.

Hughes, N.F. (1994). The enigma of angiosperm origins. Cambridge University Press, Cambridge.

Huynh, K.-L. (1972). Le problème de la polarité du pollen d'*Ephedra. Pollen et Spores* 16: 469–474.

Huynh, K.-L. (1975). Quelques phénomènes de polarité du pollen à plis multiples du genre *Spathiphyllum* (Araceae). *Beiträge zur Biologie der Pflanzen* 50: 445–456.

Kedves, M. (1994). Transmission electron microscopy of the fossil gymnosperm exines. Szeged.

Krassilov, V.A. (1997). Angiosperm origins: morphological and ecological aspects. Pensoft, Sofia and Moscow.

Liang, Y.-H. (1988). Pollen morphology of the family Zingiberaceae in China - pollen types and their significance in taxonomy. *Acta Phytotaxonomica Sinica* 26: 265–281.

Lugardon, B., Lobreau–Callen, D. and Le Thomas, A. (1978/88). Structures polliniques chez les Araceae - 1, tribu des Spathiphylleae. *Journal of Palynology* 23–24: 51–57.

Mayo, S.J., Bogner, J. and Boyce, P.C. (1997). The genera of Araceae. Royal Botanic Gardens, Kew.

van der Merwe, J.J.M., van Wyk, A.E. and Kok, P.D.F. (1988). *Dahlgrenodendron,* a remarkable new genus of Lauraceae from Natal and Pondoland. *South African Journal of Botany* 54: 80–88.

Muller, J. (1981). Fossil pollen records of extant Angiosperms. *The Botanical Review* **47**: 1–142.

Muller, J. (1984). Significance of fossil pollen for angiosperm history. *Annals of the Missouri Botanical Garden* **71**: 419–443.

Osborn, J.M., Taylor, T.N. and de Lima, M.R. (1993). The ultrastructure of fossil ephedroid pollen with gnetalean affinities from the Lower Cretaceous of Brazil. *Review of Palaeobotany and Palynology* **77**: 171–184.

Pocock, S.A.J. and Vasanthy, G. (1988). *Cornetipollis reticulata*, a new pollen with angiospermid features from Upper Triassic (Carnian) sediments of Arizona (U.S.A.), with notes on *Equisetosporites*. *Review of Palaeobotany and Palynology* **55**: 337–356.

Sampson, F.B. (1993). Pollen morphology of the Amborellaceae and Hortoniaceae (Hortonioideae: Monimiaceae). *Grana* **32**: 154–162.

Scott, R.A. (1960). Pollen of *Ephedra* from the Chinle formation (Upper Triassic) and the genus *Equisetosporites*. *Micropaleontologia* **6**: 271–276.

Takahashi, M., Takai, K. and Saiki, K. (1995). Ephedroid fossil pollen from the Lower Cretaceous (Upper Albian) of Hokkaido, Japan. *Journal of Plant Research* **108**: 11–15.

Tarasevich, V.F. (1988). Peculiarities of morphology in ridged pollen grains in some representatives of Araceae. In: A.F. Chlonova (editor) Palynology in the USSR, papers of the Soviet palynologists to the VII International Palynological Congress Brisbane, Australia, 1988. Nauka Novosibirsk, 1988, pp. 58–61.

Thanikaimoni, G. (1969). Esquisse palynologique des Aracées. Institut Français de Pondichéry, Travaux de la Section Scientifique et Technique **5**: 1–31.

Theilade, I., Mœrsk–Møller, M.–L., Theilade, J. and Larsen, K. (1993). Pollen morphology and structure of *Zingiber* (Zingiberaceae). *Grana* **32**: 338–342.

Traverse, A. (1988). Palaeopalynology. Unwin Hyman, Boston.

Trevisan, L. (1980). Ultrastructural notes and considerations on *Ephedripites*, *Eucommiidites* and *Monosulcites* pollen grains from Lower Cretaceous sediments of southern Tuscany (Italy). *Pollen et Spores* **22**: 85–132.

van der Ham, R.W.J.M., Hetterscheid, W.L.A. and van Heuven, B.J. (1998). Notes on the genus *Amorphophallus* (Araceae) - 8. Pollen morphology of *Amorphophallus* and *Pseudodracontium*. *Review of Palaeobotany and Palynology* **103**: 95–142.

van der Hammen, T. and Garcia de Mutis, G. (1966). The Palaeocene pollen flora of Colombia. *Leidse Geologische Mededelingen* **35**: 105–116.

van Konijnenburg-van Cittert, J.H.A. (1992). An enigmatic Liassic microsporophyll, yielding *Ephedripites* pollen. *Review of Palaeobotany and Palynology* **71**: 239–254.

Weber, M., Halbritter, H. and Hesse, M. (1998). The spiny pollen wall in *Sauromatum* (Araceae) - with special reference to the endexine. *International Journal of Plant Science* **159**: 744–749.

Weber, M., Halbritter, H. and Hesse, M. (1999). The basic pollen wall types in Araceae. *International Journal of Plant Science* **160**: 415–423.

Zavada, M. (1984). Angiosperm origins and evolution based on dispersed fossil pollen ultrastructure. *Annals of the Missouri Botanical Garden* **71**: 444–463.

Zavada, M. (1990). The ultrastructure of three monosulcate pollen grains from the Triassic Chinle formation, western United States. *Palynology* **14**: 41–51.

Van der Ham, R.W.J.M., Hetterscheid, W.L.A., van Heuven, B.J. and Star, W. (2000). Exine architecture in echinate pollen of *Amorphophallus* (Araceae) in relation to taxonomy. In: M.M. Harley, C.M. Morton and S. Blackmore (Editors). Pollen and Spores: Morphology and Biology, pp. 241–248. Royal Botanic Gardens, Kew.

EXINE ARCHITECTURE IN ECHINATE POLLEN OF *AMORPHOPHALLUS* (ARACEAE) IN RELATION TO TAXONOMY

RAYMOND W.J.M. VAN DER HAM[1], WILBERT L.A. HETTERSCHEID[2], BERTIE JOAN VAN HEUVEN[1] AND WIM STAR[1]

[1]Rijksherbarium / Hortus Botanicus, P.O. Box 9514, 2300 RA Leiden, The Netherlands
[2]Chrysantenstraat 28, 1214 BM Hilversum, The Netherlands

Abstract

Exine architecture in all five *Amorphophallus* species with echinate pollen has been described using light and electron microscopical data. A subdivision based on ectexine ultrastructure correlates well with the three subtypes established earlier on the basis of spine features. Together, these data sets support the preliminary taxonomical treatment of accommodating the five echinate *Amorphophallus* species in three infrageneric groups. The ectexine ultrastructure in one of the three subtypes resembles that in *Arisaema*, which genus is possibly closely related to *Amorphophallus*.

Introduction

Amorphophallus Blume is a genus of palaeotropical aroids (Araceae) occurring in Africa, Madagascar, India, continental SE Asia, Malesia and NE Australia. At present about 170 species are recognised (Hetterscheid and Ittenbach, 1996; Mayo *et al.*, 1997). Van der Ham *et al.* (1998) made an extensive light and scanning electron microscopical study of the pollen of *Amorphophallus*, including 145 species, of which 28 species were studied with transmission electron microscopy. *Amorphophallus* pollen is always inaperturate, but it is highly diverse as to size (28-88 µm) and exine surface: echinate, verrucate, areolate, striate, fossulate, scabrate and psilate, which makes *Amorphophallus* the most diverse genus in the family. The exine is also ultrastructurally remarkable. It consists of a distinct, acetolysis-resistant endexine and a usually non-acetolysis-resistant ectexine, which is a rare condition in monocotyledons. No differentiation into tectum and infratectum is apparent. Besides showing a variety of surface patterns, the ectexine of a number of species contains variously sized, shaped and distributed dark (osmiophilic) granules of still unknown composition. So far, five *Amorphophallus* species, four African and one Madagscan, were found to have echinate pollen. All five species could be studied with light and scanning electron microscopy, but only one of them with transmission electron microscopy. In the present paper the exine ultrastructure in three more species with echinate pollen is described and compared with earlier results. The aim is to obtain additional data in order to test conclusions, based on light and scanning electron microscopy and macromorphology, designating the *Amorphophallus* species with echinate pollen as a heterogeneous group.

Material and Methods

Each of the species with echinate pollen is represented by a single collection:

A. antsingyensis Bogn., Hett. et Ittenb., Hetterscheid H.AM.099 (L), Madagascar
A. aphyllus (Hook.) Hutch., Koenen s.n. (BONN), Senegal
A. doryphorus Ridl., Eggers s.n. (K), Senegal
A. dracontioides (Engl.) N.E.Br., Geerlings and Bokdam 2214 (WAG), Ivory Coast
A. elliottii Hook.f., Scott Elliott 4640 (K), Sierra Leone

All samples were processed and studied with light (LM) and scanning electron microscopy (SEM), and, except for that of *A. elliottii*, also with transmission electron microscopy (TEM). Pollen of *Amorphophallus* usually does not resist acetolysis. Therefore, the material was boiled in water for 10 minutes. Then, the pollen grains could be freed by breaking the anthers, and transferred via alcohol to a glycerin/alcohol 96% mixture (2/1). The pollen was mounted in glycerin jelly for LM observation. For SEM the pollen was mounted on stubs and sputter-coated with gold. Preparation for TEM included rehydration and fixing in 0.1% glutaraldehyde (1 week), followed by fixation with 1% OsO_4 (1 hour), prestaining with 1% uranyl acetate during dehydration, embedding in 3/7 Epon (Luft), poststaining with 5% uranyl acetate (10 minutes) and Reynolds' lead citrate (10 minutes).

Pollen grain size is given as the largest diameter (L). The ratio L/B (largest/smallest diameter) is given as an indication of pollen grain shape.

Weber *et al.* (1998) have demonstrated that the spines in the araceous genus *Sauromatum* are of amoeboid-tapetal origin, polysaccharidic in nature and do not resist acetolysis. An acetolysis-resistant ectexine was considered to be missing. Non-resistance seems to be common in *Amorphophallus*, but in some species the outer wall elements more or less withstand acetolysis (Van der Ham *et al.*, 1998), indicating at least some incorporation of sporopollenin. Homology with resistant tectate/columellate ectexines of other Araceae was assumed, regardless of low resistance to acetolysis in most cases. Data concerning acetolysis-resistance of the outer wall layer/elements in echinate *Amorphophallus* pollen are lacking so far, and as long as the non-homology of the outer layer/elements with (part of) the ectexine is not established, it is preferred to denote them as ectexinous.

Results

In contrast to most other surface types in *Amorphophallus*, the echinate type seems to be well-delimited. Occasionally, psilate pollen grains show one or a few spine-like protrusions (see Van der Ham *et al.*, 1998), but no further resemblance exists. Echinate *Amorphophallus* pollen is medium-sized (L = 34.8–46.0 μm, av. 38.9 μm) and sub-spheroidal (L/B = 1.06–1.24, av. 1.14). Variation occurs in spine features and in the ultrastructure of the exine.

A. aphyllus (Figs. 1, 2, 9)
LM: pollen grains 34.8 μm, subspheroidal (L/B = 1.24).
SEM: spines 1–2 μm long, relatively short, unstoreyed; surface between spines scabrate.
TEM: exine thickness between spines 1.4 μm; endexine 1.2 μm thick, consisting of a

homogeneous outer sub-layer, a middle sub-layer with an indistinctly radial dark pattern and a homogenous inner sub-layer; ectexine 0.2 μm thick, consisting of a homogeneous, locally undulating outer sub-layer that is continuous with the spine bases, and a thicker finely granular basal sub-layer that extends under the spines; spines homogeneous, sometimes with a dark speck near the apex.

A. dracontioides (Figs. 3, 10)
LM: pollen grains 38.2 μm, subspheroidal (L/B = 1.21).
SEM: spines 1–2 μm long, relatively short, unstoreyed; surface between spines scabrate.
TEM: exine thickness between spines 1.5 μm ; endexine 1.3 μm thick, not stratified; ectexine 0.2 μm thick, consisting of a homogeneous, locally undulating outer sub-layer that is continuous with the spine bases, and a slightly thicker finely granular basal sub-layer that extends under the spines; spines homogeneous.

A. elliottii (Fig. 4)
LM: pollen grains 35.6 μm, almost spheroidal (L/B = 1.07).
SEM: spines 1–2 μm long, relatively short, unstoreyed, connected by (standing on) a thin layer (ectexine?) that overlies a basal layer (endexine?); surface between spines scabrate.
TEM: not available.

A. doryphorus (Figs. 5, 6, 11)
LM: pollen grains 46.0 μm, almost spheroidal (L/B = 1.10).
SEM: spines 2–3 μm long, elongate, unstoreyed; surface between spines ± scabrate.
TEM: exine between spines 2.2 μm thick, almost completely consisting of endexine; endexine not stratified, though larger dark specks occur especially in the inner part, which is irregularly connected with the intine; ectexine almost absent between the spines; spines homogeneous.

A. antsingyensis (Figs. 7, 8, 12)
LM: pollen grains 39.9 μm, almost spheroidal (L/B = 1.06).
SEM: spines 3–4 μm long, elongate, 1–5-storeyed; surface between spines ± smooth.
TEM: exine thickness between spines 2.6 μm; endexine 1 μm thick, consisting of a homogeneous outer sub-layer, a spongy middle sub-layer and a homogeneous inner sub-layer; ectexine 1.6 μm thick, except for some scattered irregularly shaped darker specks ± homogeneous; spines largely to almost completely filled with a dark, 1–4-storeyed body, the convex base of which is surrounded by lighter ectexinous material.

Discussion

On the basis of spine morphology Van der Ham *et al.* (1998) distinguished three echinate subtypes:

a. short (1–2 μm), unstoreyed (*A. aphyllus*, *A. dracontioides*, *A. elliottii*)
b. elongate (2–3 μm), unstoreyed (*A. doryphorus*)
c. elongate (3–4 μm), 1–5-storeyed (*A. antsingyensis*)

243

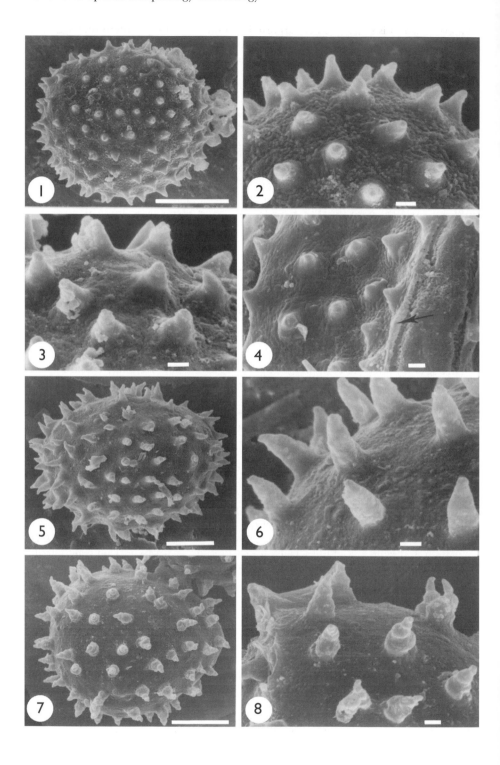

It appears that this subdivision is supported by TEM data, notably those relating to the ultrastructure of the ectexine. Subtype a (*A. aphyllus*, *A. dracontioides*) shows a thin homogeneous outer sub-layer that is continuous with the spine bases, and a finely granular ectexinous sub-layer overlying the endexine and extending under the spines (Figs. 9, 10). SEM data suggest that in *A. elliottii* the spines stand on a similar basal layer. Subtype b (*A. doryphorus*) shows spines standing more or less isolated on the endexine, a continuous ectexinous layer being hardly or not discernible (Fig. 11). Subtype c (*A. antsingyensis*) can be separated by the relatively thick ectexinous layer, while its spines are unusual by their storeyed appearance.

According to macromorphological data the *Amorphophallus* species with echinate pollen were provisionally accommodated in three predominantly African/Madagascan sections (Ittenbach, 1997; Van der Ham *et al.*, 1998). *A. aphyllus*, *A. dracontioides* and *A. elliottii* (all subtype a) constitute the section *Corynophallus* (Schott) Engl., which is a well-defined African group. *A. doryphorus* (subtype b) belongs in the African section *Brownei* Ittenb. (not yet published), together with at least one other species (pollen unknown). *A. antsingyensis* (subtype c) was placed in the African/Madagascan/Indian section *Engleri* Ittenb. (not yet published), which is a macromorphologically heterogeneous group of eight species with diverse pollen (besides echinate also striate, fossulate and psilate), and which needs further study. Obviously, the preliminary taxonomical treatment of the echinate *Amorphophallus* species, i.e. inclusion in three subgeneric groups, is well-supported by pollen morphology (SEM as well as TEM). However, pending further taxonomical and pollen morphological research, the relations of subtypes b and c with the surface types occurring in sections *Brownei* and *Engleri*, respectively, are still not clear.

In addition, within *Amorphophallus*, *A. sylvaticus* has a finely granular basal ectexine (Van der Ham *et al.*, 1998, Figs. 33–36), which suggests some relation with subtype a. However, *A. sylvaticus* has finely verrucate pollen without any trace of echinate elements. It belongs to the Indian section *Rhaphiophallus* (Schott) Engl., which is probably monophyletic (Hetterscheid *et al.*, 1994), though highly diverse as to exine surface type.

Comparison of the echinate pollen of *Amorphophallus* with that of other aroid genera studied with TEM (*Arisaema*, *Arum*, *Sauromatum*) shows that subtype c stands apart so far. Pollen of *Arum* and *Sauromatum* (Hesse, 1980; Pacini and Juniper, 1983; Pacini, 1990; Weber *et al.*, 1998) has relatively short spines more or less isolated on an endexinous base, which state, except for spine shape, resembles subtype b (*A. doryphorus*). In *Arisaema* pollen (Ohashi *et al.*, 1983; Oh *et al.*, 1990) relatively short spines are connected by a thin ectexinous layer (not distinct in Zavada, 1983). Although it is difficult to decide whether there is a finely granular basal sub-layer, this condition greatly resembles that of the echinate *Amorphophallus* subtype a. This is interesting because the genus *Arisaema*, the pollen of which is invariably echinate (Grayum, 1992), is possibly closely related to *Amorphophallus* (Hetterscheid and Ittenbach, 1996).

FIGS. 1–8. Echinate *Amorphophallus* pollen subtype a (1–4), subtype b (5–6) and subtype c (7–8). SEM; bar = 10 μm (1, 5, 7) or 1 μm (2–4, 6, 8). Figs. 1–2. *A. aphyllus*. Pollen grain and detail showing short spines. Fig. 3. *A. dracontioides*. Detail showing short spines. Fig. 4. *A. elliottii*. Detail showing short spines and cracked outer exine sub-layer (arrow). Fig. 5–6. *A. doryphorus*. Pollen grain and detail showing elongate spines. Fig. 7–8. *A. antsingyensis*. Pollen grain and detail showing elongate storeyed spines.

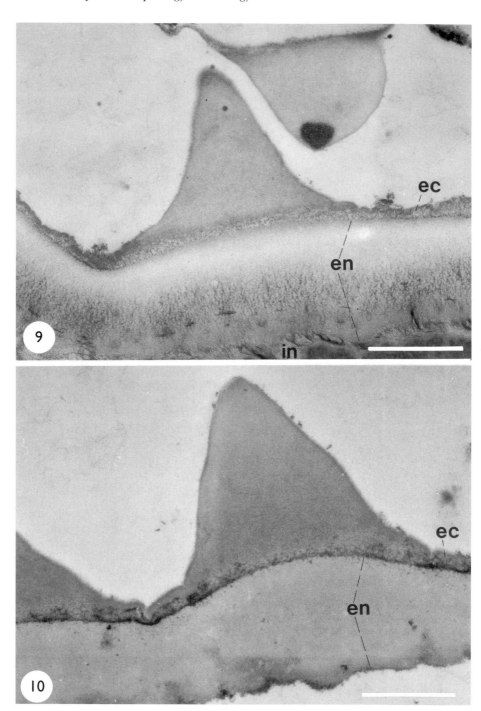

FIGS. 9–10. Echinate *Amorphophallus* pollen subtype a. TEM; bar = 1 μm. Abbreviations: ec = ectexine, en = endexine, in = intine. Fig. 9. *A. aphyllus.* Section showing intine, endexine, continuous thin ectexine and ectexinous spines; the isolated spine (upside down) shows a dark speck near its apex. Fig. 10. *A. dracontioides.* Section showing endexine, continuous thin ectexine and ectexinous spine.

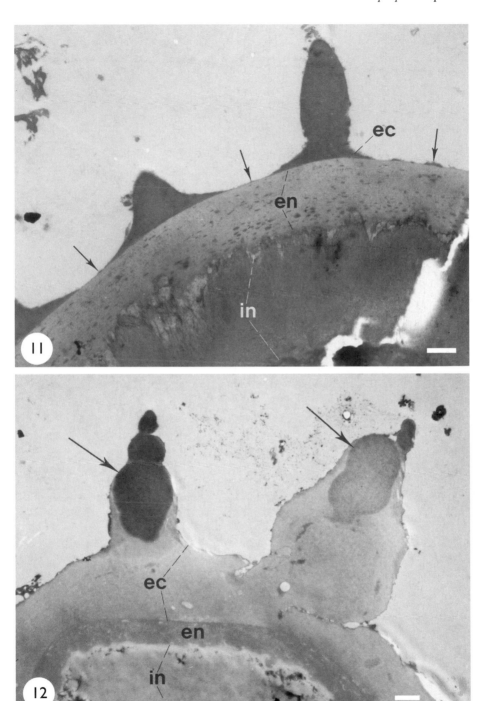

FIGS. 11–12. Echinate *Amorphophallus* pollen subtypes b(11) and c(12). TEM; bar = 1 μm. Abbreviations: ec = ectexine, en = endexine, in = intine. Fig. 11. *A. doryphorus*. Section showing intine, endexine, no continuous ectexine (arrows) between the ectexinous spines. Fig. 12. *A. antsingyensis*. Section showing intine, endexine, continuous thick ectexine and storeyed spines with dark storeyed bodies inside (arrows).

References

Grayum, M.H. (1992). Comparative external pollen ultrastructure of the Araceae and putatively related taxa. *Monographs in Systematic Botany from the Missouri Botanical Garden* 43: 167 pp.

Hesse, M. (1980). Entwicklungsgeschichte und Ultrastruktur von Pollenkitt und Exine bei nahe verwandten entomophilen und anemophilen Angiospermensippen der Alismataceae, Liliaceae, Juncaceae, Cyperaceae, Poaceae und Araceae. *Plant Systematics and Evolution* 134: 229–267.

Hetterscheid, W.L.A. and Ittenbach, S. (1996). Everything you always wanted to know about *Amorphophallus*, but were afraid to stick your nose into. *Aroideana* 19: 7–131.

Hetterscheid, W.L.A., Yadav, S.R. and Patil, K.S. (1994). Notes on the genus *Amorphophallus* (Araceae) 5. *Amorphophallus konkanensis*, a new species from India, and taxonomic reflections on *Amorphophallus* section *Rhaphiophallus*. *Blumea* 39: 289–294.

Ittenbach, S. (1997). Revision der afrikanischen Arten der Gattung *Amorphophallus* (Araceae). Thesis University Bonn.

Mayo, S.J., Bogner, J. and Boyce, P.C. (1997). The genera of Araceae. Royal Botanic Gardens, Kew.

Oh, B.U., Ko, S.C., Hong, W.P. and Kim, Y.S. (1990). A phylogenetic consideration of *Arisaema* by anatomical and palynological characters 2. Palynological characters. *Korean Journal of Plant Taxonomy* 20: 37–52.

Ohashi, H., Murata, J. and Takahashi, M. (1983). Pollen morphology of the Japanese *Arisaema* (Araceae). *Science Reports of the Tohoku University, fourth series, Biology* 38: 219–251.

Pacini, E. (1990). Tapetum and microspore function. In: S. Blackmore and R.B. Knox (editors). Microspores: evolution and ontogeny, pp. 213–237. Academic Press, London.

Pacini, E. and Juniper, B.E. (1983). The ultrastructure of the formation and development of the amoeboid tapetum in *Arum italicum* Miller. *Protoplasma* 117: 116–129.

Van der Ham, R.W.J.M., Hetterscheid, W.L.A. and Van Heuven, B.J. (1998). Notes on the genus *Amorphophallus* (Araceae) - 8. Pollen morphology of *Amorphophallus* and *Pseudodracontium*. *Review of Palaeobotany and Palynology* 103: 95–142.

Weber, M., Halbritter, H. and Hesse, M. (1998). The spiny pollen wall in *Sauromatum* (Araceae) - with special reference to the endexine. *International Journal of Plant Science*, 159: 744–749.

Zavada, M.S. (1983). Comparative morphology of monocot pollen and evolutionary trends of apertures and wall structures. *Botanical Review*, 49: 331–379.

Furness, C.A. and Rudall, P.J. (2000). Aperture absence in pollen of monocotyledons. In: M.M. Harley, C.M. Morton and S. Blackmore (Editors). Pollen and Spores: Morphology and Biology, pp. 249–257. Royal Botanic Gardens, Kew.

APERTURE ABSENCE IN POLLEN OF MONOCOTYLEDONS

Carol A. Furness and Paula J. Rudall

Royal Botanic Gardens, Kew, Richmond, Surrey, TW9 3AE, UK

Abstract

Two types of inaperturate pollen are described: functionally monoaperturate ('cryptoaperturate') and 'omniaperturate'. Their distribution in monocotyledons is discussed in the context of recent systematic investigations. The two types are probably largely non-homologous and have arisen as independent modifications, although they do occur in closely related taxa in some groups (for example, Zingiberales). Inaperturate pollen is not correlated with either successive or simultaneous microsporogenesis. Functionally monoaperturate (cryptoaperturate) pollen is closely related to the 'diffuse-sulcate' type. Omniaperturate pollen is functionally similar to spiraperturate, polyrugate, extended sulcate, or clypeate pollen (with exine plates), and is often found in taxa allied to those with these related aperture types. Taxa with inaperturate pollen occur in every major monocot group, probably related to the relatively thin exine in monocotyledons. This character has reversed or evolved several times, but its presence is a synapomorphy within some groups (for example, Zingiberales, some Liliales and some Asparagales). In many (but not all) monocotyledons the presence of inaperturate pollen is correlated with aquatic or moist habitats; another possible correlation is with the saprophytic habit.

Introduction

Inaperturate pollen and spores (with no obvious aperture) are relatively common throughout the plant kingdom. Hepatics may be primitively inaperturate with distal germination (Blackmore and Crane, 1998), and many dicotyledons have inaperturate pollen, including several 'primitive' dicotyledons and some eudicots (for example, some Euphorbiaceae and Solanaceae: Kress, 1986). In monocotyledons, inaperturate pollen is widespread and covers a diverse structural and taxonomic range. Character homologies of different types of inaperturate pollen in monocotyledons require further assessment and clarification. Since there appears to be some significance in their systematic distribution, this paper reviews their occurrence in the context of current systematic investigations of monocotyledons.

Inaperturate pollen falls into two main categories: (1) functionally monoaperturate pollen, where there is a localised thickening of the intine for pollen tube germination, although the exine is uniform (i.e. a hidden, or 'crypto-aperture'); (2) 'omni-aperturate' pollen (*sensu* Thanikaimoni, 1978), where the intine is of even thickness, and the pollen tube may germinate from anywhere on the pollen wall. Our use of the term 'cryptoaperture' thus differs from that of Thanikaimoni (1978) and Punt *et al.* (1994), where it indicates an endoaperture that is not apparent in surface view because there is no ectoaperture.

Distribution of inaperturate pollen in monocotyledons

We used the topology achieved by a combined molecular/morphological analysis of monocotyledons in general (Chase *et al.*, 1995) and a molecular analysis of multiple datasets (Chase *et al.*, in press) to define the major groupings used here: early-branching monocotyledons, Lilianae and Commelinanae. Orders follow the Angiosperm Phylogeny Group (1998).

Table 1. Summary of records of inaperturate pollen in monocotyledons (Furness and Rudall, 1999)

Unplaced taxa	Triuridaceae
Early-branching monocotyledons	Araceae (most genera) and most Alismatidae (Cymodoceaceae, Hydrocharitaceae, Juncaginaceae, Lilaeaceae, Najadaceae, Posidoniaceae, Potamogetonaceae, Scheuchzeriaceae, Zannichelliaceae and Zosteraceae)
Lilianae-Asparagales	Some genera of Convallariaceae *sensu lato* (including Ruscaceae), Iridaceae and Orchidaceae
Lilianae-Liliales	Some genera of Liliaceae, Philesiaceae, Smilacaceae, Trilliaceae and Campynemataceae
Lilianae-Dioscoreales	Some genera of Burmanniaceae and Thismiaceae
Lilianae-Pandanales	Some Cyclanthaceae, Stemonaceae and Velloziaceae
Commelinanae	Hanguanaceae (*Hanguana* Blume), some Bromeliaceae, some Xyridaceae/Abolbodaceae, most Zingiberales (Cannaceae, Heliconiaceae, Marantaceae, Musaceae, Strelitziaceae, Zingiberaceae), one species of Arecaceae

In *Acorus* L., an aquatic herb which is considered the sister taxon to all other monocotyledons, pollen is monosulcate (Grayum 1992; Rudall and Furness, 1997). *Tofieldia* Huds. has disulcate pollen, but among other early-branching monocotyledons (Alismatales *sensu* Angiosperm Phylogeny Group, 1998) the majority of taxa have inaperturate pollen (Table 1), and it is possible that this character may represent a synapomorphy for this group, with reversals back to monosulcate in some taxa. A more likely explanation, however, is that the developmental mechanisms controlling aperture form are especially labile among early-branching monocots. Many early-branching monocotyledons are water plants, and the exine is often greatly reduced, presumably because pollen is not subject to desiccation in these plants. In inaperturate taxa, the intine is often thick, either uniformly so (omniaperturate, for example, Cymodoceaceae: Pettitt, 1980, 1981) or present at one side of the grain only (cryptoaperturate, for example, *Lepilaena* J.L. Drumm. ex Harv. and *Potomogeton* L.: Wodehouse, 1935; Cranwell, 1953).

In Araceae, in general, inaperturate pollen occurs in monoecious genera while bisexual-flowered genera have aperturate pollen, although there are a few exceptions to this, including the tribe Spathiphylleae (bisexual, inaperturate), which is fairly basal (Mayo *et al.*, 1997).

Lilianae probably comprise four orders: Asparagales, Liliales, Dioscoreales and Pandanales. In Asparagales, inaperturate pollen occurs in at least three groups: (1)

250

Orchidaceae (many genera; for example, Pandolfi and Pacini, 1995; Zavada, 1983, 1990), (2) Iridaceae (*Syringodea* Hook. f., *Crocus* L. [Figs. 1-3], *Diplarrhena* Labill. [Fig. 4], *Patersonia* R. Br. and *Iris* L.: for example, Goldblatt and Le Thomas, 1992) and (3) Convallariaceae *sensu lato*, including Ruscaceae (*Aspidistra* Ker Gawl. [Figs. 5, 6], *Semele* Kunth and *Tupistra* Ker Gawl.: for example, Rudall and Campbell, 1999). Furthermore, within some of these groups inaperturate pollen has probably evolved more than once. For example, in Iridaceae, it occurs in isolated genera within subfamilies Ixioideae, Iridoideae and Nivenioideae (Furness and Rudall, 1999).

Similarly, at least four distinct groups within Liliales include taxa with inaperturate pollen: (1) *Campynemanthe* Baill., (2) some species of *Tulipa* L., (3) *Trillium* L. and possible allies, and (4) *Smilax* L. (Fig. 7) and its allies (Rudall *et al.*, in press). In Dioscoreales, inaperturate pollen occurs in Burmanniaceae (*Gymnosiphon* Blume and *Apteria* Nuttal: Chakrapani and Raj, 1971; Rübsamen, 1986) and Thismiaceae (*Thismia* Griff.: Cranwell, 1953), and there are unconfirmed reports in Dioscoreaceae (Furness and Rudall, 1999). In Pandanales, inaperturate pollen occurs in Cyclanthaceae (*Dianthovius* Hammel & Wilder and *Evodianthus* Oerst.: Harling, 1958; Hammel and Wilder, 1989), Stemonaceae (*Pentastemona* Steenis and *Stichoneuron* Hook. f.: Van der Ham, 1991) and Velloziaceae (*Vellozia* Vand.: Ayensu, 1972; Ayensu and Skvarla, 1974) and there are unconfirmed reports in Pandanaceae (Zavada, 1983; Grayum, 1992).

There are also several records of inaperturate pollen within the commelinoid clade (Commelinanae) (Table 1). Among the palms and their possible relatives (Dasypogonaceae, Chase *et al.*, 1995), there is only one record of inaperturate pollen (despite diverse aperture types), in a single species of *Pigafetta* (Blume) Becc. (Harley, 1996). Poales lack inaperturate pollen entirely, unless taxa such as *Abolboda* Bonpl. are included, but this requires further testing. In other commelinoids, inaperturate pollen occurs in (1) Xyridaceae/Abolbodaceae (three genera: *Abolboda*, *Achlyphila* Maguire & Wurdack and *Orectanthe* Maguire: Carlquist, 1960), (2) Zingiberales (most genera, with thin exine and thick, layered intine, for example Skvarla and Rowley, 1970; Kress *et al.*, 1978; Stone *et al.*, 1979; Kress and Stone, 1982, 1983a, 1983b; Kress, 1986: Figs. 8, 9), (3) *Hanguana* (Rudall *et al.*, 1999: Fig. 10) and (4) Bromeliaceae (three or more genera: for example, Halbritter, 1992).

Discussion

Inaperturate pollen may be either omniaperturate or functionally monoaperturate (cryptoaperturate). These two types can probably arise as independent modifications, although they may occur in closely related taxa within the same groups, for example, Zingiberales: Zingiberaceae (omniaperturate) and Heliconiaceae (functionally monoaperturate). Pollen of *Vallisneria* L. (Hydrocharitaceae) was described as dimorphic with inaperturate grains mixed with faintly monosulcate grains (Sharma, 1967). Functionally monoaperturate pollen occurs in *Pentastemona* and *Stichoneuron* (Stemonaceae), other Stemonaceae are monosulcate (Van der Ham, 1991; Furness and Rudall, 1999), so functionally monosulcate pollen can arise as an independent modification. Omniaperturate pollen can also evolve independently, for example, in some *Crocus* (Iridaceae) species, others are spiraperturate or polyrugate (see below). Inaperturate pollen is not correlated with either successive or simultaneous microsporogenesis, which reflects the lack of influence of tetrad orientation on development of this pollen type (Furness and Rudall, 1999). Inaperturate pollen is often found in groups with a range of different aperture types, such as Araceae, although it is rare in palms, where aperture types are also diverse.

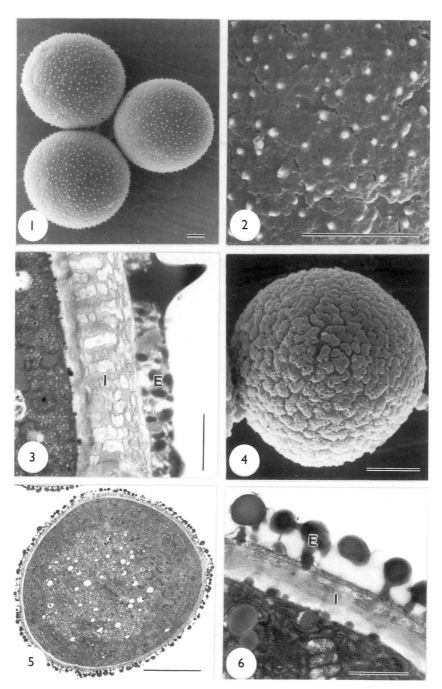

FIGS. 1–3. *Crocus* pollen. Fig. 1. *C. sativus* L. 'Saffron' (SEM). Fig. 2. *C. hadriaticus* Herb., spinulate surface (SEM). Fig. 3. *C. oreocreticus* B.L. Burtt, spinulate exine (E) and thick, channelled intine (I) (TEM). Fig. 4. *Diplarrhena latifolia* Benth., pollen grain (SEM). Figs. 5, 6. *Aspidistra lurida* Ker Gawl. Fig. 5. Pollen grain (TEM). Fig. 6. Pilate exine (E) and intine with channels (I). Scale bars: 10 μm, except 3 and 6 which are 1 μm.

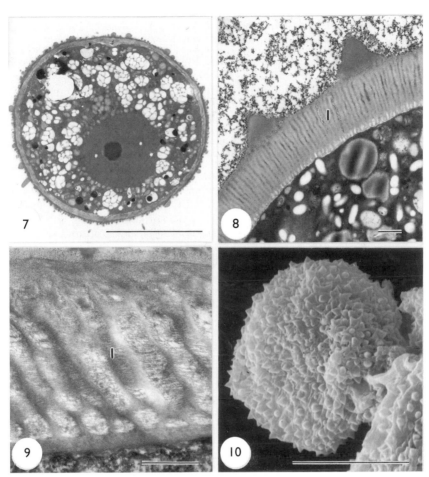

FIGS. 7–10. Fig. 7. *Smilax hispida* Muhl., pollen grain (TEM). Fig. 8. *Globba atrosanguinea* Teijsm. & Binn., thick, channelled intine (I) and spinules (TEM). Fig. 9. *Marantachloa purpurea* (Ridley) Milne-Redhead, thick, channelled intine (I) (TEM). Fig. 10. *Hanguana malayana* Merrill, spinulate pollen grain (SEM). Scale bars: 7, 10 are 10 μm; 8, 9 are 1 μm.

Functionally monoaperturate (cryptoaperturate) pollen is closely related to the 'diffuse-sulcate' type, for example in *Tupistra* (Convallariaceae: Halbritter and Hesse, 1993; Rudall and Campbell, 1999) and *Aechmea* Ruíz & Pav. (Bromeliaceae: Halbritter, 1992). Spiraperturate, polyrugate, extended sulcate, or clypeate pollen (with exine plates or shields: Halbritter and Hesse, 1995) are functionally similar to omniaperturate pollen in that the area for potential pollen tube germination is increased, and some of these aperture types may be found in taxa related to those with omniaperturate pollen; for example, *Crocus* includes some species with spiraperturate pollen, some with polyrugate pollen and some with inaperturate (omniaperturate) pollen. A number of spiraperturate and omniaperturate grains are also spinulate (Furness, 1985).

Inaperturate pollen is also fairly common among 'primitive' dicotyledons (Furness and Rudall, 1999), and Brenner (1996) proposed that the earliest angiosperm pollen was inaperturate based on Valanginian fossils, so it is not inconceivable that monocotyledons may be primitively inaperturate rather than monosulcate, although this seems unlikely in view of the monosulcate condition in *Acorus* (see above). Virtually every major monocot group includes some taxa with inaperturate pollen, indicating that this character state has reversed or evolved several times; but within some groups its presence is a synapomorphy (for example, in Zingiberales, some Liliales and some Asparagales). This abundance of inaperturate pollen in monocotyledons is probably related to the relatively thin exine. Sparseness or absence of endexine is a monocot character (Kress and Stone, 1982; Zavada, 1983; Rudall and Furness, 1997).

In many monocotyledons the presence of inaperturate pollen is correlated with aquatic or moist habitats, as in most of the early-branching taxa (an aquatic or semi-aquatic habit) and many Commelinanae, such as *Hanguana* and the taxa of Zingiberales (moist rainforests: Kress, 1986). In such environments, pollen is not subject to harmomegathic stress and can remain viable without a protective exine, as in *Heliconia* L. (Kress, 1986). However, this is by no means always the case; some Lilianae with inaperturate pollen occur in relatively dry habitats; for example, *Patersonia*, *Crocus* (Iridaceae: for example, Furness, 1985; Goldblatt and Le Thomas, 1992), *Tulipa* (Liliaceae: Schulze, 1980) and *Vellozia* (Ayensu and Skvarla, 1974).

Another possible correlation is with the saprophytic habit, possibly due to the cost of exine production, although this is speculative. At least four distinct groups of monocot saprophytes include taxa with inaperturate pollen. These include (1) Triuridaceae, a family of uncertain position possibly related to Alismatales (for example, Rübsamen-Weustenfeld, 1991), (2) the New Caledonian genus *Campynemanthe* (Liliales - Campynemataceae: Dahlgren and Lu, 1985), (3) two related families of Dioscoreales: Burmanniaceae (*Gymnosiphon* and *Apteria*: Chakrapani and Raj, 1971; Rübsamen, 1986) and Thismiaceae (*Thismia*: Cranwell, 1953) and (4) some (although not all) saprophytic orchids, such as some Vanilloideae (Cameron, 1996), Orchidoideae and Epidendroideae (Zavada, 1990, 1983). On the other hand, monosulcate pollen occurs in the Madagascan saprophyte *Geosiris* Baill. (Asparagales - Iridaceae) (Rübsamen-Weustenfeld *et al.*, 1994), and the saprophyte genus *Petrosavia* Becc. (Petrosaviaceae). According to Erdtman (1952), Rafflesiaceae (of uncertain affinities, possibly related to first-branching angiosperms: Nandi *et al.*, 1998), which are parasitic and lack chlorophyll, have inaperturate pollen (for example, *Rafflesia* R.Br. and *Rhizanthes* Dumort); also *Cytinus* L. (Cytinaceae: Angiosperm Phylogeny Group, 1998) and *Mitrastemon* Makino (Mitrastemonaceae: Angiosperm Phylogeny Group, 1998) have two or three apertures but according to Erdtman (1952) they are 'not well defined, covered by a thin membrane' . Takhtajan *et al.* (1985) reported monosulcate or monoporate pollen in *Rafflesia*, *Rhizanthes* and *Sapria* Griff., so they may in fact be cryptoaperturate. Furthermore, this correlation does not hold true for all eudicots, where thin or absent exine is in any case less common, for example, *Cuscuta* L. (Convolvulaceae) and Orobanchaceae have three or more clearly defined apertures. *Atkinsonia* F. Muell. (Loranthaceae: photosynthetic hemiparasites) is inaperturate but other Loranthaceae have distinct apertures (Feuer and Kuijt, 1980; 1985). Loss of apertures may be favoured in saprophytes and parasites, though if the shift in habit occurred first, some may still retain apertures, although this is speculative and requires testing.

Acknowledgements

We are grateful to colleagues for discussion of various aspects of this work, especially Peter Linder, Richard Bateman, Alec Pridgeon and Jim Doyle. Thanks are due to Kate Stobart (University of Leicester) for supplying the SEM of *Diplarrhena latifolia* during her work experience at Kew.

References

Angiosperm Phylogeny Group. (1998). An ordinal classification for the families of flowering plants. *Annals of the Missouri Botanical Garden* 85: 531–553.

Ayensu, E.S. (1972). Studies on pollen morphology in the Velloziaceae. *Proceedings of the Biological Society of Washington* 85: 469–480.

Ayensu, E.S. and Skvarla, J.J. (1974). Fine structure of Velloziaceae pollen. *Bulletin of the Torrey Botanical Club* 101: 250–266.

Blackmore, S. and Crane, P.R. (1998). The evolution of apertures in the spores and pollen grains of embryophytes. In: S.J. Owens and P.J. Rudall (editors). Reproductive biology, pp. 159–182. Royal Botanic Gardens, Kew.

Brenner, G.J. (1996). Evidence for the earliest stage of angiosperm pollen evolution: A paleoequatorial section from Israel. In: D.W. Taylor and L.J. Hickey (editors). Flowering plant origin, evolution and phylogeny, pp. 91–115. Chapman and Hall, New York.

Cameron, K.M. (1996). Phylogenetic relationships of the vanilloid orchids: an integration of molecular, morphological, and anatomical data. PhD thesis, University of North Carolina, Chapel Hill, North Carolina.

Carlquist, S. (1960). Anatomy of Guayana Xyridaceae: *Abolboda, Orectanthe,* and *Achlyphila. Memoirs of the New York Botanic Garden* 10: 65–117.

Chakrapani, P. and Raj, B. (1971). Pollen morphological studies in the Burmanniaceae. *Grana* 11: 164–179.

Chase, M.W., Stevenson, D.W., Wilkin, P. and Rudall, P.J. (1995). Monocot systematics: a combined analysis. In: P.J. Rudall, P.J. Cribb, D.F. Cutler and C.J. Humphries (editors). Monocotyledons: systematics and evolution, pp. 685–730. Royal Botanic Gardens, Kew.

Chase, M.W., Soltis, D.E., Soltis, P.S., Rudall, P.J., Fay, M.F., Hahn, W.J., Sullivan, S., Joseph, J., Givnish, T., Sytsma, K.J. and Price, C. (in press). Higher level systematics of the monocotyledons: an assessment of current knowledge and a new classification. In: K.L. Wilson and D.A. Morrison (editors). Systematics and evolution of monocots. CSIRO, Melbourne, Australia.

Cranwell, L.M (1953). New Zealand pollen studies. The monocotyledons. *Bulletin of the Auckland Institute and Museum* 3: 1–91.

Dahlgren, R. and Lu, A.-M. (1985). *Campynemanthe* (Campynemaceae): Morphology, microsporogenesis, early ovule ontogeny and relationships. *Nordic Journal of Botany* 5: 321–330.

Erdtman, G. (1952). Pollen morphology and plant taxonomy. Angiosperms. Almqvist and Wiksell, Stockholm.

Feuer, S.M. and Kuijt, J. (1980). Fine structure of mistletoe pollen III. Large flowered neotropical Loranthaceae and their Australian relatives. *American Journal of Botany* 67: 34–50.

Feuer, S.M. and Kuijt, J. (1985). Fine structure of mistletoe pollen VI. Small flowered neotropical Loranthaceae. *Annals of the Missouri Botanical Garden* 72: 187–212.

Furness, C.A. (1985). A review of spiraperturate pollen. *Pollen et Spores* 27: 307–320.

Furness, C.A. and Rudall, P.J. (1999). Inaperturate pollen in monocotyledons. *International Journal of Plant Sciences.* **160**: 395–414.

Goldblatt, P. and Le Thomas, A. (1992). Pollen apertures, exine sculpturing and phylogeny in Iridaceae subfamily Iridoideae. *Review of Palaeobotany and Palynology* 75: 301–315.

Grayum, M.H. (1992). Comparative external pollen ultrastructure of the Araceae and putatatively related taxa. *Monographs in Systematic Botany of the Missouri Botanical Garden* 43: 1–167.

Halbritter, H. (1992). Morphologie und systematische Bedeutung des Pollens der Bromeliaceae. *Grana* 31: 197–212.

Halbritter, H. and Hesse, M. (1993). Sulcus morphology in some monocot families. *Grana* 32: 87–99.

Halbritter, H. and Hesse, M. (1995). The convergent evolution of exine shields in angiosperm pollen. *Grana* 34: 108–119.

Hammel, B.E. and Wilder, G.J. (1989). *Dianthoveus*: a new genus of Cyclanthaceae. *Annals of the Missouri Botanical Garden* 76: 112–123.

Harley, M.M. (1996). Palm pollen and the fossil record. PhD thesis, University of East London, London.

Harling, G. (1958). Monograph of the Cyclanthaceae. *Acta Horti Bergiana* 18: 1–428.

Kress, W.J. (1986). Exineless pollen structure and pollination systems of tropical *Heliconia* (Heliconiaceae). In: S. Blackmore and I.K. Ferguson, (editors). Pollen and spores: form and function, pp. 329–345. Academic Press, London.

Kress, W.J. and Stone, D.E. (1982). Nature of the sporoderm in monocotyledons, with special reference to the pollen grains of *Canna* and *Heliconia*. *Grana* 21:129–148.

Kress, W.J. and Stone, D.E. (1983a). Pollen intine structure, cytochemistry and function in monocots. In: D.L. Mulcahy and E. Ottaviano (editors). Pollen: biology and implications for plant breeding, pp. 159–163. Elsevier, Amsterdam.

Kress, W.J. and Stone, D.E. (1983b). Morphology and phylogenetic significance of exine-less pollen of *Heliconia* (Heliconiaceae). *Systematic Botany* 8: 149–167.

Kress, W.J., Stone, D.E. and Sellers, S.C. (1978). Ultrastructure of exine-less pollen: *Heliconia* (Heliconiaceae). *American Journal of Botany* 65: 1064–1076.

Mayo, S.J., Bogner, J. and Boyce, P. (1997). The genera of Araceae. Royal Botanic Gardens, Kew.

Nandi, O.I., Chase, M.W. and Endress, P.K. (1998). A combined cladistic analysis of angiosperms using *rbcL* and non-molecular data sets. *Annals of the Missouri Botanical Garden* 85: 137–212.

Pandolfi, T. and Pacini, E. (1995). The pollinium of *Loroglossum hircinum* (Orchidaceae) between pollination and pollen tube emission. *Plant Systematics and Evolution* 196: 141–151.

Pettitt, J.M. (1980). Reproduction in seagrasses: nature of the pollen and receptive surface of the stigma in the Hydrocharitaceae. *Annals of Botany* 45: 257–271.

Pettitt, J.M. (1981). Reproduction in seagrasses: pollen development in *Thalassia hemprichii, Halophila stipulacea* and *Thalassodendron ciliatum. Annals of Botany* 48: 609–622.

Punt, W., Blackmore, S., Nilsson, S. and Le Thomas, A. (1994). Glossary of pollen and spore terminology. Laboratory of Palaeobotany and Palynology Foundation, University of Utrecht.

Rübsamen, T. (1986). Morphologische, embryologische und systematische Untersuchungen an Burmanniaceae und Corsiaceae (Mit Ausblick auf die Orchidaceae-Apostasioideae). Dissertationes Botanicae, 92. J. Cramer, Berlin.

Rübsamen-Weustenfeld, T. (1991). Morphologische, embryologische und systematische Untersuchungen an Triuridaceae. *Bibliotheca Botanica* 140: 1–113.

Rübsamen-Weustenfeld, T., Mukielka, V. and Hamann, U. (1994). Zur Embryologie, Morphologie und Systematischen Stellung von *Geosiris aphylla* baillon (Monocotyledonae - Geosiridaceae/Iridaceae). *Botanische Jahrbuch für Systematik* 115: 475–545.

Rudall, P.J. and Campbell, G. (1999). Flower and pollen structure of Ruscaceae in relation to Aspidistreae and other Convallariaceae. *Flora* 194: 201–214.

Rudall, P.J. and Furness, C.A. (1997). Systematics of *Acorus*: ovule and anther. *International Journal of Plant Sciences* 158: 640–651.

Rudall, P.J., Stevenson, D.W. and Linder, H.P. (1999). Structure and systematics of *Hanguana*, a monocotyledon of uncertain affinity. *Australian Systematic Botany*. 12: 311–330.

Rudall, P.J., Hong, W.-P., Furness, C.A., Conran, J.G., Kite, G., Stobart, K.L., and Chase, M.W. (in press). Consider the lilies - systematics of Liliales. In: K.L. Wilson and D.A. Morrison (editors). Systematics and evolution of monocots. CSIRO, Melbourne, Australia.

Schulze, W. (1980). Beiträge zur Taxonomie der Liliifloren. VI. Der Umfang der Liliaceae. *Wissenshaftliche Zeitschrift der Friedrich-Schiller Universität Jena, Mathematische-Naturwissenschaftliche* 29: 607–636.

Sharma, M. (1967). Pollen morphology of Indian monocotyledons. *Journal of Palynology. Special volume:* 1–98.

Skvarla, J.J. and Rowley, J.R. (1970). The pollen wall of *Canna* and its similarity to the germinal apertures of other pollen. *American Journal of Botany* 57: 519–529.

Stone, D.E., Sellers, S.C. and Kress, W.J. (1979). Ontogeny of exineless pollen in *Heliconia*, a banana relative. *Annals of the Missouri Botanical Garden* 66: 701–730.

Takhtajan, A.L., Meyer, N.R. and Kosenko, V.N. (1985). The pollen morphology and classification in Rafflesiaceae *s.l. Botanicheskiya Zhurnal (Moscow and Leningrad)* 70: 153–168. (In Russian).

Thanikaimoni, G. (1978). Palynological terms: proposed definitions - 1. *Proceedings of the 4th International Palynological Conference, Lucknow (1976-1977)* 1: 228–239.

Van der Ham, R.W.J.M. (1991). Pollen morphology of the Stemonaceae. *Blumea* 36: 127–159.

Wodehouse, R.P. (1935). Pollen grains. Their structure, identification and significance in science and medicine. McGraw-Hill, New York.

Zavada, M.S. (1983). Comparative morphology of monocot pollen and evolutionary trends of apertures and wall structures. *Botanical Review* 49: 331–379.

Zavada, M.S. (1990). A contribution to the study of pollen wall ultrastructure of orchid pollinia. *Annals of the Missouri Botanical Garden* 77: 785–801.

Doyle, J.A., Bygrave, P. and Le Thomas, A., (2000). Implications of molecular data for pollen evolution in Annonaceae. In: M.M. Harley, C.M. Morton and S. Blackmore (Editors). Pollen and Spores: Morphology and Biology, pp. 259–284. Royal Botanic Gardens, Kew.

IMPLICATIONS OF MOLECULAR DATA FOR POLLEN EVOLUTION IN ANNONACEAE

JAMES A. DOYLE[1], PAUL BYGRAVE[2], AND ANNICK LE THOMAS[3]

[1]Section of Evolution and Ecology, University of California, Davis, California 95616, USA;
[2]Royal Botanic Gardens, Kew, Richmond, Surrey, TW9 3AB, UK;
[3]Ecole Pratique des Hautes Etudes, Muséum National d'Histoire Naturelle, 16 rue Buffon, 75005 Paris, France

Abstract

Cladistic analyses based on morphology suggested that pollen characters are among the most valuable in systematics of Annonaceae and tended to confirm earlier hypotheses on pollen evolution. Molecular (*rbc*L) analyses of the same taxa and combined analyses of morphology and *rbc*L strengthen some previous inferences but refute others. These analyses confirm that granular monosulcate pollen is primitive (as in the basal genus *Anaxagorea*) and that groups with tetrads and inaperturate monads form a derived clade. However, *rbc*L implies that evolution of exine structure was more homoplastic than thought, although trees based on the combined data set are more consistent with this character, implying that some of the conflicting molecular data are weak. The worst conflict concerns the xylopioids, which have granular tetrads: *Cananga* is linked with the near-basal ambavioids (consistent with its three integuments and irregular endosperm ruminations), while *Xylopia* and *Neostenanthera* are separate lines within the inaperturates. Re-examination suggests that recognition of the xylopioids was due in part to misinterpretation of stamen and pollen morphology and reveals features that may be more consistent with *rbc*L, such as a vestigial sulcus in the pollen of *Cananga*.

Introduction

Over the past three decades, palynology has played a prominent role in systematics of Annonaceae, the largest family of primitive angiosperms ("Magnoliidae"). This is because of their remarkable pollen diversity, indications that pollen may provide a long-sought key to relationships within the family, and general interest in the original pollen morphology of angiosperms as a whole.

Pre-cladistic ideas on pollen evolution

The first major palynological survey of Annonaceae was by Walker (1971a, 1971b, 1972). At the base, he placed several predominantly American genera with mono-sulcate pollen and finely reticulate, columellar exine structure, which he called the *Malmea* tribe. Monosulcate pollen occurs in other magnoliids and is believed to be primitive for angiosperms as a whole, but Walker thought that Annonaceae (except

259

Pseudoxandra) differed from other magnoliids in having a proximal rather than a distal sulcus. He postulated several "trends" to more advanced pollen types. "Cataulcerate" pollen (with a round proximal thin area), often shed as tetrads, arose in two lines: the *Annona* subfamily, with columellar exine structure, and the *Fusaea* subfamily, with what Walker called "microtectate" structure; in the latter, most genera (such as *Cananga* and *Xylopia*) have tetrads, but *Anaxagorea* (the only Asian-American genus) has monosulcate monads. Pollen became inaperturate in the *Uvaria* tribe, sometimes with echinate sculpture (mostly African-Asian lianas). Inaperturates gave rise to sulculate pollen in the *Guatteria* tribe (now known to be zonasulculate: Morawetz and Waha, 1985), in which the exine is reduced almost to nothing.

Le Thomas and Lugardon (Le Thomas and Lugardon, 1974, 1975, 1976; Lugardon and Le Thomas, 1974; Le Thomas, 1980-1981) extended and modified these results by SEM and TEM studies, especially of African taxa, many of which are more primitive than anticipated. First, they refuted the concept that the sulcus is proximal in Annonaceae; the thin area in pollen of the cataulcerate type is proximal, but it is not homologous with the sulcus. This is especially clear in the Madagascan genus *Ambavia*, which has both a distal sulcus and a proximal thin area (Le Thomas and Lugardon, 1975). They interpreted the cataulcerate tetrads as basically inaperturate. Second, they showed that Walker's (1971a) microtectate exines, seen in such African genera as *Piptostigma* and *Polyceratocarpus*, really have what Van Campo and Lugardon (1973) described as granular structure and interpreted as primitive in angiosperms. Granular structure also occurs in *Anaxagorea* (Maas and Westra, 1984; Hesse *et al.*, 1985). Le Thomas (1980-1981) recognised intermediate states between granular and columellar, in which the tectum is often microverrucate. The whole series occurs in the large African-Asian genus *Polyalthia* Blume (then including *Greenwayodendron*). The echinate inaperturates show reduction of the tectum, leaving only the spines. Third, Lugardon and Le Thomas (1974) and Le Thomas (1980-1981) recognised a new trend for multiplication of foliations in the nexine, both in columellar tetrads (for example, *Annona*) and inaperturate monads (for example, *Uvaria*). Fourth, *Xylopia* and other taxa with granular tetrads have a thick basal layer that Le Thomas (1980-1981) suggested was derived from fused granules, based on comparison with granular monosulcate groups such as *Cleistopholis* and *Polyceratocarpus*, which have enlarged inner granules.

Cladistic analyses based on morphology

The advent of cladistics provided a method for testing these ideas more explicitly: amassing as many characters as possible and using the principle of parsimony to find the tree(s) most consistent with these. Doyle and Le Thomas (1994, 1995, 1996) presented a series of analyses based on 79 morphological characters, 11 of which were palynological. Annonaceae were rooted by including the five families of Magnoliales that were their closest relatives in an analysis of primitive angiosperms by Donoghue and Doyle (1989), a result confirmed by molecular work (Chase *et al.*, 1993; Soltis *et al.*, 1997; Savolainen *et al.*, in press). All these taxa were rooted by a hypothetical ancestor based on Canellaceae, Winteraceae, and *Austrobaileya* C.T.White, the closest relatives of Magnoliales in Donoghue and Doyle (1989), and Bennettitales, Gnetales, and *Caytonia* H.H.Thomas, the closest relatives of angiosperms in Doyle and Donoghue (1986, 1992). Molecular analyses have confirmed that Winteraceae and Canellaceae may be the sister group of Magnoliales, but *Austrobaileya* is more distant, and Magnoliales are too far from the base of angiosperms for other seed plants to be relevant.

Results of the most recent analysis (Doyle and Le Thomas, 1996) are summarised in Fig. 1. Outgroup relationships are poorly resolved, with *Degeneria*, Myristicaceae, and Magnoliaceae linked with Annonaceae in various arrangements. However, ingroup relationships are remarkably insensitive to these variations; *Anaxagorea* is always at the base, based on its petiole trace with an adaxial plate of xylem, laminar stamens, inner staminodes, and irregular endosperm ruminations, all features found in other Magnoliales but unusual in Annonaceae. Since *Anaxagorea* and the next lines have granular monosulcate pollen, these results confirm that this pollen type is primitive and homologous with similar pollen in other Magnoliales. The analysis of Donoghue and Doyle (1989) implied that it is also primitive for angiosperms as a whole, since Magnoliales were basal in their trees, but in the molecular analyses cited Magnoliales are nested among columellar groups, implying that their granular structure is instead a reversal (Doyle, 1998).

Above *Anaxagorea* is the ambavioid clade, consisting of the relict genera *Ambavia* (Madagascar), *Tetrameranthus* (Brazil), and *Cleistopholis* (Africa), united by the chromosome number n = 7 and three integuments. The ambavioids also retain granular monosulcate pollen, relatively primitive stamens, and irregular endosperm ruminations. Above two granular monosulcate lines (*Greenwayodendron*, *Polyalthia stuhlmannii*) is a larger clade, including *Piptostigma* and two other granular genera (piptostigmoids) and Walker's (1971a) *Malmea* tribe (malmeoids), united by the origin of reticulate-columellar exine structure. Next is the miliusoid clade, which shows trends to intermediate and columellar structure and to disulculate apertures; this includes disulculate members of *Polyalthia* and other Asian-Australian genera, but not the columellar-monosulcate *Polyalthia* group, which is nested among malmeoids with similar pollen. Disulculate pollen originated independently in *Sapranthus* (piptostigmoids), zonasulculate pollen in the *Guatteria* group (malmeoids).

The rest of the family, the inaperturate clade, consists of groups with inaperturate tetrads and monads. Besides loss of the aperture, this clade is united by proliferation of nexine foliations. The inaperturates are linked with the miliusoids by a shift from boat-shaped to round pollen. At the base are *Uvaria* and other lianas with granular monads (uvarioids), including the echinate *Monanthotaxis* group. The remaining taxa are united by tetrads, *Xylopia*, *Cananga*, and *Neostenanthera*, with granular tetrads and a nexine thought to be derived from fused granules, form the xylopioid clade. This is linked with *Toussaintia* and four taxa with pseudosyncarpous fruits, a feature that arises independently in the *Annona* group. The pseudosyncarps include *Fusaea*, which Walker (1971a) associated with the xylopioids because it has tetrads, but the other members revert to monads.

The annonoid clade, essentially Walker's (1971a) *Annona* subfamily, is united by a third origin of reticulate-columellar structure, often very coarse. Besides *Annona*, this includes the North American genus *Asimina*; the *Cymbopetalum* group, with extremely large polyads; and two African genera united by parasyncarpous fruits, *Monodora* and *Isolona*. *Isolona* has single grains; its position implies that these are secondary (*cf.* Le Thomas *et al.*, 1986). Consistent with this, its exine is thinner on one side, presumably a vestige of the proximal thinning in the tetrads. Another member is the African-Asian liana *Artabotrys*, with hook-like inflorescences for climbing, which has a thin area that Walker (1971a) and Le Thomas (1980-1981) interpreted as a reduced sulcus; however, its position in the annonoids suggests that the thin area may actually be a proximal thinning like that of *Isolona* (Doyle and Le Thomas, 1996).

Doyle and Le Thomas (1997) did experiments to estimate the contribution of pollen and other characters to this result (i.e., their relative systematic value). When pollen characters are removed, the basal branch is not *Anaxagorea* but the annonoid genus *Uvariopsis*. This rooting requires a highly implausible scenario of pollen evolution -

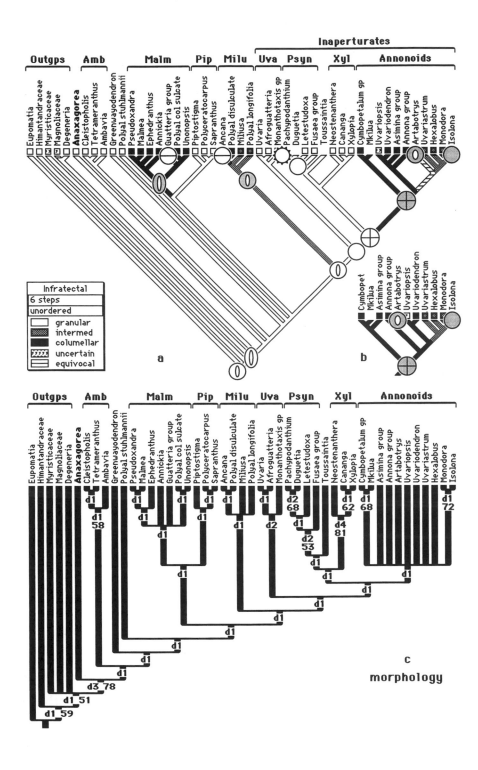

from granular monosulcates to columellar tetrads at the base of the family, then back to granular monosulcates in *Anaxagorea* - and is much less parsimonious when pollen characters are included. Based on consistency indices, which are inverse measures of homoplasy, pollen and fruit and seed characters are more informative than vegetative and floral characters.

Unfortunately, it was clear from the beginning that these results were uncertain. First, the trees changed considerably from one analysis to the next, with addition of taxa, refinement of character definitions, and scoring of previously unknown states, although many groups remained intact and only changed relations with each other. Second, many such alternative arrangements differ by only one or two steps in terms of the same data set. Third, consistency indices are much lower than average for this number of taxa (Sanderson and Donoghue, 1989), indicating that homoplasy is unusually rampant. To resolve these problems, there may be new characters that can be extracted from morphology, and many known anatomical and cytological characters remain to be studied in many taxa. However, the greatest source of new characters is molecular data, which have overturned or proved beyond doubt many previous ideas on plant relationships (*cf.* Doyle, 1998).

Molecular analyses

The first molecular study of Annonaceae was by van Zuilen (1996), based on spacer sequences between the *trn*L 3' intron and the *trn*F gene in 19 genera. This confirmed some of the main results of morphological analyses: *Anaxagorea* was basal, and taxa with inaperturate tetrads and monads (including *Uvaria* and *Artabotrys*) formed a clade, as did the pseudosyncarps. Some relationships in the inaperturates and the position of *Sapranthus* were different. However, it was hard to say whether these differences were due to conflicts between the two kinds of data or to the smaller sampling of taxa.

A much larger data set (over 150 species) has been amassed by one of us (Bygrave) for the chloroplast gene *rbc*L, which has been widely used in higher-level phylogenetic studies (Chase *et al.*, 1993). Taxon sampling was designed to include the taxa used by Doyle and Le Thomas (1996), so that the two data sets could eventually be compared and combined, both to study the congruence and implications of the two kinds of data and to obtain a better estimate of the phylogeny. The present article concentrates on implications for the evolution and systematic value of pollen characters, based on their behaviour on trees derived from *rbc*L and the combined data set. These comparisons in turn suggest some general lessons on the interpretation and use of pollen characters and ways in which molecular systematics may suggest new avenues of palynological research.

FIG. 1. (a, b) Representative most parsimonious trees from the two islands found in the morphological analysis (Doyle and Le Thomas, 1996), with sketches of pollen types superimposed on lines shaded with MacClade (Maddison and Maddison, 1992) to indicate exine structure. Sketches show pollen unit (monads, tetrads), shape, aperture condition (monosulcate, sulculate, inaperturate), exine structure (granular vs. columellar or intermediate), and echinate sculpture. (c) Strict consensus of most parsimonious trees, with decay and bootstrap values. Outgps = magnolialian outgroups, Amb = ambavioids, Malm = malmeoids, Pip = piptostigmoids, Milu = miliusoids, Uva = uvarioids, Psyn = pseudosyncarps, Xyl = xylopioids, Polyal = *Polyalthia*, col sulcate = columellar sulcate.

Materials and Methods

A major issue in combining molecular and morphological data sets is whether to represent clades by single species (exemplars) or treat them as units (compartmentalisation: Mishler, 1994). Each approach has strengths and weaknesses (Yeates, 1995; Bininda-Emonds *et al.*, 1998). When states vary within a taxon, compartmentalisation requires assumptions on which states are ancestral, preferably based on an analysis of relationships within the taxon. A problem is that this increases the number of unknown or polymorphic scorings, which in combination with homoplasy may lead not only to lower resolution but also to incorrect relationships (Nixon and Davis, 1991). However, the use of exemplars may also lead to incorrect relationships, because most exemplars have apomorphies that arose within the taxon, and some of these may be convergences with other taxa (Bininda-Emonds *et al.*, 1998). Experiments by Bininda-Emonds *et al.* (1998) suggest that reconstruction of ancestral states gives more accurate results than use of majority states or exemplars.

For practical reasons, molecular systematists have usually relied on exemplars (an exception is Doyle *et al.*, 1994), but morphologists have used both approaches. Doyle and Le Thomas (1996) used a pragmatic mixture of approaches. They represented some large, heterogeneous groups, such as the malmeoids and *Polyalthia*, with several exemplars, chosen to cover the morphological spectrum in the group. Others they treated as units, such as the *Cymbopetalum* group (= Tribe Bocageeae), for which the analysis of Johnson and Murray (1995) provided trees that allowed reconstruction of ancestral states. Except in this case, these groups correspond closely to the genus after which they are named, and for simplicity we will therefore refer to them as *Annona*, *Asimina*, *Fusaea*, *Guatteria*, and *Monanthotaxis*.

In the present study, we used one exemplar species in the *rbc*L data set to represent each taxon of Doyle and Le Thomas (1996), rather than undertaking the laborious task of reconstructing ancestral states (see Appendix for species and voucher specimens). Fortunately, the complete *rbc*L analysis confirms that all of the morphological taxa represented by more than one species are monophyletic, with one exception: Doyle & Le Thomas assumed that the African genus *Anonidium* Engl. & Diels belonged to the *Annona* group, but in the complete *rbc*L trees it is a nearby line. To reduce the risk of incorrect placement due to convergent apomorphies, we usually chose the member of a clade that formed the shortest branch on an *rbc*L tree (based on 76 species corresponding to the Doyle and Le Thomas taxon set). Where there were several short branches, we used secondary criteria such as sequence completeness or morphological advancement. To root Magnoliales as a whole, we used *Tasmannia* R.Br., in the Winteraceae, which (together with Canellaceae) are the closest outgroup in the molecular analyses of Chase *et al.* (1993) and Savolainen *et al.* (in press).

Data sets might also be combined by including all species in the molecular data set that belong to a given taxon in the morphological data set, and scoring these species the same for morphological characters (Linder and Kellogg, 1995). However, this procedure would greatly increase the number of taxa and the computation time. The two data sets might be compared by "pruning" the complete *rbc*L tree down to the taxa in the morphological analysis. This might give a better indication of what *rbc*L says about relationships of these taxa, but it would be difficult to say whether conflicts between trees based on morphology and *rbc*L were due to differences between the two types of data, or to more or less complete taxon sampling. Actually, trees based on the complete and reduced *rbc*L taxon sets are encouragingly similar. Where they do differ, it is usually relationships based on the reduced set that appear more anomalous in terms of morphology, some of which will be mentioned below.

Asian Annonaceae are under-represented in this study, relative to their high species diversity, since the morphological analysis concentrated on taxa for which exine structure is known, most of which are African and American. However, this should not have a major effect on inferences on pollen evolution, since most Asian taxa appear to belong to two clades, the miliusoids and uvarioids, which have fairly uniform pollen.

Sequences of the *rbc*L exon were obtained using total DNA (tDNA) extraction methods, PCR, primers, and sequencing methods described in Fay *et al.* (1997). Some tDNAs were kindly provided by C.M. van Zuilen (for extraction methods, see van Zuilen, 1996). Because the tDNA extracts were degraded, we obtained no sequences for *Afroguatteria* and *Uvariastrum* (although we did obtain a partial *trn*L-F sequence for the former) and only a partial sequence for *Unonopsis*. In the combined analyses, *Afroguatteria* and *Uvariastrum* were scored as unknown for *rbc*L characters, so their position is determined by morphology. In figures showing the evolution of morphological characters on molecular trees, they are inserted at consistent positions. *Letestudoxa* could not be amplified using the 1F/724R primers, so its *rbc*L sequence is also incomplete. Sequences for *Asimina, Cananga, Isolona, Mkilua, Sapranthus, Degeneria, Eupomatia, Galbulimima, Knema, Magnolia,* and *Tasmannia* were obtained from the Jodrell archive (Qiu *et al.,* 1993; Chase and Schatz, unpublished data) and Genbank. All tDNA extracts and sequence data are available upon request to P. Bygrave or M.W Chase, The Jodrell Laboratory, Royal Botanic Gardens, Kew.

To find most parsimonious trees, we used PAUP (version 3.1.1; Swofford, 1990), conducting many heuristic search replicates with TBR branch swapping and MULPARS to find different "islands" (Maddison, 1991), or families of most parsimonious trees that can be transformed into each other by moving one taxon at a time. We used MacClade (Maddison and Maddison, 1992) to study character evolution and character support for clades. To evaluate the relative parsimony of alternative hypotheses, we used the constraints option in PAUP to force particular taxa together.

To evaluate the robustness of clades, we also conducted bootstrap and decay analyses (Felsenstein, 1985; Bremer, 1988; Donoghue *et al.,* 1992). Bootstrap analyses were based on 200 replicates, each using closest addition sequence and TBR branch swapping, holding 5 trees at each step, and retaining up to 1000 trees. Decay analyses were performed by searching for trees equal to, or less than, one and two steps longer than the shortest trees and observing which clades remain in the strict consensus. To determine the decay value of clades remaining in the "two-off" search, some of which may decay in trees of this length that were not found because the search was not exhaustive, we searched for shortest trees not compatible with a constraint tree in which the group in question forms a clade.

Results and Discussion

The 180 trees based on morphology (425 steps; CI = .27, RI = .55, RC = .15) form two islands, which differ in relationships in the annonoids. Representative trees from the two islands are illustrated in Figs. 1a and 1b. Fig. 1c is a strict consensus tree, with decay and bootstrap values (the latter not calculated by Doyle and Le Thomas, 1996).

The *rbc*L analysis yielded 15,172 trees (709 steps; CI = .53, RI = .63, RC = .34), all belonging to one island. A strict consensus is shown in Fig. 2a. The consistency indices are not directly comparable to those for morphology, since autapomorphies were excluded from the morphological data set but not the *rbc*L data set. The large number of trees reflects many equally parsimonious arrangements within the malmeoid-piptostigmoid-miliusoid (MPM) and inaperturate clades, in which many internodes have lengths of 0 or 1.

The 780 trees based on the combined data set (1191 steps; CI = .41, RI = .56, RC = .23) form three islands, consisting of 42, 38, and 700 trees. A strict consensus is shown in Fig. 2b, while consensus trees for each of the three islands are shown in Fig. 3. The islands differ most significantly in outgroup relationships (position of *Eupomatia*) and the positions of the *Cananga*-ambavioid clade, *Annickia*, and the *Artabotrys-Guatteria* clade. Relationships at the piptostigmoid-malmeoid grade and among the inaperturates appear to be poorly resolved. However, much of the lack of resolution is due to "jumping" of single taxa between widely separated positions: *Annickia* between the base of the MPM clade, just below the malmeoids, and within the malmeoids; *Afroguatteria* between the uvarioids and *Guatteria*; *Letestudoxa* between the *Artabotrys-Guatteria* clade and the pseudosyncarps. *Uvariastrum* alternates among various positions above the *Annona-Asimina* line. These uncertainties reflect the fact that we have no *rbc*L data for *Afroguatteria* and *Uvariastrum* and only half a sequence for *Letestudoxa*. The association of *Afroguatteria* with *Guatteria* can be eliminated from further discussion because *trn*L-F data (Bygrave, unpublished) place it in the uvarioids, as do morphological data.

Implications of these results will be illustrated by plotting characters on trees based on morphology, *rbc*L, and the combined data. These comparisons reveal several relatively minor conflicts between the two kinds of data, plus one major conflict, where morphology (including pollen) seems to strongly support one alternative, but *rbc*L data are so much stronger that they leave little doubt that the morphological result is incorrect. We will consider whether this error arose because morphology is intrinsically misleading, or because the characters involved were misinterpreted in ways that might have been avoided. Finally, we will assess how these results affect the generalisations of Doyle and Le Thomas (1997) concerning the relative value of pollen and other characters.

General implications

Fig. 4 shows an *rbc*L tree with aperture evolution traced on it. As with morphology (Fig. 1), *Anaxagorea* is basal, confirming that its granular monosulcate pollen is primitive. Above this are three major clades. The MPM clade includes the malmeoids, piptostigmoids, and miliusoids of Doyle and Le Thomas (1996), plus *Greenwayodendron* and *Polyalthia stuhlmannii* (previously lower), but minus *Guatteria*. This clade also retains monosulcate pollen, replaced by disulculate in the miliusoids. These taxa (with the differences noted) form two adjacent lines in Fig. 1, but they were united in earlier analyses (Doyle and Le Thomas, 1994, 1995), and in terms of the Doyle and Le Thomas (1996) data set it "costs" only 2 extra steps to force them together as a clade. Hence this is only a weak contradiction. The second clade includes the ambavioids (previously just above *Anaxagorea*). However, these are linked with *Cananga*, which was in the xylopioid clade in the morphological analyses; this is the major conflict mentioned. The third clade consists of the inaperturates, also united by multiple nexine foliations, plus *Guatteria*. The uvarioids with monad pollen are not basal, but nested within the clade, linked with *Hexalobus*. Furthermore, the two remaining xylopioids, *Xylopia* and *Neostenanthera*, are also separated from each other. In analyses of the complete data set (Bygrave, unpublished), *Neostenanthera* is linked with *Anonidium* (which Doyle and Le Thomas assumed was in the *Annona* group), on or next to a line leading to *Annona, Asimina, Diclinanona, Disepalum*, and *Goniothalamus* (the last three not included in the present analysis). Interestingly, *Neostenanthera* and *Goniothalamus* share "winged" seeds, the unique feature of indument on the seed (van Setten and Koek-Noorman, 1992), and similar pollen (Walker, 1971a).

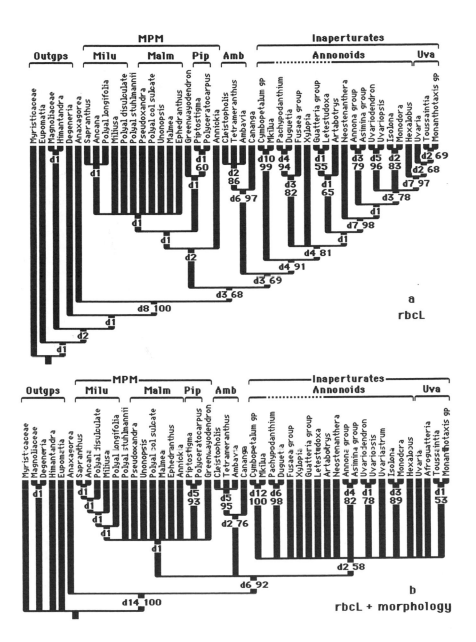

Fig. 2. (a) Strict consensus of most parsimonious trees based on *rbc*L, with decay and bootstrap values. (b) Strict consensus of most parsimonious trees based on the combined data set (*rbc*L plus morphology), with decay and bootstrap values. Brackets are dotted above taxa not included in the annonoids by Doyle and Le Thomas (1996). Abbreviations as in Fig. 1, except MPM = malmeoid-piptostigmoid-miliusoid clade.

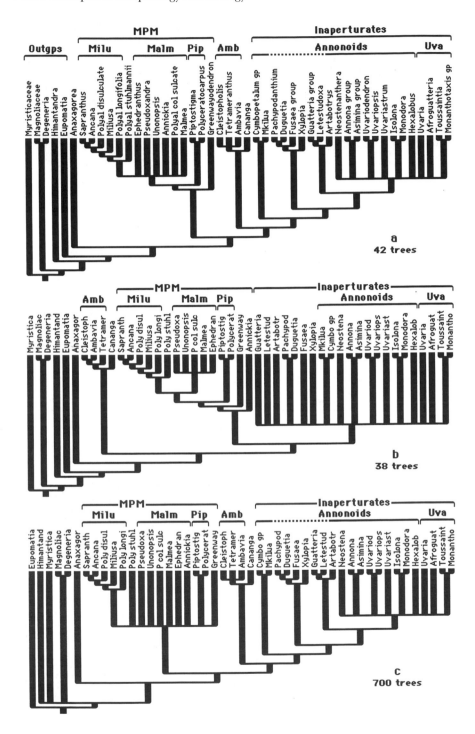

FIG. 3. Strict consensus of each of the three islands of trees based on the combined data set.

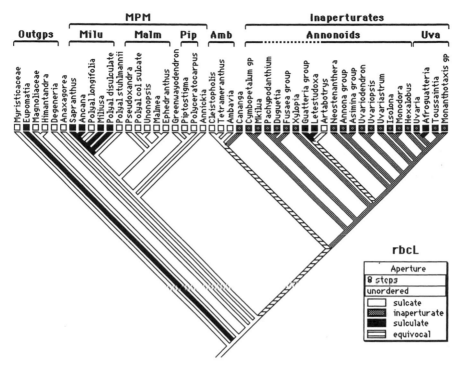

FIG. 4. Representative tree based on *rbc*L, showing evolution of the pollen aperture character.

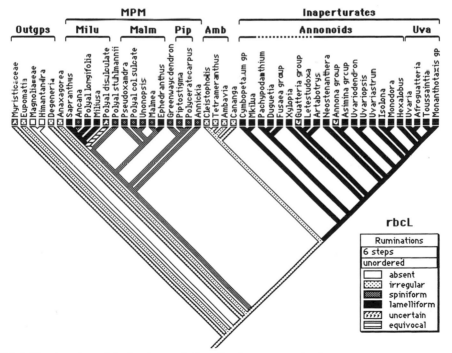

FIG. 5. Tree based on *rbc*L, showing evolution of endosperm ruminations.

Fig.5 shows a non-palynological character that is actually more consistent with *rbc*L than it is with the morphological data set, namely endosperm ruminations, which are irregular in *Anaxagorea* and the ambavioids but become spiniform in the MPM clade and lamelliform in the inaperturates, with some convergences in the miliusoids. It undergoes 6 steps, rather than 8 on the morphological trees (Doyle and Le Thomas, 1996, Fig. 18). Other cases where *rbc*L supports the systematic value of seed and fruit characters will be seen below.

So far, molecular data seem fairly consistent with previous ideas on pollen evolution. However, exine structure presents a more confused picture. On the morphological trees (Fig. 1a), columellar structure is derived from granular three times, with the intermediate type transitional to columellar in the miliusoids, derived from columellar in the annonoids, so the character undergoes 6 steps (including one in Magnoliaceae). However, on the *rbc*L trees (Fig. 6a), the three states are intermixed in both the MPM clade and the inaperturates, implying a total of 12-13 steps. A critical taxon is the African columellar genus *Annickia*, which is at the base of the MPM clade, below the granular piptostigmoids, rather than in the malmeoids.

These results suggest that exine structure is much less reliable in Annonaceae than previously thought (consistent with molecular analyses of angiosperms as a whole, which imply that the granular structure of Magnoliales is itself a reversal, not primitive: Doyle, 1998). However, closer examination suggests that some of the contradictory molecular data are weak. This fits the fact that the MPM clade is the part of the *rbc*L tree where relationships are most poorly resolved, as seen in the consensus tree and the low bootstrap and decay values in Fig. 2a.

Trees based on the combined data (Figs. 2b, 3a-c) also indicate that the strength of the molecular results varies in different parts of the tree. In most respects, the combined trees are like both morphological and *rbc*L trees where these agree (basal position of *Anaxagorea*, with stronger bootstrap support; unity of the ambavioids; unity of the inaperturates), but like the *rbc*L trees where the two differ (breakup of the xylopioids, association of *Cananga* with the ambavioids, unity of the MPM clade, nesting of the uvarioids within the annonoids), indicating that the *rbc*L data overwhelm morphology. However, in some trees (Fig. 3b) the *Cananga*-ambavioid line is just above *Anaxagorea*, as in morphological trees. This implies that the *rbc*L evidence linking it with the inaperturates is weak, as does the low bootstrap value for this node in Fig. 2a (69).

Relationships in the combined trees are also more consistent with pollen characters, especially in the MPM clade. In *rbc*L trees, *Polyalthia longifolia*, which has a reduced sulcus, is nested among disulculate taxa, implying a reversal to monosulcate (Fig. 4), and the disulculate *Polyalthia* group is separated from the other miliusoids in some trees. However, in the combined trees (Fig. 2b), the miliusoids always form a clade, with *P. longifolia* at the base, so their disulculate pollen is derived once from monosulcate, with no reversals.

The combined trees also suggest that exine structure is less homoplastic than *rbc*L would imply. This is not surprising, but it is not inevitable either; conceivably, the *rbc*L data could be so strong that exine data would have no effect. In some trees (for example, Fig. 3b), *Annickia* is still basal in the MPM clade, implying that its columellae are not homologous with those of other malmeoids, and columellae arise three times in the MPM clade. However, in Fig. 6b, *Annickia* is in the malmeoids, so that columellae arise only once here (directly from granular) and once in the miliusoids (via intermediate). In Fig. 6c, *Annickia* is just below the other malmeoids, consistent with anything from one origin (plus reversals) to three. As a result, the character undergoes 4 or 5 steps in the MPM clade, rather than 6 steps on the *rbc*L trees. In other words, here it may be exine structure that refutes molecular data, rather than vice versa.

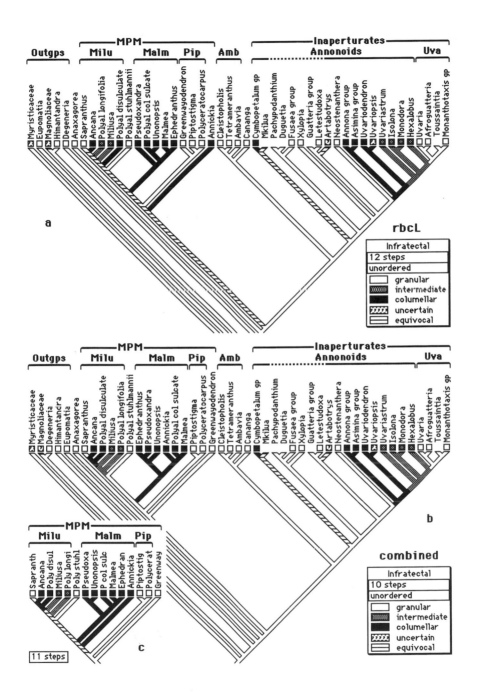

FIG. 6. Evolution of infratectal structure on trees based on (a) *rbcL* and (b, c) the combined data set.

In contrast, relationships in the inaperturates are similar in the combined and *rbc*L trees, with separate origins of columellae in the *Cymbopetalum* group, the core annonoids, and *Artabotrys*. As a result, the character as a whole is only 1-3 steps less homoplastic than it was on the *rbc*L trees (10-11 steps rather than 12-13). This suggests that origin of columellae and reversion to granular structure were indeed easier than previously inferred, though not as much so as *rbc*L alone would imply.

A character that might help resolve relationships in the MPM clade is chromosome number, which like ruminations is more consistent with the *rbc*L and combined trees than those based on morphology. On the morphological trees (Doyle and Le Thomas, 1996, Fig. 19), the basic number is n = 8, which gives rise to n = 9 four times (in *Sapranthus*, some malmeoids, the miliusoids, and the *Cymbopetalum* group) and n = 7 three times, for a total of 7 steps. However, on the *rbc*L and combined trees (Fig. 7), where the malmeoids and miliusoids are consolidated, n = 9 originates only twice, so the number of steps is only 5 (or 6 if *Annickia*, with n = 8, is nested among the malmeoids, as in Fig. 6b). Published data are lacking for the piptostigmoids and most malmeoids, but Fig. 7 predicts that the former should have n = 8, the latter n = 9. Future counts could therefore strengthen or weaken this arrangement.

Specific conflicts

Several conflicts between trees based on morphology and *rbc*L are relatively minor, in the sense that they involve alternatives that are almost as parsimonious in terms of morphology, or have little effect on hypotheses concerning the course of pollen evolution.

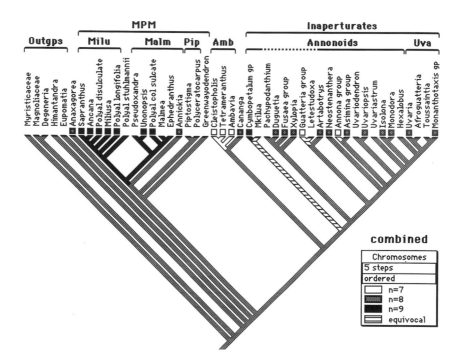

FIG. 7. Tree based on the combined data set, showing evolution of chromosome number.

First, *Letestudoxa* is dissociated from the other pseudosyncarps in the *rbcL* trees. This appears to be an artifact, first of the fact that we obtained only half of its *rbcL* sequence, and second of taxon sampling; in the complete analysis (Bygrave, unpublished), which includes a fifth genus of pseudosyncarps, *Pseudartabotrys* Pellegr., *Letestudoxa* is in the pseudosyncarp clade. Furthermore, with the present *rbcL* data set, bootstrap support for the *Artabotrys-Letestudoxa-Guatteria* clade is only 65, and it costs only 2 steps to place *Letestudoxa* in the pseudosyncarps. Finally, all four pseudosyncarps are united in some combined trees (Fig. 3b).

Second, *Sapranthus* is in the piptostigmoids in the morphological trees, based on percurrent tertiary venation and thick-walled, sessile fruits, but in the miliusoids in the *rbcL* and combined trees, as argued by Morawetz (1988). There is evidence for this in the morphological data set: it costs only 1 step to move *Sapranthus* to the miliusoids, and although its pollen is like that of the piptostigmoids in being large and granular, it is disulculate, as in the miliusoids. Conversely, in the *rbcL* and combined trees, *Greenwayodendron* moves up onto or next to the piptostigmoid line; this is consistent with pollen morphology, since *Greenwayodendron* resembles *Piptostigma* and *Polyceratocarpus* in having granular monosulcate pollen.

Third is the case of the African liana *Toussaintia*, which converges with *Magnolia* L. in having an elongate receptacle and numerous petals. In Doyle and Le Thomas (1996), *Toussaintia* was linked with the pseudosyncarps, based on large sepals and imbricate petals, but in the *rbcL* and combined trees it is in the uvarioids, consistent with its liana habit and stipitate fruits. *Toussaintia* "jumped" between these two positions in different morphological studies; in Doyle and Le Thomas (1994, 1995), it was at the base of the uvarioids.

An unexpected result concerns *Guatteria*, with zonasulculate pollen and an extremely reduced exine (Morawetz and Waha, 1985), which is nested in the malmeoids in morphological trees. In the *rbcL* and combined trees, it is in the inaperturates, linked with *Artabotrys* (with or without *Letestudoxa* and/or *Afroguatteria*, discussed above as probably unrelated). This seems to be a strong result; in terms of morphology, it costs 5 steps to link *Guatteria* with *Artabotrys*, but in terms of *rbcL* it costs 14 steps to nest it in the malmeoid-miliusoid clade. *Guatteria* also shares at least one morphological advance with *Artabotrys*, numerous bracteoles, although this arises in several other taxa. This result does not contradict pollen data as much as it does other characters (silica cells, complex midrib histology, one basal ovule), since *Guatteria* had to be scored as unknown for most exine characters; essentially, its pollen could be derived from anything. In Doyle and Le Thomas (1996), *Guatteria* was linked with *Unonopsis* and the columellar-sulcate *Polyalthia* group by crossed mesotesta fibres, but these also occur in most inaperturates. Endosperm ruminations are usually spiniform in the MPM clade and lamelliform in the inaperturates, but *Guatteria* is variable (van Setten and Koek-Noorman, 1992). Perhaps only molecular data can reveal its true relationships.

A more severe conflict involves the uvarioids, which are at the base of the inaperturates in the morphological trees, but nested within them in the *rbcL* and combined trees, linked with the African annonoid *Hexalobus* (or *Uvariastrum* and/or *Hexalobus* in some combined trees). This implies that the granular monads of the uvarioids are derived from columellar tetrads. In the morphological trees, *Hexalobus* is linked with the parasyncarps *Monodora* and *Isolona*, based on fused petals, a feature restricted to these taxa (and the apparently related genus *Asteranthe*, not included in this analysis). The *rbcL* data confirm that the parasyncarps are nearby, but by placing the uvarioids between them and *Hexalobus* they imply that sympetaly arose twice, or that the petals of the uvarioids are secondarily free.

Morphological evidence that the uvarioids do not belong here seems strong: it costs 6 steps to force them together with *Hexalobus*, and there is not a single morphological character in our data set that would link the two groups. Placing the uvarioids here requires several reversals - not only from columellar tetrads to granular monads, but also from capitate to sessile stigmas, and from sessile to stipitate fruits. We see no way to reinterpret these characters. However, the opposing molecular evidence is much stronger - the uvarioids are linked with *Hexalobus* at a bootstrap level of 97, and it costs 23 steps to move them to the base of the inaperturates. Furthermore, neither of the pollen reversals is unusual. Secondary monads also occur in the pseudosyncarps, *Artabotrys-Guatteria*, and *Isolona*, and one of the uvarioids, *Toussaintia*, has tetrads, which may or may not be primitive. With the arrangement found, intermediate exine structure forms a transitional stage in the reversal from columellar to granular (Fig. 6). Finally, some morphological characters suggest that the uvarioids belong somewhere above *Artabotrys-Guatteria*, particularly oil cells in the nucellus and the endosperm. Like the case of ruminations (Fig. 5), this may be evidence that seed characters are especially useful in Annonaceae. On the other hand, it is less certain that the parasyncarps are separated from *Hexalobus*, since in trees based on *trn*L-F these groups form a clade and the uvarioids are slightly lower (Bygrave, unpublished).

The xylopioid problem

The worst conflict is the breakup of the xylopioids, with granular tetrad pollen, in the *rbc*L and combined analyses. This is especially disconcerting because the xylopioids are the most strongly supported clade in the morphological analysis (Fig. 1b), with decay and bootstrap values of 4 and 81, respectively. This is not just a conflict with the Doyle and Le Thomas data set; the xylopioids were also grouped by Walker (1971a) and by van Heusden (1992), working on floral morphology, though not by van Setten and Koek-Noorman (1992), working on fruits and seeds.

There can be little doubt in this case that the morphological result is incorrect. In the *rbc*L analysis (Fig. 2a), bootstrap support for the *Cananga*-ambavioid clade is 97. In terms of morphology, it costs 9 steps to move *Cananga* to the ambavioids and 5 more to separate *Xylopia* and *Neostenanthera*, but in terms of *rbc*L it costs 24 steps to associate the three genera, and there is not a single *rbc*L synapomorphy that would unite them. In contrast, *Cananga* does share several morphological features with the ambavioids. The most remarkable is presence of a third (middle) integument, a feature restricted to Annonaceae (Corner, 1949; Christmann, 1986; Svoma, 1998). Another is apical articulation of the monocarp stipe in *Cananga* and *Cleistopholis* (*Ambavia* and *Tetrameranthus* were scored as unknown because they have sessile monocarps); this may reflect origin of the stipe from the receptacle rather than the carpel (*cf.* Johnson, 1989). A synapomorphy that shows more homoplasy across the family is low vessel density. *Cananga* is also like the ambavioids in having irregular endosperm ruminations - this is a reversal when *Cananga* is in the xylopioids, but primitive when it is linked with the ambavioids. *Cananga* is not a member of the ambavioids; they are united by two lateral ovules, tuberculate seeds, and the chromosome number n = 7, as well as by *rbc*L data. According to the complete *rbc*L analysis (Bygrave, unpublished), two additional Asian tritegmic genera also belong here: *Cyathocalyx* Champ. ex Hook.f. & Thomson is linked with *Cananga*, while the ambavioids include *Mezzettia* Becc., also with n = 7. Christmann (1986) also reported three integuments in *Artabotrys* and other genera not included in this analysis, but these reports should be confirmed with developmental studies.

This case might be taken as evidence that pollen and other morphological characters are inherently misleading, compared to molecular data. However, the conflict may be due in part to potentially avoidable misinterpretations of morphology. This may be seen by critically examining the five characters that unite the xylopioids, plus some that place them in the inaperturates as a whole.

The first character is stamen morphology (Fig. 8): (0) laminar, as in *Anaxagorea* (Fig. 8a) and other Magnoliales; (1) narrower but with a tongue-like connective extension, as in *Ambavia* (Fig. 8b), *Tetrameranthus*, and *Greenwayodendron*; (2) peltate-truncate, with a cap covering the sporangia, as in most of the family; or (3) peltate but with an apiculus on the cap, the xylopioid state. Since this seemed like a logical series, Doyle and Le Thomas (1996) ordered the character, so there were two steps between tongue-like and peltate-apiculate. However, the concept of a peltate-apiculate state appears to be invalid. It was based on species such as *Xylopia quintasii* Engl. & Diels (Fig. 8c) and extended to *Cananga* and *Neostenanthera*. However, other *Xylopia* species, such as *X. staudtii* Engl. & Diels (Fig. 8d), have normal peltate-truncate stamens. *Xylopia* was therefore scored as 2/3, but when it is located between *Cananga* and *Neostenanthera*, the algorithm treats the clade as having state 3. *Cananga* (Fig. 8e) has a pointed apex, but the subapical part is more elongate than the cap of typical peltate taxa, like the tongue-like extension of the ambavioids. Endress (1975) also noted that the apex is stiffened by a sclerenchyma layer, which is lacking in peltate stamens. The pointed apex in *Neostenanthera* (Fig. 8f) is a small, tilted cap that does not cover the sporangia; it

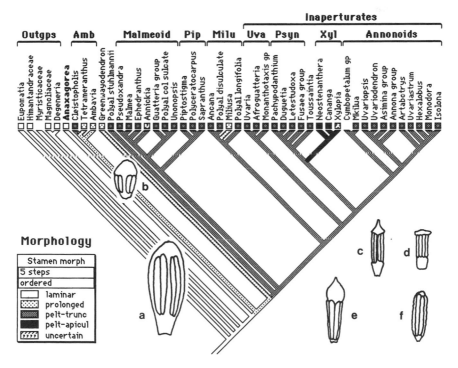

FIG.8. Tree based on morphology (Doyle and Le Thomas, 1996), showing evolution of stamen morphology. Sketches of stamens: (a) *Anaxagorea acuminata* (Dunal) A.DC. (after Maas and Westra, 1984), laminar; (b) *Ambavia gerrardii* (Baill.) Le Thomas (Le Thomas 117), prolonged; (c) *Xylopia quintasii* (after Le Thomas, 1969), peltate-apiculate; (d) *X. staudtii* Engl. & Diels (Letouzey 13300), peltate-truncate; (e) *Cananga odorata* (Lam.) Hook.f. & Thomson (Deroin s.n.); (f) *Neostenanthera myristicifolia* (Oliv.) Exell (Letouzey 13025).

might be reduced from cap-like or tongue-like. Ordering this character acts as an additional force "attracting" *Cananga* to the xylopioids: it means the character undergoes two extra steps when *Cananga* is linked with the ambavioids, when actually there may be no steps at all.

Another relevant character is stigma form: (0) sessile, (1) capitate (often like a match head), or (2) borne on a definite style, where state 2 unites the xylopioids and pseudosyncarps. *Xylopia* and *Neostenanthera* have long, thin, tapering styles (Le Thomas, 1969), but *Cananga* has an obconical stigma (van Heusden, 1992), more like an elongate version of the capitate type. As in *Cyathocalyx* section *Drepananthus* Scheff. *sensu* J. Sinclair, the stigmas of *Cananga* form a disk-like compitum (Sinclair, 1955; Endress, 1994).

Two other xylopioid synapomorphies are narrow petals and concave petal base (which are not always correlated). We see no way to reject these *a priori*; the petal base in *Cananga* is hard to categorise, but for this reason it was already scored 0/1. However, it is suggestive that both characters are highly homoplastic even on the morphological trees.

The xylopioids are also united by two pollen characters. One is basal layer composed of fused granules (Fig. 9), equated with the state found in the monosulcate genera *Cleistopholis* and *Polyceratocarpus* (Fig. 10; Le Thomas, 1980-1981), but coarser. The other (Fig. 15) is loss of multiple nexine foliations, which arose at the base of the inaperturates. This character has an especially strong effect, since it has three states (absent, few and discontinuous, multiple), which were ordered. Thus if the xylopioids are nested in the inaperturates, origin of their nexine structure involves three steps (fusion of granules; reduction and then loss of foliations), the same as three characters "attracting" them to each other. However, closer examination of related taxa suggests that other scenarios should be considered. In annonoid tetrads (Fig. 11), the bases of the columellae are often enlarged, producing an undulating layer suggestive of the nexine of *Xylopia* (Fig. 9), but with foliations below. Considering the positions of *Xylopia* and *Neostenanthera* in the *rbc*L trees, perhaps their nexines were derived from this type by massive deposition of sporopollenin, obscuring the foliations (Doyle and Le Thomas, 1995). If so, the transformation would involve not three steps, but only one. In addition, there are subtle variations in the nexine of the xylopioids; it seems most clearly composed of large granules in *Cananga* (Fig. 12), which makes sense if *Cananga* is related to the ambavioids.

At the very least, these observations suggest that previous character definitions were too heavily biased toward one rather speculative scenario of exine evolution. Bias against the alternative "annonoid" hypothesis could be avoided by treating foliations as unordered and scoring *Xylopia* and *Neostenanthera* as unknown for fused granules. Ontogenetic studies might help resolve this problem - for example, if they showed that foliations are present early in development in *Xylopia* but not in *Cananga*.

Finally, in all tetrads of *Cananga* examined with SEM (Fig. 14), each monad has an indentation at the distal pole, like a vestigial sulcus. This means *Cananga* is more like *Ambavia* (Fig. 13), with its distal sulcus and thin proximal exine (Le Thomas and Lugardon, 1975), than previously recognised. Doyle and Le Thomas (1994, 1996) argued that the thin proximal exine in *Isolona* is a vestige of a tetrad ancestry, but not that of *Ambavia*, which was not nested among tetrad groups. However, if *Cananga* and the ambavioids are linked with the inaperturates, as in *rbc*L and most combined trees, tetrads may have arisen in the common ancestor of the two lines and reverted to monads in the ambavioids.

To test whether these questionable interpretations could be responsible for the conflict between morphology and *rbc*L, we eliminated the peltate-apiculate stamen state; scored *Cananga* as having tongue-like stamens, capitate stigmas, and a sulcus;

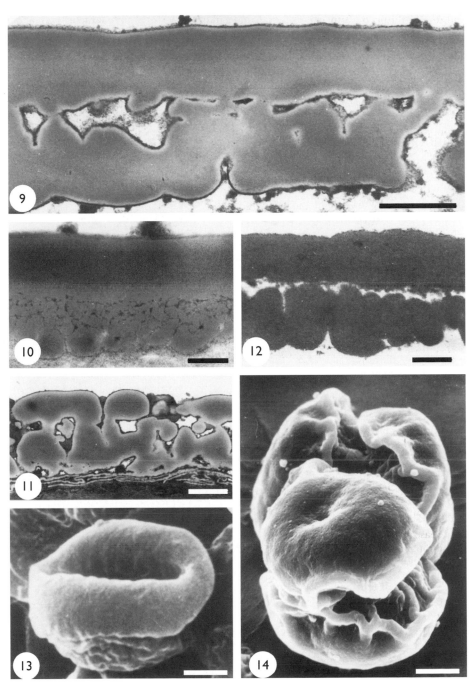

FIGS. 9–14. Pollen of Annonaceae. Figs. 9–12. Sections of exines, TEM. Fig. 9. *Xylopia staudtii* (Letouzey 11854), 20,000 ×. Scale bar = 1 μm. Fig. 10. *Polyceratocarpus pellegrinii* Le Thomas (Letouzey 10248), 50,000 ×. Scale bar = 0.2 μm. Fig. 11. *Uvariastrum pierreanum* Engl. & Diels (Letouzey 10225), 5000 ×. Scale bar = 2 μm. Fig. 12. *Cananga odorata* (Raynal 16312), 10,000 ×. Scale bar = 1 μm. Fig. 13. *Ambavia gerrardii* (Capuron 12583), SEM, 2200 ×. Scale bar = 5 μm. Fig. 14. *Cananga odorata* (Raynal 16312), SEM, 1100 ×. Scale bar = 10 μm.

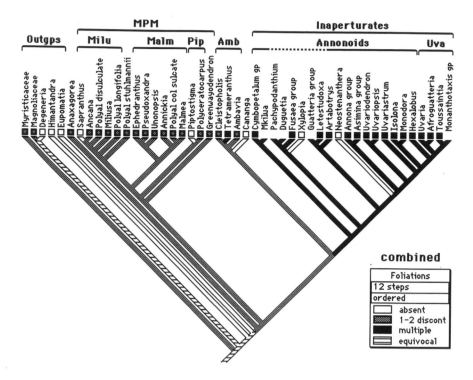

FIG. 15. Tree based on the combined data set, showing evolution of nexine foliations.

scored *Neostenanthera* as having either tongue-like or peltate stamens; redefined foliations as unordered; and scored *Xylopia* and *Neostenanthera* as unknown for fused granules. In the resulting trees, the xylopioids still form a clade, but the difference in parsimony between these trees and trees in which *Cananga* is linked with the ambavioids falls from 9 steps to only 2. The number of steps in the modified characters decreases on the *rbc*L and combined trees (on the latter, stamen form by 3 steps, foliations by 2), implying that they are more consistent with *rbc*L. In other words, what seemed to be a major conflict becomes a minor one. There are still too many convergences among the xylopioids that cannot be eliminated *a priori* (for example, petal form), and molecular data may be needed to infer their true relationships, but morphology is not as bad as it seemed.

Closer examination of exine structure might also be useful in the *Cymbopetalum* group, to see if it has features consistent with its apparent separation from other columellar "annonoids."

Consistency and homoplasy of character sets

Doyle and Le Thomas (1997) assessed the relative value of different kinds of morphological characters by examining three consistency indices for subsets of characters (Table 1). On the morphological trees, indices for vegetative and floral characters are low, while those for pollen and fruits and seeds are higher. On the *rbc*L

278

TABLE 1. Consistency indices for subsets of characters on trees based on morphology (Fig. 1b), *rbc*L (Fig. 5a), and the combined data (Fig. 5b). CI, consistency index; RI, retention index; RC, rescaled consistency index. Numbers in parentheses were found after reinterpretation of stamen, pollen, and stigma characters.

		Vegetative	Floral	Pollen	Fruit & Seed
	CI	24 (24)	24 (24)	26 (26)	33 (33)
Morphology	RI	43 (43)	44 (43)	73 (72)	62 (62)
	RC	10 (10)	11 (10)	19 (19)	21 (21)
	CI	22 (22)	21 (20)	17 (18)	33 (33)
*rbc*L	RI	35 (35)	31 (32)	55 (57)	62 (62)
	RC	8 (8)	6 (6)	10 (11)	20 (20)
	CI	23 (23)	22 (21)	18 (20)	33 (33)
Combined	RI	39 (39)	34 (35)	58 (60)	62 (62)
	RC	9 (9)	7 (7)	11 (12)	20 (20)

trees, indices for vegetative and floral characters fall slightly, but those for pollen decrease much more, while indices for fruits and seeds remain at almost the same high level. However, pollen characters still have a relatively high retention index; this means that although derived states arise more than once, they still unite several taxa in one or more of these cases (i.e., they are local synapomorphies rather than autapomorphies). On the combined trees, indices for most characters increase slightly, including those for pollen, reflecting the effect of exine structure on inferred relationships in the MPM clade, where *rbc*L data are weak. With the modified data set (in parentheses), indices for pollen characters on the *rbc*L and combined trees increase somewhat, implying that previous character definitions and state assignments exaggerated the level of homoplasy in pollen evolution.

Conclusion

These analyses provide a mixed assessment of previous inferences on pollen evolution in Annonaceae. They indicate that some results, such as the unity of the xylopioids with granular tetrads, were incorrect because of misinterpretation of some characters and undetectable homoplasy in others. As a general lesson, they warn against character definitions that depend too heavily on one scenario of morphological evolution, such as origin of the xylopioid nexine from fused granules, and against ordering states unless their interrelations are very clear. At the same time, our analyses confirm the primitive status of granular monosulcate pollen and the relationship of inaperturate monads and tetrads, and they suggest that pollen may complement *rbc*L in sorting out relationships in the MPM clade. They also show that molecular systematics can suggest new avenues of palynological research, such as developmental studies that might clarify homologies of the xylopioid nexine.

Acknowledgements

We thank Peter Endress for leaf material of *Ancana* and discussions of character evolution, Thierry Deroin for preserved floral material, Karen van Zuilen for tDNAs, David Johnson and Peter Linder for valuable comments on the manuscript, and the Ecole Pratique des Hautes Etudes for financial support.

References

Albert, V.A., Williams, S.E. and Chase, M.W. (1992). Carnivorous plants: phylogeny and structural evolution. *Science* 257: 1491–1495.

Bininda-Emonds, O.R.P., Bryant, H.N. and Russell, A.P. (1998). Supraspecific taxa as terminals in cladistic analysis: implicit assumptions of monophyly and a comparison of methods. *Biological Journal of the Linnean Society* 64: 101–133.

Bremer, K. (1988). The limits of amino acid sequence data in angiosperm phylogenetic reconstruction. *Evolution* 42: 795–803.

Chase, M.W. and 42 others (1993). Phylogenetics of seed plants: an analysis of nucleotide sequences from the plastid gene *rbcL*. *Annals of the Missouri Botanical Garden* 80: 526–580.

Christmann, M. (1986). Beiträge zur Histologie der Annonaceen-Samen. *Botanische Jahrbücher für Systematik* 106: 379–390.

Corner, E.J.H. (1949). The annonaceous seed and its four integuments. *New Phytologist* 48: 332–364.

Donoghue, M.J. and Doyle, J.A. (1989). Phylogenetic analysis of angiosperms and the relationships of Hamamelidae. In: P.R. Crane and S. Blackmore (editors). Evolution, systematics, and fossil history of the Hamamelidae, vol. 1, pp. 17–45. Clarendon Press, Oxford.

Donoghue, M.J., Olmstead, R.G., Smith, J.F. and Palmer, J.D. (1992). Phylogenetic relationships of Dipsacales based on *rbcL* sequences. *Annals of the Missouri Botanical Garden* 79: 333–345.

Doyle, J.A. (1998). Phylogeny of vascular plants. *Annual Review of Ecology and Systematics* 29: 567–599.

Doyle, J.A. and Donoghue, M.J. (1986). Seed plant phylogeny and the origin of angiosperms: an experimental cladistic approach. *Botanical Review* 52: 321–431.

Doyle, J.A. and Donoghue, M.J. (1992). Fossils and seed plant phylogeny reanalyzed. *Brittonia* 44: 89–106.

Doyle, J.A., Donoghue, M.J. and Zimmer, E.A. (1994). Integration of morphological and ribosomal RNA data on the origin of angiosperms. *Annals of the Missouri Botanical Garden* 81: 419–450.

Doyle, J.A. and Le Thomas, A. (1994). Cladistic analysis and pollen evolution in Annonaceae. *Acta Botanica Gallica* 141: 149–170.

Doyle, J.A. and Le Thomas, A. (1995). Evolution of pollen characters and relationships of African *Annonaceae*: implications of a cladistic analysis. In: A. Le Thomas and E. Roche (editors). 2e Symposium de Palynologie africaine, Tervuren (Belgique), pp. 241–254. Centre International pour la Formation et les Echanges Géologiques, Publication Occasionnelle, 1995/31, Orleans.

Doyle, J.A. and Le Thomas, A. (1996). Phylogenetic analysis and character evolution in Annonaceae. *Bulletin du Muséum National d'Histoire Naturelle, section B, Adansonia* 18: 279–334.

Doyle, J.A. and Le Thomas, A. (1997). Significance of palynology for phylogeny of *Annonaceae*: experiments with removal of pollen characters. *Plant Systematics and Evolution* 206: 133–159.

Endress, P.K. (1975). Nachbarliche Formbeziehungen mit Hüllfunktion im Infloreszenz- und Blütenbereich. *Botanische Jahrbücher für Systematik* 96: 1–44.

Endress, P.K. (1994). Diversity and evolutionary biology of tropical flowers. Cambridge University Press, Cambridge.

Fay, M.F., Swensen, S.M. and Chase, M.W. (1997). Taxonomic affinities of *Medusagyne oppositifolia* (*Medusagynaceae*). *Kew Bulletin* 52: 111–120.

Felsenstein, J. (1985). Confidence limits on phylogenies: an approach using the bootstrap. *Evolution* 39: 783–791.

Hesse, M., Morawetz, W. and Ehrendorfer, F. (1985). Pollen ultrastructure and systematic affinities of *Anaxagorea* (Annonaceae). *Plant Systematics and Evolution* 148: 253–285.

Johnson, D.M. (1989). Revision of *Disepalum* (Annonaceae). *Brittonia* 41: 356–378.

Johnson, D.M. and Murray, N.A. (1995). Synopsis of the tribe Bocageeae (Annonaceae), with revisions of *Cardiopetalum, Froesiodendron, Trigynaea, Bocagea,* and *Hornschuchia*. *Brittonia* 47: 248–319.

Le Thomas, A. (1969). Annonacées. Flore du Gabon. Muséum National d'Histoire Naturelle, Paris.

Le Thomas, A. (1980-1981). Ultrastructural characters of the pollen grains of African Annonaceae and their significance for the phylogeny of primitive angiosperms. *Pollen et Spores* 22: 267–342, 23: 5–36.

Le Thomas, A. and Lugardon, B. (1974). Quelques types de structure grenue dans l'ectexine de pollens simples d'Annonacées. *Comptes Rendus de l'Académie des Sciences Paris*, série D, 278: 1187–1190.

Le Thomas, A. and Lugardon, B. (1975). Ultrastructure d'un pollen original parmi les Annonacées (*Ambavia*). *Bulletin de la Société Botanique de France* 122: 109–111.

Le Thomas, A. and Lugardon, B. (1976). De la structure grenue à la structure columellaire dans le pollen des Annonacées. *Adansonia*, série 2, 15: 543–572.

Le Thomas, A., Morawetz, W. and Waha, M. (1986). Pollen of palaeo- and neotropical Annonaceae: definition of the aperture by morphological and functional characters. In: S. Blackmore and I.K. Ferguson (editors). Pollen and spores: form and function, pp. 375–388. Academic Press, London.

Linder, H.P. and Kellogg, E.A. (1995). Phylogenetic patterns in the commelinid clade. In: P.J. Rudall, P.J. Cribb, D.F. Cutler and C.J. Humphries (editors). Monocotyledons: systematics and evolution, pp. 473–496. Royal Botanic Gardens, Kew.

Lugardon, B. and Le Thomas, A. (1974). Sur la structure feuilletée de la couche basale de l'ectexine chez diverses Annonacées. *Comptes Rendus de l'Académie des Sciences Paris*, 279: 255–258.

Maas, P.J.M. and Westra, L.Y.T. (1984). Studies in Annonaceae. II. A monograph of the genus *Anaxagorea* A. St. Hil., Part 1. *Botanische Jahrbücher für Systematik* 105: 73–134.

Maddison, D.R. (1991). The discovery and importance of multiple islands of most-parsimonious trees. *Systematic Zoology* 40: 315–328.

Maddison, W.P. and Maddison, D.R. (1992). MacClade: analysis of phylogeny and character evolution, version 3. Sinauer Associates, Sunderland, Mass.

Mishler, B.D. (1994). The cladistic analysis of molecular and morphological data. *American Journal of Physical Anthropology* 94: 143–156.

Morawetz, W. (1988). Karyosystematics and evolution of Australian *Annonaceae* as compared with *Eupomatiaceae, Himantandraceae,* and *Austrobaileyaceae*. *Plant Systematics and Evolution* 159: 49–79.

Morawetz, W. and Waha, M. (1985). A new pollen type, C-banded and fluorochrome counterstained chromosomes, and evolution in *Guatteria* and related genera (Annonaceae). *Plant Systematics and Evolution* 150: 119–141.

Nixon, K.C. and Davis, J.I. (1991). Polymorphic taxa, missing values and cladistic analysis. *Cladistics* 7: 233–241.

Qiu, Y.L., Chase, M.W., Les, D.H. and Parks, C.R. (1993). Molecular phylogenetics of the Magnoliidae: cladistic analyses of nucleotide sequences of the plastid gene *rbc*L. *Annals of the Missouri Botanical Garden* 80: 587–606.

Sanderson, M.J. and Donoghue, M.J. (1989). Patterns of variation in levels of homoplasy. *Evolution* 43: 1781–1795.

Savolainen, V., Chase, M.W., Hoot, S.B., Morton, C.M., Soltis, D.E., Bayer, C., Fay, M.F., De Bruijn, A., Sullivan, S. and Qiu, Y.L. (in press). Phylogenetics of flowering plants based upon a combined analysis of plastid *atp*B and *rbc*L gene sequences. *Systematic Biology*.

Sinclair, J. (1955). A revision of the Malayan Annonaceae. *Gardens' Bulletin Singapore* 14: 149–516.

Soltis, D.E., Soltis, P.S., Nickrent, D.L., Johnson, L.A., Hahn, W.J., Hoot, S.B., Sweere, J.A., Kuzoff, R.K., Kron, K.A., Chase, M.W., Swenson, S.M., Zimmer, E.A., Chaw, S.M., Gillespie, L.J., Kress, W.J. and Sytsma, K.J. (1997). Angiosperm phylogeny inferred from 18S ribosomal DNA sequences. *Annals of the Missouri Botanical Garden* 84: 1–49.

Svoma, E. (1998). Seed morphology and anatomy in some *Annonaceae*. *Plant Systematics and Evolution* 209: 177–204.

Swofford, D.L. (1990). PAUP: phylogenetic analysis using parsimony, version 3.0. Illinois Natural History Survey, Champaign, Illinois.

Van Campo, M. and Lugardon, B. (1973). Structure grenue infratectale de l'ectexine des pollens de quelques Gymnospermes et Angiospermes. *Pollen et Spores* 15: 171–187.

Van Heusden, E.C.H. (1992). Flowers of Annonaceae: morphology, classification, and evolution. *Blumea Supplement* 7: 1–218.

Van Setten, A.K. and Koek-Noorman, J. (1992). Fruits and seeds of Annonaceae. Morphology and its significance for classification and identification. Studies in Annonaceae XVII. *Bibliotheca Botanica*, Stuttgart, 142: 1–101.

Van Zuilen, C.M. (1996). Patterns and affinities in the *Duguetia* alliance (Annonaceae). Molecular and morphological studies. Doctoral thesis, Utrecht University.

Walker, J.W. (1971a). Pollen morphology, phytogeography, and phylogeny of the Annonaceae. *Contributions from the Gray Herbarium* 202: 1–131.

Walker, J.W. (1971b). Contributions to the pollen morphology and phylogeny of the Annonaceae. I. *Grana* 11: 45–54.

Walker, J.W. (1972). Contributions to the pollen morphology and phylogeny of the Annonaceae. II. *Botanical Journal of the Linnean Society* 65: 173–178.

Yeates, D.K. (1995). Groundplans and exemplars: paths to the tree of life. *Cladistics* 11: 343–357.

Appendix

Taxa used in morphological analyses, exemplar species and voucher specimens representing them in the molecular analyses, and publication where the *rbc*L sequence was first reported (when not indicated, in this paper).

Outgroups: *Tasmannia insipida* DC. (Winteraceae), A.N. Rodd 5540, NSW (Albert *et al.*, 1992)

Eupomatia R.Br. (Eupomatiaceae): *E. bennettii* F.Muell., Qiu 90022, NCU (Qiu *et al.*, 1993)

Himantandraceae: *Galbulimima belgraveana* (F.Muell.) Sprague, P.H. Weston 929, NSW (Qiu *et al.*, 1993)

Degeneria I.W.Bailey & A.C.Sm.: *D. roseiflora* J.M. Miller (Degeneriaceae), J.M. Miller 1189-63, SUVA (Qiu *et al.*, 1993)

Myristicaceae: *Knema latericia* Elmer, Qiu 91041, NCU (Qiu *et al.*, 1993)

Magnoliaceae: *Magnolia salicifolia* (Siebold & Zucc.) Maxim., Qiu 82, NCU (Qiu *et al.*, 1993)

Afroguatteria Boutique: not sequenced

Ambavia Le Thomas: *A. gerrardii* (Baill.) Le Thomas, Rabevohitra 2035, P

Anaxagorea A.St.-Hil.: *A. dolichocarpa* Sprague & Sandwith, P.J. Maas 7080, U

Ancana F.Muell.: *A. hirsuta* Jessup, L.W. Jessup 706, K

Annickia Setten & Maas: *A. kummeriae* (Engl. & Diels) Setten, D.M. Johnson 1942, K

Annona L. group: *A. muricata* L., Qiu 90031, NCU (Qiu *et al.*, 1993)

Artabotrys R.Br.: *A. hexapetalus* (L.f.) Bhandari, M. Chase 1203, K

Asimina Adans. group: *A. triloba* (L.) Dunal, Qiu 15, NCU (Qiu *et al.*, 1993)

Cananga (DC.) Hook.f. & Thoms.: *C. odorata* (Lam.) Hook.f. & Thoms., M. Chase 219, NCU (Qiu *et al.*, 1993)

Cleistopholis Pierre ex Engl.: *C.* sp., Wieringa & van Nek 3278, WAG

Cymbopetalum Benth. group: *C. brasiliense* Benth., P.J. Maas *et al.* 7778, U

Duguetia A.St.-Hil.: *D. lanceolata* A.St.-Hil., P.J. Maas *et al.* 8043, U

Ephedranthus S.Moore: *E. parviflorus* S.Moore, J.A. Ratter *et al.* 4402, K

Fusaea (Baill.) Saff. group: *F. longifolia* Saff., M. Jansen-Jacobs *et al.* 2827, U

Greenwayodendron Verdc.: *G. suaveolens* (Engl. & Diels) Verdc., Semsei 2376, K

Guatteria Ruiz & Pav. group: *G. punctata* (Aubl.) R.A.Howard, J.F. Molino 1593, K, U

Hexalobus A.DC.: *H. monopetalus* (A.Rich.) Engl. & Diels, Kisena & Ruffo 90, K

Isolona Engl.: *I.* sp., G. Schatz 3364, WIS

Letestudoxa Pellagr.: *L.* sp., F.J. Breteler 12858, WAG

Malmea R.E.Fr.: *M.* sp. nov., L. Chatrou *et al.* 8, U

Miliusa Lesch. ex A.DC.: *M. horsfieldii* (Benn.) Baill. ex Pierre, M. Chase 2084, K

Mkilua Verdc.: *M. fragrans* Verdc., RBG Kew cult. 1990-1129, K

Monanthotaxis Baill. group: *M. filamentosa* (Diels) Verdc., Etuge 2377, K

Monodora Dunal: *M. myristica* Dunal, RBG Kew cult. 1973-12039, K

Neostenanthera Exell: *N. hamata* (Benth.) Exell, P. Bamps 2247, K

Pachypodanthium Engl. & Diels: *P.* sp., Wieringa & van Nek 3290, WAG

Piptostigma Oliv.: *P. pilosum* Oliv., M. Cheek 7869, K

Polyalthia stuhlmannii (Engl.) Verdc., Luke & Robertson 1424, K

Polyalthia longifolia (Sonn.) Thwaites, D.M. Johnson 1965, K

Polyalthia columellar sulcate group: *P. sumatrana* (Miq.) Kurz, D. Duling 35, K, BRUN

Polyalthia disulculate group: *P.* sp. aff. *rumphii* (Blume) Merr., A.P. Davis 788, K

Polyceratocarpus Engl. & Diels: *P. parviflorus* (Bak.f.) Ghesq., D. Thomas 9737, K

Pseudoxandra R.E.Fr.: *P. polyphleba* (Diels) R.E.Fr., P.J. Maas *et al.* 8227, U

Sapranthus Seem.: *S.* sp., G. Schatz 1199, WIS

Tetrameranthus R.E.Fr.: *T.* sp., D. Simpson, C.A. Cid *et al.* 8084, U
Toussaintia Boutique: *T. orientalis* Verdc., Luke 1609, K
Unonopsis R.E.Fr.: *U. floribunda* Diels, P.J. Maas *et al.* 8225, U
Uvaria L.: *U. lucida* Bojer ex Sweet, A.G.M. 75, K
Uvariastrum Engl. & Diels: not sequenced
Uvariodendron (Engl. & Diels) R.E.Fr.: *U.* sp., S. Cable 2187, K
Uvariopsis Engl.: *U. globiflora* Keay, Utrecht cult. 84GR00381, U
Xylopia L.: *X. nitida* Dunal, M.F. Prevost 3213, K, U, CAY

Borsch, T. and Wilde, V. (2000). Pollen variability within species, populations, and individuals, with particular reference to *Nelumbo*. In: M.M. Harley, C.M. Morton and S. Blackmore (Editors). Pollen and Spores: Morphology and Biology, pp. 285–299. Royal Botanic Gardens, Kew.

POLLEN VARIABILITY WITHIN SPECIES, POPULATIONS, AND INDIVIDUALS, WITH PARTICULAR REFERENCE TO *NELUMBO*

THOMAS BORSCH[1] AND VOLKER WILDE[2]

[1]Botanisches Institut und Botanischer Garten, Friedrich-Wilhelms-Universität Bonn, Meckenheimer Allee 170, 53115 Bonn, Germany
[2]Forschungsinstitut Senckenberg, Paläobotanik, Senckenberganlage 25, 60325 Frankfurt a.M., Germany

Abstract

During a study of the infrageneric systematics of modern and fossil *Nelumbo* an extreme variability in pollen aperture condition as well as in exine sculpturing was found. A comprehensive analysis of the variation of pollen characters between and within different American, European, and Asian populations of *Nelumbo* revealed the regular existence of a variability pattern, rather than the occurrence of single grains deviating in aperture condition, as reported by earlier authors. However, the variability of pattern found in *Nelumbo* cannot be explained by any of the known phenomena such as successiformy or incompatibility systems. Therefore, pollen variability phenomena were considered throughout the angiosperms, to get a better picture of the kind and extent of variation below the species level. It appears that infraspecific pollen variation is at least determined by seven different biological syndromes. A very important phenomenon is polymorphism of the exine sculpture within anthers. This continuous variation can be explained by hypothesising self-assembly processes that act within a genetically fixed framework in pattern formation. It is suggested that in certain taxonomic groups the overall reliability of pollen morphological data can be enhanced by considering variability phenomena.

Introduction

The sporoderm is generally composed of an intine or endospore (the inner pecto-cellulosic part) and an exine or exospore (the outer part, predominantly consisting of sporopollenin), (Blackmore and Barnes, 1987). A major differentiation of the sporoderm within single individuals occurred with the rise of heterospory (Chaloner and Pettitt, 1987) and the development of morphologically distinct mega- and microspores. This paper is concerned with the microspores of angiosperms, and in particular with the exine.

Variation in angiosperm pollen within species was known for a long time as part of the heterostylous syndrome (Hildebrand, 1863; Darwin, 1877). In many species, which exhibit two or three floral morphs, pollen from each of these morphs can be distinguished by size differences and often also by differences in the sculpture of the ectexine. In this case pollen variability follows a highly determined pattern. Van Campo (1976) drew attention to the fact that pollen varies below the species level in a far higher number of taxa, and even within individuals and single pollen sacs. She recognised three distinct evolutionary patterns (successiformy, spiralisation, breviaxy), which describe a defined spectrum of variability of aperture configurations. For example, in successiformy tricolpate, pericolpate, and periporate grains occur, and the concept is that the pericolpate and periporate conditions are derived from the tricolpate. In their analysis of feeding pollen in *Lagerstroemia indica* L. Pacini and Bellani (1986) listed other pollen polymorphisms in angiosperms that are mostly connected to specialised pollination syndromes.

Work on the systematics of *Nelumbo* (Borsch and Barthlott, 1996) indicated an extreme case of pollen variability. What was found exceeds the earlier known presence of deviating monosulcate grains (Kuprianova, 1979) in this genus, which is commonly known for its tricolpate pollen. Because of the importance of pollen characters for the systematics of modern and fossil *Nelumbo* (several fossil taxa are based on pollen, for example, Kuprianova and Tarasevich, 1983), an extensive survey was undertaken to get a more precise idea of the variation of pollen characters between and within extant American, European, and Asian populations (Borsch, Wilde, Neinhuis; work in progress). In addition to the apertures the sculpture of the ectexine was found to be extremely variable, even among grains from individual anthers. However, it appeared that none of the currently known variability phenomena could explain this variation. Another project on the systematics of the so-called 'Messel-waterlily' from the Eocene of Messel, Germany, (Wilde, Borsch, Collinson, Goth, Schaarschmidt, work in progress) further strengthened our interest in pollen variability below the species level. In this case, monosulcate and trichotomosulcate pollen co-occur within single anthers.

Because pollen variability is a very important issue when applying pollen characters to systematics, palaeobotany, and pollen analysis, some key questions became apparent: 1. What is the extent of pollen variation below the species level? 2. Is it possible to relate this variation to particular biological phenomena? 3. Which pollen characters vary in such phenomena? 4. In which angiosperm lineages do those phenomena occur? The present paper should draw attention to variation patterns without questioning the great achievements in applying pollen morphology to systematics, palaeobotany and pollen analysis that developed after the pioneering works by Wodehouse (1935) and Erdtman (1952). On the contrary, it aims at outlining the kind and amount of variation existing within species, populations, and individuals, and should thus stimulate consideration of variation phenomena in interpreting and using pollen data.

Material and Methods

Pollen was removed from mature anthers with open pollen sacs. Grains were acetolysed for 3 minutes according to the method of Erdtman (1960), and re-suspended twice in distilled water. Samples were then dried on glass slides, and transferred to aluminium stubs, previously covered with a carbon sticky pad (Plano GmbH, Marburg, FRG) for scanning electron microscopy (SEM). Specimens were covered with gold (ca. 25 nm) in a sputter coater (Balzers Union SCD 040, Balzers GmbH, Wiesbaden, FRG). The investigations were carried out with a Cambridge S 200 scanning electron microscope, equipped with a LaB_6-cathode for high resolution. SEM micrographs of the exine were taken from a central position within a mesocolpium at standard magnifications of 5000 or 20000, depending on the scale of the pattern. The surface was positioned perpendicular to the electron beam, since this gives the best view of little cavities and is easily reproducible. Approximate grain sizes were recorded to cover possible grain-size dependent sculpture variation.

Pollen was sampled from the following herbarium specimens: *Nelumbo nucifera* Gaertner subsp. *lutea* (Willdenow) Borsch & Barthlott: Britton, N.L. 1491, Jamaica, (NY); Hankenson, E.L. s.n., USA, New York, (NY); Massey, J.R. & Thomas, W. 3295, USA, North Carolina, (NY). *Quercus ilex* L. subsp. *ilex*: Borsch 3268, France, (FR); 3279, France (FR). *Tilia platyphyllos* Scop.: Holm-Nielsen, L. & Jeppesen, S. 549, Denmark, (FR). Fossil pollen was photographed from '*Messel-waterlily*' (ined.) = Me 3014 from the Eocene of Messel, Germany, (Senckenberg).

Terminology

Both terms heteromorphism and polymorphism have been widely used to describe the occurrence of different pollen types within species, or at lower levels, within a population or within a pollen sac. For stringency we only use 'polymorphism' since this term better reflects the overall range of variability (discrete characters *and* gradual differences). Pollen variation may then be further characterised by variation phenomena. This has the advantage that information about the biological context is provided, which automatically includes the level(s) of variation. The distinction of a more restricted term "pollen heteromorphism" (i.e. the production of different fertile pollen types by a single plant, each of which must be present in all anthers, and thus all flowers, throughout the life cycle or at least most of it, Till-Bottraud *et al.* 1995) is not necessary in this context.

Results

Variation phenomena below the species level

The analysis of selected taxa and a screening of the palynological literature revealed an astonishing number of cases of pollen variability below the species level. However, aberrant grains (those obviously suffering from irregular defects in ontogenetic processes) have to be distinguished from 'regular' variability. The main criterion for 'regular' variability, and for being included in our study, is that it must be a characteristic part of the reproductive system of a taxon. Viability is not sufficient as a criterion, because there are plants in which sterile grains are produced by specialised anthers on a regular basis (see below). Hybridisation can affect pollen variability in the same way as differentiation of populations, and therefore has to be considered although more than one species is involved.

For a better understanding of aspects of function, inheritance etc. of the variability phenomena it might be helpful to distinguish the following levels: polymorphism at the population level (= different populations of a species have different pollen characters); polymorphism at the level of individuals (= different individuals within a population exhibit deviating pollen characters); polymorphism at the level of different inflorescences/ flowers (= different inflorescences or flowers of an individual differ in pollen characters); polymorphism at the anther/pollen sac level (= different anthers/pollen sacs of a flower possess different pollen types); polymorphism below the pollen sac level (= single pollen sacs produce different types of pollen). For each of the phenomena it is important to know the lowest level at which variation occurs, because the same variability can be observed when comparing samples from higher taxonomic levels.

In the following we describe seven different phenomena of pollen variability below the species level. The theoretical concept that led to this approach is that pollen variability results from different underlying biological phenomena. Therefore, the best way to analyse this variability appeared to be in trying to recognise the underlying phenomena and to regard the actual differences in pollen characters as a consequence of them. The different phenomena may be synapomorphies for a number of taxa, autapomorphies for

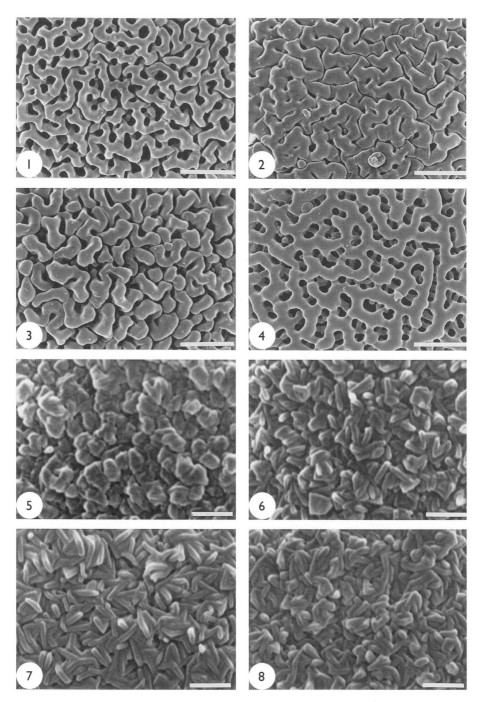

FIGS. 1–8. Figs. 1–4. Exine of *Nelumbo nucifera* subsp. *lutea*. Figs. 1–3. Different grains from the same flower, Hankenson s.n. Fig. 4. Grain from a different population, Britton 1491. Figs. 5–11. Exine of *Quercus ilex* subsp. *ilex*. Figs. 5–7. Different grains from the same inflorescence, Borsch 3268. 8. Grain from a different tree, Borsch 3279. Scale bars: Figs. 1–4 = 5 μm, 5–8 = 1 μm.

single taxa, may have evolved independently, or may be plesiomorphic and very widespread. A lot more research is necessary on their evolution, on the cytological and molecular mechanisms involved, and on the relative influence of genetical versus physical control (self-assembly) in pattern formation. However, the approach presented may be useful in separating the mechanisms regarded, and inspiring for further work. The first four variability phenomena are independent from any evolutionary differentiation of populations into distinct lineages. The fifth type of polymorphism relates to an evolutionary pattern, and can occur at all levels (see also below), whereas the last two result from a differentiation of populations (shifts in allele frequencies, speciation). For those, the amount of pollen variability depends on the species concept.

1. Polymorphism of the exine sculpture below the pollen sac level

Among all plants that have been studied more thoroughly for their exine microsculpture, *Nelumbo* was found to be extremely variable (apertures are discussed separately). The tectum is composed of ± isolated strands, which are undulated, branched and three-dimensionally intertwined. Within certain limits these basic sculpture elements vary continuously in size, degree of branching, and degree of fusion (Figs. 1–4). Individual pollen sacs usually produce grains that differ morphologically (Figs. 1–3, from a population close to Lake Ontario, state of New York), and the proportions of these morphotypes may differ from pollen sac to pollen sac. An individual (in this case an individual may be equivalent to a population because of a clonal structure) has a certain range of morphological variability, whereas grains deviating more strongly may be found in different populations (Fig. 4, from a population from Jamaica). Different populations may also have some grains with identical morphology, but there is no regularity. Similar variation of the exine sculpture is present in *Quercus ilex* (Fagaceae, Fig. 5–8). Differences can be found in the size, number and aggregation pattern of the rodlets. In addition, the coverage of the rodlets by 'masking sporopollenin' (Rowley 1996), differs. This receptor-independent sporopollenin seems to account for a considerable part, but not all, of the variation. In *Nelumbo* and *Quercus* pollen grains exhibit the same exine pattern over all their surface. In the trizonocolporate grains of *Tilia platyphyllos* (Tiliaceae) the muri of the coarsely patterned pertectate exine (Christensen and Blackmore, 1988) are larger at the poles compared to the equatorial region. However, a comparison of an identical polar position reveals clear differences in the shape and size of lumina, and in the ratio of large versus small lumina (Figs. 9, 10) in grains of the same anther. In *Amorphophallus titanum* Becc. (Araceae) exine sculpture variability from punctate to reticulate occurs among flowers from the same inflorescence (Borsch and Wolter, 1998). Polymorphism of the exine sculpure, which is continuous within certain limits, has mostly been overlooked. However, it seems likely that it is a rather widespread phenomenon.

2. Polymorphism related to a functional specialisation within an individual

Several unrelated taxa have flowers with two different types of anthers, feeding anthers and fertilisation anthers (Faegri and van der Pijl, 1966), which often produce different pollen. The known examples are rather few, and it is difficult to estimate how widely distributed the phenomenon is in angiosperms. Mattsson (1976) shows for *Tripogandra* (Commelinaceae) that two types of pollen (a spheroidal and fertile type, and an elongated and sterile type) are produced in different whorls of stamens. Pacini and Bellani (1986) described blue pollen serving pollination from large anthers and yellow feeding pollen from small anthers of *Lagerstroemia indica* L. (Lythraceae). In flowers of *Couroupita guianensis* Aubl. (Lecythidaceae) the androecium is differentiated into two parts, one of which has large showy, and the other small anthers (Mori *et al.*, 1980).

The pollen produced by the large anthers is rugose, sterile and remains in tetrads, whereas the pollen produced by the small anthers is smooth and fertile. The functional specialisation in both taxa is to provide pollen as reward for pollinators, which is kept separate from pollen serving in pollination. The function of a pollen dimorphism occuring in *Amorphophallus titanum* (Araceae) and related species is less clear. *Amorphophallus* produces pollen strands that consist of small sterile, intineless grains and large, fertile grains (Borsch and Wolter, 1998). Polymorphism related to the possession of chasmogamous and cleistogamous flowers may be included here. Data for a pollen morphological differentiation between the two flower types are provided by Lord and Eckard (1984, 1986).

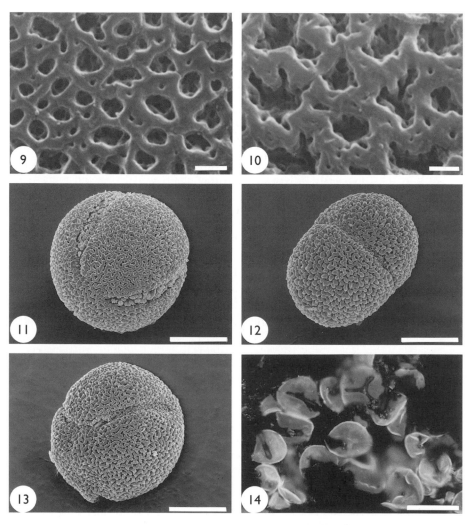

FIGS. 9–14. Figs. 9–10. Exine of *Tilia platyphyllos*, grains from the same flower, Holm-Nielsen & Jeppesen 549. Figs. 11–13. Grains of *Nelumbo nucifera* subsp. *lutea*. Figs. 11–12. From Hankenson s.n. Fig. 13. From Massey & Thomas 3295. 14. Grains *in situ* from *'Messel Waterlily'* (ined.). Scale bars: Figs. 9–10 = 1 µm, Figs. 11–13 = 20 µm, Fig. 14 = 50 µm.

3. Polymorphism related to a heterostyly/incompatibility syndrome
Heterostyly is a type of genetic polymorphism in which plant populations are composed of two (distyly) or three (tristyly) morphs that differ reciprocally in the heights of stigmas and anthers in flowers. This style-stamen polymorphism is usually accompanied by a sporophytically controlled self-incompatibility syndrome (Barrett, 1992). Heterostyly is a well-documented phenomenon and will therefore not be discussed here in detail (for reviews see Barrett, 1992; Dulberger, 1992; Lloyd and Webb, 1992; Bir Bahadur *et al.*, 1984). It occurs in approximately 30 angiosperm families (Barrett, 1992, Bir Bahadur *et al.*, 1984) and in many cases it includes a stigma and pollen polymorphism that plays a role in preventing inbreeding through incompatibility (Dulberger, 1992). One classical example is *Armeria maritima* Willd.: long-styled flowers produce pollen with a coarsely reticulate exine, whereas in short-styled flowers the exine reticulation is considerably finer (for example Baker, 1953). In many taxa pollen varies in size, and in some it differs also in aperture number and exine pattern.

4. Polymorphism related to dioecy
Dimorphic pollen can arise in the evolution of dioecy. Anderson (1979) and Anderson and Symon (1989) have shown for *Solanum appeniculatum* H. et B. ex Dun that functionally male flowers have tricolporate pollen whereas functionally female flowers have inaperturate pollen. The inaperturate grains retain their viability but cannot produce pollen tubes. The production of a protoplast without the ability of the pollen grain to germinate may be understood as a nutritional reward for pollinators (Levine and Anderson, 1986). NcNeill and Crompton (1978) reported the presence of reticulate vs. punctate pollen in different populations of *Silene alba* (Miller) E.H.L. Kranse. However, this polymorphism was not unequivocally proven to belong to the dioecious syndrome and could also be caused by cryptic heterostyly (McNeill and Crompton, 1978), or be the result of an unrecognised taxonomic problem.

5. Polymorphism related to an evolutionary pattern
When Van Campo (1976) established successiformy, spiralisation, and breviaxy, she remarked that, "these series of pollen apertural configurations obey a limited number of rules which produce a limited number of patterns of variation". Each of these series resembles an evolutionary pattern. This means that all the different aperture configurations within such a pattern are homologues. In cladistic terminology they resemble different character states which can be transformed into each other in progenitor-descendant relationships. Van Campo (1976) also showed that the same evolutionary pattern may appear in different, unrelated lineages of angiosperms. As has been discussed by Borsch and Barthlott (1998) for the evolution of metareticulate pollen within the successiform pattern, this independent appearance is to be considered parallel, not convergent. It may be assumed that the genetic information for a whole set of characters is present in a clade (for example, in Caryophyllales for tricolpate, pericolpate, periporate, and metareticulate pollen). In some taxa of this clade only one aperture configuration is expressed, in others several are expressed. This would also explain the occurrence of evolutionary patterns like successiformy at all levels below the species and above. However, there are theoretical problems with this hypothesis, given the idea is correct that homologues may not co-occur in a single organism (Patterson, 1982), but the discussion of this issue goes beyond the present paper.

A further pattern, not mentioned by Van-Campo (1976) is the sulcate-trichotomosulcate pattern. The co-occurrence of mono- and trichotomosulcate grains was repeatedly demonstrated, in extant (for example, Canellaceae, Wilson, 1964; various basal angiosperms, Walker, 1974; palms, Harley, 1999; and Asparagales, Rudall *et al.*, 1997) and fossil taxa (for example, *Anacostia*, Friis *et al.*, 1997; the 'Messel-waterlily', Fig. 14). Interestingly, additional deviating aperture configurations were never reported in this pattern. Walker and Doyle (1975) inferred trichotomosulcate pollen to be derived from monosulcate. Rudall *et al.* (1997) showed that all taxa examined in Asparagales with trichotomosulcate pollen have simultaneous microsporogenesis, although not all taxa with simultaneous microsporogenesis have trichotomosulcate pollen. They further suggest the presence of a developmental basis for the correlation between trichotomosulcate pollen and simultaneous microsporogenesis.

Nelumbo pollen is extremely variable not only in the primary pattern, as discussed above, but also in the secondary pattern. The major problem is that the homology of deviant aperture configurations (single apertures that may be straight, irregularly curved, circular, or two apertures; Figs. 11-13) occurring in addition to the tricolpate is not resolved. Different possibilities for an interpretation are: 1. The monoaperturate condition in *Nelumbo* is not homologous to monosulcate or zonosulcate pollen in other taxa. In this case the assumption of Kuprianova (1979) that *Nelumbo* has pollen which is transitional between that of the monosulcate and tricolpate taxa would not be valid. *Nelumbo* would not provide such spectacular evidence for the evolution of tricolpate pollen by a simple triplication of a single aperture. The variability in *Nelumbo* could either represent an early state in the evolution of tricolpate pollen at which the control of aperture formation was not very strict, or an autapomorphy. 2. The monoaperturate grains in *Nelumbo* are homologous to monosulcate and/or zonosulcate pollen. This would imply that tricolpate pollen in *Nelumbo* is derived from monosulcate, since the monosulcate condition can be considered plesiomorphic in angiosperms (Walker, 1974, Doyle, 1987-88, Crane, 1990). Variability was generated because the derived character state did not completely replace the plesiomorphic one in *Nelumbo*. Evidence is provided for the evolutionary origin of tricolpate pollen from monosulcate, in line with Kuprianova (1979).

The recent cladistic analyses of angiosperm phylogeny, largely based on molecular data, place *Nelumbo* in Proteales, a clade consisting of Nelumbonaceae, Platanaceae, and Proteaceae (Chase, *et al.*, 1993; APG, 1998). In this topology the Proteales are sister to the core eudicots, with the Ranunculales basal to both of them. However, an appearance of monosulcate pollen in *Nelumbo* either would have to be interpreted as a reversal, or a parallel evolution of the tricolpate condition in Ranunculales, in Proteales, and in core eudicots would have to be assumed. Current knowledge allows no assignment of *Nelumbo* pollen varibility to any of the recognised evolutionary patterns. Pending on the homology of monoaperturate grains in *Nelumbo* being proven, *Nelumbo* might represent a condition plesiomorphic to the successiform pattern, and would show how the successiform pattern is connected to other patterns comprising monocolpates. Flynn and Rowley (1971) and Rowley (1975) reported that aperture formation in *Nelumbo* does not follow the prevailing model, according to which, "primexine is not formed at sites of future germinal apertures". A uniform primexine template with probacules over the entire surface of the grain was found instead. Recent studies on pollen development by Kreunen and Osborn (1999) confirm this and indicate a post-tetrad establishment of apertures in *Nelumbo*. However, if aperture formation in *Nelumbo* follows a unique way this would point to an autapomorphy. The study of Kreunen and Osborn (1999) further revealed the co-occurrence of tetragonal and tetrahedral tetrads. From tetragonal tetrads the presence of successive, and from tetrahedral the presence of simultaneous microsporogenesis, may be concluded. Successive microsporogenesis may be interpreted as a retained plesiomorphic state, since it seems to be plesiomorphic in the angiosperms (Rudall *et al.*, 1997). Interestingly, both successive and simultaneous microsporogenesis also occur in Proteaceae, for which Blackmore and Barnes (1995) showed simultaneous microsporogenesis to produce triporate pollen conforming to Garside's rule (*Grevillea rosmariniifolia* A. Cunningham) and successive microsporogenesis to produce biporate pollen (*Dryandra polycephala* Benth.). However, information on the type of microsporogenesis does not provide sufficient evidence for elucidating the homology of the aperture configurations in *Nelumbo*. Simultaneous microsporogenesis seems to produce a variety of different aperture configurations, monosulcate (for example, in palms, Harley, 1999; in Asparagales, Rudall *et al.*, 1997) and tricolpate (eudicots, Rudall *et al.*, 1997). On the other hand the presence of successive microsporogenesis does not necessarily imply that pollen with a distal sulcus is formed, as *Dryandra* (biporate), or *Ceratophyllum* (no aperture region differentiated, Takahashi, 1995) show.

Thus, a lot more research is required to understand the precise homologies of aperture configurations in *Nelumbo*. Most remarkably, Borsch, Wilde, and Neinhuis (pers. obs.) could not find any trichotomosulcate grains in *Nelumbo* although large numbers of grains in many populations were screened. In an extremely variable taxon such as *Nelumbo* in which a monosulcate condition is believed to occur, occasional trichotomosulcate grains should be expected as well. Obviously, the monosulcate-trichotomosulcate pattern is not present, perhaps an argument for a non-homology of monoaperturate grains in *Nelumbo* with monosulcate pollen. Moreover, the tetrad illustrated by Kuprianova (1979) shows within-tetrad variation(!): one grain has three meridional colpi, one has a single equatorial, and two single meridional apertures each. The possibility may therefore not be excluded that the determination of apertures in *Nelumbo* is governed by different processes which are stochastic in part, and that the tricolpate grains in *Nelumbo* are not identical to tricolpate pollen present in other eudicots. Circular apertures in *Nelumbo* may rather be explained by a fusion of two apertures at their ends since the positions of meridional apertures in relation to the poles seem to vary as well. Moreover, Doyle and Hotton (1991) showed aperture variation in tricolpates (radially 'twisted', spiraperturate, with a single sulcus) from the Lower Cretaceous of Gabon which might have some similarity with the phenomenon in *Nelumbo*, but further study is required.

6. Polymorphism related to ploidy

Variation related to different levels of ploidy occurs in grain size and aperture numbers. An increase in pollen size, aperture number, and size of the meshes of the ectexine reticulum was reported in artificially induced autopolyploids of *Arabidopsis thaliana* (L.) Heynh. by Bronckers (1963). The correlation of ploidy level and pollen morphology was found in several taxa, whereas in other taxa no significant correlation exists (for example, Gould, 1957; Mignot *et al.*, 1994). Rogers and Harris (1969) further observed various smaller grains within anthers of triploid hybrids in *Linum*, which were formed around chromosomes that were not included into the reforming nuclei at the end of the telophase ('stray chromosomes').

7. Polymorphism related to the differentiation of populations (geographical variation)

A nice example is presented by Muller (1979) for *Allophylus cobbe* Blume (Sapindaceae), which shows different types of exine sculpture in different parts of Asia. Another example is the variation in *Picea abies* L. as described by Birks (1978). However, there is not much known on geographical variation of pollen characters below the species level. But with geographic variation there is the inherent problem of unrevealed taxa, or introgression from other taxa.

Implications for the understanding of exine sculpture determination

The study of variation patterns in relation to the underlying biological phenomena may provide insight into the processes involved in generating this variation, and thus, in generating morphological patterns. Pattern formation in pollen takes place at two levels as pointed out by Dickinson and Sheldon (1986) and Schmid et al. (1996). The primary pattern is the overall micro-architecture of the pollen wall, whereas the secondary pattern is the system of apertures.

There is strong evidence that the secondary pattern is microtubule-dependent and created by the meiotic cells (Heslop-Harrison, 1971; Sheldon and Dickinson, 1986; Schmid et al., 1996 for review). This paragraph focuses on the primary pattern and the mechanisms determining it, which are far less understood. In principle pattern development is under genetic control (Blackmore and Crane, 1988). However, the extent to which genetic information is responsible, and the levels at which genetic control operates during the development of complicated exines is not clear, and may be different in different taxa. Ontogenetic studies show that pollen exines are initiated as patterned primexines composed of polysaccharides, proteins, and cellulose, during the tetrad stage, before the meiotic division (Dickinson and Heslop-Harrison, 1977). These primexines become impregnated with sporopollenin later on (Dickinson and Heslop-Harrison, 1977; Takahashi, 1989 a, b). Sheldon and Dickinson (1983), and Dickinson and Sheldon (1986) suggested for reticulate pollen of *Lilium* that the exine pattern is generated in the plasmalemma by organised segregation of membrane components and the formation of platelets in the plasmalemma. These authors proposed the involvement of biophysical re-organisation (by addition of hydrophobic constituents) or reaction diffusion events (Lacalli, 1981), and thus of self-assembly in pattern generation.

Heslop-Harrison (1972) was the first to suggest that self-assembly may be involved in pollen pattern formation. Van Uffelen (1991) considered self-assembly processes to play the dominant role in spore wall formation, with the living system contributing only by supplying material of a particular quality and concentration, by pre-patterning the condensation surface, and by governing the physico-chemical environment. Based more on theoretical considerations Gabarayeva (1993) and Scott (1994) pointed to a major participation of self-assembly processes also in pollen. Nowadays, lycopsid megaspores are much better understood than pollen exines of seed plants. Megaspore walls are thicker (up to 50 μm) and consist of a system of closely packed mono-dispersed particles (Hemsley et al., 1992; Hemsley et al., this volume). Hemsley et al. (1994) developed a hypothesis of aggregation of regular particles by depletion flocculation, and later provided experimental evidence in favour of this (Hemsley et al., 1998).

The above described polymorphism in exine sculpture provides evidence for the participation of self-assembly in primary pattern determination from a different perspective. Self-assembly processes are characterised by the fact that a pattern is not reproduced absolutely identical, but according to particular rules. The continuous variation that is based on a modification of sculpture elements as found in exines of *Tilia*, *Nelumbo*, and *Quercus* corresponds to this. A genetic control of this polymorphism is not very likely because it would require an enormous effort, which seems not to be justified by any functional constraints (see Hemsley et al., 1994). The coarse reticulate pattern in *Tilia* (Figs. 9–10) may be determined by a mechanism that involves a patterned primexine acting as a template, and since variation occurs at the level of the coarse reticulum, the template should vary as well. In *Nelumbo* the architecture of the ectexine is more complicated. Apart from a template at the primexine level, other processes may be involved. This is also in agreement with the observation frequently made in developmental studies that an initial pattern established during the tetrad stage may be modified in various ways during the free microspore stage (Schmid et al., 1996). An alternative explanation for the polymorphism in *Nelumbo* would be a system in which different grains arrest in different ontogenetic stages. When these grains are compared at the same time after the pollen sac has opened they will differ in exine morphology. Kreunen and Osborn (in press) have shown how the ectexine is completed during the development, and that aborted grains do exist in *Nelumbo*. However, the total of infra-individual ectexine variation in *Nelumbo* exhibits numerous patterns that cannot be transformed into other patterns simply by addition of material (Figs. 1-4). Thus, ontogenetic phenomena may account for some of the polymorphism, but additional mechanisms are required to explain the variation of the ectexine pattern. In *Quercus* (Figs. 5-8) the ectexine resembles an aggregation of rodlets, but the formation of rodlets as basic structural elements, with a very low level of complexity, can hardly be explained with a plasmalemma-based template. Microsculptures as in *Quercus* require the operation of self-assembly processes at a different level. Based on the described ectexine variation phenomena it must be postulated that self-assembly operates at different levels within the system leading to the formation of the primary pattern. The latest stage of modification in *Quercus* may be the condensation of 'receptor-independent sporopollenin' (Rowley, 1996), which seems to vary to a higher degree than the aggregation of rodlets.

Implications for systematics, palaeopalynology and pollen analysis

The approach presented here, of distinguishing particular variability phenomena will help to understand the variation that occurs in pollen. This is of particular importance in palaeopalynological studies exclusively dealing with dispersed pollen, but also for any other application of pollen characters. Therefore, it will be extremely important to assess pollen variability in any taxon under study. More extensive studies taking the underlying biological principles into consideration will enable evaluation of pollen character variation and its systematic occurrence. The results may be used to make generalisations and predictions, and thus may enhance the overall reliability in the application of pollen morphological data.

Acknowledgements

We would like to express our gratitude to the directors and curators of the herbaria of the Forschungsinstitut Senckenberg (FR) and the New York Botanical Garden (NY) for loaning herbarium specimens and permission to study pollen. We appreciate that M. Hesse, Vienna, spent his time with us for extensive discussions. M.E. Collinson, Egham, J.A. Doyle, Davis, A.R. Hemsley, Cardiff, and J.M. Osborn, Kirksville, provided valuable comments on earlier drafts of the manuscipt. For various kinds of help we are grateful to W. Barthlott, C. Neinhuis, R. Pretscher, Bonn, and K. Schmidt (Frankfurt). We further thank two anonymous reviewers for their valuable suggestions.

References

APG (1998). An ordinal classification for the families of flowering plants. *Annals of the Missouri Botanical Garden* **85**: 531–553.

Anderson G.J. (1979). Dioecious *Solanum* of hermaphroditic origin is an example of a broad convergence. *Nature* 282 (5741): 836–838.

Anderson, G.J. and Symon, D.E. (1989). Functional dioecy and andromonoecy in *Solanum. Evolution* **43**: 204–219.

Baker, H.G. (1953). Dimorphism and monomorphism in the Plumbaginaceae. II. Pollen and stigmata in the genus *Limonium*. *Annals of Botany* **17**: 433–445.

Barrett, S.C.H. (1992) Heterostylous genetic polymorphisms: model systems for evolutionary analysis. In: S.C.H. Barrett (editor). Evolution and function of heterostyly. *Monographs on Theoretical and Applied Genetics* **15**, pp. 1–29.

Bir Bahadur, Laxmi S.B. and Rama Swamy N. (1984). Pollen morphology and heterostyly. A systematic and critical account. *Advances in Pollen and Spore Research* **12**: 79–126.

Birks, H.J.B. (1978). Geographic variation of *Picea abies* (L.) Karsten pollen in Europe. *Grana* **17**: 149–160.

Blackmore, S. and Barnes, S.H. (1987). Embryophyte spore walls: origin, development and homologies. *Cladistics* **3**: 199–209.

Blackmore, S. and Barnes, S.H. (1995). Garside's rule and the microspore tetrads of *Grevillea rosmariniifolia* A. Cunningham and *Dryandra polycephala* Bentham (Proteaceae). *Review of Palaeobotany and Palynology* **85**: 111–121.

Blackmore, S. and Crane, P.R. (1988). The systematic implications of pollen and spore ontogeny. In: C.J. Humphries (editor), Ontogeny and systematics, pp. 85–115, Columbia University Press, New York.

Borsch, Th. and W. Barthlott (1996). Classification and distribution of the genus *Nelumbo* Adans. (Nelumbonaceae). *Beiträge zur Biologie der Pflanzen* **68**: 421–450.

Borsch, Th. and W. Barthlott (1998). Structure and evolution of metareticulate pollen. *Grana* 68–78.

Borsch, Th. and Wolter, M. (1998). Pollenmorphologie. In: W. Barthlott and W. Lobin (editors). *Amorphophallus titanum. Tropische und Subtropische Pflanzenwelt* 99, pp. 135–145.

Bronckers, F. (1963): Variations polliniques dans une série d'autopolyploïdes artificiels d'*Arbidopsis thaliana* (L.) Heynh. *Pollen et Spores* **5**: 233–238.

Chaloner, W.G. and Pettitt, J.M. (1987). The inevitable seed. *Bulletin de la Societé Botanique de France (Actualité Botanique)* **134**: 39–49.

Chase, M.W., Soltis, D.E., Olmstead, R.G., Morgan, D., Les, D.H., Mishler, B.D., Duvall, M.R., Price, R.A., Hills, H.G., Qiu, Y.-L., Kron, K.A., Rettig, J.H., Conti, E., Palmer, J.D., Manhart, J.R., Sytsma, K.J., Michaels, H.J., Kress, W.J., Karol, K.G., Clark, W.D., Hedrén, M., Gaut, B.S., Jansen, R.K., Kim, K.-J., Wimpee, C.F., Smith, J.F., Furnier, G.R., Strauss, S.H., Xiang, Q.-Y., Plunkett, G.M., Soltis, P.S., Swensen, S.M., Williams, S.E., Gadek, P.A., Quinn, C.J., Eguiarte, L.E., Golenberg, E., Learn, G.H. Jr., Graham, S.W., Barrett, S.C.H., Dayanandan, S., and Albert, V.A. (1993). Phylogenetics of seed plants: an analysis of nucleotide sequences from the plastid gene *rbc*L. *Annals of the Missouri Botanical Garden* **80**: 528–580.

Christensen, P.B. and Blackmore, S. (1988). The Northwest European pollen flora 40: Tiliaceae. *Review of Palaeobotany and Palynology* **57**: 33–43.

Crane, P.R. (1990). The phylogenetic context of microsporogenesis. In: Blackmore, S. and Knox, R.B. (editors). Microspores: ontogeny and evolution, pp. 11–41, Academic Press, London.

Darwin, C. (1877). The different forms of flowers on plants of the same species. Murray, London.

Dickinson, H.G. and Heslop-Harrison, J. (1977). Ribosomes, membranes and organelles during meiosis in angiosperms. *Philosophical Transactions of the Royal Society, London, Biology* **277**: 327–342.

Dickinson, H.G. and Sheldon, J.M. (1986). The generation of patterning at the plasma membrane of the young microspores in *Lilium*. In: Blackmore, S. and Ferguson, I.K. (editors). Pollen and spores: form and function, pp. 1–17. Academic Press, London.

Doyle, J.A. (1987-88). Pollen evolution in seed plants: a cladistic perspective. *Journal of Palynology* **23–24**: 7–18.

Doyle, J.A. and C.L. Hotton (1991). Diversification of early angiosperm pollen in a cladistic context. In: Blackmore, S. and Barnes, S.H. (editors). Pollen and spores, patterns of diversification, pp. 169–196. Oxford Science Publications, Oxford.

Dulberger, R. (1992). Floral polymorphisms and their functional significance in the heterostylous syndrome. In: S.C.H. Barrett (editor). Evolution and Function of Heterostyly. *Monographs on Theoretical and Applied Genetics* **15**, pp. 41–48.

Erdtman, G. (1952). Pollen morphology and plant taxonomy. Angiosperms. - Almqvist & Wiksell, Stockholm.

Erdtman, G. (1960). The acetolysis method. *Svensk Botanisk Tidskrift* **54**: 561–564.

Faegri, K. and van der Pijl, L. (1966). The principles of pollination ecology. Pergamon Press, Toronto.

Flynn, J.J. and Rowley, J.R. (1971). The primexine of *Nelumbo nucifera*. *Experientia* **27**: 227–228.

Friis, E.M., Crane, P.R. and Pedersen, K.R. (1997). *Anacostia*, a new basal angiosperm from the Early Cretaceous of North America and Portugal with tricho-tomocolpate/monocolpate pollen. *Grana* **36**: 225–244.

Gabarayeva, N.I. (1993). Hypothetical ways of exine pattern determination. *Grana* **33**, suppl. 2: 54–59.

Gould, F.W. (1957). Pollen size as related to polyploidy and speciation in the *Andropogon saccaroides- A. barbinodis*-complex. *Brittonia* **9**: 71–75.

Harley, M.M. (1999). Tetrad variation: influence on pollen form and systematics in the Palmae. In: M.H. Kurmann and A.R. Hemsley (editors). Evolution of plant architecture, pp. 289–304. Royal Botanic Gardens, Kew.

Hemsley, A.R., Collinson, M.E. and Brain, A.P.R. (1992). Colloidal crystal-like structure of sporopollenin in the megaspore walls of recent *Selaginella* and similar fossil spores. *Botanical Journal of the Linnean Society* **108**: 307–320.

Hemsley, A.R., Collinson, M.E., Kovach, W.L., Vincent, B. and Williams, T. (1994). The role of self-assembly in biological systems: evidence from iridescent colloidal sporopollenin in *Selaginella* megaspore walls. *Philosophical Transactions of the Royal Society, London, B* **345**: 163–173.

Hemsley, A.R, Vincent, B., Collinson, M.E., and Griffiths, P.C. (1998). Simulated self-assembly of spore exines. *Annals of Botany* **82**: 105–109.

Hemsley, A.R., Collinson, M.E., Vincent, B., Griffiths, P.C. and Jenkins, P.D. (2000): Self-assembly of colloidal units in exine development. In: M.M. Harley, C.M. Morton and S. Blackmore (editors). Pollen and spores: morphology and biology, pp. 31–44, Royal Botanic gardens, Kew.

Heslop-Harrison, J. (1971). Wall pattern formation in angiosperm microsporogenesis. *Symposium Series of the Society for Experimental Biology* **25**: 277–300.

Heslop-Harrison, J. (1972): Pattern in plant cell walls: morphogenesis in miniature. *Proceedings of the Royal Institution of Great Britain* **45**: 335–352.

Hildebrand, F. (1863). Dimorphismus von *Primula sinensis*. *Verhandlungen des Naturhistorischen Vereins der Rheinlande und Westfalens, Sitzungsbericht* **20**: 183–184.

Kreunen, S.S. and Osborn, J.M. (1999). Pollen and anther development in *Nelumbo*. (Nelumbonaceae). *American Journal of Botany* **86**: 1662–1676.

Kuprianova, L.A. (1979). On the possibility of the development of tricolpate pollen from monosulcate. *Grana* **18**: 1–4.

Kuprianova, L.A. and Tarasevich, V.F. (1983). Morphologiya pyl'tsy sovremennykh iskopaenykh vidov roda *Nelumbo* (Nelumbonaceae): *Botanicheskii Zhurnal (Moscow & Leningrad)* **68**: 137–147.

Lacalli, T.C. (1981). Dissipative structures and morphogenetic pattern in unicellular algae. *Philosophical Transactions of the Royal Society, London, Biology* **294**: 547–588.

Levine, D.A. and Anderson, G.J. (1986). Evolution of dioecy in *American Solanum*. In: W.G. D'Arcy (editor). Solanaceae: Biology and Systematics, pp. 264–273. Columbia University Press, New York.

Lloyd, D.G. and Webb, C.J. (1992). The evolution of heterostyly. In: S.C.H. Barrett (editor). Evolution and function of heterostyly. *Monographs on Theoretical and Applied Genetics* **15**, pp. 151–178.

Lord, E.M. and Eckard, K.J. (1984). Incompatibility between the dimorphic flowers of *Collomia grandiflora*, a cleistogamous species. *Science* **223**: 695–696.

Lord, E.M. and Eckard, K.J. (1986). Ultrastructure of the dimorphic pollen and stigmas of the cleistogamous species *Collomia grandiflora* (Polemoniaceae). *Protoplasma* **132**: 12–22.

Mattsson, O. (1976). The development of dimorphic pollen in *Tripogandra* (Commelinaceae). In: Ferguson, I.K. and Muller, J. (editors). The evolutionary significance of the exine, pp. 163–183. Academic Press, London.

McNeill, J. and Crompton, C.W. (1978). Pollen dimorphism in *Silene alba* (Caryophyllaceae). Canadian Journal of Botany **56**: 1280–1286.

Mignot, A., Hoss, C., Dajoz, I., Leuret, C., Henry, J.P., Dreuillaux, J.M., Heberle-Bors, E. and Till-Bottraud, I. (1994). Pollen aperture polymorphism in the angiosperms: importance, possible causes and consequences. *Acta Botanica Gallica* **141**: 109–122.

Mori, S.A., Orchard, J.E. and Prance, G.T. (1980). Intrafloral pollen differentiation in the New World Lecythidaceae, subfamily Lecythidoideae. *Science* **209**: 400–403.

Muller, J. (1979). Form and function in angiosperm pollen. *Annals of the Missouri Botanical Garden* **66**: 593–632.

Pacini, E. and Bellani, L.M. (1986). *Lagerstroemia indica* L. pollen: form and function. In: Blackmore, S. and Ferguson, I.K. (editors). Pollen and spores: form and function, pp. 347–357. Academic Press, London.

Patterson, C. (1982). Morphological characters and homology. In: Joysey, K.A. and Friday, A.E. (editors). Problems of phylogenetic reconstruction, pp. 21–74. Academic Press, London.

Rogers, C.M. and Harris, B.D. (1969). Pollen exine deposition: a clue to its control. *American Journal of Botany* **56**: 1209–1211.

Rowley, J.R. (1975). Germinal apertural formation in pollen. *Taxon* **24**: 17–25.

Rowley, J.R. (1996). Exine origin, development and structure in pteridophytes, gymnosperms and angiosperms. In: Jansonius, J. and MacGregor, D.C. (editors). Palynology: principles and applications, vol. 1, pp. 443–462, American Association of Stratigraphic Palynologists Foundation.

Rudall, P.J., Furness, C.A., Chase, M.W. and Fay, M.F. (1997). Microsporogenesis and pollen sulcus type in Asparagales (Lilianae). *Canadian Journal of Botany* **75**: 408–430.

Schmid, A.-M.M., Eberwein, R.K. and Hesse, M. (1996). Pattern morphogenesis in cell walls of diatoms and pollen grains: a comparison. *Protoplasma* **193**: 144–173.

Scott, R.J. (1994). Pollen exine - the sporopollenin enigma and the physics of pattern. In: Scott, R.J. and Stead, A.D. (editors). Molecular and cellular aspects of plant reproduction, pp. 49–82. Cambridge University Press.

Sheldon, M.J. and Dickinson, H.G. (1983). Determination of patterning in the pollen wall of *Lilium henryi*. *Journal of Cell Science* **63**: 191–208.

Sheldon, M.J. and Dickinson, H.G. (1986). Pollen wall formation in *Lilium*: The effect of chaotropic agents, and the organization of the microtubular cytoskeleton during pattern development. *Planta* **186**: 11–23.

Takahashi, M. (1989a). Pattern determination of the exine in *Caesalpinia japonica* (Leguminosae: Caesalpinioideae). *American Journal of Botany* **76**: 1615–1626.

Takahashi, M. (1989b). Development of the echinate pollen wall in *Farfugium japonicum* (Compositae: Senecioneae). *Botanical Magazine, Tokyo* **102**: 219–234.

Takahashi, M. (1995). Development of structureless pollen wall in *Ceratophyllum demersum* L. (Ceratophyllaceae). *Journal of Plant Research* **108**: 205–208.

Till-Bottraud, I., Mignot, A., De Paepe, R., and Dajoz, I. (1995). Pollen heteromorphism in *Nicotiana tabacum* (Solanaceae). *American Journal of Botany* **82**: 1040–1048.

Uffelen, G.A. van (1991). The control of spore wall formation. In: Blackmore, S. and Barnes, S.H. (editors) Pollen and spores: patterns of diversification, pp. 89–102. Clarendon Press, Oxford.

Van Campo, M. (1976). Patterns of pollen morphological variation within taxa. In: Ferguson, I.K. and Muller, J. (editors). The evolutionary significance of the exine, pp. 125–137. Academic Press, London.

Walker, J.W. (1974). Aperture evolution in the pollen of primitive angiosperms. *American Journal of Botany* **61**: 1112–1136.

Walker, J.W. and Doyle, J.A. (1975). The bases of angiosperm phylogeny: palynology. *Annals of the Missouri Botanical Garden* **62**: 664–723.

Wilson, T.K. (1964). Comparative morphology of the Canellaceae. III. Pollen. *Botanical Gazette* **125**: 192–197.

Wodehouse, R.P. (1935). Pollen grains. McGraw-Hill, New York.

Lobreau-Callen, D., Malecot, V. and Suarez-Cervera, M. (2000). Comparative study of pollen from apetalous Crotonoideae and some other uniovulate Euphorbiaceae: exine ultrastructure at the aperture. In: M.M. Harley, C.M. Morton and S. Blackmore (Editors). Pollen and Spores: Morphology and Biology, pp. 301–324. Royal Botanic Gardens, Kew.

COMPARATIVE STUDY OF POLLEN FROM APETALOUS CROTONOIDEAE AND SOME OTHER UNIOVULATE EUPHORBIACEAE : EXINE ULTRASTRUCTURE AT THE APERTURE

DANIELLE LOBREAU-CALLEN[1],[2], VALERY MALECOT[1] AND MARIA SUAREZ-CERVERA[3]

[1]FR3-CNRS, Institut d'Ecologie, Paléobotanique et de Paléoécologie, Université Pierre et Marie Curie, 12 Rue Cuvier, F-75005 Paris
[2]Phytomorphologie de l'EPHE, Muséum national d'Histoire naturelle, 16 Rue Buffon, F-75005 Paris, France
[3]Department of Botany, Faculty of Pharmacy, University of Barcelona, s.n. Av. Juan XXIII, E-08028 Barcelona, Spain

Abstract

A phylogenetic analysis of the Euphorbiaceae has been undertaken using pollen and macromorphological characters. It includes representatives of each of the five subfamilies. Results show the family to be monophyletic. The uniovulate subfamilies constitute a clade which has the grade of the biovulate Phyllanthoideae-Oldfieldioideae as sister group. The Phyllanthoideae are polyphyletic and the monophyletic Oldfieldioideae are rooted in the apomorphic members of Phyllanthoideae. The Crotonoideae are monophyletic and have character states which place them near the Phyllanthoideae. Within subfamily Crotonoideae, there is a tendency towards the inaperturate condition, however, the pollen of the five apetalous tribes of Crotonoideae has either simple or composed apertures. TEM shows that, in several genera, endexine is visible in the endoaperture, and in a few others the apertures are covered by an operculum in the endexine (*Manihot*), or an operculum comprising both ectexine and endexine (*Hevea*). Compared with the pollen ultrastructure of the other two uniovulate subfamilies of Euphorbiaceae, which are sister groups, the apetalous tribes of Crotonoideae show some affinity with the subfamily Acalyphoideae, in particular with Acalypheae (*Claoxylon*), (endoapertural endexine), but differ from Alchorneae (*Bocquillonia*) which has an ectexinous operculum. The apertural ultrastructure of the three studied species of Euphorbioideae (Hippomaneae), shows rare characters also observed in the pollen of Crotonoideae. This study indicates that apertural exine character states in the Euphorbiaceae are often homoplasic and autapomorphic.

Introduction

The Euphorbiaceae are a large, monophyletic family comprising about 325 genera and 7950 species. Among the five subfamilies (Webster, 1994) three, Acalyphoideae, Crotonoideae and Euphorbioideae, have a single ovule and form a clade according to the strict consensus cladogram based on vegetative as well as reproductive characters (Levin and Simpson, 1994), and also based on molecular sequences (Wurdack and Chase, 1996).

Of the twelve tribes of Crotonoideae, seven (59 genera) are predominantly tropical, and have flowers which in general possess petals, and produce inaperturate pollen (Nowicke, 1994; Lobreau-Callen and Suárez-Cervera, 1997). The remaining five tribes (15 genera) are primarily sub-tropical: Adenoclineae, Gelonieae, Micrandreae, Manihoteae and Elateriospermeae, and have apetalous flowers which produce aperturate pollen. The pollen of the Crotonoideae is characterised by crotonoid ornamentation (Erdtman, 1952; Lobreau-Callen and Suárez-Cervera, 1997) and crotonoid ultrastructure (Lobreau-Callen and Suárez-Cervera, 1997). The ornamentation and, in most cases, the ultrastructure are similar throughout the subfamily. The infratectum is columellate or granular, the foot layer continuous, fragmented or absent, while the endexine differs in texture from the foot layer, being compact and lamellate (Thanikaimoni et al., 1984; Nowicke, 1994; Lobreau-Callen and Suárez-Cervera, 1997).

In most of the Acalyphoideae the pollen is usually 3-6-colporate (Punt, 1962; Gillespie, 1994; Lobreau-Callen and Suárez-Cervera, 1994; Takahashi et al., 1995; Fernández-González and Lobreau-Callen, 1996; Nowicke et al., 1998), while in Euphorbioideae it is generally tricolporate with a well-defined smooth margin (except in Stomatocalyceae) (Punt, 1962; El-Ghazaly and Chaudhari, 1993; Suárez-Cervera et al., 1995; Malécot and Lobreau-Callen, unpublished).

Exine characters of the mesocolpium of Crotonoideae are much more similar to those of Phyllanthoideae, one of the two non-uniovulate subfamilies, than to those of the other two uniovulate subfamilies (Lobreau-Callen and Suárez-Cervera, 1997). However, using mesocolpial exine characters, Acalyphoideae and Euphorbioideae show clear affinities to one another (Fernández-González and Lobreau-Callen, 1996). In Acalyphoideae and Euphorbioideae the non-apertural exine is often tectate-perforate, the infratectum columellar, the foot layer and endexine irregular, and the boundary between these two layers is sometimes delimited by the presence of microfibrils (Lobreau-Callen and Suárez-Cervera, 1994).

In order to understand how variation in ultrastructural characters might have lead to the inaperturate condition, the different types of apertures found in Crotonoideae (Lobreau-Callen and Suárez-Cervera, 1997) have been studied using electron microscopy. Furthermore, because certain apetalous tribes of Crotonoideae and several Euphorbioideae are occasionally referred to Acalyphoideae (Lobreau-Callen and Suárez-Cervera, 1994, 1997; Webster, 1994; Fernández-González and Lobreau-Callen, 1996), tricolporate pollen of some representative genera were also studied: Acalyphoideae group: *Claoxylon* (Acalypheae) and *Bocquillonia* (Alchorneae) as well as *Excoecaria*, *Gymnanthes* and *Omalanthus* (Hippomaneae) from subfamily Euphorbioideae. A comparison of the aperture characters of these pollen types should make it possible to establish relationships that might exist among these groups.

Materials and Methods

Pollen preparation
Pollen for both scanning and transmission electron microscopy (SEM and TEM) was removed either from herbarium specimens in the Paris Herbarium (P), or from living material cultivated in the Arboretum de Chèvreloup (Appendix 1). For SEM study, samples were acetolysed then coated with gold and examined using a Jeol 840

microscope. For TEM work unacetolysed pollen was fixed in 2.5% glutaraldehyde, in cacodylate sodium buffer, at 0.025 M at 4°C. Herbarium material was post-fixed in 1% osmium tetroxide. A sample taken from living material of *Manihot esculenta* was fixed in 2% osmic acid added to a mixture of 0.8% potassium ferricyanide K3Fe(CN)6 – phosphate buffer. In all cases, after dehydration in acetone, pollen was embedded in Spurr resin. Sections were stained with uranyl acetate and lead citrate (AU/CP), and observations were made using an Hitachi 200 SEM, or a Philips 200 or 301 transmission electron microscope (TEM) (Lobreau-Callen and Suárez-Cervera, 1994, 1997).

Cladistic analysis (Fig. 37; Appendix 2; Table 1).

The outgroups (Appendix 1) selected were *Staphylea pinnata* (Staphyleaceae, Rosids: Geraniales), *Celastrus paniculatus* (Celastraceae, Eurosids I: not placed in an order), and four members of Malpighiaceae: *Acridocarpus smeathmannii*, *Byrsonyma lucida*, *Echinopterys eglandulosa*, *Hiptage benghalensis* (Eurosids I: Malpighiales). The last two families are sister groups to the Euphorbiaceae (Bremer *et al.*, 1998). For the cladistic analysis PAUP (3.1.1) was used. The tree (Fig. 37) was produced using McClade (3.0.7).

Terminology

In general the terminology follows Punt *et al.* (1994), however, a few terms follow Erdtman (1952), Thanikaimoni (1978), or Lobreau-Callen and Suárez-Cervera (1997).

Results

I. Crotonoideae [Figs. 1–19]
Pollen aperture type in the apetalous tribes is diverse (SEM). Two major types of aperture are recognised:
1. Tricolporate or tricolpate
1A. Apertural membrane scabrous or granular
1Aa – Endoaperture sub-elliptic and generally poorly delimited laterally [Figs. 2–5]

TEM. In the mesocolpium of the pollen of *Micrandra elata* the crotonoid tectum rests on a columellar infratectum (Fig. 2, left); the foot layer is thick and the endexine irregular with microchannels. Above and below the equatorial region (Fig. 2, right), the ectexine is in irregular masses. At the margin of the furrow, the endexine is composed of many small, irregular granules. Under the furrow it is thicker and more compact, a few short, irregular lamellae are visible. At the endoaperture (Fig. 3) the ectexine is reduced to a few granulae and clavae. The endexine is spongy and consists of irregular, poorly consolidated globules (Fig. 4), while under the furrow endexine is absent. At the lateral extremities of the endoaperture, the spongy endexine terminates abruptly (Figs. 4–5). Throughout the aperture, the pollenkitt is appressed to the endexine or to the intine; it is particularly dense, and includes irregular crystalline electron transparent elements (Figs. 3–5).
Taxon examined: Micrandra elata (Micrandrineae (Micrandreae))

Key to symbols in figs. 1–36: Am = aperture margin; C = columella; c = costae; cl = clavae E = endexine; Ec = compact endexine; ES = spongy endexine; Ex = lateral extremity of an endoaperture; F = foot layer; G = granular infratectum; GE = ectexinous granules; Gr = pollen grain; I = intine; K = pollenkitt; M = mesocolpium; O = operculum; OpE = endexinous operculum; T = tectum.

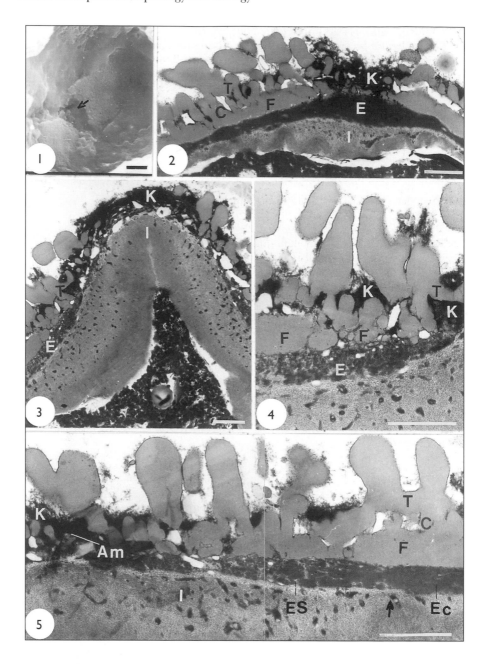

FIGS. 1–5. Fig. 1. *Glycydendron amazonicum*, inner view of a broken grain showing endoaperture (arrow) SEM. Figs. 2–5. *Micrandra elata*, transverse sections through apertures. Fig. 2. Subequatorial section showing abundant pollenkitt on the ectoaperture. TEM. Fig. 3. Section through an endoaperture. TEM. Fig. 4. Detail of the spongy endexine in the endoaperture, and under the margin of the ectoaperture. TEM. Fig. 5. Section through an ectoaperture margin (left) and one side of an endoaperture (right), note compact, spongy endexine. TEM. Scale: Fig. 1 = 1 μm; Figs. 2–5 = 0. 1 μm.

1Ab – Endoaperture H-shape, well-delimited [Fig. 6]

TEM. In the mesocolpium of the pollen of *Klaineanthus gaboniae* the structure of the tectum is consolidated, the infratectum is columellar, and the foot layer is irregular and continuous (Fig. 6, left).The endexine comprises two parts, one continuous and contiguous with the foot layer and the other discontinuous as a result of overlapping of short lamellae and irregular masses. At the edge of the ectoaperture the tectum is discontinuous and irregular, and ends abruptly (Fig. 6 right). The infratectum and the foot layer are disorganised and reduced to irregular masses which terminate at the aperture margin. The endexine is reduced to irregular masses formed of short lamellae, it is absent in the endoaperture region.

Taxon examined: Klaineanthus gaboniae (Adenoclineae)

1B. Ectoaperture ornamented with free elements

1Ba – Apertural membrane poorly delimited [Figs. 1, 7–9]

SEM. Pollen ornamented with clavae (*Cladogelonium, Endospermum, Glycydendron, Tetrorchidium*) or prisms (*Adenocline, Ditta*). In pollen with tricolpate apertures (*Adenocline*), the endexine is either interrupted by endocracks in the mesocolpium (Fig. 7) or smooth (*Cladogelonium*). While tricolporate pollen has an endocingulum (*Ditta, Endospermum, Tetrorchidium*) or a small endoaperture (*Glycydendron* - Fig. 1).

TEM. In the pollen of *Tetrorchidium congolense* the meridian section through the mesocolpium transects the endocingulum (Fig. 8) which corresponds to the interruption of the endexine. At the apocolpia irregular endexine is present. In polar section the margin of a furrow near a polar extremity (Fig. 9) shows the tectum to be thinner than in the mesocolpium, while the infratectum shows some subspheroidal elements, occasionally these are more or less columellate. The foot layer is formed of numerous irregular granular masses fused one to another. In the ectoaperture, there are free ectexinous elements, and the foot layer is comprised of numerous contiguous, irregular masses.

Taxa examined: Adenocline acuta, Ditta myricoides, Endospermum chinense, E. ovalifolium, Glycydendron amazonicum, Tetrorchidium congolense (Adenoclineae); *Cladogelonium madagascariensis* (Gelonieae).

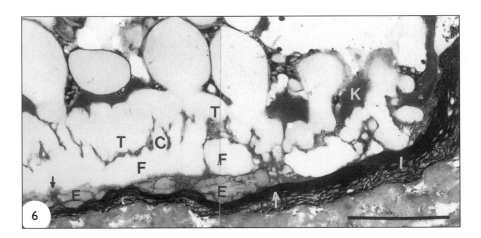

FIG. 6. *Klaineanthus gaboniae.* Transverse section of an ectoaperture margin, and one side of the endoaperture (arrow). TEM. Scale = 0. 1 μm.

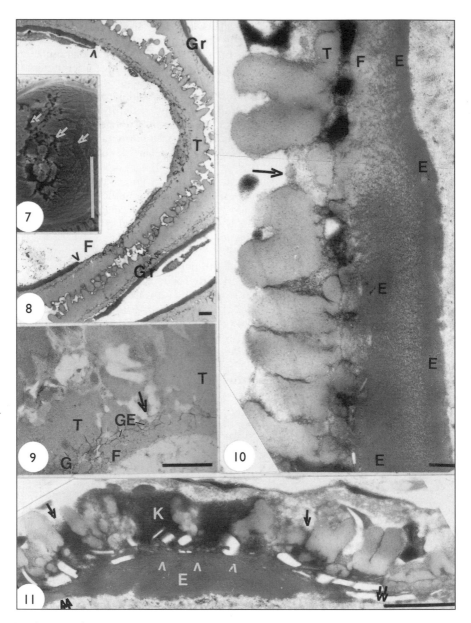

FIGS. 7–11. Fig. 7. *Adenocline acuta,* inside of broken grain, note endocracks below a furrow and at the mesocolpium (arrows). SEM. Figs. 8–9. *Tetrorchidium congolense.* Fig. 8. Meridian section through the endocingulum at a mesocolpium: arrowheads indicate the limits of the endocingulum. TEM. Fig. 9. Ectoaperture (arrow) transverse section through the endocingulum, near a polar extremity. TEM. Figs. 10–11. *Hevea kunthiana.* Fig. 10. Meridian section through the operculum: detail of the polar extremity of the operculum with a commissure (arrow) in the ectexine between operculum (below) and apocolpium (above), note the continuous endexine between the operculum and the apocolpium (pontoperculum). Fig. 11. Transverse section through an operculum (single arrows = commissure of the ectexine at the lateral sides of the operculum; double arrows = commissure of endexine at the lateral side of the operculum; white arrowheads = short irregular lamellae, contiguous with the foot layer). TEM. Scale: Fig. 7 = 10 μm; Figs. 8–9, 11 = 0. 1 μm; Fig. 10 = 0. 01 μm.

1Bb – Aperture membrane with a distinctly delimited structure [Figs.10–11]

TEM. In the pollen of *Hevea kunthiana* the tectum is interrupted abruptly along the edge of the furrow. The infratectum is reduced to a few irregular, scattered granulae, and the foot layer is reduced, and traversed by microfibrils. The endexine is interrupted, and replaced by a few dense flattened masses. At the polar end (Fig. 10), and at the lateral edges of the furrow (Fig. 11), the ectexine is clearly delimited, discontinuous and replaced by groups of clavae that are fused at the base (operculum). Under the furrow, the endexine is thicker and changes in texture, becoming heterogeneous, with short, irregular lamellae in the upper quarter, contiguous with the foot layer (Fig. 11, white arrowheads). At the polar end of the furrow, the endexine is continuous with the endexine of the apocolpium; on the lateral edges of the furrow, the endexine is clearly delimited and shows a discontinuity with the mesocolpial endexine (Fig. 11, double arrows). The endexine consists of a pontoperculum. Throughout the aperture, the base of the pollenkitt includes crystalline elements that are electron transparent.

Taxon examined: Hevea kunthiana (Heveineae, Micrandreae)

2. Periporate or cryptoporate

2A. Apertural membrane smooth or granular [Figs. 12–15]

SEM. In *Manihot* the aperture is obstructed by the clearly delimited and discontinuous operculum - endexine (Fig. 12). At the centre of the aperture, the endexine forms a thick and compact lens or operculum not visible on the surface (Fig. 14). However, in the pollen of Elateriospermeae (*Elateriospermum*), a similar apertural endexinous operculum does not exist, although endexine seems to be present under the apertures (Fig. 15).

TEM. In pollen of *Manihot esculenta* around the aperture (Fig. 13) the tectum has clavate ornamentation. The clavae are very short and sharply interrupted. The granular infratectum is absent, and the foot layer is very reduced. The attenuated endexine is continuous and then abruptly interrupted.

Taxa examined: Manihot esculenta, M. tripartita (Manihoteae) and *Elateriospermum tapos* (Elateriospermeae)

2B. Apertural membrane clavate [Figs. 16–19]

SEM. The apertural membrane of the pollen of *Suregada chauvetia* is sculptured by short clavae (Figs. 16, 17).

TEM. In the pollen of *Suregada chauvetia*, the tectum around the apertures comprises closely packed interrupted elements which end abruptly (Fig. 18). The bases of the clavate elements rest on a few ectexinous granules or, more frequently, directly on the endexine (Fig. 19). The endexine is irregular and lamellate, as in the mesocolpium (Fig. 19). The lamellae anastomose and include dispersed ectexinous masses. Heterogeneous and fibrillar masses of lipoprotein that are as electron dense as the endexine are present in all apertures either at the centre of the heads of the clavae, at the bases of the clavae or, forming globular elements between them.

Taxon examined: Suregada chauvetia (Gelonieae)

II. Acalyphoideae [Figs. 20–27]

Acalypheae - endoapertural endexine present [Figs. 20–23]

SEM. In the endoaperture, endexine present near the margin disappeared under the furrow (Fig. 23).

TEM. In the pollen of *Claoxylon insulanum* the tectum is perforated at the edge of the furrow (Fig. 20), the infratectum is columellar, and the foot layer is fragmented into irregular masses while, in the mesocolpium, it is continuous and crossed by microchannels. The apertural membrane is ectexinous and reduced to granules (Fig. 21). Away from the equatorial region, below the electron dense endexine layer,

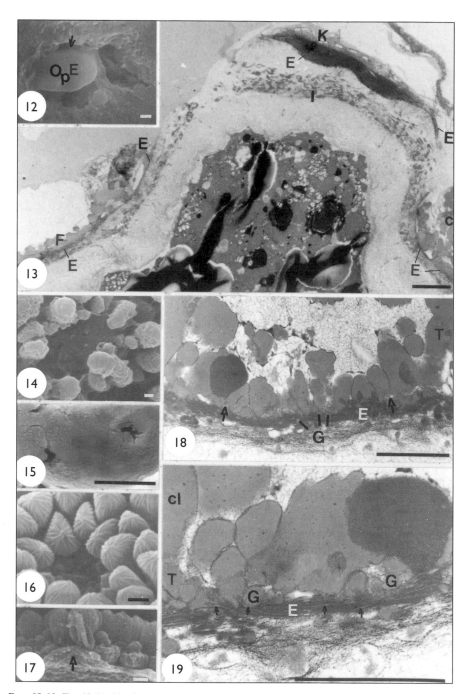

FIGS. 12–19. Figs 12–14. *Manihot esculenta.* Fig. 12. A broken grain, showing innermost side of the endexinous operculum (OpE). SEM. Fig. 13. Endexinous operculum on a tri-layered intine. TEM. Fig. 14. Ectoaperture. SEM. Fig. 15. *Elateriospermum tapos,* a broken grain showing inner view of apertures. SEM. Figs. 16–19. *Suregada chauvetia.* Fig. 16. Ectoaperture. SEM. Fig. 17. Section through the exine showing endexine in aperture (arrow). SEM. Fig. 18. Transverse section through an aperture. TEM. Fig. 19. Detail of clavae, and endexine lamellae (arrows). TEM. Scale: Fig. 15 = 10 µm; Figs. 12, 14, 16–17 = 1 µm; Figs. 13, 18–19 = 0. 1 µm.

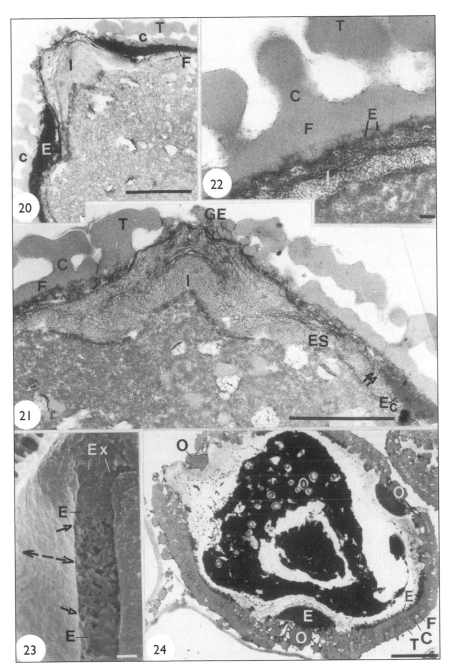

FIGS. 20–24. Figs. 20–23. *Claoxylon insulanum.* Fig. 20. Meridian section through an aperture showing costae. TEM. Fig. 21. Equatorial section through aperture with compact endexine at margin of endoaperture (Ec and double arrow), note spongy endexine (ES) present at endoaperture margin, but absent below the ectoaperture. TEM. Fig. 22. Detail of the spongy endexine in the endoaperture margin opposite. TEM. Fig. 23. Endoaperture: note endexine near the lateral extremity, costae (arrows), axis of the ectoaperture (arrows). SEM. Fig. 24. *Bocquillonia sessiliflora*, sub-equatorial section of a whole pollen grain showing the opercula in an endoaperture (left) and above the endoapertures (below and right). TEM. Scale: Figs. 20, 22, 24 = 0. 1 μm; Fig. 21 = 0. 01 μm; Fig. 23 = 1 μm.

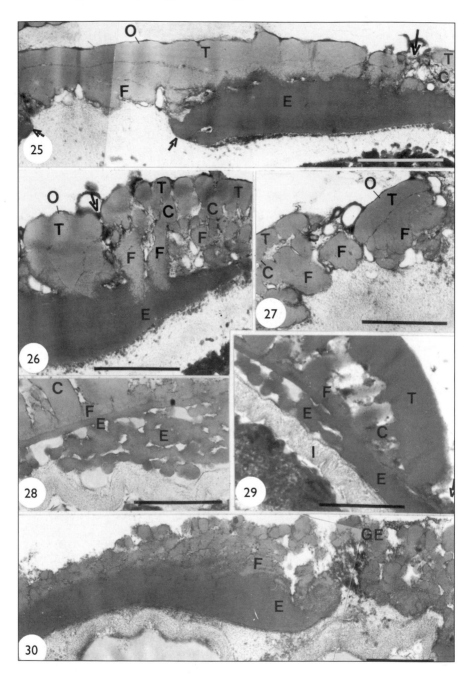

FIGS. 25–30. Figs. 25–27. *Bocquillonia sessiliflora.* Fig. 25. Meridian section through the operculum from above the endoaperture (left arrows) to the polar end (right): note commissure in the ectexine (right arrow). TEM. Figs. 26–27. Transverse sections of operculum at the lateral margin of a furrow. Fig. 26. Endexine thicked under the furrow margin and continuous with that of the mesocolpium. TEM. Fig. 27. Section through an endoaperture. Figs. 28–30. *Omalanthus nutans.* TEM. Fig. 28. Granular-lamellate endexine at apocolpium. TEM. Fig. 29. Granular endexine below the ectoaperture margin; arrow (bottom right) indicates the ectoaperture. TEM. Fig. 30. Meridian section through aperture. TEM. Scale (all figs.): 0.1 μm.

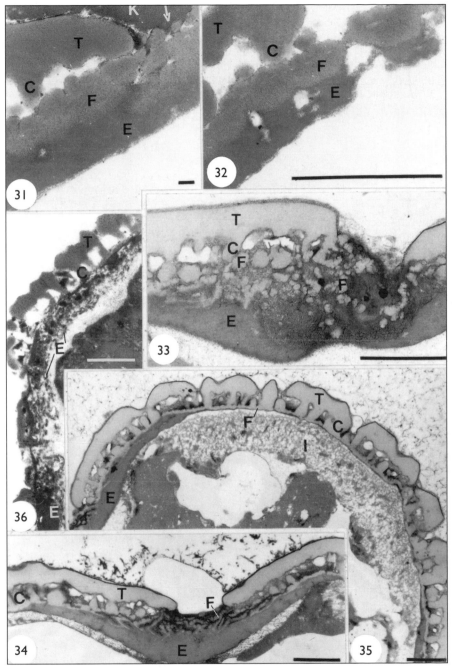

FIGS. 31–36. Figs. 31–32. *Excoecaria agallocha.* Fig. 31. Transverse section through the ectoaperture margin. TEM. Fig. 32. Detail of the foot layer and the endexine in mesocolpium. TEM. Figs. 33–35. *Gymnanthes elliptica.* Fig. 33. Sub-equatorial section through a composite aperture. TEM. Fig. 34. Transverse section across aperture and adjacent mesocolpia. TEM. Fig. 35. Transverse section through mesocolpium and adjacent aperture (left). TEM. Fig. 36. *Savia danguiana.* To show endexine in an endoaperture. TEM. Scale: Fig. 31 = 0. 01 μm; Figs. 32–36 = 0. 1 μm.

globular masses are apparent, particularly under the costae perpendicular to the furrow (Figs. 21, 22). Near the lateral extremities of the endoaperture, the endexine is thin and has a spongy or even coarsely lamellate appearance. The foot layer-endexine boundary is diffuse (Fig. 22). Below the furrow, and in the ectoaperture, the endexine is absent (Fig. 22). In equatorial section (Fig. 21) the endoapertures are bordered by well-delimited, compact endexine which changes abruptly to a spongy texture. Seen in meridian section (Fig. 20) the thick costae terminate abruptly.

Taxa examined: Acalypha pancheriana, Claoxylon insulanum and *C. humbert ii.*

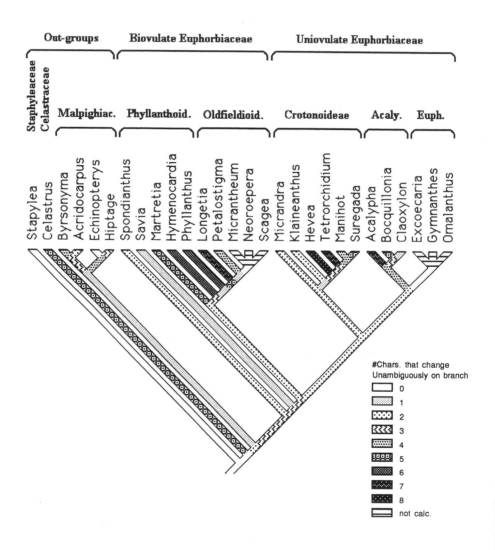

FIG. 37. Strict consensus tree based on 6 trees of 261 steps. Acaly. = Acalyphoideae; Euph. = Euphorbioideae; Malpighiac. = Malpighiaceae; Oldfieldioid. = Oldfieldioideae; Phyllanthoid. = Phyllanthoideae.

Alchorneae - ectexinous operculum [Figs. 24–27]

TEM. In the pollen of *Bocquillonia sessiliflora* the three apertures are operculate (Fig. 24) the ectoaperture being covered by a clearly delimited ectexinous structure (Fig. 25). The ectoaperture is bordered by a microperforate tectum, a columellar infratectum, and a foot layer comprised of irregular masses, while, in the mesocolpium, the foot layer is thick and continuous (Fig. 24). Below the foot layer, the endexine is irregular and somewhat granular. Above or below the equatorial region, the foot layer-endexine boundary is diffuse (Fig. 26). On the edge of the furrow, above the endoaperture, only the ectexine is present, and the foot layer comprises coarsely sub-ellipsoid or cylindrical masses (Fig. 27).

Taxon examined: Bocquillonia sessiliflora

III. Euphorbioideae

Apertures in all three taxa examined are tricolporate, the aperture membrane is ectexinous [Figs. 28–35]

TEM. At the aperture, the exine in the three species examined is similar with a smooth tectate margin, an infratectum of short columellae, a foot layer formed of irregular masses, and a thick endexine (Figs. 28–35). In the pollen of *Omalanthus nutans* (Figs. 28–30), above and below the equator, and at the aperture margins, the tectum is traversed by microchannels, and the foot layer is densely perforated. The endexine is thin in the mesocolpial area, numerous masses of more or less spheroidal layered endexinous globules are apparent in the apocolpial area (Fig. 28). In the furrow the aperture membrane comprises ectexinous granulae and lamellae, and the endexine is compact and thick (Fig. 29, extreme bottom right). At the equator the endoaperture is bordered by an endexine thickening, while around the endoaperture, a few short foot layer lamellae are interbedded with endexine (Fig. 30). The endoaperture membrane is composed of irregular ectexine masses. At the aperture margin in the pollen of *Excoecaria agallocha* the endexine is micro-sculptured, as in the apocolpium (Fig. 31). In the mesocolpium, however, between the extremities of the ectoapertures the foot layer is formed of contiguous irregular masses, and the endexine comprises a discontinuous layer with a spongy appearance (Fig. 32). At the equator the endoaperture membrane (Fig. 31, extreme bottom right) comprises numerous endexinous masses, interspersed with short ectexinous lamellae. In *Gymnanthes elliptica* the foot layer at the aperture margin is irregular and the surface comprised of contiguous masses (Fig. 33). Around the pore (Fig. 34), the foot layer is reduced to irregular masses or granulae and irregular short lamellae. Adjacent to the endoaperture the endexine, which is absent in the mesocolpium, forms a costa interspersed with short foot layer lamellae (Figs. 34–35). At the centre of the pore the membrane is formed of ectexinous granulae supported by an aceto-soluble micro-fibrillar layer (Fig. 35).

Taxa examined: Excoecaria agallocha, Gymnanthes elliptica, and *Omalanthus nutans* (Hippomaneae)

Cladistic analysis (Fig. 37; Table 1; Appendix 2)

For aperture ultrastructure 30 pollen characters were identified (characters 1–30). In addition 17 floral, fruit, and vegetative characters have been included (characters 31–47). All the characters are unordered and polymorphism is indicated by the two states co-occurring in a taxon (Table 1). Six trees were obtained with 261 steps (CI = 0,464; RI = 0,572), these are summarised in the strict consensus tree (Fig. 37).

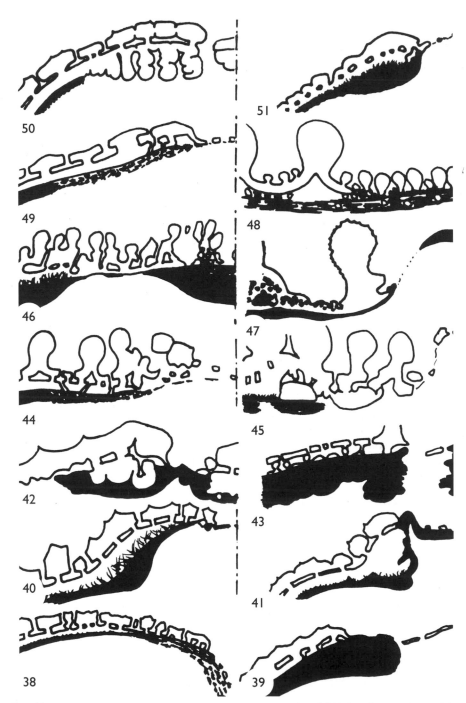

FIGS. 38–51. Diagrams to show exine ultrastructure at aperture margin, solid black indicates endexine. (Figs. 38–42, based on Lobreau-Callen and Suárez-Cervera, 1993, 1994). Figs. 38–41. Phyllanthoideae. Fig. 38. *Savia danguiana.* Fig. 39. *Spondianthus preusii.* Fig. 40. *Martretia quadricornis.* Fig. 41. *Hymenocardia acida.* Figs. 42–43. Oldfieldioideae. Fig. 42. *Longetia buxoides.* Fig. 43. *Micrantheum hexandrum.* Figs. 44–48. Crotonoideae. Fig. 44. *Micrandra elata.* Fig. 45. *Klaineanthus gaboniae.* Fig. 46. *Hevea kunthiana.* Fig. 47. *Manihot esculenta.* Fig. 48. *Suregada chauvetia.* Figs. 49–51. Acalyphoideae. Fig. 49. *Claoxylon insulanum.* Fig. 50. *Boquillonia sessiliflora.* Fig. 51. *Acalypha pancheriana.*

TABLE 1. Data Matrix for the 30 pollen and 17 macromorphological characters.

	1	2	3	4	5	6	7	8	9	10	11	12	13	14	15	16	17	18	19	20	21	22	23	24	25
Phyllanthoideae																									
Hymenocardia	2	0	0	2	1	0	0	1	0	4	1	1	0	1	2	3	1	0	0	2	0	0	1	0	4
Martretia	0	0	0	0	1	1	0	1	0	2	2	1	0	1	0	1	1	1	0	1	0	0	8	0	4
Phyllanthus	0	2	2	2	2	0	0	1	0	2	1	1	1	1	0	4	1	2	0	2	0	0	0	0	5
Savia	0	0	0	0	1	2	0	0	0	1	1	0	0	0	0	0	2	0	0	1	0	0	0	?	0,1
Spondianthus	0	0	0	0	0	2	0	0	2	2	1	0	0	0	0	0	1	0	0	1	0	0	6	?	1
Crotonoideae																									
Hevea	0	0	0	2	2	1	0	1	0	2	0	1	0	1	0	1	0	0	2	6	0	0	3	0	3
Klaineanthus	0	0	0	0	1	1	0	1	2	2	0	0	0	1	0	0	1	0	0	2	0	0	3	0	3
Manihot	0	2	2	2	2	0	0	3	1	2	0	0	0	0	0	1	0	2	3	2	0	0	3	0	3
Micrandra	0	0	0	0	0,1	0	0	1	1	2	1	0	0	1	0	0	1	0	0	0	0	0	3	0	3
Tetrorchidium	0	0	0	0	2	1	0	3	0	2	0	0	0	1	0	4	0	0	0	4	0	1	3	2	4
Suregada	0	2	2	2	2	0	0	3	1	2	0	0	0	2	0	2	0	1	0	2	0	0	3	1	3
Acalyphoideae																									
Acalypha	0	0	0	1	1	0	2	0	0	4	1	1	0	1	2	4	1	0	0	1	0	0	1	0	4
Bocquillonia	0	0	0	0	0	1	0	0	1	2	1	1	0	2	0	0	1	0	1	3	0	0	1	2	6
Claoxylon	0	0	0	0	1	1	0	0	1	2	1	1	0	1	0	0	2	0	0	5	0	0	1	2	1
Euphorbioideae																									
Excoecaria	0	0	0	0	1	1	1	1	1	2	0	0	0	1	0	0	1	1	0	1	0	2	0	0	2
Gymnanthes	0	0	0	0	1	1	1	1	0	3	1	0	0	1	0	0	1	1	0	1	0	0	0	0	2
Omalanthus	0	0	0	0	1	1	1	1	3	2	1	0	0	1	0	0	1	1	0	1	0	2	0	0	2
Oldfieldioideae																									
Longetia	2	1	0	1	1	0	0	2	2	2	1	1	0	1	1	3	2	0	0	2	1	1	2	0	4
Micrantheum	0	2	2	2	4	0	0	1	2	2	1	0	1	1	0	4	1	0	0	2	0	0	2	0	4
Neoroepera	0	2	2	2	1	0	0	1	2	0	1	0	0	1	0	4	1	0	0	2	0	0	2	0	4
Petalostigma	0	2	2	2	3	0	0	1	0	0	1	0	0	1	0	5	2	0	0	2	0	0	2	0	4
Scagea	0	2	2	2	1	0	0	1	2	0	1	0	0	1	0	4	1	0	0	2	0	0	2	0	4
Malpighiaceae																									
Acridocarpus	0	0	0	0	1	2	0	0	2	1	1	0	0	2	0	1	1	0	0	1	0	0	6	0	7
Byrsonyma	0	0	0	0	0	1	0	1	0	1	1	1	0	2	0	1	1	0	0	1	0	0	6	0	7
Echinopterys	0	1	1	0	0	1	0	4	0	1	0	0	0	2	0	0	1	0	0	2	0	0	8	0	4
Hiptage	1	1	3	2	0	0	0	4	0	0	0	0	0	2	0	2	1	0	1	2	0	0	5	0	6
Celastraceae																									
Celastrus	0	0	0	0	1	2	0	0	1	0	3	0	1	2	1	2	1	0	0	1	0	0	6	0	0
Staphyleaceae																									
Staphylea	0	0	0	0	0	1	0	1	1,2	0	2	1	1	2	0	2	0	0,1	0	1	0	0	0	0	1

Table 1 continued

	26	27	28	29	30	31	32	33	34	35	36	37	38	39	40	41	42	43	44	45	46	47
Phyllanthoideae																						
Hymenocardia	0	1	1	1	0	0	1	1	2	?	1	4	0	0	0	2	1	0	0	0	2	0
Martretia	0	1	1	1	0	0	1	1	0	0	0	4	1	0	0	2	0	1	0	0	2	0
Phyllanthus	0	1	1	1	0	0	1	1	1	0	0	2	1	0	0	2	0	1	0	0	4	0
Savia	0	1	1	0	0	0	1	0	0	0	0	2	0	0	0	2	0	0	0	0	2	0
Spondianthus	0	1	1	0	0	0	1	0	0	0	0	1	0	0	0	2	0	0	0	0	0	0
Crotonoideae																						
Hevea	0	1	1	1	0	0	1	1	0	1	1	2	1	1	0	1	0	0	1	1	0	1
Klaineanthus	0	1	6	1	0	0	1	1	0	0	0	2	0	0	0	1	0	0	1	2	0	0
Manihot	2	1	5	2	0	0	1	1	0	0	1	2	0	0	1	1	0	0	1	1	2	1
Micrandra	0	1	5	0	0	0	1	1	0	0	0	2	0	0	1	1	0	0	1	2	0	0
Tetrorchidium	2	0	4	0	0	0	1	1	0	0	0	5	0	0	0	1	0	0	1	2	2	0
Suregada	2	3	3	3	0	0	1	1	0	0	0	3	0	0	0	1	0	0	0	1	2	0
Acalyphoideae																						
Acalypha	0	1	5	0	0	0	1	1	1	1	0	4	0	1	1	1	0	0	0	?	4	0
Bocquillonia	0	1	1	0	0	0	1	1	1	1	1	5	1	1	1	1	0	2	0	?	4	0
Claoxylon	0	1	1	0	0	0	1	1	1	1	1	3	0	1	0	1	0	0	0	?	4	0
Euphorbioideae																						
Excoecaria	0	4	8	0	0	0	1	1	1	2	0	5	0	0	0	1	0	1	1	2	0	0
Gymnanthes	0	1	0	4	0	0	1	1	1	2	0	5	0	0	1	1	0	1	1	2	2	0
Omalanthus	0	1	7	0	0	0	1	1	1	2	1	3	0	1	1	1	0	0	1	2	2	0
Oldfieldioideae																						
Longetia	0	2	4	0	0	0	1	1	1	0	?	3	0	0	1	2	0	0	0	0	4	0
Micrantheum	0	3	2	0	1	0	2	2	2	0	0	3	0	0	1	2	0	2	0	0	5	0
Neoroepera	0	0	2	0	1	0	1	1	1	0	0	3	0	0	1	2	0	2	0	0	2	0
Petalostigma	0	2	4	0	0	0	1	0	2	0	0	3	1	0	1	2	0	2	0	0	2	0
Scagea	0	0	2	0	1	0	1	1	1	0	0	4	0	0	1	1	0	0	0	0	4	0
Malpighiaceae																						
Acridocarpus	0	1	1	0	0	1	0	0	0	0	1	0	0	0	0	0	1	0	0	0	2	0
Byrsonyma	0	1	2	0	0	0,1	0	0	0	1	1	0	0	1	0	0	0	?	0	0	0,2	0
Echinopterys	1	3	2	0	0	1	0	0	0	0	0	1	0	0	0	0	1	0	0	0	?	0
Hiptage	2	2	2	0	0	1	0	0	0	0	0	1	0	0	0	0	1	0	0	0	1	0
Celastraceae																						
Celastrus	0	1	1	0	0	0	0	0	0	0	1	1	0	0	0	0	0	2	0	0	3	0
Staphyleaceae																						
Staphylea	0	1	1	0	2	0	0	0	0	0	0	1	0	0	1	0	0	?	0	0	2	1

Discussion

Pollen characteristics of uniovulate Euphorbiaceae

In the uniovulate Euphorbiaceae, the ectoaperture is usually covered by a fragmented ectexine, on a primarily endexinous membrane. While in the endoaperture endexine is not always present. In most cases the ectexine and endexine of the mesocolpia are continuous into the apertural region. In three genera, however, the endoaperture is covered by a distinctly delimited exine, the composition of which varies between the three genera. In *Bocquillonia* it is strictly ectexinous, as in all Alchorneae (Takahashi *et al.*, 1995); in *Manihot* it is endexinous, as in other Manihoteae (Lobreau-Callen and Suárez-Cervera, 1997), while in *Hevea* it comprises both ectexine and endexine. Diversity in endoaperture membrane structure is also observed in the biovulate Euphorbiaceae, especially the presence of endexine, for example in *Savia danguyana* (Phyllanthoideae) (Fig. 36).

According to Wodehouse (1935), Iversen and Troels-Smith (1950), Erdtman (1952), and Thanikaimoni (1978) the pollen of all three taxa is operculate, the definition of an operculum, accepted by the above authors, being a thickening of measurable bulk clearly delimited by a commissure in the exine of the aperture membrane, with a structure similar to that of the remaining exine of the pollen grain. However, according to the more restricted definition of Punt *et al.* (1994), only the pollen of *Bocquillonia* and, in part, that of *Hevea* are operculate, as an operculum is "A distinctly delimited ectexinous/sexinous structure which covers part of an ectoaperture".

The variation in apertural exine between taxa make it possible to define two pollen groups for the apetalous Crotonoideae which correspond with those proposed by Webster (1994). Furthermore, these groups correspond with those defined from mesocolpial exine ultrastructure (Lobreau-Callen and Suárez-Cervera, 1997).

Group 1. In the endoaperture region the endexine is abruptly modified in texture or shows discontinuity with that of the rest of the exine, whether or not there is an endexinous operculum *sensu* Wodehouse (1935), Iversen and Troel-Smith (1950), Erdtman (1952) and Thanikaimoni (1978) (Micrandreae, *Glycydendron*, Manihoteae, Elateriospermeae).

Group 2. The endexine is either absent from the endoaperture (colporate pollen) or is continuous and uniform in texture throughout the pollen grain (porate pollen) (Adenoclineae, Gelonieae: *Cladogelonium*).

Polymorphic characters

Polymorphism, which is found in the few taxa in this study, consists of a continuous range, rather than one discrete, character state for a particular character. For example, sculpture of the apertural membrane: smooth to granular; exine: reticulate to perforate; endexine of the apertural margin: granular to irregular masses. Nevertheless, according to theoretical studies on the use and coding of polymorphic character states in cladistic analysis, such characters can be phylogenetically informative. The number of polymorphic character states is low in our analysis. We would not expect significant differences in our tree topology by using other coding methods (Wiens, 1995).

Relationships of the Crotonoideae with other subfamilies of Euphorbiaceae

The monophyletic Euphorbiaceae (Levin and Simpson 1994) are defined by unisexual flowers and ovule number. The only pollen character with unequivocal character states is the fragmented ectexine at the aperture membrane. The family consists of two major groups of species: those which are biovulate, and those which are uniovulate. The two groups were first distinguished using ovule characters in the analysis of Webster (1994).

This analysis, based on a wide range of macromorphological characters, supports the classification of Webster (1994), as does the analysis of Wurdack and Chase (1996), based on *rbc*L sequence data. Within the family, although only a small number of taxa has been considered in the present study, subfamilies Acalyphoideae, Crotonoideae, Euphorbioideae, and Oldfieldioideae appear to be monophyletic, while the Phyllanthoideae are apparently polyphyletic. The distinction of the two sister groups Acalyphoideae-Euphorbioideae is based on the reduction in sepal number, and variation in stamen number.

Within the Phyllanthoideae the genera *Savia* and *Spondianthus* form two branches (Fig. 37) at the base of the family (Phyllanthoideae I), suggesting that their tricolporate, tectate-perforate pollen is plesiomorphic within the Euphorbiaceae. The remaining taxa (Phyllanthoideae II) including *Hymenocardia*, which is sometimes referred to the monogeneric family Hymenocardiaceae, are grouped in a grade in which the Oldfieldioideae are rooted. Phyllanthoideae II and Oldfieldioideae forming a clade (Fig. 37), are characterised by a diffuse foot layer-endexine boundary. The costae, simple apertures, and complex exine associated with the pollen of the genera included in this clade are apomorphic character states. The biovulate Oldfieldioideae are characterised by four unambiguous pollen synapomorphies: echinate tectum, foot layer reduced or absent, endexine thick and continuous, and a distinct foot layer-endexine boundary. Although *Scagea*, the only uniovulate genus of Oldfieldioideae is distinct from other members of the subfamily, and sometimes placed within subfamily Crotonoideae; like the Phyllanthoideae II, it shows many of the apomorphic character states shared by Oldfieldioideae and Phyllanthoideae II.

Crotonoideae, are characterised by the non-ambiguous synapomorphy of crotonoid exine structure (character states 23: [3], 25: [5]). Within the subfamily the operculum (character 15) is a diverse, homoplasious character with a number of autapomorphic character states. Less complex apertures, with endexine present in the endoaperture, correspond with a number of apomorphic character states: reduced endoaperture, general lack of differentiation of ornamentation between mesocolpium and aperture membrane, reduction of exine structure, and a granular infratectum. These particular exine features apparently lead to the inaperturate condition (Lobreau-Callen and Suárez-Cervera, 1997).

In the strict consensus tree the Crotonoideae are located just above the biovulate Phyllanthoideae (Phyllanthoideae I), and the pollen of the Crotonoideae reflects this relationship. In the mesocolpium some of the tectum characteristics are similar: tectum either supra-reticulate or with funnel-shaped perforations, either striate clavae or verrucate, and foot layer-endexine boundary diffuse (Lobreau-Callen and Suárez-Cervera, 1997). The ultrastructure of the exine at the aperture (Figs. 38–48) also supports the affinity of these two groups. In a few genera such as *Savia* (Fig. 36, 38), *Hymenocardia* (Lobreau-Callen and Suárez-Cervera, 1994; Fig. 41), endexine is present in the endoaperture. In several other taxa (for example *Martretia, Spondianthus*, Figs. 39–40) foot layer masses are apparent at the aperture, while endexine is absent and, at the foot layer-endexine boundary, the aperture margin may be diffuse (*Martretia*), or not (*Spondianthus*) (Lobreau-Callen and Suárez-Cervera, 1993). It is noteworthy that the presence of endexine in the endoaperture also occurs in some Acalyphoideae, for example, in tribe Plukenetieae (Suárez-Cervera *et al.*, 1997) as well as in Euphorbioideae, Euphorbieae (Suárez-Cervera *et al.*, 1995). However, the endexine structure is similar to that of the Phyllanthoideae, tribes Acalypheae and Micrandreae, suggesting that this is a homoplasious feature.

Acalyphoideae and Euphorbioideae, defined by two unequivocal synapomorphies, are sister groups. Acalyphoideae (Figs. 49–51) are characterised by variation of the infratectum at the ectoaperture margin, a usually distinct foot layer-endexine boundary, a micro-echinate tectum and a shrubby plant habit. Within Acalyphoideae, tribe Alchorneae (*Bocquillonia*) is characterised by generally homogeneous pollen, but with an autapomorphic character state for the operculum (Punt, 1962; Takahashi *et al.*, 1995).

Conclusion

In this study the two uniovulate groups of Webster (1994), which are supported by some palynological characters (Lobreau-Callen and Suárez-Cervera, 1997) are not reflected by the characters of exine ultrastructure at the aperture. In phylogenetic analyses, variations in apertural exine morphology of euphorbiacious pollen (Figs. 38–51) are often autapomorphic or homoplasious characters. Furthermore, the cladistic analysis using only 47 characters, has produced clades which are frequently defined by two or three unequivocal synapomorphies. This suggests that the clades are probably not sufficiently robust (Doyle and Le Thomas, 1996). In this phylogenetic analysis only a few pollen characters have been used as unambigous synapomorphies; but even if more characters were used few of the pollen characters would lend support to the clades as defined here. Indicative that within the Euphorbiaceae many of the pollen characters or character states for example, granular infratectum, operculum, and endoapertural endexine, are homoplasious, apomorphic characters. Some character states, such as those of the operculum, are autapomorphic and characteristic at tribal level (Manihoteae). In order to achieve better resolution from phylogenetic analyses of the family, it will be necessary to consider a much larger number of taxa for each tribe, to provide a better understanding of both macromorphological and pollen characters and their variability (character states).

Acknowledgements

The authors gratefully acknowledge the Directors of the herbaria, from which pollen material was obtained for this study, in particular the Paris Herbarium. For the English translation we thank P. Lowry (Missouri Botanical Garden, USA). G. MacPherson, Euphorbiaceae systematist (Missouri Botanical Garden, USA) re-read the draft manuscript, and J.A. Doyle (University Davis of California, USA) who suggested improvements in the cladistic section. Technical assistance for SEM preparation work was provided by N. D'Amico (Laboratoire de Palynologie, EPHE, Paris). We thank CIME (Université Pierre-et-Marie Curie, Paris) for use of SEM facilities, and the University of Barcelona for use of TEM facilities. Special thanks are expressed to the editors and to the two anonymous reviewers for their improvements to the manuscript.

References

Bremer, K., Bremer, B. and Thulin, M. (1998). Classification of flowering plants. Internet http://www. systbol. uu. sc/classification/overview. html.

Doyle, J. and Le Thomas, A. (1996). Phylogenetic analysis and character evolution in Annonaceae. *Bulletin du Museum national d'Histoire naturelle, Paris, 4 ser., sect. B, Adansonia* 18: 279–334.

El-Ghazaly, G. A. and Chaudhary, R. (1993). Pollen morphology of some species of the genus *Euphorbia* L. *Review of Palaeobotany and Palynology* 78: 293–319.

Erdtman, G. (1952). Pollen morphology and plant taxonomy. Angiosperms. Almqvist and Wiksell, Stockholm.

Fernández-González, D. and Lobreau-Callen, D. (1996). Le pollen de la tribu des Acalypheae (Acalyphoideae, Euphorbiaceae). *Grana* 35: 266–284.

Gillespie, L. (1994). Pollen morphology and phylogeny of the Plukenetieae (Euphorbiaceae). *Annals of the Missouri Botanical Garden* 81: 317–348.

Iversen, J. and Troels-Smith, J. (1950). Pollenmorphologiske definitioner og typer. *Danmarks Geol. Undersögelse* 4. Raekke **3**: 1–52.

Levin, G. A. and Simpson, M. G. (1994). Phylogenetic implications of pollen ultrastructure in the Olfieldioideae (Euphorbiaceae). *Annals of the Missouri Botanical Garden* **81**: 203–238.

Lobreau-Callen, D. and Suárez-Cervera, M. (1993). Variations morphologiques et ultrastructurales du pollen de *Bischofia* et d'autres Phyllanthoideae (Euphorbiaceae). *Palynosciences* **2**: 223–237.

Lobreau-Callen, D. and Suárez-Cervera, M. (1994). Pollen ultrastructure of *Hymenocardia* Wallich ex Lindley and comparison with other Euphorbiaceae. *Review of Palaeobotany and Palynology* **81**: 257–278.

Lobreau-Callen, D. and Suárez-Cervera, M. (1997). Le pollen des Crotonoideae Apétales (Euphorbiaceae): ultrastructure de l'exine. *Review of Palaeobototany and Palynology* **98**: 257–291.

Nowicke, J. W. (1994). A palynological study of Crotonoideae (Euphorbiaceae). *Annals of the Missouri Botanical Garden* **81**: 245–269.

Nowicke, J. W., Takahashi, M. and Webster, G. L. (1998). Pollen morphology, exine structure and systematics of Acalyphoideae (Euphorbiaceae). Part 1. Tribes Clutieae (*Clutia*), Pogonophoreae (*Pogonophora*), Chaetocarpeae (*Chaetocarpus, Trigonopleura*), Pereae (*Pera*), Cheiloseae (*Cheilosa, Neoscortechinia*), Erismantheae pro parte (*Erismanthus, Moultonianthus*), Dicoelieae (*Dicoelia*), Galearieae (*Galearia, Microdesmis, Panda*) and Ampereae (*Amperea, Monotaxis*). *Review of Palaeobototany and Palynology* **102**: 115–152.

Punt, W. (1962). Pollen morphology of the Euphorbiaceae with special reference to taxonomy. *Wentia* **7**: 1–116.

Punt, W., Blackmore, S., Nilsson, S. and Le Thomas, A. (1994). Glossary of pollen and spore terminology. *Laboratory of Palynology and Palaeobotany Contribution Series 1.* Utrecht, LPP Foundation.

Suárez-Cervera, M., Gillespie, M., Fernández-González, D. and Marquez, J. (1997). Pollen grain ultrastructure of two African species of the subtribe Tragiinae (Euphorbiaceae). International Association of African Palynology (IAAP). 14–19 September 1997, University of Witwatersrand, Johannesburg, South Africa, Abstracts: 43.

Suárez-Cervera, M., Marquez, J., Martin, J., Molero, J. and Seoane-Camba, J. (1995). Structure of the apertural sporoderm of pollen grains in *Euphorbia* and *Chamaesyce* (Euphorbiaceae). *Plant Systematics and Evolution* **197**: 11–122.

Takahashi, M., Nowicke, J. W. and Webster, G. L. (1995). A note on remarkable exines in Acalyphoideae (Euphorbiaceae). *Grana* **34**: 282–290.

Thanikaimoni, G. (1978). Palynological terms: proposed definitions-1. In: D.C. Bharawaj, K.M. Lele, R.K. Kar, H.P. Singh, R.S. Tiwari, Vishnu-Mittre and H.K. Maheshwari (editors). Proceedings of the IV International Palynological Conference, pp. 228–239. Birbal Sahni Institute of Palaeobotany, Lucknow, India.

Thanikaimoni, G., Caratini, C., Nilsson, S. and Grafström, E. (1984). Omniaperturate Euphorbiaceae pollen with striate spines. *Bulletin du Jardin Botanique National de Belgique* **54**: 105–125.

Webster, G. L. (1994). Synopsis of the suprageneric taxa of Euphorbiaceae. *Annals of the Missouri Botanical Garden* **81**: 33–144.

Wodehouse, R. P. (1935). Pollen grains. McGraw-Hill, New York.

Wurdack, K. and Chase M. W. (1996). Molecular systematics of Euphorbiaceae *sensu lato* using *rbc*L sequence data. *American Journal of Botany* Abstracts 580, 83(6): 203.

Wiens, J.J. (1995). Polymorphic characters in phylogenetic systematics. Systematic Biology **44**: 482–500

APPENDIX 1. Species examined. With the exception of a few specimens, pollen material is from the collections in the Paris Herbarium (P). Other institutions are indicated where appropriate. Living material is from plants cultivated at the Arboretum de Chèvreloup (Paris Museum National d'Histoire Naturelle). Species examined with TEM are indicated.

EUPHORBIACEAE
ACALYPHOIDEAE Ascherson (1864)
ACALYPHEAE Dumortier (1829)
Acalypha pancheriana Baill. : Mackee 27235, New Caledonia (TEM).
Claoxylon humbertii Leandri: Capuron 22729, Madagascar.
Claoxylon insulanum Müll. Arg. : MacKee 27681, New Caledonia (TEM).
ALCHORNEAE
Bocquillonia sessiliflora Baill. : MacKee 33629, New Caledonia (TEM).

CROTONOIDEAE Pax (1884)
ADENOCLINEAE (Müll. Arg.) Webster (1975)
Adenocline acuta (Thunb.) Baill. : Schlechter 2732, South Africa (TEM).
Ditta myricoides Griseb. : Howard & Newling 15745, Puerto Rico (TEM).
Endospermum chinense Benth. : Poilane 23827, Vietnam (TEM).
Endospermum ovalifolium Pax: Phusomsaeng 212, Thailand (TEM).
Glycydendron amazonicum Ducke: Black 47–1082 F24, Brazil (TEM).
Klaineanthus gaboniae Pierre: Klaine 426, Gabon; s. n. Gabon; Tisserant 2359, Central Africa (TEM).
Tetrorchidium congolense Léonard: Sita 2657, Congo (TEM).
Tetrorchidium didymostemon (Baill.) Pax & Hof. : Descoing 6497, Gabon; Tisserant 208, Congo.
Tetrorchidium macrophyllum Poeppig: Asplund 18608, Ecuador.
Tetrorchidium parvulum Müll. Arg. : Glaziou 8087, Brazil.
Tetrorchidium rubrivenium Poeppig and Endl. : Smith & Reitz 12727, Brazil.
GELONIEAE (Müll. Arg.) Pax (1890)
Cladogelonium madagascariensis Leandri: SF. 13254, Madagascar (TEM).
Suregada baronii (S. Moore) Croizat: Capuron, SF. 11919, Madagascar.
Suregada chauvetia R. Smith: Capuron, SF. 12818, Madagascar (TEM).
Suregada occidentale (Hoyle) Croizat: Vigne 4092, Ghana.
MANIHOTEAE (Müll. Arg.) Pax (1890)
Cnidoscolus urens (L.) Arthur: Harley 16484, South America.
Manihot esculenta Crantz: cultivated, Arboretum de Chèvreloup (TEM).
Manihot tripartita (Sprengel) Müll. Arg. : Glaziou 18470, Brazil.
MICRANDREAE (Müll. Arg.) Webster (1975)
Hevea kunthiana (Baill.) Huber: Jenman 7578, British Guiana (TEM).
Micrandra elata (Didrichs.) Müll. Arg. : Tawjoeran 14647P, Surinam (TEM).
Micrandra lopezii Schultes: Schultes & Lopez 9725, Brazil: Amazonia.
Micrandra rossiana Schultes: Richard s. n., Brazil.
Micrandra siphonoides Benth. : Wurdack & Adderley 43250, Venezuela.
Micrandra spruceana (Baill.) Schultes (= *Cunuria spruceana* Baill.): Krukoff 8858, Brazil, Amazonia.
Micrandra crassipes (Müll. Arg.) Schultes (= *Cunuria sprucei* Müll. Arg.): Schultes & Lopez 9668, Brazil, Amazonia.
ELATERIOSPERMEAE Webster (1975)
Elateriospermum tapos Blume: Yates 2569, Sumatra (TEM).

APPENDIX 1 contd.

EUPHORBIOIDEAE
HIPPOMANEAE
Excoecaria agallocha L. : MacKee 18918, New Caledonia (TEM).
Gymnanthes elliptica Swartz (= *Ateramnus elliptica* Browne): Harris 10387, Jamaica (TEM).
Omalanthus nutans (Forst.) Guill. : MacKee 15758, New Caledonia (TEM).

PHYLLANTHOIDEAE Ascherson (1864)
PHYLLANTHEAE Dumortier (1829)
Martretia quadricornis Beille: Louis 15082 BR, Congo (TEM).
Phyllanthus subglomeratus Poiret: Hahn 322, Martinique.
Savia danguiana Lovell 77 (TEM).
Spondianthus preussii Engl. var preusii: Jalelple 6483B, Ivory Coast (TEM).
HYMENOCARDIEAE (Müll. Arg.) Hutchinson J. (1969)
Hymenocardia acida Tulasne: Léonard 2BR, Zaïre.
Hymenocardia ripicola Léonard: Rewere 727BR, Zaïre; Vanderijst 9516 BR, Zaïre.
Hymenocardia ulmoides Oliver: Gerard 1042BR, Gabon.

OLDFIELDIOIDEAE Köhler & Webster (1969)
CALETIEAE Müll. Arg. (1865)
Longetia buxoides Baill. : MacKee 12824, New Caledonia (TEM).
Micrantheum hexandrum Hooker: Leonhardt & Mossen s. n., Australia (TEM).
Neoroepera banksii Benth. : Johnson 533A BRI , Australia (TEM).
Petalostigma quadricularis K. Domin. : Mueller 90, Australia (TEM).
Scagea oligostemon (Guill.) McPherson: McKee 41283, New Caledonia (TEM).

CELASTRACEAE
Celastrus paniculatus Willd. : Leschenault 27, India.

MALPIGHIACEAE
Acridocarpus smeathmannii (DC.) Guill. & Per. : Lobreau-Callen s. n., Ivory Coast.
Byrsonyma lucida A. Rich. : Sagot 96, French Guiana.
Echinopterys eglandulosa (Juss.) Small: Diguet s. n., Mexico.
Hiptage benghalensis (L.) Kurz: Rao 38961 BSI (NC), India.

STAPHYLEACEAE
Staphylea pinnata L. : cultivated, Arboretum de Chèvreloup.

APPENDIX 2. Pollen characters and character states.

[? = character is unknown in study group]
1. **Form**: subspherical [0]; cuboid [1]; oblate [2].
2. **Aperture number**: 3–4 [0]; 5–7 [1]; > 7 [2].
3. **Aperture position**: zonate [0]; spiral [1]; panto-aperturate [2]; hexaporate arrangment, positioned at variance with overall pollen grain symmetry [3].
4. **Ectoaperture form**: composite aperture, long furrow [0]; composite aperture, short furrow [1]; simple aperture, pore or furrow [2].
5. **Sculpture of the apertural membrane**: smooth [0]; granular or verrucate [1]; clavate [2].
6. **Margin of the ectoaperture**: wide (porus) [0]; parallel [1]; constricted [2].
7. **Aperture margin, tectum**: absent [0]; distinct [1]; diffuse [2].
8. **Aperture margin, infratectum**: columellae shorter than those of mesocolpium [0]; columellae as in the mesocolpium [1]; granular, an narrower than in mesocolpium [2]; granular, as wide as in mesocolpium [3]; granular-columellate structure [4].
9. **Aperture margin, endexine**: smooth [0]; granular [1]; irregular masses [2]; fused granulae [3].
10. **Aperture margin, foot layer**: absent [0]; continuous, thin [1]; irregular masses [2], 3 = lamellae and irregular masses [3]; thicker than in mesocolpium [4].
11. **Aperture margin, endexine thickening**: absent [0]; present but thin [1]; costate [2]; conduplicate [3].
12. **Aperture margin, foot layer/endexine boundary**: absent [0]; distinct [1]; diffuse, microfibrillar [2].
13. **Aperture margin, panto-aperturate thinning**: absent [0]; present [1].
14. **Aperture membrane, ectexine**: absent or reduced to a few small granulae [0]; discontinuous, reduced to fragmented elements [1]; continuous [2].
15. **Endexine, composition surrounding endoaperture**: as at furrow margin [0]; thicker than at furrow margin [1]; thick and fibrillar where it intergrades with foot layer [2].
16. **Endexine, centre of aperture membrane**: irregular, channelled [0]; compact or slightly lamellate [1]; lamellate-granular [2]; reduced to a few granulae [3]; absent [4].
17. **Endexine, distribution in aperture membrane**: regular across membrane [0]; absent in centre (endoaperture) [1]; modified, spongy in centre (endoaperture) [2]; absent [3].
18. **Foot layer-endexine boundary in the membrane**: clearly defined [0]; interbedded [1]; absent [2].
19. **Operculum**: absent [0]; present, ectexinous [1]; present, comprises ectexine and endexine [2]; present, endexinous [3].
20. **Endoaperture**: indistinct or with extremity poorly defined [0]; elliptic [1]; circular [2]; H-shaped [3]; an endocingulum [4]; furrow-like, equatorially elongated [5]; absent [6].
21. **Non-apertural exine, distribution of the infratectum at the meso- and the apocolpium (pantoporate grains excepted)**: present [0]; absent at the mesocolpium [1].
22. **Non-apertural exine, distribution of endexine**: similar in meso- and apocolpium (-porium) [0]; apocolpium only [1]; mesocolpium only [2].
23. **Non-apertural exine, tectum sculpture**: smooth [0]; microechinate [1]; echinate [2]; clavate [3]; micro-rugulate [4]; areolate [5]; crested [6]; microstriate [7]; microrugulate-echinate [8].

24. **Non-apertural exine, free supratectal elements sculpture**: smooth [0]; foveolate [1]; microstriate [2].
25. **Non-apertural exine, tectum structure**: reticulate [0]; perforate [1]; with funnel-shaped perforations [2]; tilioide structure (supratectal reticulum as in *Tilia*) [3]; microperforate [4]; reduced-clavate [5]; continuous [6].
26. **Non-apertural exine, infratectum**: columellate [0]; granular with fused elements [1]; free-granular [2].
27. **Non-apertural exine, foot layer**: absent or not distinct from infratectum [0]; continuous [1]; thin [2]; discontinuous with irregular masses [3]; continuous in apocolpium, discontinous in mesocolpium [4].
28. **Non-apertural exine, endexine**: absent [0]; thin, and more or less granular [1]; thick, smooth and compact [2]; lamellate [3]; thin, smooth and regular [4]; irregular with channels [5]; thin with irregular masses [6]; thickened at apocolpium [7]; disontinuous in mesocolpium [8].
29. **Non-apertural exine, foot layer-endexine boundary**: distinct [0]; diffuse, fibrillar [1]; comprising interbedded lamellae [2]; comprising interbedded granulae [3]; absent [4].
30. **Electron density of the endexine**: denser than ectexine [0]; similar [1].

Macromorphology
31. **Flower, symmetry**: actinomorphic [0]; zygomorphic [1].
32. **Flower, sexuality**: bisexual [0]; unisexual [1].
33. **Flower, petals**: present [0]; absent [1].
34. **Staminate flowers, number of sepals**: 5 [0]; less than 5 [1]; more than 5 [2].
35. **Staminate flowers, sepals**: imbricate [0]; valvate [1].
36. **Staminate flowers, sepals**: free [0]; connate [1].
37. **Staminate flowers, number of stamens**: two cycles (10) [0]; one cycle (5) [1]; very variable, maximum of 10 [2] more than 10 [3]; very variable with a maximum of 8 [4]; 2–3 [5].
38. **Staminate flowers, stamens**: free [0]; connate [1].
39. **Pistillate flowers, sepals**: free [0]; connate [1].
40. **Pistillate flowers styles**: free [0]; connate [1].
41. **Pistillate flowers, number of ovule per carpel**: (2)–3 [0]; 2 [1]; 1 [2].
42. **Fruit wing**: absent [0]; present [1].
43. **Fruit, caruncle or aril**: absent [0]; present [1]; aril [2].
44. **Latex**: absent [0]; present [1].
45. **Laticifer channels**: absent [0]; articulate [1]; inarticulate [2].
46. **Growth habit**: tree [0]; liana [1]; upright shrub [2]; scandent shrub [3]; shrubby tree [4]; herbaceous [5].
47. **Leaf**: entire [0]; palmate [1].

Ladd, P.G., Parnell, J.A.N. and Thomson, G. (2000). The morphology of pollen and anthers in an unusual myrtaceous genus (*Verticordia*). In: M.M. Harley, C.M. Morton and S. Blackmore (Editors). Pollen and Spores: Morphology and Biology, pp. 325–347. Royal Botanic Gardens, Kew.

THE MORPHOLOGY OF POLLEN AND ANTHERS IN AN UNUSUAL MYRTACEOUS GENUS (*VERTICORDIA*)

P.G. Ladd[1], J.A.N. Parnell[2] and G. Thomson[3]

[1]School of Environmental Science, [3]School of Biological Sciences and Biotechnology, Murdoch University, Murdoch, Western Australia 6150
[2]Herbarium, School of Botany, Trinity College, Dublin 2, Ireland

Abstract

Pollen and anther morphology and structure of *Verticordia* are unusual in the Myrtaceae and extremely varied. *Verticordia* pollen grains are small, isopolar, tricolporate with the colpi brevicolpate or brevissimicolpate or porate, flattened and triangular in polar view with acute or blunt angles at the apertures. Sides of the polar outline (amb) are normally concave but may be straight or convex. Surface patterning varies from psilate to strongly scabrate. Pollen in most Myrtaceae is dry, but in *Verticordia* pollen is shed in a fluid (pollenkitt). In some *Verticordia* species the pollen and pollenkitt combines with oil produced from a gland located in the anther connective or at the apex of the filament. In species with a pollen presenter and gland (mostly in subgenus *Eperephes*) the pollen and oil mixes before anthesis, and is deposited onto the pollen presenter hairs below the stigma, to be carried away from the anthers as the style elongates during flower opening. Species in the subgenus *Chrysoma* lack a pollen presenter and the pollen and oil mixes in a depression or cowl-like structure at the apex of the anther. In other species (mostly in subgenus *Verticordia*) the anther gland is small and non-functional or absent and, depending on the species, the pollen-pollenkitt mixture is deposited on a pollen presenter or (in only a few species) directly dispersed from the front of the anther loculi. Pollen and anther morphology are discussed in relation to pollination and taxonomy in the genus.

Introduction

The Myrtaceae is an extremely important family in the Southern Hemisphere. For example, tree species in the family dominate the forests of Australia, there are many taxa which form important components of the forests of South America, and of the islands of Oceania (Johnson and Briggs, 1981).

In many species of Myrtaceae the anthers perform a dual function. Not only are they responsible for pollen production and dispersal, but also for pollinator attraction through the formation of brush blossoms (Johnson and Briggs, 1981). In these cases anthers, therefore, carry out the normal male role and the role of the perianth. Despite the prevalence of this trait in the family, there are also many taxa which do not have brush blossom inflorescences. In many of these, the anthers outdo their brush blossom relatives in their specialisation of androecial form and function.

It has been proposed that the ancestral form of the Myrtaceae flower had a prominent androecium (Johnson and Briggs, 1984), somewhat similar to the brush blossom type. Thus the non brush blossom type of flower seems a derived state in the family. These flower forms are mainly found in the *Leptospermum* and *Chamelaucium* informal taxonomic groups of Johnson and Briggs (1984), and are located as sister groups in their phylogram outlining relationships in the Myrtaceae. It appears, therefore, that floral charcteristics clearly reflect phylogeny. In the development of flower form it is likely that plant-pollinator relationships will be one of the main factors influencing evolution in the group.

The two informal taxonomic groups mentioned above form a species-rich part of the highly diverse sclerophyll flora growing on nutrient poor soils on the Australian continent. It is suspected that accompanying this floral diversity is a considerable diversity of mutualistic, highly specific interactions between flora and pollinators (for example Houston *et al.*, 1993, but see Bell, 1994 for an alternative view).

While there has been some concentration on characteristics of the androecium in the Myrtaceae, with particular attention to the genus *Eucalyptus* (Bentham, 1867), in smaller genera there has been little detailed study. George (1991) certainly emphasised the importance of the anthers in his treatment of *Verticordia* but did not examine the pattern of forms in detail.

Pollen studies in the Myrtaceae have been even less extensive. Apart from a survey of the family (Pike, 1956) there have been few detailed studies except for examination of taxa likely to be important in fossil pollen studies (for example Chalson and Martin, 1995; Pickett and Newsome, 1997), one detailed study of the pollen in the tribe Metrosiderinae (Gadek and Martin, 1981) and one preliminary one by Chantaranothai (1989) of *Eugenia*. Pike (1956) recognised three pollen types in the Myrtaceae from the southwestern Pacific region, separated from each other on the basis of their colpi: longicolpate, syncolpate or parasyncolpate, brevicolpate or brevissimicolpate. The reliability of these groups appeared to be confirmed by Patel *et al.* (1984).

The genus *Verticordia* is restricted to Western and Northern Australia and is unusual in the family for a number of reasons:

- the flowers of many species form spectacular inflorescences with the attractive parts being composed mainly of sepals

- there is a diversity of anther form in the genus

- the pollen is unusual in the family as it is colporate to porate rather than parasyncolporate

- the pollen is shed in a fluid, rather than dry, as is the case in the majority of the Myrtaceae

Verticordia currently comprises 99 species and is divided into three subgenera, each with a number of sections (George, 1991). The characteristic feature of the flowers is the intricately divided perianth, particularly the sepals. The flowers of some species form tightly packed, brilliantly coloured inflorescences which are widely used in dried floral art.

The androecium in most species remains within the flower at anthesis and pollen is deposited onto the pollen presenter before the flower opens (Fig. 1). The gynoecium is inferior with a simple style. In most species the style bears a series of unicellular hairs placed subapically below the stigma. The anthers dehisce onto these hairs and pollen is presented to pollinators from this structure (Fig. 2), rather than from the anthers in the majority of species (George, 1991; Ladd, 1994). Some species lack the pollen presenter and pollen in these species is dispensed directly from the anthers. Flower colour varies between species from white through pink and mauve to red and even bright yellow.

This study of anther form and pollen in *Verticordia* aims to relate the forms to both the systematics of the genus and to the likely pollination specialisations in the different species.

Materials and Methods

Pollen for scanning electron microscopy (SEM) was prepared following a modification of the procedure outlined in Parnell (1991). Anthers from mature buds preserved in 3% glutaraldehyde in 0.025M phosphate buffer (pH 7.0) and stored in 75% ethanol, were excised, placed in a plastic centrifuge tube containing ca. 5cc volume of 50% acetone and crushed using a glass rod and suspended using a whirli-mix. After 10 minutes the pollen was centrifuged down at 2,500rpm for 8 minutes. Pollen was then dehydrated through a graded series of acetone solutions and filtered through Endecott steel sieves (90μm), spun down as before, decanted and the pollen re-suspended in the remaining minimal volume of acetone. The pollen-containing solution was then dropped directly onto an SEM stub and allowed to dry overnight in a dessicator. The stubs were gold-coated in a Polaron SC500 sputter coater for viewing in a Leica Cambridge Stereoscan 360. The advantage of this method over conventional acetolysis is principally that it cleans the pollen grains usually as well as acetolysis does, is less destructive of 3-D structure of thin-walled pollen and is very much quicker. Some species were treated by the normal acetolysis method (Moore and Webb, 1978) in order to examine the internal wall structure, particularly the pore region. Pollen of 53 species was examined.

For pollen description we follow the terminology of Moore and Webb (1978), Erdtman (1952) and Punt *et al.* (1994). Terms for the components of the anther follow Green (1980).

The anthers of 57 species from most sections of the genus were examined. Wherever possible fresh material was examined and was collected from natural populations or from cultivated plants. Voucher specimens for species collected from the field are lodged in the herbarium of the School of Biological Sciences, Murdoch University. Anthers from at least two inflorescences from each of two different plants were morphologically examined, if there were sufficient plants in flower.

Material for histological examination was collected as immature or mature buds, fixed in 3% glutaraldehyde in 0.025M phosphate buffer (pH 7.0) and embedded in Spurr's resin after dehydration through a graded series of acetone solutions. Light microscope sections were stained with 1% methylene blue and 1% azur II in 1% sodium tetraborate (Richardson *et al.* 1960). Anthers prepared for the SEM were dehydrated through a graded series of ethanol solutions, transferred to amyl acetate, critical point dried, then gold coated.

FIGS 1–2. *Verticordia*. Fig. 1. *V. huegelii.* bud after pollen release onto the pollen presenter hairs. S - stigma, A - anther. Scale = 1 mm. Fig. 2. *V. penicillaris* pollen on pollen presenter hairs (from an open flower). Scale = 0.5 mm.

Results

Pollen in *Verticordia* is generally small and diameter measurements in polar view range between 13 μm to 46 μm (Table 1). Considerable difficulty was experienced in obtaining equatorial measurements and, therefore, these are only rarely given. The mean grain size in subgenus *Verticordia* (27.2 ± 1.8μm) appears significantly larger than in the other subgenera (*Eperephes* - 22.8 ± 1.1 μm; *Chrysoma* -21.9 ± 1.4 μm) (F=3.5; $p \leq 0.05$; Duncan's Multiple Range Test; $p \leq 0.05$ for *Verticordia* vs. *Chrysoma* and for *Verticordia* vs. *Eperephes*). The grains are isopolar, flattened (prolate or peroblate) and triangular with three apertures. The angles tend to be rounded and the sides of the amb in most species are concave (Figs. 3, 7) but may be straight (Fig. 5) to slightly convex. The apertures are basically colporate with the colpi short (Fig. 7, b1 in Table 1) to very short (Fig. 5, b2 in Table 1) though they are sometimes non-existant, in which case the apertures are functionally porate (Fig. 4). Subgenus *Eperephes* has a higher proportion of species with brevicolporate grains than do the other two subgenera. The wall in section is relatively thin with little distinction between the endexine and ectexine. At the pores the ectexine may thicken slightly and the endexine diverges from the ectexine to form a vestibulum. In most species the pore is so large that there is little overarching of the ectexine so there is virtually no roof to the structure, unlike the situation in many other Myrtaceae, such as *Eucalyptus*.

The structure of the tectum ranges from psilate to microfossulate (*sensu* Gadek and Martin, 1981). The tectum has sparse to very dense microperforations (Fig. 6), and in some species is surmounted by microscabrate elements on the surface (Figs. 7, 8).

In *Verticordia* the stamen is demarcated clearly into the microsporangia (anther) attached on the dorsal side to a discrete filament. Unlike many other Myrtaceae (for example *Eucalyptus*), anthers on most *Verticordia* species are not versatile and the distinction between the connective of the anther and the filament is not well demarcated. In transverse section the anthers may be arcuate, rectangular or trapezoidal, and the anther form and position is generally lingulate but may be filamentous (*sensu* Hufford, 1996). In all species so far examined there are four microsporangia.

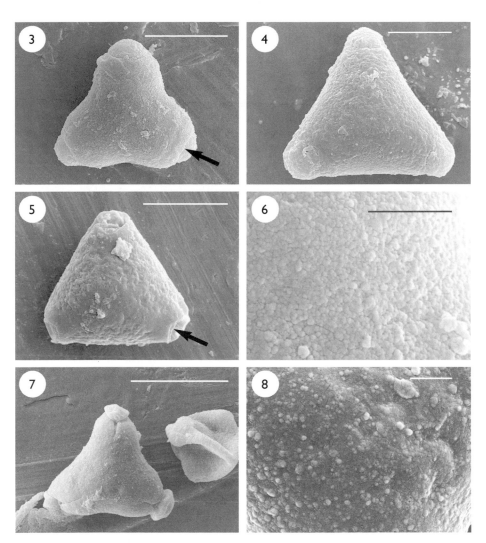

FIGS 3–8 *Verticordia*. Fig. 3. *V. brevifolia* pollen showing very short colpi (arrow). Scale = 10 μm. Fig. 4 *V. insignis* porate pollen without any colpi. Scale = 10 μm. Fig. 5. *V. huegelii* pollen showing extremely short colpi (arrow). Scale = 20 μm. Fig. 6. *V. brevifolia* pollen surface showing micropunctate tectum. Scale = 2 μm. Fig. 7. *V. pholidophylla* pollen grain with very long colpi and microscabrate sculpture. Scale = 10 μm. Fig. 8. *V. pholidophylla* pollen surface showing microscabrate sculpture. Scale = 2 μm.

As in many other Myrtaceae species the gland is located in the upper part of the stamen. The conventional terms of "anther", "connective" and "filament" have tended to be ambiguous (Hufford, 1996) as the terms are usually used functionally rather than ontogenetically. In *Verticordia* the connective can be considered as the upper part of the filament. This is supported by the fact that in several species (*V. ovalifolia*, *V. nobilis*, *V. staminosa*) a gland similar to that in the anther occurs in the apex of the staminodes. The filament has a single vascular trace which may split in the connective with branches going to the microsporangia and the gland (Fig. 10).

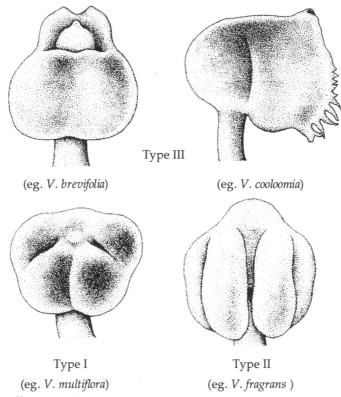

Type III

(eg. *V. brevifolia*)   (eg. *V. cooloomia*)

Type I   Type II
(eg. *V. multiflora*)   (eg. *V. fragrans*)

FIG. 9. Three *Verticordia* anther types.

Anther dehiscence is introrse and, in the case of species with a pollen presenter, occurs before anthesis, with pollen deposited on the hairs of the pollen presenter below the stigma (Fig. 2). In all species the pollen is extruded as a sticky mass of pollen and tapetal material to which, in those with functional glands, is added the fluid from the gland. This fluid may be yellow or clear and is oily in appearance. The pollen is extruded from the anthers through pores or slits depending on the type of anther.

The anthers of *Verticordia* can be classified into three main types (Fig. 9, Table 1), which to a large extent accord with the taxonomic classification of the species.

Type I   Rectangular
Type II   Oblong
Type III   Saccate

Excluding the glandular part, anther anatomy is similar to that of a generalised dicotyledon anther form. The epidermis of the microsporangium consists of a single layer of lozenge-shaped cells underlain by the block-like, thin-walled endothecial cells. Endothecial thickenings are columnar (*sensu* Manning, 1996) in most species but are absent in a few. Underlying the endothecium is one (to several) layer(s) of cells with very thin walls and small lumina. The tapetal cells are internal to these and are isodiametric, stain heavily and have thin walls. The cells of the connective are also approximately isodiametric and are thin walled and often contain crystals in the lumina similar to crystals which occur under the epidermis of the hypanthium.

TABLE 1. Taxonomic subgenera and sections. C - Chrysoma, section Chrysoma; J - Chrysoma, section Jugata; U - Chrysoma, section Unguiculata; Si - Chrysoma, section Sigalantha; Ch - Chrysoma, section Chrysorhoe; Co - Chrysoma, section Cooloomia; Sy - Chrysoma, section Synandra; E cory - Eperephes, section Corynatoca; E integ - Eperephes, section Integripetala; E t - Eperephes, section Tropica; E p - Eperephes, section Pennuligera; E v - Eperephes, section Verticordella; V pen - Verticordia, section Penicillaris; V intric - Verticordia, section Intricata; V cat - Verticordia, section Catocalypta; V inf - Verticordia, section Infuscata; V mic - Verticordia, section Micrantha; V piloc - Verticordia, section P locosta; V plat - Verticordia, section Platandra; V cory - Verticordia, section Corymbiformis; V ela - Verticordia, section Elachoschista; Vv - Verticordia, section Verticordia. Aperture type - Brevicolporate, (b1) Brevissimicolporate, (b2) Porate (p). ne - not examined.

Verticordia species	Section	Pollen amb	Pollen surface	Pollen size, polar (pore to pore) × equatorial (μm)	Aperture type	Pollen presenter	Anther type
Subgenus Verticordia							
V. brachypoda Turcz.	V piloc	Straight to concave	Scabrate to pustulate	46 × 16	b2	Very long (>1mm) hairs with forked ends	Type I, no gland
V. brownii (Desf.) DC.	V cory	ne	ne	ne	ne	Hairs (<0.5 mm) with forked ends	Type I, no gland
V. capillaris A.S. George	V cory	ne	ne	ne	ne	Simple hairs (<0.2 mm)	Type I, no gland, possibly only two locelli
V. crebra A.S. George	Vv	Concave	Scabrate	31 × 15	b2	Simple hairs (<0.5 mm) longer on one side	Type I, small non-functional gland
V. densiflora Lindley	V cory	ne	ne	ne	ne	Hairs (<0.5 mm) with forked ends	Type I, no gland
V. eriocephala A.S. George	V cory	± Straight to weakly concave	Scabrate to rugulate	19	b2	In most specimens hairs absent, in some few hyaline hairs (<0.2 mm) below stigma	Type I, very small gland
V. fastigiata Turcz.	V. mic	Concave	± Smooth	19	b1	ne	ne
V. fimbrilepis Turcz.	Vv	Slightly concave	Weakly scabrate	16	?b2	ne	ne

Species							
V. gracilis A.S. George	V plat	Concave	Scabrate	19 × 14	b2/p	Hairs (<0.5 mm) with forked ends	Type I, large gland over the top of the thecae but globular overall
V. habrantha Schauer	V cat	± Straight	Weakly scabrate	15	b2	Absent	Type I, prominent small apical appendage but no gland
V. harveyi Benth.	Vv	Slightly concave	Weakly scabrate	24	b1	ne	ne
V. helichrysantha F. Muell.	Vv	Slightly convex	Weakly scabrate	28 × 19	b1	Sparse simple hairs (<0.6 mm)	Type I with a small non-functional gland with orange contents
V. hugelii Endl.	V piloc	Straight	Weakly scabrate (psilate) to smooth	30	b2	Simple hairs (<0.5 mm) below very large (1.5 mm) stigma	Type I, small non-functional gland with brown fluid
V. insignis A.S. George	V cat	± Straight to slightly concave	Weakly scabrate	30 × 20	b2/p	Absent	Type I, no gland - very fimbriate staminodes
V. minutiflora F. Muell.	V mic	ne	ne	ne	ne	Clavate hairs (<0.3 mm)	Type I, no gland, generally globular shape
V. mitchelliana C. Gardner	V intric	Concave	Scabrate	35	b2	Simple hairs (<0.5 mm)	Type II, no gland
V. monadelpha Turcz.	V intric	± Straight	± Smooth	35	b2/p	Absent	Type I, no gland but apical solid umbo
V. multiflora Turcz. ssp. multiflora	V piloc	Convex	Weakly scabrate to smooth	36 × 19	b2	Long hairs (>1 mm) with forked bulbous ends	Type I, no gland
V. oxylepis Turcz.	V inf	Weakly concave to ± smooth	Weakly scabrate	26 × 12	b1	Simple hairs (<0.5 mm)	Type I, very minute cavity visible in cleared anthers

Species	Code	Shape	Surface	Size	Style / hairs	Gland / hairs	Anther type
V. penicillaris F. Muell.	V pen	Convex	Scabrate	18	b2 s[i]mple hairs	Very long (3 mm) gland with sticky orange contents - non functional	Type II, very reduced
V. plumosa (Desf.) Druce	Vv	Slightly concave	Weakly scabrate	23 × 14	t2 (<0.5 mm)	Simple hairs non-functional gland	Type I, no gland
V. polytricha Benth.	V cory	ne	ne	ne	ne (<0.5 mm) on one side of style below stigma	Simple hairs	Type I, no gland
V. pulchella A.S. George	V intric	Convex	Scabrate	ne	b2 (<0.2 mm) below stigma	Few thin hairs	Type II, no gland
V. roei Endl.	V cat	Concave	Weakly scabrate	30	b2	Absent	Type I, no gland
V. stenopetala Diels	Vv	± Straight	Weakly scabrate	27 × 18	ne (<0.5 mm)	Simple hairs functional gland	Type I, small non
V. verticordina (F. Muell.) A.S. George	V ela	Very strongly concave	Smooth	33 × 13	b2/p (<0.5 mm)	Simple hairs (<0.5 mm)	Type I, no gland
Mean ± SE			16.0 ± 0.9	28.1 ± 1.9;			
Subgenus *Eperephes*							
V. aereifolia E.A. George & A.S. George	Ep	Concave	Rugulate	28	b1 on one side of style	Dense simple hairs (<0.5 mm)	Type II
V. blepharophylla A.S. George	Ev	Straight to slightly concave	Psilate to finely scabrate	23	b1	Forked hairs (<0.5 mm)	Type II
V. carinata Turcz.	Ev	Weakly concave	± Smooth	19	b1 (<0.5 mm)	Forked hairs	ne
V. cunninghamii Schauer	E t	Straight	± Smooth	28 × 16	b1	Simple hairs (<0.5 mm) in single row below stigma	Type II, small gland

Species		Shape	Surface	Size		Hairs	Type
V. dichroma A.S. George	Ep	Concave	Rugulate	31	b2	Simple hairs (<0.2 mm) on one side of style	Type II
V. etheliana C. Gardner	Ep	Concave	Weakly scabrate	24 × 13	b1	Simple hairs (<0.5 mm) slightly longer on one side of style	Type II
V. fragrans A.S. George	Ep	Slightly concave	Weakly scabrate	19	b1/b2	Forked hairs (<0.5 mm) on one side of style	Type II
V. grandis J.L. Drumm.	Ep	Concave	Weakly scabrate	24	b1	Simple hairs (<0.5 mm)	Type II
V. helmsii S. Moore	E integ	ne	ne	ne	ne	single row of crooked hairs (<0.5 mm)	Type II, no gland
V. hughanii F. Muell.	Ev	Concave	Scabrate	25	b1	Simple hairs (<0.2 mm)	Type II
V. lepidophylla F. Muell.	Ep	ne	ne	ne	ne	Long and short hairs with forked ends (<0.5 mm)	Type II
V. lindleyi Schauer ssp. *purpurea* A.S. George	Ev	Slightly concave	Scabrate to pustulate	20	b2	Forked hairs (<0.5 mm)	Type II
V. mitodes A.S. George	Ev	Concave	Weakly scabrate	22 × 12	b1	Forked hairs (<0.6 mm)	Type II
V. muelleriana E. Pritzel	Ep	Convex	Strongly scabrate	21	b1	Intricate hairs (~ 1 mm)	Type II
V. oculata Meissner	Ep	Strongly convex	Weakly scabrate	27 × 18	b2	Intricate hairs (> 1 mm)	Type II
V. ovalifolia Meissner	E cory	± Straight	Rugulate	13 × 9	b1	Simple hairs (<0.2 mm)	Type II, with large cavity, staminodes also have a gland
V. pennigera Endl.	Ev	Concave	Psilate to weakly scabrate	19	b1	Forked hairs (<0.5 mm)	Type II
V. pholidophylla F. Muell.	Ev	Very strongly concave	Scabrate (almost pustulate)	18	b2	Forked hairs (<0.5 mm)	Type II

V. picta Endl.	E integ	Concave	Scabrate	17	b?	Reflexed forked hairs (<0.2 mm)	Type II, no gland
V. rennieana F. Muell.	E integ	Slightly concave	Scabrate	23 × 13	b2/p	Clavate hairs in a single row (<0.5 mm)	Type II, large rounded apex
V. tumida A.S. George	Ev	ne	ne	ne	ne	Hairs (<0.5 mm) some bifurcate with forked ends	Type II
V. venusta A.S. George	Ep	Concave	Rugulate	32	b2	Hairs (<0.5 mm) with forked ends	Type II
Mean ± SE				22.8 ± 1.1; 13.5 ± 1.3			
Subgenus *Chrysoma*							
V. acerosa Lindley	C	Straight	± Smooth, weakly scabrate	25	b2/p	Absent	ne
V. amphigia A.S. George	J	Concave	± Smooth, weakly scabrate	22 × 7	b1	Absent	Type III, small hood
V. brevifolia A.S. George	J	Concave	Scabrate	17	b1	Absent	Type III, flaps over top of loculi, but not peaked
V. chrysantha Endl.	J	= Straight	Weakly scabrate	19	b2	Absent	Type III, moderate sized hood
V. citrella A.S. George	C	Concave	Scabrate	20	b2	Absent	Type III, very small hood
V. cooloomia A.S. George	Co	Concave	Psilate	16	b2	Absent	Type III, large gland and enveloping appendage
V. endlicheriana Schauer	C	Slightly concave	Almost smooth	28 × 14	b2	Absent	Type III, quite small, entire appendage, gland protrudes prominently
V. grandiflora Endl.	U	± Straight	Scabrate to rugulate	29	b1	Absent	ne

V. integra A.S. George	Si	Concave	Scabrate	20	b2	Absent	Type III, large tubular flange at top of anther
V. nitens (Lindley) Endl.	Ch	Concave	Weakly scabrate	16	b2	Absent	Type III, large cavity and appendage with tubular peak
V. nobilis Meissner	U	Slightly concave	Smooth	32	b2/p	Absent	Type III, large cavity, elongate bifurcate appendages, gland also in staminodes
V. patens A.S. George	Ch	Concave	Scabrate	15	b2/p	Absent	Type III, small cavity at apex, small hood
V. rutilastra A.S. George	U	ne	ne	ne	ne	Absent	Type III, large cavity, elongate bifurcate appendages
V. serrata (Lindley) Schauer	Si	Concave	Scabrate	21	b2	Absent into a tubular flange	Type III, small apical appendage formed
V. staminosa C. Gardner & A.S. George	Sy	Slightly concave	± Smooth	26 × 12	b2	Absent gland also in staminodes	Type III, projection absent to very short,
Mean ± SE				21.9 ± 1.4; 11 ± 2.1			

Gland size and structure vary between species. The gland is initially schizogenous but lysigenous at maturity with the breakdown of all the cells within the gland cavity. The gland wall structure in some groups is similar to that of the microsporangia. Sometimes glands are absent.

The outer wall of the gland consists of the epidermis which is continuous with the epidermis of the rest of the anther but may have a thicker outer wall than occurs over the microsporangia. Underlying the epidermis are one to two layers of thin-walled, blocky cells which may contain polyphenol material. These cells grade into the endothecial cells of the microsporangial wall (Fig. 11). Underlying these cells are one to two layers of squamous, thin-walled cells with very narrow lumina. In species with small, non-functional glands the gland cells stain heavily and are relatively large and thin-walled (Fig. 12). There may be small intercellular spaces between these cells. In species with large glands, the glands are lined by one to two layers of small elongate densely staining cells which are continuous with a three dimensional network of cells delimiting oil-containing voids (intercellular spaces) (Fig. 10). The cells are thin-walled, have no large vacuoles but contain many small, clear inclusions (Fig. 13). Ultimately the network of cells collapses. In these species squeezing anthers from mature buds causes the gland initially to express a clear oil, followed by a milky slurry consisting of the remains of the network of oil-producing cells which had settled to the floor of the cavity.

Description of anthers

Type I

Type I anthers are rectangular. Morphologically and structurally they appear to be the simplest type, and consist of four microsporangial sacs that are more or less isodiametric. This anther type is closer to the filamentous form (Hufford, 1996) than any other in the genus. However, there is often a small umbonate projection at the apex in some species (for example *V. monadelpha*). In other species (for example *V. gracilis*, *V. oxylepis*) two of the locelli (*sensu* Green, 1980) are reduced in size, and the outline of the sacs is only slightly lobed. This gives the anthers a globular, rather than rectangular form. Dehiscence is porate.

Most species with this anther type either lack an anther gland or only have a small gland (Fig. 12) which does not extrude its contents (i.e. it is non-functional). Of 19 species examined, about two thirds have no gland, one species, *V. gracilis*, has a relatively large oil-filled cavity from which oil is extruded (Fig. 16), whereas the remainder have a rudimentary gland which may contain clear or brown viscous fluid. The gland in *V. eriocephala* is unusual because it is formed by the close conjunction of several glands (Fig. 14) in the upper part of the connective/filament, virtually identical with glands in other parts of the flower (Fig. 15).

In *V. helichrysantha* the gland is small and contains large, thin-walled cells with densely staining contents and scattered small vacuoles (Fig. 12). It is surrounded by cells similar to those in the connective. While the epidermal cells over most of the anther are small with reduced lumina, the epidermal cells are 4 - 5 times larger with correspondingly large lumina over the adaxial surface of the gland.

This anther type is restricted to the subgenus *Verticordia*; only four of the species in the subgenus, examined in this study, do not have this form.

Type II

In contrast to Type I the microsporangia are elongated, being twice as long as wide and giving the anther an oblong outline (Fig. 17). The anthers are lingulate, with the microsporangia facing inward, and may be trapezoidal to arcuate in transverse section.

Dehiscence is by a slit at the intersection of the two lobes of each microsporangium.

Microsporangial wall structure does not differ greatly from that of Type I. However, the larger connective and gland means there is a greater amount of parenchyma tissue than in Type I anthers.

A gland located at the top of the connective is present in most species and usually results in a rounded apex to the anther. Oil is expressed from a pore on the adaxial side of the anther, at the time of dehiscence, and mixes with the pollen.

The difference between the structures around the gland in *V. pennigera* and *V. grandis* is worth noting. In *V. grandis* the gland is very large and the cells surrounding it are similar to those over the microsporangia. However, at the region of the pore the endothecium is absent and the epidermal cells are relatively enlarged. The gland cells stain heavily with methylene blue and toluidine blue and form a network of thin cells throughout the gland cavity. In *V. pennigera* the gland is somewhat smaller and the cells surrounding it are different from those of the microsporangial wall. Anthers of *V. ovalifolia* (section *Corynatoca*) show some anatomical differences from the other species examined. Morphologically the anthers are similar to those of *V. grandis*. However, anatomically there are differences, with the epidermal cells being larger than in other species and lacking polyphenol contents. The endothecial cells are smaller and more equi-dimensional than in *V. grandis* but the gland wall is similar to that of the microsporangia. The network of gland cells is similar to that in other species examined in the subgenus (Fig. 18). In *V. ovalifolia* the staminodes also have a gland similar to that in the anther (Fig. 19).

All the species examined in subgenus *Eperephes* have this type of anther and virtually all have a large functional gland. Exceptions to this are *V. helmsii* and *V. picta*, in section *Integripetala*, which seem to lack a functional gland. In addition two species in subgenus *Verticordia*, section *Intricata* (*V. pulchella*, *V. mitchelliana*), have oblong anthers without a gland, whereas *V. penicillaris* (section *Penicillaris*) has oblong anthers with a very reduced gland with sticky orange contents.

Type III

Type III anthers are saccate. They are morphologically distinguished from Type I anthers by the trapezoidal transectional form (Fig. 20), and from Type II anthers by the poor external demarcation of the microsporangial lobes, and porate dehiscence. In addition, in this form, there has been elaboration of the apex of the anther in a number of species (Fig. 21).

Two elaborations of the apex of the anther can be recognised. In one the apical projection is produced into two horns which form a lunate structure enabling pollen and oil to be held by surface tension between them. In the other, the projection expands from the adaxial side of the anther into a cowl-like structure which forms a reservoir to contain pollen and oil once this is expressed at anthesis. Apart from this apical development the form of the anther is uniform for all species.

FIGS 10–15. *Verticordia*. Fig. 10. *V. cooloomia* anther with large gland (G), locellus (L) and branching vascular trace (t). Scale = 100 μm. Fig. 11. *V. patens* anther showing transition from the microsporangial wall to the gland wall. Locellus - L, gland - G. Scale = 20 μm. Fig. 12. *V. helichrysantha* non-functional anther gland in which the gland cells (dark cells) show globular clear inclusions but do not form a network. Scale = 50 μm. Fig. 13. *V. renniana* anther gland with a network of glandular cells with many clear globular inclusions. Scale = 100 μm. Fig. 14. *V. eriocephala* anther with gland (G) similar to glands on the hypanthium. Arrow - single layer epidermis, L - locellus. Scale = 50 μm. Fig. 15. *V. eriocephala* hypanthium glands. Scale = 50 μm.

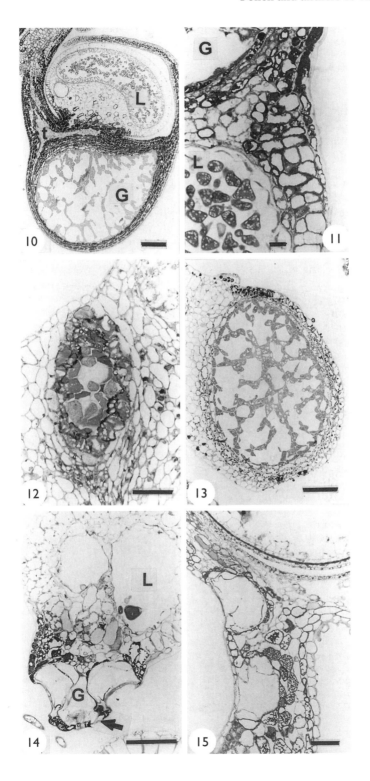

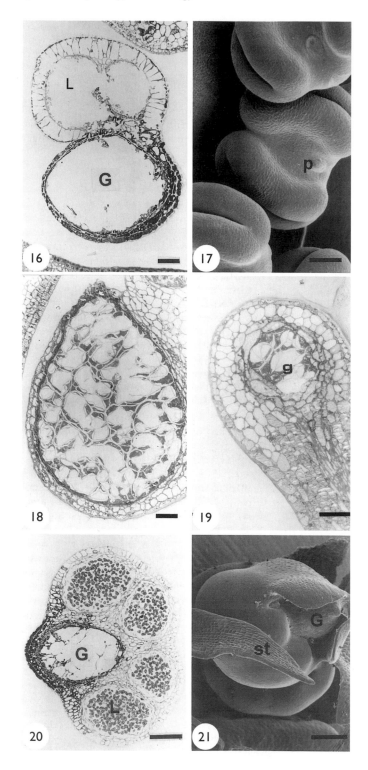

The structure of the microsporangial wall and gland in Type III anthers is similar to the structure of Type II anthers. The epidermis consists of small narrow cells with a thick outer wall and very small lumina. The cells surrounding the gland may be rather different from the cells of the microsporangial wall. In *V. cooloomia* the cells of the hypodermis around the gland stain heavily indicating polyphenol contents (Fig. 9). However, in other species (for example *V. nitens*) there are few polyphenol-containing cells.

The cells of the apical/abaxial projection are similar to those of the adaxial side of the anther. In *V. cooloomia* they stain strongly with methylene blue, indicating that they are full of polyphenols but in other taxa the cells have few, non-staining, contents.

Discussion and Conclusions

Pickett and Newsome (1997) indicated that pollen of the Myrtaceae is considered distinctive, usually being syncolpate or parasyncolpate. By contrast, as first noted by Pike (1956), *Verticordia* pollen is usually brevicolporate to brevissimicolporate and, as identified here, grains of some species lack any colpus and are thus porate. *Actinodium* is the only genus in the Myrtaceae that Pike identified with porate pollen and this genus is quite closely related to *Verticordia*. In Pike's (1956) key to pollen types, *Verticordia* is in the group identified by short colpi and grains greater than 16 μm. However, with a larger sample of species this is shown to be somewhat unreliable, as at least two species have smaller pollen (*V. ovalifolia*, *V. patens*), and several species have triporate grains with no evidence of colpi. Nevertheless, the majority of species examined here would be assigned to the correct group using Pike's key.

The present study supports Pike's (1956) views, which were based on only one species of *Verticordia*. Particular details of pollen grains such as the colpi are consistent within taxonomic categories, at least on a broad scale. The pollen wall structure is consistent with that from more detailed examination of distantly related taxa in the family (for example *Tristania*, *Eucalyptus*; Gadek and Martin, 1982) leading to the prediction that wall structure in the Myrtaceae may be fairly uniform. In addition, grains in some *Verticordia* species show sculptural elements on the surface of the tectum which have not been specifically identified in any other taxa in the family.

Although there seems to be no detailed correlation between anther and pollen types, there is a general relationship in which the largest mean pollen diameter occurs in subgenus *Verticordia*. This group has the smallest anthers in which the gland is either non-functional or absent. It therefore appears that reduction of anther volume is related to increase in pollen size.

The pollen grains are not strongly ornamented, which is consistent with many other Myrtaceae. The lack of variation in the pollen grains makes the Myrtaceae a difficult group to distinguish at a lower taxonomic level in fossil pollen studies (Pickett and Newsome, 1997). In *Verticordia* there is a correlation between subgenus and aperture type. Most species in the subgenera *Chrysoma* and *Verticordia* have brevissimicolporate to porate grains whereas over half the species examined in subgenus *Eperephes* have longer colpi (brevicolporate grains).

Figs 16–21. *Verticordia*. Fig. 16. *V. gracilis* anther, locellus (L), and gland (G) bounded by polyphenol-containing cells. Scale = 50 μm. Fig. 17. *V. muelleriana* anthers showing gland pore (p). Scale = 100 μm. Fig. 18. *V. ovalifolia* anther gland with network of glandular cells. Scale = 50 μm. Fig. 19. *V. ovalifolia* staminode gland with network of glandular cells (g). Scale = 25 μm. Fig. 20. *V. patens* transection of anther showing gland (G) and locelli (L). Scale = 100 μm. Fig. 21. *V. brevifolia* anther with gland (G) and staminode (st). Scale = 200 μm.

The presence and restriction of colporate to porate apertures in members of the *Chamelaucium* group in general is worthy of note. It is generally considered that the apertures of pollen, in particular colpi, are involved in the function of the grains (Muller, 1980). Muller noted that in oblate grains syncolpate apertures are common. Presumably this type of colpal arrangement assists in accommodating stresses involved in water loss from the grain during pollen tranfer. It is notable that most of the non-syncolpate taxa, which Pike (1956) listed, have pollen which is shed in a fluid. This would diminish a need for adjustments of pollen size due to dessication while pollen is transferred from the anthers to a stigma.

The three anther types identified in *Verticordia*, can be related to the three subgenera identified by George (1991). In particular the saccate anthers of subgenus *Chrysoma* are structurally uniform, although there is a sequence of morphological changes which is related to specialisation in dispensing pollen. The rectangular form is predominant in, and restricted to, subgenus *Verticordia*, whereas the oblong form is found mainly in subgenus *Eperephes*.

The anthers of some species, however, do not fit the general pattern of relationship between form and the taxonomic classification. In particular *V. gracilis*, with a large gland on a small globular anther, is not consistent with the rest of the subgenus *Verticordia*. In terms of anther shape *V. pulchella* and *V. mitchelliana*, in section *Intricata*, do not match with those of *V. monadelpha* which has an anther form similar to the majority of other species in the subgenus.

In subgenus *Eperephes* the anthers of species in sections *Verticordella* and *Corynotheca* are quite uniform with oblong microsporangia and large glands. However the anthers of two species in section *Integripetala* lack an anther gland and show some similarity to *V. pulchella* and *V. mitchelliana* in subgenus *Verticordia*.

There has been some discussion of whether *Verticordia* is monophyletic or not. Johnson and Briggs (1984) tentatively recognised "*Catocalypta*" (equals subgenus *Eperephes*; George, 1991) as a distinct entity from *Verticordia*. George (1991) commented that if *Verticordia* was to be split there would need to be more than two genera. On the basis of our study of anther morphology subgenus *Chrysoma* appears to be very uniform, and clearly separate from the other two subgenera, thereby supporting George's views. However the anther forms in *Eperephes* and subgenus *Verticordia* show much variation, as indicated above, and reassignment of some species within the subgenera could be warranted.

The anatomical structure of the microsporangial wall is generally similar across all the anther types, although there is some variation in cell size between species. Type I anthers are structurally simpler and smaller. The only other detailed anatomical study of Myrtaceae anthers has been of *Thryptomene calycina* (Beardsell *et al.*, 1989). The microsporangial wall of *Verticordia* anthers is similar to that of *T. calycina* but in *Verticordia* the epidermal cells of the microsporangia are much smaller. Furthermore, the lumina are smaller and have thicker outer walls than is the case in *T. calycina*. In *V. ovalifolia* the endothecial cells are similar to those of *T. calycina* but those of other species are larger (for example *V. gracilis*, *V. eriocephala*).

Oil glands are characteristic of the Myrtaceae and have been anatomically described for *Eucalyptus* as a schizogenous cavity lined with secretory cells (Carr and Carr, 1970). The cavity is formed because the enlargement and differentiation of the surrounding cells exceeds that of the secretory cells. This seems to be the usual manner whereby the anther glands form but there are several variations of this pattern of development.

Of all the species examined, the gland of *V. eriocephala* shows the greatest similarity with the structure of glands in its primary tissues. The anther gland consists of a coalescence of several small glands into a single feature.

In species with small non-functional glands, development is slightly different. In these, the cells of the gland are large, densely stained and have globular vacuoles, but there is little formation of intercellular space. It appears that the gland cells proliferate to keep pace with the filament expansion so that few or no intercellular spaces are formed. In contrast, in species with large glands, gland cells initially fill the immature gland cavity. Later, as the cavity expands, the cell network forms and oil is secreted into the intercellular spaces. As the anthers mature this cell network breaks down and the cavity is filled with oil and debris from the disintegrated cells.

The gland structure of *Thryptomene calycina* is very similar to that of species of *Verticordia* with large glands. The sections (Figs. 13 and 14; Beardsell *et al.*, 1989) from immature *T. calycina* anthers accord most closely with the anatomy of mature *Verticordia* anthers, especially in the presence of a network of cells enclosing large intercellular spaces. However in *T. calycina* Beardsell *et al.* (1989) noted that from an intermediate stage of development the proportion of the intercellular space decreased as the anther matured. This seems the reverse of what would be expected if the spaces are to act as reservoirs for oil production from the gland cells before the oil is expressed at maturity. The gland cells in sections taken from more mature anthers of *T. calycina* (20 days prior to anthesis) are similar to the cells in the non-functional glands (for example *V. helichrysantha*) of the species examined in this study. It may be that gland and microsporangial development in *T. calycina* is not synchronised and thus sections of glands, from what appeared to be mature anthers, had immature glands.

Oil production by flower organs occurs in a number of taxa (for example Buchmann, 1987; Bussell *et al.*, 1995) but is rarely produced by the androecium. In some species of *Mouriri* (Melastomataceae) each anther bears a concave oil gland (Buchmann and Buchmann, 1981) and the connective of *Cyphomandra* anthers produces scent compounds (Sazima *et al.*, 1993) some of which can be classed as oils. In neither of these taxa is the structure of the oil producing glands similar to those in the Myrtaceae. The contents of the *Thryptomene calycina* anther gland were demonstrated to be lipids by colour reaction to a number of stains (for example oil red O, Beardsell *et al.*, 1989). Analysis of two closely related *Verticordia* species in subgenus *Chrysoma* (*V. nitens*, *V. aurea*) showed that a number of oils were present and that despite their close relationship the most abundant compounds were different between the species (Houston *et al.*, 1993).

When pollen is shed from anthers in most angiosperms it is relatively dry. In animal pollinated species the grains may be stuck together to various degrees by tapetal material (pollenkitt Knox, 1984). However, having pollen shed in a viscous fluid is uncommon. The addition to this pollen-pollenkitt mixture of oily glandular material is an even more unusual trait. In *Verticordia* it has been proposed that extrusion of the pollen is "effected by contraction of the locules" (George, 1991). However, there seems to be no mechanism that could cause this. Endothecial thickenings have been considered to promote rupture of the stomium and hence pollen shedding (Esau, 1977; D'Arcy, 1996). However, Manning (1996) considered thickenings were more related to phylogeny than function. In general, anther dehiscence seems to be related to dessication (Schmid, 1976), and the endothecium (in particular the thickenings), rather than acting to open the stomium, may ensure a regular collapse of the cells so the stomium opens evenly. In most *Verticordia* species dessication is inapplicable as the pollen is shed before the flower opens, and hence, in a very moist micro-environment. It seems more likely that pollen shedding is related to osmotic changes in the anther. Possibly break down of the tapetal cells at anther maturity leads to a decreased water potential in the loculus. This causes water to be drawn into the loculus from the filament vascular trace increasing the volume of the contents which eventually rupture the stomium. The pollenkitt fluid, pollen and oil (if present) then adheres to the pollen presenter or, in subgenus *Chrysoma,* mixes at the apex of the anther.

The ancestral condition in the Myrtaceae is for species to have versatile, dorsifixed anthers with an anther gland (Johnson and Briggs, 1984). *Verticordia* anthers could be derived from a versatile form with an increase in the extent of attachment of the microsporangia to the filament. In Type II and III anthers the microsporangia are fused to the filament throughout their length, and appear to be basally attached. However, the dorsal attachment is still clear in many species (for example *V. citrella*, *V. serrata*). Rectangular anthers tend to be dorsally to apically attached.

If the ancestral type of Myrtaceae anther, as suggested by Johnson and Briggs (1984), is taken as the progenitor form in *Verticordia*, a number of evolutionary changes in this basic form can be recognised within the genus (Fig. 22).

The general trend is of fusion of the anther and filament, reduction in microsporangial size and loss of the gland, and increase in pollen grain size. Reduction in microsporangium size and stomium length, and thickening of the connection between the filament and connective, leads to Type I anthers. Further reduction in the size of the proximal locelli leads to the more globular type of the rectangular anther as is found in *V. gracilis*. Loss of the gland in both lines leads to the Type I anther without a gland. Increased fusion between the filament and connective produces Type II anthers and loss of the gland occurs in some species. To derive the Type III anther involves the trends seen in the other two groups - increase in the fusion of the filament and connective, and decrease in the stomium to form a pore for dehiscence, plus elaboration of the anther apex.

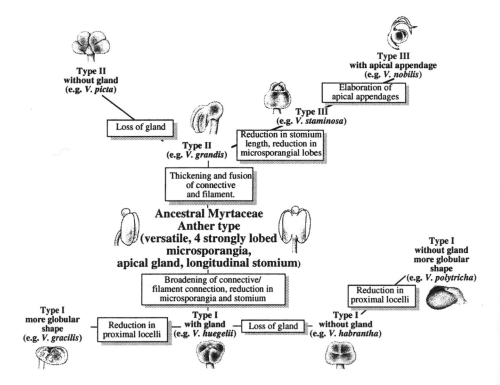

FIG. 22. Summary diagram of the changes necessary to form the various types of anthers in *Verticordia* from a hypothetical ancestral Myrtaceae anther type.

Ecologically, the importance of the anther gland relates to aspects of pollination biology. In subgenus *Chrysoma* there is indication that the oils act as pollinator attractants and may ensure reproductive isolation of different plant species. Pollination in two species (*V. nitens, V. aurea*) has been studied in some detail (Houston *et al.*, 1993). Particular bee (*Euryglossa* spp.) species visit the different *Verticordia* species and appear to be oligolectic. In addition, few or no other potential pollinators visit the flowers. *Verticordia cooloomia* also seems to have a different *Euryglossa* species to *V. nitens* or *V. aurea* acting as principal pollinator (T. Houston pers. comm.). The females of these bees collect nectar from the top of the ovary and also probe the spout-like extensions of the anther to collect the pollen-oil mixture. Analysis of the contents of the crop of the species which visits *V. nitens* showed that the oil contents of the crop were very similar to the contents of the anthers. The fact that the predominant oils in the anther glands of the two species are different (Houston *et al.*, 1993) provides a mechanism for selection of specific pollinators which can utilise the food source and are also specifically attracted to the plant species.

There have been no other such detailed studies of pollinators in the other two subgenera and there are few records of visitors to different *Verticordia* species. The oil from anther glands may assist adhesion of pollen to the pollen presenter. Clearly, having a nectar gland is not necessary for the effective operation of a pollen presenter as most species in subgenus *Verticordia* do not have a functional anther gland. However, within that subgenus all the species which lack a pollen presenter also seem to lack an anther gland. In contrast, in subgenus *Eperephes* most species have large anther glands. The loading ability of forked, intricately branched or clavate, pollen presenter hairs is likely to be superior to that of simple hairs. In subgenus *Eperephes* the larger anther glands do correlate with more intricate pollen presenter hairs because the proportion of species with non-simple pollen presenter hairs is much higher (68%) than in subgenus *Verticordia* (21%).

Pollen in most *Verticordia* species is small (*sensu* Walker and Doyle, 1975) and oily, and the oil may act to protect the pollen from dessication (O'Brien, 1996) or increase pollen longevity (Tyagi *et al.*, 1991). Another possibility is that the oils act as attractant scent chemicals such as advocated for anther glands in some legumes (Lucknow and Grimes, 1997). The more brightly coloured species of *Verticordia* have either a sweet scent or none at all (George, 1991). However the location of scent production is not known. Some of the oils from the gland may be aromatic compounds, and be effective when deposited on the exerted pollen presenter. However this possibility remains to be tested.

Acknowledgements

We would like to thank P. and N. Moyle for supplying fresh material for a number of species from their living collection, J. Macklin for preparing material for SEM and for some of the plates, H. Hunter for the drawings and A. George for helpful discussion. The Western Australian Herbarium supplied pickled material of *Verticordia capillaris* and *V. cunninghamii*. PGL thanks Professor M. B. Jones for access to facilities at Trinity College.

References

Beardsell, D.V., Williams, E.G. and Knox, R.B. (1989). The structure and histochemistry of the nectary and anther secretory tissue of the flowers of *Thryptomene calycina* (Lindl.) Stapf (Myrtaceae). *Australian Journal of Botany* 37: 65–80.

Bell, D.T. (1994). Plant community structure in southwestern Australia and aspects of herbivory, seed dispersal and pollination. In: M. Arianoutsou and R.H. Groves (editors). Plant-animal interactions in Mediterranean-type ecosystems. pp. 63–72. Kluwer, Dordrecht.

Bentham, G. (1867). Flora Australiensis Vol 3., L. Reeve and Co., London.

Buchmann, S.L. (1987). The ecology of oil flowers and their bees. *Annual Review of Ecology and Systematics* **18**: 343–369.

Buchmann, S.L. and Buchmann, M.D. (1981). Anthecology of *Mouriri myrtilloides* (Melastomataceae: Memecyleae), an oil flower in Panama. *Biotropica* **13** (2, suppl.): 7–24.

Bussell, B.M., Considine, J.A. and Spadek, Z.E. (1995). Flower and volatile oil ontogeny in *Boronia megastigma. Annals of Botany* **76**: 475–463.

Carr, S.G.M. and Carr, D.J. (1970). Oil glands and ducts in *Eucalyptus* L'Herit. 3. Development and structure of oil glands in the embryo. *Australian Journal of Botany* **18**: 191–212.

Chalson, J.M. and Martin, H.A. (1995). The pollen morphology of some co-occurring species of the family Myrtaceae from the Sydney region. *Proceedings of the Linnean Society of New South Wales* **115**: 163–191.

Chantaranothai, P. (1989). The taxonomy of *Eugenia* L. *sensu lato* (Myrtaceae) in Thailand. Unpublished Ph.D. thesis. Trinity College, Dublin.

D'Arcy, W.G. (1996). Anthers and stamens and what they do. In: W.G. D'Arcy and R.C. Keating (editors). The anther, pp. 1–24. Cambridge University Press, Cambridge.

Erdtman, G. (1952). Pollen Morphology and Plant Taxonomy. Almqvist and Wiksell.

Esau, K. (1977). Anatomy of seed plants. Second edition, Wiley, New York.

Gadek, P.A. and Martin, H.A. (1981). Pollen morphology in the subtribe Metrosiderinae of the Leptospermoideae (Myrtaceae) and its taxonomic significance. *Australian Journal of Botany* **29**: 159–184.

Gadek, P.A. and Martin, H.A. (1982). Exine ultrastructure of myrtaceous pollen. *Australian Journal of Botany* **30**: 75–86.

George, A.S. (1991). New taxa, combinations and typifications in *Verticordia* (Myrtaceae: Chamelaucieae). *Nuytsia* **7**: 321–394.

Green, J.W. (1980). A revised terminology for the spore-containing parts of anthers. *New Phytologist* **84**: 401–406.

Hufford, L. (1996). The origin and early evolution of angiosperm stamens. In: W.G. D'Arcy and R.C. Keating (editors). The anther, pp. 58–91. Cambridge University Press, Cambridge.

Houston, T.F., Lamont, B.B., Radford, S. and Errington, S.G. (1993). Apparent mutualism between *Verticordia nitens* and *V. aurea* (Myrtaceae) and their oil-ingesting bee pollinators (Hymenoptera: Colletidae). *Australian Journal of Botany* **41**: 369–380.

Johnson, L.A.S. and Briggs, B.G. (1981). Three old southern families – Myrtaceae, Proteaceae and Restionaceae. In: A. Keast (editor). Ecological Biogeography of Australia Vol. 1, pp. 429–469. W. Junk, The Hague.

Johnson, L.A.S. and Briggs, B.G. (1984). Myrtales and Myrtaceae - a phylogenetic analysis. *Annals of the Missouri Botanical Garden* **71**: 700–756.

Knox, R.B.(1984). The pollen grain. In: B.M. Johri (editor). Embryology of Angiosperms pp. 197–271. Springer-Verlag, Berlin.

Ladd, P.G. (1994). Pollen presenters in the flowering plants - form and function. *Botanical Journal of the Linnean Society* **115**: 165–195.

Lucknow, M. and Grimes, J. (1997). A survey of anther glands in the mimosoid legume tribes Parkieae and Mimoseae. *American Journal of Botany* **84**: 285–297.

Manning, J.C. (1996). Diversity of endothecial patterns in the angiosperms. In: W.G. D'Arcy and R.C. Keating (editors). The anther, pp 136–158. Cambridge University Press, Cambridge.

Moore, P.D. and Webb, J.A. (1978). An Illustrated guide to pollen analysis. Hodder and Stoughton, London.

Muller, J. (1980). Form and function in angiosperm pollen. *Annals of the Missouri Botanical Garden* **66**: 593–632.

O'Brien, S.P. (1996). Timetable of stigmatic receptivity and development and pollen tube growth in *Chamelaucium uncinatum* (Myrtaceae). *Australian Journal of Botany* **44**: 649–659.

Parnell, J.A.N. (1991). Pollen morphology of *Jovibarba* Opiz and *Sempervivum* L. (Crassulaceae). *Kew Bulletin* **46**: 733–738.

Patel, V.C., Skvarla, J.J. and Raven, P.H. (1984). Pollen characteristics in relation to the delimitation of the Myrtales. *Annals of the Missouri Botanical Garden* **71**: 858–969.

Pickett, E.J. and Newsome, J.C. (1997). *Eucalyptus* (Myrtaceae) pollen and its potential role in investigations of Holocene environments in southwestern Australia. *Review of Palaeobotany and Palynology* **98**: 187–205.

Pike, K.M. (1956). Pollen morphology of Myrtaceae from the south-west Pacific area. *Australian Journal of Botany* **4**: 13–53.

Punt, W., Blackmore, S., Nilsson, S. and Le Thomas, A. (1994). Glossary of pollen and spore terminology. Laboratory Paleobotany and Palynology Contributions Series 1, LPP Foundation, Utrecht.

Richardson, K.C., Jarret, L. and Finke, E.H. (1960). Embedding in epoxy resin for ultrathin sectioning in electron microscopy. *Stain Technology* **35**: 313–323.

Sazima, M., Vogel, S., Cocucci, A. and Hausner, G. (1993). The perfume flowers of *Cyphomandra* (Solanaceae): pollination by euglossine bees, bellows mechanism, osmophores, and volatiles. *Plant Systematics and Evolution* **187**: 51–88.

Schmid R. (1976). Filament histology and anther dehiscence. *Botanical Journal of the Linnean Society* **73**: 303–315.

Tyagi, A., Considine, J. and McComb, J. (1991). Germination of *Verticordia* pollen after storage at different temperatures. *Australian Journal of Botany* **40**: 151–155.

Walker, J.W. and Doyle, J.A. (1975). The bases of angiosperm phylogeny: palynology. *Annals of the Missouri Botanical Garden* **62**: 664–723.

Graham, S.A. and Graham, A. (2000). The Evolution of diaperturate pollen in the eurypalynous genus *Cuphea* (Lythraceae). In: M.M. Harley, C.M. Morton and S. Blackmore (Editors). Pollen and Spores: Morphology and Biology, pp. 349–364. Royal Botanic Gardens, Kew.

THE EVOLUTION OF DIAPERTURATE POLLEN IN THE EURYPALYNOUS GENUS *CUPHEA* (LYTHRACEAE)

SHIRLEY A. GRAHAM AND ALAN GRAHAM

Department of Biological Sciences, Kent State University, Kent, Ohio 44242, USA

Abstract

Cuphea is remarkable for having extensive variation in pollen structure and sculpturing at the species level. One of the several forms is diaperturate pollen which is confirmed in 23 of the ca. 250 species of this typically triaperturate genus. Diaperturate pollen is found in three of 13 taxonomic sections. The pores develop at the opposing poles of the grain where three colpi converge; functional equatorial pores are not apparent. Morphologically-based cladistic parsimony analyses employing 37 taxa and 42 characters indicate a single origin for all species with diaperturate pollen from an ancestor with trisyncolporate pollen with a striate exine. The 60 most parsimonious trees generated using PAUP had t.l.=205, ci=0.40, ri=0.57, but short branch lengths and low bootstrap values. The hypothesised phylogenetic relationships partially conflict with the current classification. Sections *Pseudocircaea* and *Melvilla* subsect. *Pachycalyx*, composed of species with diaperturate pollen, are nested within the large sect. *Euandra* which is consequently paraphyletic. The results are consistent with other morphological data in supporting the need for a revised classification. According to the analyses, in South America small green- to purple-flowered species with diaperturate pollen were ancestral to the large, red-flowered species which in turn gave rise to the red-flowered, triaperturate species in North America.

Introduction

The Lythraceae are palynologically more diverse than any other family in the order Myrtales (Patel *et al.* 1984; Graham, S. and Graham, A. 1971; Graham, A. *et al.* 1985, 1987, 1990). Within the family the genus *Cuphea*, a group of herbaceous perennials of the New World tropics, is unique in having exceptional pollen diversity at the species level. An early light microscope pollen survey of 135 of the ca. 250 species of *Cuphea* revealed extensive variation in such pollen features as shape, aperture number, colpi length, and sculpturing (Graham, S. and Graham, A. 1971). Subsequent scanning electron microscope observations of *Cuphea* pollen have added greater details that are proving highly informative in systematic and phylogenetic studies (Graham, S. 1988, 1989, 1998a, 1998b).

One of the most unusual pollen types in *Cuphea* is diaperturate pollen which appears in some species among the triaperturate types more typical of the family and of dicotyledonous angiosperms generally. The triaperturate grains have three colpi associated with three equatorial pores (tricolporate; Figs. 1–4). The diaperturate pollen in *Cuphea* has two topographically polar pores through which run three colpi (Figs. 5, 6). *Cuphea* is one of only a few genera of dicotyledons in which diaperturate grains are

FIGS. 1–6. Pollen of *Cuphea*. Fig. 1. *C. riparia,* Ekman 10242 (A). Fig. 2. *C. micropetala,* Graham 757 (KE-G). Fig. 3. *C. schumannii,* Tenorio 10697 (KE-G). Fig. 4. *C. parsonsia,* Schafer 11666 (NY). Fig. 5. *C. annulata,* Daley *et al.* 1715 (KE-G). Fig. 6. *C. gardneri,* Anderson 7336 (KE-G). Scale bar = 7 μm.

characteristic of certain species and not merely aberrant forms mixed with predominantly triaperturate forms. Among other examples is the genus *Fuchsia* L. in the Onagraceae, sister family to the Lythraceae, where diaperturate pollen is the rule; triaperturate grains in *Fuchsia* occur only in some polyploid species (Nowicke *et al.* 1984). In the unrelated Proteaceae, three of 50 genera (*Banksia* L.f., *Dryandra* R. Br. and some species of *Embothrium* J. R. Forst. & G. Forst.) have consistently diaperturate pollen, and in the triaperturate Moraceae, Ulmaceae, and Urticaceae a few genera, including *Ficus* L., *Trema* Lour. and *Sarcopilea* Urban, have one or a few species characterised by diaperturate pollen (Erdtman 1952). *Faramia* Aubl. in the Rubiaceae is also known to have species with either di- or triaperturate pollen. On the whole, however, species with consistently diaperturate pollen are rare in genera with characteristically triaperturate pollen.

Reports of 2-, 3-, and/or 4-aperturate grains within a genus often do not make clear if a single form is constant for the different species or if one form predominates while the others occur as aberrant grains (see, for example, the citation of mixed forms within a species in the Gentianaceae, Capparidaceae, and Euphorbiaceae in Erdtman 1952). In most instances, 2- or 4-aperturate grains are found either as aberrant grains within a species with triaperturate pollen or the pore number is variable (for example, in *Alnus* Miller, *Ulmus* L.). Four- (or more) aperturate grains within taxa commonly having triaperturate pollen are often indicative of polyploidy.

In *Cuphea* diaperturate pollen is confirmed in 23 of ca. 250 species and these are classified in three of the 13 sections of the genus (Table 1; Koehne 1903). All are of South American, primarily Brazilian, distribution. Section *Pseudocircaea* Koehne is the only section that is uniformly diaperturate. The section is generally recognised by the synapomorphy of persistent petals. In addition, other features unite the species and suggest the section is monophyletic, i.e. leaves are typically thin and long-petioled, stems are very viscous-haired, and the plants inhabit disturbed sites.

Table 1. Species of *Cuphea* with diaperturate pollen arranged by taxonomic section.

Sect. *Pseudocircaea* (4 spp./4 diaperturate)
  C. impatientifolia A. St.-Hil., *C. lutescens* Koehne,
  C. persistens Koehne, *C. sessiliflora* A. St.-Hil.

Sect. *Melvilla* (ca. 43 spp. in 6 subsections/10 diaperturate)
  Subsect. *Pachycalyx* (12 spp./10 diaperturate)
    C. andersonii S. Graham, *C. annulata* Koehne,
    C. cylindracea S. Graham, *C. egleri* Lourt., *C. froesii*
    Lourt., *C. fuchsiifolia* A. St.-Hil., *C. gardneri*
    Koehne, *C. rubro-virens* T. B. Cavalc., *C. sabulosa* S.
    Graham, *C. teleandra* Lourt.

Sect. *Euandra* (ca. 65 spp. in 5 subsections/ 9 diaperturate)
  Subsect. *Platypterus* (ca. 15 spp./4 diaperturate)
    C. ingrata Cham. & Schltdl., *C. glutinosa* Cham. &
    Schltdl., *C. acinifolia* A. St.-Hil., *C. thymoides*
    Cham. & Schltdl.
  Subsect. *Hilariella* (ca. 20 spp./3 diaperturate)
    C. pseudovaccinium A. St.-Hil., *C. pohlii* Lourt.,
    C. adenophylla T. B. Cavalc.
  Subsect. *Oidemation* (ca. 22 spp./2 diaperturate)
    C. tuberosa Cham. & Schltdl., *C. stenopetala* Koehne

In section *Melvilla* Koehne, one-quarter of the species are diaperturate. They occur only in the South American subsection *Pachycalyx* Koehne in 10 of the 12 species. Floral morphology in the two triaperturate exceptions, *C. grandiflora* Koehne and *C. pulchra* Moric., suggests the species may be incorrectly classified in subsect. *Pachycalyx* (Graham, S. 1990). Section *Melvilla* comprises North and South American species whose specialised long, dorsally convex, red/yellow floral tubes are correlated with bee and hummingbird pollinators. Subsection *Pachycalyx* is defined within the section primarily by reduced epicalyx segments (the "appendages").

The remaining species of *Cuphea* with diaperturate pollen are found in the large section *Euandra* where just nine of 65 species have diaperturate grains. They are classified in three of the five subsections. Section *Euandra* is defined by stamens extending to the distal margin of the floral tube and the subsections are mostly based on differences in the seed margin (for example, thick rounded margins vs. thinned wing-like ones).

The presumed basal species in *Cuphea* have triaperturate pollen, as do the other 30 genera of the family, so diaperturate grains must represent a derived condition within the genus. Transformation to the diaperturate state from a triaperturate one in the genus has happened infrequently, possibly only once. Given this rarity, knowledge of the evolution of diaperturate *Cuphea* pollen could provide information useful in establishing phylogenetic relationships among species with this feature. Further, in the context of a range of vegetative, floral, and seed morphological changes, it could help clarify phylogenetic relationships at the sectional level and assist in the construction of a more natural classification in this taxonomically difficult genus. Two hypotheses for the evolution of diaperturate pollen in *Cuphea* are: 1) the form evolved just once in the genus, and all species with diaperturate grains are related by this common diaperturate ancestor; or 2) diaperturate pollen evolved more than once, perhaps a minimum of three times, meaning that it developed at least once in each taxonomic section in which it is found.

In this study we use morphologically-based cladistic parsimony analysis to investigate the evolution of diaperturate pollen in *Cuphea* and to answer the following questions: 1) how many times has diaperturate pollen evolved in *Cuphea*; 2) what monophyletic group or groups are supported by this pollen state; 3) how do the phylogenetic relationships of species with diaperturate pollen compare to their present taxonomic classification?

Materials and Methods

Pollen of previously unstudied species was obtained from and vouchered by collections at KE-Graham. Pollen of species from other herbaria in the permanent pollen reference collection at KE is vouchered at one of several herbaria (A, GH, MICH, NY, US; Graham, S. and Graham, A. 1971). Pollen samples for LM and SEM were processed by the standard acetolysis method of Erdtman (1960). Samples for SEM were sputter-coated with gold-palladium and examined and photographed with a Cambridge Stereoscan Mark IIA microscope. Cladistic analyses of morphological characters employed PAUP vers. 3.1.1 (Swofford 1993), and character evolution was examined using MacClade vers. 3.07 (Maddison and Maddison 1992).

Description of Cuphea Pollen

The basic pollen form in *Cuphea* is oblate, tricolporate (flattened from pole to pole with the pores equidistantly spaced at the equator), amb triangular to oval-triangular in polar view, tectate, and 14–40 μm in diameter (Figs. 1–4). The pores vary from non-protruding to strongly protruding (cf. Figs. 1–4). Syncolpate grains are most common but species with non-syncolpate grains occur in at least four sections (sections

Archocuphea, Cuphea, Heteranthus, and *Brachyandra*). Sculpture patterns range from psilate to rugulate to striate and vary in the degree of striation density, height of the sculpture elements, and by different organisation of the elements at the poles versus at the equator. Interaperturate wall thickenings, apparent in LM, characterise triaperturate pollen in species of one section, sect. *Trispermum* (Graham, S. and Graham, A. 1971).

A range of pollen size is found within all triaperturate pollen types. Although there are a number of species with grains less than 21 μm in diameter, and a few reach diameters of 30–40 μm, most are within 22–28 μm. Grains may expand to different degrees depending on treatment, mounting medium, age of the slide preparation, fresh versus herbarium samples, and age of the herbarium collection. The sizes coded in the data matrix (Table 3) are an average taken from glycerin-mounted acetolysed pollen, except in a few cases where lactic acid mounts were necessary because of the rarity of the material.

Diaperturate grains are much less variable. They are prolate to prolate-spheroidal with two circular pores protruding to varying degrees beyond the margin of the grain (Figs. 5–8). The pores are connected by three colpi (Fig. 8), the exine is rugulate to short-striate with the sculpture elements oriented perpendicular to the colpi, and the size ranges from 21 to 30 μm long including the protruding pores. No pores are evident in the colpi at the equator (Fig. 7). The pollen develops in tetrahedral tetrads and these show that the two functional pores are positioned at the true poles of the grains.

FIGS. 7–8. Pollen of *Cuphea fuchsiifolia* (Cavalcanti & Graham 2308 (KE-Graham), illustrating absence of equatorial pores. Fig. 7. Equatorial view. Fig. 8. Oblique equatorial view with three colpi visible. Scale bar = 7 μm.

Within the anthers of species with prolate diaperturate-pollen a few oblate-spheroidal grains are found that have two polar pores and a modification of the exine at the mid-point of the colpi where pores would be expected (poroidate?). Detailed developmental investigations will be needed to determine the ontogeny of the diaperturate forms, but one possibility suggested by current observations is that they result from the arrested development of the three equatorial pores and the *de novo* formation of pores in the syncolpate region of the poles. In a 5% sucrose solution, the protoplasm of pollen of the diporate pollen species *Cuphea fuchsiifolia* A. St. Hil.,

Table 2. List of characters used for cladistic analysis. The list was constructed for all species of *Cuphea* and therefore includes some characters invariant for the subset of species considered in this analysis (char. 1, 2, 21, 22, 25, 36, 39, 50). These invariant characters were ignored in the analyses.

1. HABIT: 0=perennial; 1=annual.
2. ROOTSTOCK INTERIOR COLOUR: 0=clear; 1=red.
3. HEIGHT: 0=0.75 m or more; 1=less than 0.75 m.
4. STATURE: 0=erect; 1=decumbent, procumbent, vine-like.
5. MALPIGHIACEOUS HAIRS: 0=absent; 1=present.
6. PETIOLE LENGTH OF LARGEST LEAVES: 0=2 mm or more; 1=less than 2 mm.
7. LEAVES: 0=opposite; 1=whorled.
8. LEAF BASE: 0=acute to decurrent; 1=acute to rounded; 2=rounded to cordate.
9. LARGEST LEAF LENGTH: 0=40 mm or more; 1=less than 40 mm.
10. STEM LEAF/INFLORESCENCE BRACT TRANSITION: 0=gradual; 1=abrupt.
11. FLOWER ARRANGEMENT: 0=opposite; 1=alternate; 2=whorled; 3=umbelliform clusters.
12. PEDICEL ATTACHMENT: 0=basal-near basal; 1=ventral.
13. INFLORESCENCE: 0=distinct, small-bracteate, terminal; 1=indistinct, leafy, flowers distributed along the stem.
14. INFLORESCENCE: 0=simple (raceme, spike); 1=compound (thyrse).
15. PEDICEL LENGTH: 0=avg.0–4 mm; 1=avg.greater than 4 mm.
16. BRACTEOLES: 0=leaf-like; 1=reduced to hairs; 2=absent.
17. FLORAL TUBE LENGTH: 0=consistently less than 14 mm; 1=variably 14–24 mm; 2=consistently more than 20 mm.
18. FLORAL TUBE COLOUR: 0=green to purple-red; 1=red or red-orange.
19. FLORAL TUBE DORSAL SIDE: 0=horizontal to concave; 1=convex.
20. FLORAL TUBE BASE: 0=spurless, no measurable extension beyond the pedicel at 40×; 1=rounded to elongate, horizontal to descending or slightly ascending.
21. SEPALS: 0=equal to subequal; 1=unequal, the dorsal lobe longer than the other 5 lobes.
22. LOBE MARGIN: 0=non-ciliated; 1=ciliated or ciliolated.
23. FLORAL TUBE MARGIN: 0=blunt, dorsal and ventral sides equally extended; 1=oblique, ventral side extended farther than the dorsal.
24. APPENDAGES: 0=absent; 1=present as thickenings, only margins free; 2=present as free-standing cylinders or leaf-like growths 1/2 as long as the calyx lobes or longer.
25. INTERNAL ALAE: 0=absent; 1=present.
26. PETALS: 0=caducous or absent; 1=persistent.
27. PETAL LENGTH: 0=all equal or subequal; 1=unequal, dorsals larger than the ventrals; 2=absent; 3=unequal, ventrals larger than dorsals.
28. DORSAL PETAL COLOUR: 0=light to deep purple or fuchsia; 1=deep maroon or purple-black; 2=white; 3=red-orange; 4=petals none.
29. DORSAL AND VENTRAL PETAL COLOUR: 0=same; 1=different; 2=ventral or dorsal and ventral petals absent.
30. STAMEN BASE: 0=lower half of the floral tube near ovary; 1=above the middle of the floral tube.
31. STAMEN INSERTION: 0=all at same level; 1=2 dorsal-most deeper.
32. STAMEN NUMBER: 0=8; 1=11.
33. ANTHER POSITION: 0=5-11 exserted; 1=anthers reaching sinuses; 2=anthers deep in tube.
34. STAMEN POSITION: 0=evenly distributed; 1=ventrally gathered.

35. POLLEN PORE NUMBER: 0=3; 1=2.
36. POLLEN INTERAPERTURATE THICKENINGS: 0=absent; 1=present.
37. POLLEN COLPI: 0=non-syncolpate; 1=syncolpate.
38. POLLEN EXINE: 0=psilate to scabrate; 1=rugulate to uniformly striate; 2=coarsely striate only at the pores.
39. POLLEN EXINE AT POLES: 0=psilate, striate, or rugulate; 1=irregular mesh.
40. POLLEN DIAMETER: 0=21 μm or less; 1=22 μm or greater.
41. OVARY LOCULES: 0=equal in size; 1=ventral locule reduced.
42. OVULES: 0=more than 25; 1=(10-)13-25; 2=3-10(-12).
43. PLACENTA: 0=retained in capsule; 1=emergent.
44. SEED LENGTH: 0=less than 2.0 mm long; 1=2.0 mm long or longer.
45. SEED SHAPE: 0=elongated pyramidal; 1=bilateral compressed; 2=inflated, minimally compressed.
46. SEED MARGIN: 0=thick, rounded, non-winged; 1=thinned, narrowly winged.
47. SEED HAIRS: 0=straight; 1=spirally twisted; 2=hairs absent or minimally developed, non-emergent.
48. SEED OIL DOMINANT FATTY ACID: 0=C18:2; 1=C14:0; 2=C12:0; 3=C10:0.
49. DISC: 0=cup-shaped; 1=unilateral, broad to massive, triangular to oblong, erect to deflexed in the spur; 2=unilateral, slender, deflexed or inrolled; 3=absent.
50. STEM GLAND PATCHES: 0=absent; 1=present.
51. FLORAL TUBE WIDTH: 0=1-4 mm; 1=greater than 4 mm.
52. VESICLES: 0=absent; 1=present.
53. SEED APEX: 0=obtuse; 1=retuse; 2=carunculate.
54. STIGMA: 0=surpassing anther level during anthesis, long-exserted at floral maturity; 1=remaining at or below anther level during anthesis, not or scarcely exserted at floral maturity.

---

emerged only through the two polar pores, both in the strictly diporate grains and in the polar-pored plus equatorial poroidate grains. No protoplasm emerged from the equatorial regions of the poroidate pollen in any grains observed, suggesting there is a loss of both structural and functional equatorial pores in the transitional forms to diporate pollen.

Fresh pollen of *Cuphea fuchsiifolia* was tested for viability with cotton blue/lactic acid. Fifty-nine percent of 300 grains counted stained, indicating viability. These included both strictly diaperturate and diaperturate plus poroidate grains.

Cladistic Analyses

Parsimony analysis was performed on 37 taxa including three outgroups used to polarise the characters. For the outgroups *Cuphea denticulata* Kunth was selected as a representative of the basal species in the genus, *Pleurophora* because it is sister genus to *Cuphea*, and *Adenaria* as an earlier member of the lythracean clade containing *Cuphea* (Graham, S. *et al.* 1993). Seventeen of 23 species with diaperturate pollen were chosen for analysis. Excluded were two diaperturate species from Section *Euandra* and four from Section *Melvilla* subsect. *Pachycalyx* that were morphologically closely similar to one or more selected species. The full set of 23 species was not included in order to reduce the size of the matrix for greater ease of analysis. Seventeen species with triaperturate pollen were selected as place holders for the sections of the genus that could possibly have phylogenetic connections to the species with diaperturate pollen. Excluded from the analysis were representatives of the derived North American sections (*Heterodon*, *Leptocalyx*, and *Diploptychia*) and the highly specialised South American sections (*Archocuphea*, *Heteranthus*, *Melicyathium*, *Trispermum*, and *Amazoniana*). From a matrix of 54 morphological and anatomical characters for 100 species of *Cuphea* that was previously generated for revisionary studies, a subset of species was extracted for this study. Eight of the 54 characters were invariant (char. 1, 2, 21, 22, 25, 36, 39, 50) leaving 46 characters for analysis. The eight invariant characters, included in Table 2, were ignored in the analyses. Three characters (char. 5, 32, 41) were uninformative autapomorphies (Table 3).

Parsimony analyses using various addition options were performed. Simple addition of taxa followed by successive *a posteriori* weighting based on a reconstituted consistency index (rc) provides the basis for discussion. In all analyses, characters were unordered and optimised with ACCTRAN, multiple states were treated as polymorphisms, and the options for saving all minimal trees (MULPARS) and tree-bisection reconnection (TBR) were chosen. Steepest descent was in effect in the initial search as a means of sampling islands of trees, and not selected during analysis of re-weighted characters. The initial heuristic search was repeated employing the same options but excluding the pollen aperture character (character 35) in order to examine how its absence changed the cladistic pattern generated. A search for trees one step longer than the most parsimonious trees of the initial analysis was made to estimate the robustness of the clades using simple addition, MULPARS, and steepest descent options. A bootstrap analysis (Felsenstein 1985) was conducted heuristically to assess branch support using 100 replicates.

The analysis with simple addition and characters equally weighted resulted in 60 most parsimonious trees with tree length (t.l.)=205, consistency index (ci)=0.40, and retention index (ri)=0.57. The strict consensus and one randomly selected tree from the 60 trees are illustrated in figures 9 and 10, respectively. All species with diaperturate grains (clade CAMP-ADEO) arise from a single node along one of three branches. All species of sect. *Pseudocircaea*, all species sampled from sect. *Melvilla*, and nine of the 17 species sampled from sect. *Euandra* (including some tri- and all diaperturate species) evolve from this node. Branch lengths of the cladogram are very short for all internal branches of the ingroup indicating weak support for all clades of the ingroup (Fig. 10). With the exception of branches from the outgroup with 93% support and the MELV-SCHU branch with 58% support, bootstrap values for all other branches had poor support of < 50%. The morphological data, as indicated by the low consistency index, have a high degree of homoplasy.

Two rounds of successive weighting obtained four equally most parsimonious trees each time. Thirty-one of the 46 variable characters contributed some weight to determining tree structure; characters 12, 18, 19, 20, 30, 43, and 51 were fully weighted. Eleven other characters on a 10 point scale were weighted at 5, 4, or 3 (including pollen aperture number weighted at 3). The remaining characters had weights of 2 or 1 and were highly homoplastic.

One of the four equally most parsimonious trees is used as a basis for examination of evolution of pollen pore number (Fig. 11) and for comparison of phylogenetic and taxonomic relationships of species with diaperturate pollen (Fig. 12). The same diaperturate monophyly is present after character re-weighting as was present in the initial search with equally weighted characters. Diaperturate pollen evolved once from a triaperturate, syncolporate, rugulate or striate ancestor. The diaperturate condition subsequently was lost twice, along the branch to *C. campylocentra* Griseb. and at the base of the most highly derived representatives of section *Melvilla*, the clade MELV-PARX (Fig. 11). The gain of syncolpy and the move from psilate to rugulate or striate exine in pollen structure (characters 35, 37, 38) occurred prior to the gain of diaperturate pollen on the preceeding branch (Fig. 12). Also in Fig. 12 *Cuphea confertiflora* A. St.-Hil. (with triaperturate pollen) has moved from a diaperturate clade in the unweighted consensus (Fig. 9) to a position as sister to the diaperturate clade CAMP-ADEO.

Table 3. Data matrix of 37 taxa by 54 characters. The invariant characters 1, 2, 21, 22, 25, 36, 39, and 50 are ignored in the analysis. Species of *Cuphea* are listed by specific epithets. Specific epithets are abbreviated to the first four letters on the cladogram; CAME=*C. campestris;* CAMP=*C. campylocentra.* Symbols used for multistate taxa are: a=0&1; c=0&2. Missing data=?

	5	1 0	1 5	2 0	2 5	3 0	3 5	4 0	4 5	5 0
Adenaria	00000	00000	30101	20000	00000	00200	00000	00000	00000	02030	0000
denticulata	00100	10010	00000	20000	00010	03001	11100	00000	11101	01010	0001
melanium	00110	10a10	11100	00000	00010	00001	11200	00000	12101	01220	0001
carthagenensis	00100	00010	11100	00001	00010	00001	01200	01201	12111	11220	0001
campestris	00100	10000	11100	00001	00120	00001	11100	01201	12111	11?10	0101
urbaniana	00100	00010	11100	00001	00120	00001	11100	01001	12111	01?10	0120
lysimachioides	00000	11100	21001	00000	00110	00001	11100	01201	12111	11?10	0001
strigulosa	00100	10010	11100	00001	00010	00001	11100	01201	12111	11110	0000
campylocentra	00100	10010	11000	00001	00020	00001	11000	01101	10101	01?10	0100
melvilla	00011	00101	11001	02111	00a20	02421	11000	01101	10101	02210	1000
schumannii	00000	00101	11011	a2111	00120	01021	11000	01101	11111	01310	1000
micropetala	00000	00000	11001	02111	00a20	00301	11000	01101	10112	02110	1000
watsoniana	00010	00000	11101	02111	00a20	0c201	11000	01100	11111	01?10	0000
paradoxa	00110	00000	11101	02111	00120	02421	11010	00101	12111	01?10	0000
gardneri	00000	00a00	1110a	01111	00120	10201	11011	01101	12111	01?10	0000
egleri	00000	10100	11110	02111	00110	00101	11011	01101	12111	01?10	0010
teleandra	00000	10010	11100	01111	00110	00001	11011	01101	12111	01?10	0010
cylindracea	00000	00210	11100	02011	00020	00301	11011	01101	12111	01?10	0010
sabulosa	00100	10210	11100	00?11	00020	02421	11001	01100	12111	01?10	0010
glutinosa	00110	10110	11100	00001	00110	00001	11001	01101	12101	11210	0000
tuberosa	00100	00201	110a1	00001	00010	01001	11001	01101	11111	11?10	0000
lutescens	00000	00100	11110	a0001	00020	10c01	11001	01001	12111	01210	0000
sessiliflora	00110	00210	11110	00001	00010	10c01	11a01	01001	12111	01210	0001
persistens	00100	00000	11110	00001	00010	10001	11101	01100	11111	01?10	000?
impatientifolia	?0100	00000	11010	00001	00010	10001	11001	01101	12111	01?10	000?
fuchsiifolia	00100	00100	11101	01001	00010	10001	11011	01101	12111	01?10	0000
ingrata	00110	10?10	11100	00001	00020	00c01	11101	01101	11111	01210	0011
thymoides	00110	10010	11100	00001	00010	00001	11101	01101	12101	01210	0001
pseudovaccinium	00000	10110	11100	00001	00010	00001	11101	01101	12111	01210	0001
stenopetala	00000	10100	11100	00001	00010	?0001	11101	01101	111?1	11?10	0001
adenophylla	00000	10010	11100	00001	00010	00001	11101	01101	12111	01?10	0001
confertiflora	00100	10201	110a0	00001	00a10	00001	11100	01101	12111	01210	0000
disperma	00110	10010	11100	00001	00010	00001	11100	00000	12111	01?10	0000
acinos	00110	10110	01101	00001	00010	00001	11100	00000	12111	01210	0001
sclerophylla	00000	10001	01110	00001	00020	00001	11100	00000	12111	01210	0001
ferruginea	00100	11100	11011	00001	00010	00001	11100	0000?	12111	01?10	0011
Pleurophora	00000	10010	10100	00000	00020	00000	01000	00001	10001	01000	0000

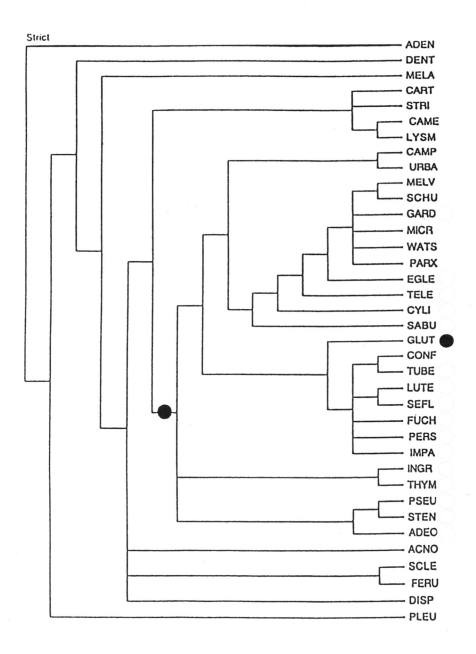

FIG. 9. Strict consensus of 60 most parsimonious trees; tree length=205 steps, consistency index = 0.40, retention index = 0.57. The left dot on the cladogram signifies point of origin of the diaperturate pollen form; all species with diaperturate grains are marked by dots. Species epithets, provided in the text, are reduced to four letters.

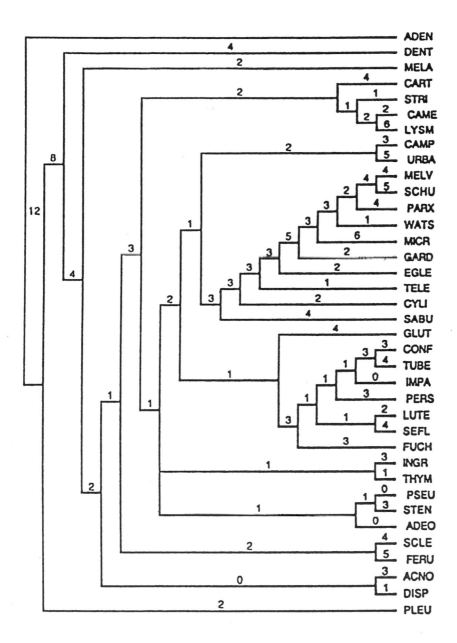

FIG. 10. One of 60 most parsimonious trees; branch lengths (the amount of character change) indicated for each branch.

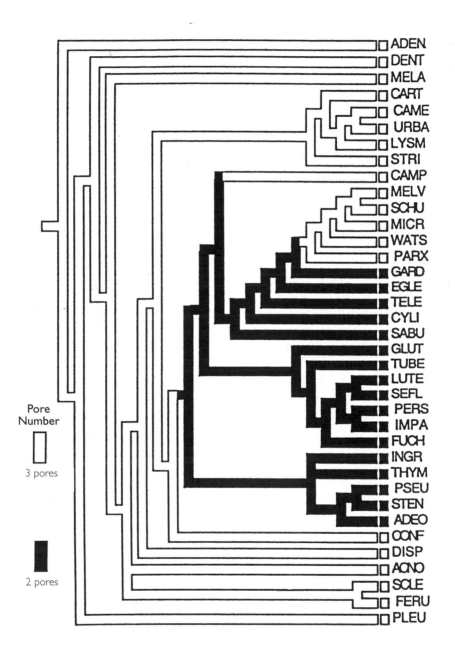

FIG. 11. Evolution of diaperturate pollen on one of the four most parsimonious trees obtained after *a posteriori* weighting of 60 trees from the initial analysis. Reversals occur on the lineages to *C. campylocentra* and to the MELV-PARX grade.

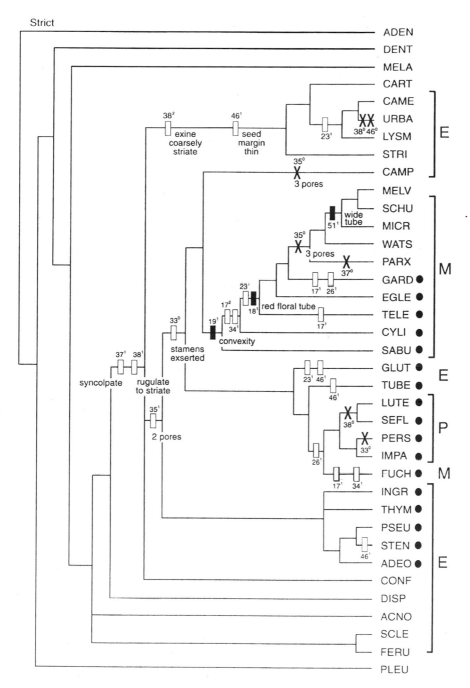

FIG. 12. One of the four most parsimonious trees obtained after *a posteriori* weighting of 60 trees from the initial analysis (Fig. 9). Sections are indicated as: E=*Euandra*, M=*Melvilla*, P=*Pseudocircaea*. States of selected characters are portrayed as superscripted numbers (see Tables 2 & 3). Solid bar = unique synapomorphy; open bar = parallelism; X=reversal to a previous character state.

Among the most informative and least homoplastic characters with respect to the evolution of species with diaperturate pollen are the following:

Character 17 - Floral tubes longer than 14 mm and presumed to have larger, possibly more specialised pollinators, occur in clade MELV-CYLI and arise independently in *C. fuchsiifolia* A. St.-Hil.

Character 18 - Red floral tubes are uniquely synapomorphic for MELV-TELE.

Character 19 - Convex floral tubes are uniquely synapomorphic for MELV-SABU.

Character 26 - Persistent petals evolved twice according to this hypothesis, in *C. gardneri* Koehne and in the clade LUTE-FUCH.

Character 33 - Exserted stamens, which imply that pollination is primarily accomplished by outcrossing, characterise the large, mainly diaperturate pollen clade CAMP-FUCH. Exserted stamens are derived from an ancestor with stamens reaching only as far as the margin of the floral tube opening.

Characters 37 and 38 - The syncolporate condition and striate exine evolved prior to diaperturate pollen and are present in most diaperturate pollen. The exceptions are found in *C. lutescens* Koehne, and *C. sessiliflora* A. St.-Hil. where the exine is psilate.

Among the diaperturate species the clade CAMP-FUCH, which had evolved long-exserted stamens, further diverged by gain in clade MELV-SABU of floral tubes that are longer (also wider in MELV-WATS), intense red, red-orange, or red/yellow, and dorsally curved. The divergence clearly relates to selection for outcrossing and to greater correlation with specialised pollinators. The red colour attracts bees and the dorsal convexity accommodates th long tongues of bees or curved bills of hummingbirds. These floral changes did not occur in the other two clades with diaperturate pollen, GLUT-IMPA and INGR-ADEO. However, *C. fuchsiifolia*, sister to clade LUTE-IMPA (taxonomically sect. *Pseudocircaea*), has independently acquired similar features attractive to pollinators, i.e., a longer floral tube which is intensely coloured and long-exserted stamens that emerge more or less collectively from the ventral side of the floral tube.

The stability of the diaperturate clade was examined by removing the aperture character from the matrix and re-analysing the data as in the initial analysis. Results were 838 most parsimonious trees of t.l.=200, ci=0.40 and ri=0.56. The branch on which diaperturate pollen evolved and nearly all other branches collapsed to form a single polycotomy of all ingroup members. Clearly, the numerous other characters used in the analysis, including the other pollen characters, are too widely and randomly distributed to maintain the branching pattern in the absence of the aperture character. Thus, change in pore number provides a good phylogenetic signal in the genus. Other outcomes of the re-analysis were retention of clade MELV-SABU on one branch of the polycotomy; and, in the only other line with resolution, the sister species relationship between *C. confertiflora* and *C. tuberosa* Cham. & Schltdl.

The search for trees one-step longer found 20,001 trees before the analysis was aborted. A strict consensus of 19,941 trees of tree length 206, obtained by filtering trees, was unresolved for the ingroup with the exception of the clade MELV-CYLI which is the same clade, minus *C. sabulosa* S. Graham, to survive in the analysis that excluded the aperture character.

The low lack of resolution in cladistic analysis is typical of *Cuphea* analyses. However, the relationships expressed in the initial heuristic search and especially in the results after successive weighting are consistent with intuitive views based on careful comparison of the morphology, cytology, palynology, and seed oil composition for the genus (for example S. Graham 1988, 1989, 1990, 1998a). In the following discussion, the taxonomic implications of our results are examined based on these preliminary phylogenies.

Comparison with Taxonomic Classification

According to the phylogeny generated after successive weighting (Fig. 12), sect. *Melvilla* is monophyletic if *C. fuchsiifolia* is reclassified in sect. *Pseudocircaea*. The placement of the species there is warranted not only by possession of the apomorphic persistent petal state and purple floral tube but also by possession of the same base chromosome number as the other members of the section, x = 7. The base chromosome number in section *Melvilla* is x = 8 with most species being tetraploids with n = 16.

The members of sect. *Melvilla* subsect. *Pachycalyx* form a grade basal to the rest of the section (SABU-GARD in Fig. 12). The three Mexican species included in this study, *C. watsoniana* Koehne, *C. micropetala* Kunth, and *C. schumannii* Koehne, are derived from Brazilian ancestors, an origin in agreement with the general understanding that the Brazilian Highlands are the primary centre of speciation for the genus, and that the western Sierra Madre of Mexico, with a distinctly different set of species, is a secondary centre of diversity (Graham 1992). The sister species to the *Melvilla* clade is a member of section *Euandra* with triaperturate pollen.

The monophyletic section *Pseudocircaea* (including *C. fuchsiifolia*) is nested within a larger clade whose basal members are classified in section *Euandra*. This study cannot resolve the relationships among species of section *Euandra* or between *Euandra* and sections *Melvilla* and *Pseudocircaea* because the large size of *Euandra* precludes treatment of sufficient taxa to answer those questions. However, both sections *Melvilla* and *Pseudocircaea* appear to be highly specialised clades within sect. *Euandra*. Section *Euandra* is thus paraphyletic and it may ultimately prove to be polyphyletic although this analysis does not test that hypothesis. If either hypothesis is substantiated after analyses on additional subsets of species, and the addition of information from molecular data sets, the classification now in use will require major renovation to reflect the true relationships among the taxa.

Conclusions

Based on morphological analyses we conclude that diaperturate pollen had a single origin from a South American ancestor with tricolporate, striate pollen like that occurring in some members of sect. *Euandra*. Section *Euandra* is seen as paraphyletic with some descendants currently recognised as sections *Melvilla* and *Pseudocircaea*. The *Melvilla* grade of South American species with diaperturate pollen (subsect. *Pachycalyx*) gave rise to species adapted for outcrossing by specialised pollinators. Some members of this lineage, following a reversal to triaperturate pollen or retention of triaperturate pollen from triaperturate ancestors, radiated into North America. The investigation of the evolution of pollen pore number has provided insight into the relationships among members of sections *Euandra*, *Melvilla*, and *Pseudocircaea*, and will be useful in continuing inquiries into the phylogeny of the genus and its correlation with taxonomy.

Acknowledgements

The support of the United States National Science Foundation through grant DEB-9509524 to S. Graham is gratefully acknowledged.

Literature

Erdtman, G. (1952). Pollen morphology and plant taxonomy. Almqvist & Wiksell, Stockholm.

Erdtman, G. (1960). The acetolysis method. A revised description. *Svensk Botanisk Tidskrift* 54: 561–564.

Felsenstein, J. (1985). Confidence limits on phylogenies: an approach using the bootstrap. *Evolution* 39: 783–791.

Graham, A., Nowicke, J., Skvarla, J.J., Graham, S.A., Patel, V. and Lee, S. (1985). Palynology and systematics of the Lythraceae. I. Introduction and genera *Adenaria* through *Ginoria. American Journal of Botany* 72: 1012–1031.

Graham, A., Nowicke, J., Skvarla, J.J., Graham, S.A., Patel, V. and Lee, S. (1987). Palynology and systematics of the Lythraceae. II. Genera *Haitia* through *Peplis. American Journal of Botany* 74: 829–850.

Graham, A., Graham, S., Nowicke, J.W., Patel, V. and Lee, S. (1990). Palynology and systematics of the Lythraceae. III. Genera *Physocalymma* through *Woodfordia,* addenda, and conclusions. *American Journal of Botany* 77: 159–177.

Graham, S.A. (1988). Revision of *Cuphea* section *Heterodon* (Lythraceae). *Systematic Botany Monographs* 20: 1–168.

Graham, S.A. (1989). Revision of *Cuphea* sect. *Leptocalyx* (Lythraceae). *Systematic Botany* 14: 43–76.

Graham, S.A. (1990). New species of *Cuphea* section *Melvilla* (Lythraceae) and an annotated key to the section. *Brittonia* 42: 12–32.

Graham, S.A. (1992). New chromosome counts in Lythraceae-Systematic and evolutionary implications. *Acta Botánica Mexicana* 17: 45–51.

Graham, S.A. (1998a). Revision of *Cuphea* sect. *Diploptychia* (Lythraceae). *Systematic Botany Monographs* 53: 1–96.

Graham, S.A. (1998b). Relacionamentos entre as espécies autógamas de *Cuphea* seção *Brachyandra* (Lythraceae). *Acta Botanica Brasileira* (in press).

Graham, S.A. Crisci, J.V. and Hoch, P.C. (1993). Cladistic analysis of the Lythraceae *sensu lato* based on morphological characters. *Botanical Journal of the Linnean Society* 113: 1–33.

Graham, S.A. and Graham, A. (1971). Palynology and systematics of *Cuphea* (Lythraceae). II. Pollen morphology and infrageneric classification. *American Journal of Botany* 58: 844–857.

Koehne, E. (1903). Lythraceae. In: A. Engler (editor). *Das Pflanzenreich* IV. 216: 1–326. Wilhelm Englemann, Leipzig, Germany.

Maddison, W. and Maddison, D. (1992). MacClade. Vers. 3.07. Sinauer Assoc. Inc., Sunderland, Massachusetts.

Nowicke, J.W., Skvarla, J.J., Raven, P.H. and Berry, P.E. (1984). A palynological study of the genus *Fuchsia* (Onagraceae). *Annals of the Missouri Botanical Garden* 71: 35–91.

Patel, V.C., Skvarla, J.J. and Raven, P.H. (1984). Pollen characters in relation to the delimitation of Myrtales. *Annals of the Missouri Botanical Garden* 71: 858–969.

Swofford, D.L. (1993). PAUP. Vers. 3.1. Smithsonian Institution, Washington, D.C.

68 TAXA AND 32 CHARACTERS: RESOLVING SPECIES RELATIONSHIPS USING MORPHOLOGICAL DATA

MARK A. CARINE AND ROBERT W. SCOTLAND

Department of Plant Sciences, University of Oxford, South Parks Road, Oxford. OX1 3RB. UK

Abstract

Thirty two morphological characters, including eight pollen characters, were scored for sixty eight species of *Strobilanthes sensu lato* from southern India and Sri Lanka. Parsimony analyses were performed using the total data set, using pollen characters only, and using macromorphological characters only. The results of the analysis of the pollen data alone were incongruent with the results of the analysis performed on the total data set. Pollen characters therefore do not estimate the topology of the tree resulting from the analysis of all characters. All eight pollen characters were homoplastic in the total analysis, but all contain significant levels of synapomorphy as evidenced by their Retention Index values. The strict consensus from the total analysis of 68 taxa and 32 characters was well resolved even though 88% of characters were homoplastic. It is suggested that this is because homoplastic characters can provide structure to the tree when coupled with dense sampling of taxa. The study suggests that morphological data can resolve species relationships in large taxa such as *Strobilanthes* that are characterised by high levels of character conflict.

Introduction

Strobilanthes Blume *sensu lato* (Acanthaceae), comprising 350–450 species, is distributed throughout south and south-east Asia and Melanesia, where it forms a dominant component of the vegetation in many upland areas. Species delimitation, species relationships and generic circumscription within the group remain ambiguous.

Within *Strobilanthes sensu lato* there is considerable variation in pollen morphology. This was first documented by Radlkofer (1883), and subsequently investigated by Lindau (1893, 1895), Clarke (1907), Bremekamp (1944), Vishnu-Mittre and Gupta (1966) Terao (1982, 1983), Scotland (1993) and Carine and Scotland (1998). While the earliest classifications of *Strobilanthes sensu lato* by Nees (1832, 1847), Anderson (1867) and Clarke (1885) used gross morphological characters exclusively, subsequent attempts to classify the group have made use of variation in pollen morphology.

Lindau (1895), Clarke (1907) and Ridley (1923) recognised two pollen types in *Strobilanthes sensu lato*, and distinguished genera on that basis. Lindau (1895) distinguished *Lamiacanthus* Kuntze and *Pseudostenosiphonium* Lindau with '*stachelpollen*' (spheroidal pollen with spines) from *Strobilanthes* which he characterised by *'rippenpollen'* (ellipsoidal pollen with longitudinal ribs). Clarke (1907) and Ridley (1923) distinguished *Acanthopale* C. B. Clarke with spheroidal pollen from *Strobilanthes* with ellipsoidal pollen. However, Lindau's sampling of pollen morphology in the group was limited, and both *rippenpollen* and *stachelpollen* are present in *Strobilanthes sensu* Lindau (Bremekamp 1944; Vishnu-Mittre and Gupta 1966; Terao 1982, 1983; Scotland 1993). Further, the conceptualisation of pollen morphology as two shape classes by Lindau (1895), Clarke (1907) and Ridley (1923) failed to accurately describe the diversity of pollen morphology within *Strobilanthes* (Carine and Scotland,1998).

Bremekamp (1944) split *Strobilanthes sensu lato* into 54 segregate genera based on pollen morphology and other morphological characters. In contrast to other classifications of the group incorporating pollen morphology, Bremekamp's system has been widely followed in many Flora accounts of the region (see Carine and Scotland,1998). However, in reviewing the pollen morphology of the southern Indian and Sri Lankan species of *Strobilanthes,* Carine and Scotland (1998) highlighted several inadequacies in Bremekamp's classification:

- The diversity of pollen morphology is much greater than Bremekamp described. While Bremekamp recognised seven pollen types in the southern Indian and Sri Lankan *Strobilanthes,* Carine and Scotland (1998) recognised twenty two.
- Bremekamp's classification was not comprehensive. Of the 74 species of *Strobilanthes* in southern India and Sri Lanka, 22% (16 species) were left unaccounted for by Bremekamp (*Genus adhuc incertum*).
- Bremekamp's descriptions were inaccurate. The pollen morphologies of fourteen species of *Strobilanthes* from southern India and Sri Lanka were incorrectly described, and their taxonomic position in the Bremekamp system is therefore highly questionable.

Earlier classifications, irrespective of the data used, have proved unsatisfactory for a number of reasons. Classifications were often proposed based on limited sampling: Bremekamp, by his own admission, had not seen material of many species of *Strobilanthes* because of the political situation at the time (Bremekamp 1944: 10-11); Lindau (1893, 1895) surveyed the pollen morphology of a limited number of species of *Strobilanthes.* Without adequate sampling, the nature and distribution of character variation cannot be assessed accurately. Thus, the distribution of Lindau's (1893, 1895) pollen types in *Strobilanthes sensu lato* does not correspond with the delimitation of the genera he proposed, even though pollen morphology was the 'character' used for generic delimitation. Descriptions of the variation have proven inaccurate, as Carine and Scotland (1998) showed for Bremekamp (1944). Finally, earlier classifications have relied on *a priori* weighting of characters. *A priori* character weighting was adopted to delimit groups in *Strobilanthes* because there is no obvious congruence between characters. This is indicative of homoplasy, which has effectively concealed species relationships and thus inhibited attempts to understand intuitively the patterns of character distribution and the delimitation of groups.

Parsimony analysis provides an explicit methodology with which patterns of character distribution and character congruence may be explored. In this paper, cladistic analyses of pollen and other morphological characters in the southern Indian and Sri Lankan *Strobilanthes sensu lato* are presented, the patterns of pollen character distributions in the group discussed, and the taxonomic significance of pollen data evaluated.

Strobilanthes sensu lato from southern India and Sri Lanka is taken to include *Strobilanthes* and the long-recognised genus *Stenosiphonium* Nees. *Stenosiphonium* has traditionally been distinguished from *Strobilanthes* solely on the basis of ovule number -

Stenosiphonium has 6-8 ovules in each ovary, while *Strobilanthes* has 2-4. The southern Indian and Sri Lankan species of *Strobilanthes sensu lato* investigated in this paper form a putative monophyletic group, geographically distinct from the rest of *Strobilanthes sensu lato*. None of Bremekamp's (1944) genera or groups comprise both southern Indian species and species from the rest of *Strobilanthes sensu lato*, and Wood (1994) considered that, 'The *Strobilanthes* flora of southern India [and Sri Lanka] is rich and varied but has no connection with that of the Himalaya'. The monophyly of the southern Indian and Sri Lankan *Strobilanthes sensu lato* is currently under investigation. Therefore, in all analyses of character distribution in the southern Indian and Sri Lankan *Strobilanthes sensu lato* presented in this paper the cladograms are unrooted.

Pollen morphological variation

Carine and Scotland (1998) recognised twenty two pollen types in the southern Indian and Sri Lankan *Strobilanthes* to aid identification.

The pollen types recognised may be grouped into two shape classes, ellipsoidal (Figs. 1-9) and spheroidal (Figs. 10-17). Ellipsoidal grains have P/E ratios in the range 1.3 – 1.8. All are tricolporate and have pseudocolpi between the apertures which divide the grain into a number of longitudinal ribs (for example Fig. 1). The number of longitudinal ribs is variable between species, and in some cases within species.

The longitudinal ribs of most species with ellipsoid pollen coalesce at the poles (Fig. 1). In *S. anceps* Nees, *S. lupulina* Nees, *S. perrottetiana* Nees, *S. punctata* Nees, *S. tristis* (Wight) T. Anderson and *S. zeylanica* T. Anderson, however, the ribs completely encircle the poles (Fig. 2), although in *S. lupulina* this character is polymorphic, with the longitudinal ribs of some grains from a single anther coalescing at the poles and others completely encircling the poles.

Species with ellipsoidal pollen show variation in the ornamentation of the longitudinal ribs between pseudocolpi. The ribs of many species have a punctate tectum lacking tectal ornamentation (Fig. 1). Tectal ornamentation, when present, may take the form of a continuous strip along the centre of each rib (Fig. 3), which in some species has prominent side projections (Fig. 4), discontinuous aggregations along the centre of the rib (Fig. 5), coarse reticulation (Fig. 6), or spines (Fig. 7).

The Sri Lankan endemic *Strobilanthes stenodon* C. B. Clarke is polymorphic for spines. *Strobilanthes stenodon* has a disjunct distribution with populations in Ritigala and Matale. Pollen grains from the Ritigala population are unornamented (Fig. 8), while those from Matale, have spines along the longitudinal ribs (Fig. 9). The two populations also differ in other morphological characters and are treated as separate taxa in this paper pending a revision of the taxonomy of this species (in preparation).

All spheroidal pollen grains have spines. The spheroidal pollen types vary in the structure of the sexine, and in the number, distribution and type of aperture. The sexine in some grains comprises prominently raised areas supported by columellae, with a perforate tectum, and terminating in a spine (Fig. 10). The raised areas are discrete in some species, while in others the columella layer is greatly reduced between the raised areas. In other species, the columella layer is of a continuous thickness (Fig. 11), or slightly undulating.

The number of apertures in species with spheroidal pollen grains is variable. The majority have three apertures, but four-, five- and seven-aperturate grains are also found. The apertures are equatorially distributed, or distributed over the entire grain (pantoaperturate). Ectoapertures may be porate (Fig. 12), brevicolpate (Fig. 13), colpate (Fig. 14) or cryptoaperturate (Fig. 15).

Spine morphology varies considerably. For example, the spines of *S. gardneriana* (Nees) T. Anderson (Fig. 14) are short (c. 3μm) and rounded; those of *S. exserta* C.B. Clarke (Fig. 15) short (c. 3 –6μm), amorphous and one to several are present on each raised area; those of *S. lurida* Wight (Fig. 16) are long (c. 15μm) and tapering to a rounded point; spines of *S. calycina* Nees (Fig. 17) are c. 6.5μm long, prominently striated and tapering from a broad base.

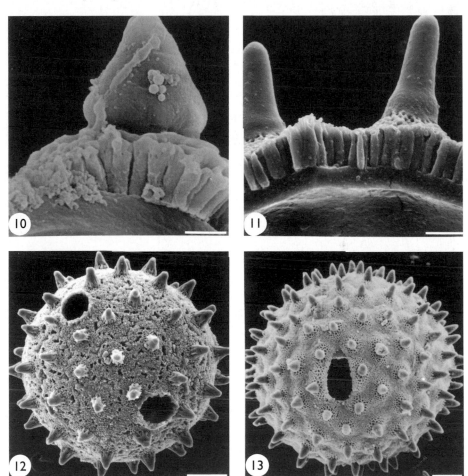

FIGS. 10–13. *Strobilanthes* pollen morphology. Fig. 10. *Strobilanthes viscosa* (Arnott ex Nees) T. Anderson, sexine section (scale bar = 1 μm). Fig. 11. *Strobilanthes lurida* Wight, sexine section (scale bar = 3 μm). Fig. 12. *Strobilanthes laxa* T. Anderson (scale bar = 10 μm). Fig. 13. *Strobilanthes rhamnifolia* (Nees) T. Anderson (scale bar = 10 μm)

FIGS. 1–9. *Strobilanthes* pollen morphology. Scale bar = 10 μm. Fig. 1. *Strobilanthes neilgherrensis* Beddome, mesocolpial view. Fig. 2. *Strobilanthes anceps* Nees, polar view. Fig. 3. *Strobilanthes helicoides* (Nees) T. Anderson, mesocolpial view. Fig. 4. *Strobilanthes sexennis* Nees, mesocolpial view. Fig. 5. *Strobilanthes anamallaica* J.R.I. Wood, mesocolpial view. Fig. 6. *Strobilanthes pulcherrima* T. Anderson, mesocolpial view. Fig. 7. *Strobilanthes urceolaris* Gamble, mesocolpial view. Fig. 8. *Strobilanthes stenodon* C.B. Clarke, Ritigala population (*nov. sp.*, see text), mesocolpial view. Fig. 9. *Strobilanthes stenodon*, Matale population, mesocolpial view

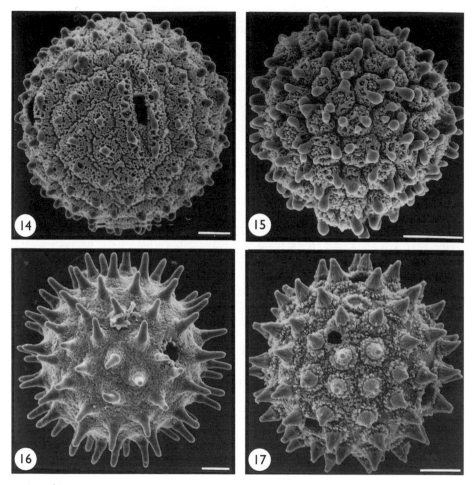

FIGS. 14–17. *Strobilanthes* pollen morphology. Scale bar = 10 μm. Fig. 14. *Strobilanthes gardneriana* (Nees) T. Anderson. Fig. 15. *Strobilanthes exserta* C.B. Clarke. Fig. 16. *Strobilanthes lurida* Wight. Fig. 17. *Strobilanthes calycina* Nees.

Full descriptions of the pollen morphologies of the southern Indian and Sri Lankan *Strobilanthes* are given in Carine and Scotland (1998).

Systematic analysis

Matrix design

Thirty two morphological characters (Appendix 1), including eight pollen characters (characters 1 - 8) have been coded for sixty eight species of *Strobilanthes sensu lato* from southern India and Sri Lanka. The following species have not been included in the analysis due to lack of pollen material and/or herbarium material: *S. canarica* Beddome, *S. circarensis* Gamble, *S. dupeni* Beddome ex C.B. Clarke, *S. hypericoides* J.R.I. Wood, *S. minor* Talbot, *S. membranacea* Talbot, *S. pentandra* J.R.I. Wood, *S. thwaitesii*

T. Anderson, and *S. warreensis* Dalzell. A full list of taxa, authorities and vouchers for pollen material is given in Carine and Scotland (1988).

Twenty seven of the characters scored are binary characters and five are multi-state characters, of which three (characters 2, 4 and 8) have four states and two (characters 17 and 22) have three states. Thus, there are 40 apomorphic character states.

The character coding strategy follows that advocated by Hawkins *et al.* (1997) for characters that may be absent or present, and if present may include further discrete character state variation. These were coded as two characters rather than as a multistate character. Thus, tectal ornamentation, which may be absent, or present as spines, a solid strip, a discontinuous strip or reticulation, is scored as two characters: character 7, tectal ornamentation presence/absence and character 8, type of tectal ornamentation. Taxa lacking tectal ornamentation are scored inapplicable for character 8.

An initial matrix was compiled by the first author. The characters and coding protocol for a sample of the total taxa were checked against specimens and dissections by the second author and Jon Bennett, who also works on *Strobilanthes* systematics. This process highlighted several problematic characters which were subsequently eliminated from the matrix to leave the final matrix comprising 68 taxa and 32 characters. Characters such as bract shape, used by earlier authors, cannot be divided into discrete character states when the variation across the whole of the southern Indian and Sri Lankan species of *Strobilanthes sensu lato* is taken into account. Such characters have not been included in the matrix. Only meristic characters and continuous characters that can be divided into discrete states have been included. The data matrix is given in Appendix 2.

Cladistic analysis

Parsimony analyses using PAUP version 3.1.1 (Swofford 1993) were performed on the following:
1. The complete data set (characters 1 – 32).
2. All characters excluding pollen characters (characters 9 – 32).
3. Pollen characters only (characters 1 – 8).
All analyses were performed with character states unordered.

1. Analysis of complete data set

Three species were excluded from the analyses as redundant taxa (*S. lanata*, *S. nigrescens*, *S. pulcherrima*) with identical coding to other species (*S. kunthiana*, *S. diandra* and *S. rubicunda* respectively).

The search strategy, based on that of Catalán *et al.* (1997) was as follows:
1. A search compising 5000 random-order-entry starting trees (TBR, MULPARS), saving no more than ten trees at each replicate of length more than or equal to 2 steps was performed. This search resulted in 150 trees of length 148 steps.
2. To find all trees compatible with the consensus tree from (1), the strict consensus tree was loaded as a constraint, and a search consisting of 5000 replicates of random addition sequence performed, (TBR, MULPARS), saving all trees *compatible* with the constraint tree. 6862 trees of length 148 steps were saved. One of the 6862 optimal trees is presented in Fig. 18, with the redundant taxa that were excluded from the analysis incorporated. Nodes not present in the strict consensus tree are indicated by arrows.
3. To search for other islands of equally parsimonious trees, or for islands of shorter trees, a third round of searches was implemented. The strict consensus tree from the second round was loaded as a constraint, and a search consisting of 5000

replicates of random addition sequence performed (TBR, MULPARS), saving not more than 10 trees at each replicate of length equal to, or more than, 2 steps and *not* compatible with the constraint tree. Thus trees equal in length but not compatible with the constraint from (1), or trees shorter than those found in (1) could be recovered by this search. This search failed to find any trees of length 148 steps or less (shortest tree = 149 steps). From this, we concluded that the search implemented in (1) had discovered all islands of optimal trees, and that the trees recovered in (2) represented all most parsimonious trees.

2. Analysis excluding pollen data

An analysis was performed using characters 9–32 only. Five taxa were excluded as redundant taxa (*S. nigrescens* T. Anderson, *S. lanata* Nees, *S. lurida*, *S. micrantha* Wight, *S. rubicunda* (Nees) T. Anderson), with identical coding to other species (respectively: *S. diandra* (Nees) Alston, *S. kunthiana* (Nees) T. Anderson ex Bentham, *S. laxa* T. Anderson (both *S. lurida* and *S. micrantha*), and *S. pulcherrima* T. Anderson).

The search strategy implemented in PAUP was identical to that used in the analysis of the total data set. 672 trees of length 99 steps were recovered. The strict consensus tree is presented in Fig. 19.

3. Analysis of pollen data

Analysis of pollen characters only (characters 1–8) was performed using the branch and bound search option in PAUP. Fifteen species were included in the analysis. All other species (53) were excluded from the analyses as redundant taxa, with identical coding to one of the 15 included species.

5439 most parsimonious trees of length 18 steps were found. The strict consensus, incorporating redundant taxa excluded from the analysis, is presented in Fig. 20. Taxa that were included in the analysis are highlighted in bold text.

Discussion

Comparison of data sets

Figure 20 represents the strict consensus tree from the analysis performed using pollen characters only (characters 1-8). As can be seen from the cladogram, the species can be separated about the branch marked 'A' into those with spheroidal pollen grains, and those with ellipsoidal pollen grains. The unrooted tree from parsimony analysis of pollen data alone is therefore compatible with the classifications of Lindau (1895), Clarke (1907) and Ridley (1923) whose major divisions within *Strobilanthes sensu lato* were based upon pollen shape class.

However, the results of the analysis of pollen data are incongruent with the results from the analysis of the total data set (Fig. 18). Only two of the clades present in Fig. 20 are present in the strict consensus of the analysis of the total data set (Fig. 18). These are indicated by solid circles. In both cases, the clades comprise taxa with identical coding. Thus, the pollen data alone cannot be used to estimate the topology of the trees obtained using a combination of micro- and macromorphological characters.

Figure 19 shows the strict consensus tree from the analysis of characters 9-32 (i.e. excluding pollen). In the strict consensus from this analysis there are 12 nodes that are also present in the strict consensus from the analysis of the total data set (see Fig. 18). These nodes are indicated by solid circles. However, while there is similarity between the two results, there are also significant differences. For example, while group A is

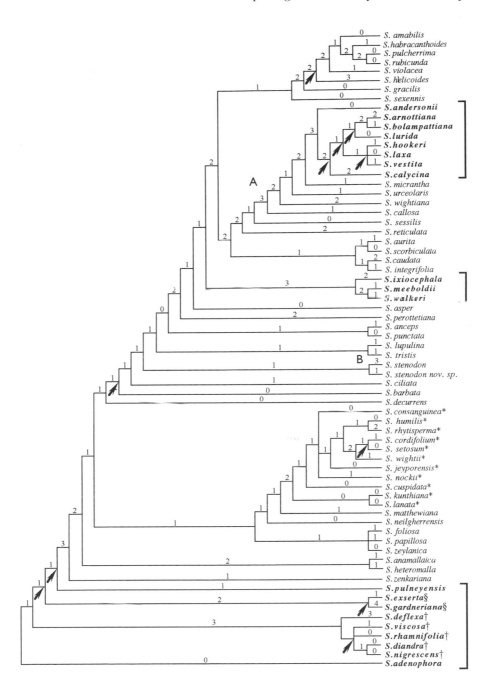

FIG. 18. One of 6862 optimal unrooted trees of length 148 steps from the analysis of the total data set (characters 1-32). Arrows indicate nodes not present in the strict consensus tree. Taxon names in bold denote taxa with spheroidal pollen grains. Square brackets [ ], denote clades in which all species have spheroidal pollen grains. Letters A and B refer to clades discussed in the text. Symbols *, § and † denote taxa from each of three clades.

sister to group B in Fig. 19, in the analysis of the total data set, the two groups are widely separated (see Fig. 18). Group C in Fig. 19 represents an amalgam of three clades distinguished in Fig. 18 (the symbols *, †, and § distinguish taxa from each of the three clades present in Fig. 18, that constitute the single group C in Fig. 19). Thus, in common with the pollen character sub-set, the sub-set of macromorphological characters fails to accurately estimate the topology of the tree obtained using the total data set. The combination of pollen and macromorphological characters in the analysis of the total data set resulted in most parsimonious trees in which the majority of nodes were not present in the results from either sub-set analysed individually.

The pattern of distribution of pollen characters

On Fig. 18, taxa with spheroidal pollen are indicated in bold text, and those with ellipsoidal pollen in plain text. Square brackets, [ ], indicate groups distinguished by their spheroidal pollen shape. As can be seen from Fig. 18, three groups are distinguished by spheroidal pollen shape. Thus, while this character state is homoplastic, it does have taxonomic value in delimiting groups.

Figure 21 illustrates the distribution of all pollen characters (characters 1-8) on the most parsimonious tree depicted in Fig. 18. From this, it can be seen that pollen characters are all homoplastic, but do delimit groups. Thus, character 7, the presence/absence of tectal ornamentation, maps on to the cladogram at four nodes (indicated by arrows on Fig. 21). Character 5, columella layer continuous/discontinuous maps on to the cladogram twice (indicated by *); the presence of a continuous columella layer defines one clade, but is also present in *S. meeboldii* Craib.

As can be seen from Table 1, c.i. for all characters ranges from 0.111 to 1.00, and r.i. ranges from 0.375 to 1.00. Pollen characters (characters 1 - 8) have c.i. values ranging from 0.250 to 0.600. All pollen characters are therefore homoplastic. However, from Table 1 it can also be seen that pollen characters have r.i. values which range from 0.667 to 0.889. Thus, whilst all pollen characters are homoplastic, they do contain significant levels of synapomorphy.

Tree statistics

The tree presented in Fig. 18 is one of 6862 optimal trees from the analysis of the complete data set (characters 1-32). This result was obtained from an extensive set of heuristic searches involving 10, 000 random order replicates. Arrows on Fig. 18 represent nodes absent in the strict consensus tree. In a fully resolved cladogram, the number of nodes is $n - 2$, where n is the number of taxa. In a fully resolved cladogram of 68 taxa, there are, therefore, 66 nodes. In the strict consensus of the analysis of the total data set (see Fig. 18), there are 51 nodes: the strict consensus can be said to be 77% resolved.

The high degree of resolution for sixty eight taxa and thirty two characters is surprising. However, the sixty eight taxa are represented in the matrix by sixty five unique combinations of characters. This suggests a high level of character conflict, a characteristic of the variation has thwarted previous attempts to understand intuitively the patterns of character distribution and the relationships of taxa in the group.

In the recent debate surrounding the use of molecular methods to resolve phylogenies, Hillis (1996) showed that complex phylogenies could be inferred by parsimony, even if the phylogenetic signal is weak. He suggested that this was dependent upon homoplasy (or 'noise') being distributed across the many branches of the tree and co-varying patterns of homoplasy between any two taxa being relatively rare. From his simulation studies using molecular data, he concluded that incorporating large numbers of terminal taxa may increase the accuracy of the estimated trees.

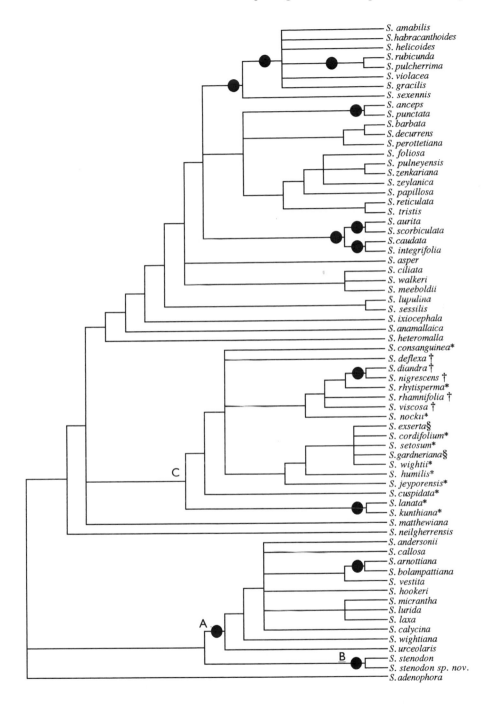

FIG. 19. Strict consensus tree of 672 optimal unrooted trees of length 99 steps from the analysis of characters 9-32 (i.e. excluding pollen data). Solid circles indicate nodes also present in the strict consensus from the analysis of the total data set. Letters A, B, and C refer to clades discussed in the text. Symbols *, § and † distinguish taxa from each of three clades present in Fig 18.

TABLE 1. c.i. and r.i. values for pollen characters from the analysis of the total data set.

CHARACTER	Tree steps	c.i.	r.i.
1. Pollen shape	3	0.333	0.889
2. Pollen aperture number	5	0.600	0.667
3. Pollen aperture distribution	3	0.333	0.667
4. Ectoaperture type	7	0.429	0.733
5. Columella layer continuity	2	0.500	0.883
6. Columella layer structure	3	0.333	0.818
7. Presence of tectal ornamentation	4	0.250	0.880
8. Type of tectal ornamentation	5	0.600	0.857
9. Sessile gland dots	1	1.000	1.000
10. Flower pedicels	1	1.000	1.000
11. Presence of bracteoles	6	0.167	0.375
12. Presence of secondary flower buds	3	0.333	0.714
13. Calyx division	4	0.250	0.893
14. Calyx lobe inequality	3	0.333	0.500
15. Calyx venation	7	0.143	0.600
16. Calyx apex	3	0.333	0.500
17. Corolla shape	6	0.333	0.875
18. Presence of twist in corolla tube	6	0.167	0.722
19. Presence of hairs to retain style	2	0.500	0.500
20. Length of row of hairs to retain style	8	0.125	0.611
21. Hairs to retain style borne on papillae	3	0.333	0.600
22. Stamen number	8	0.250	0.750
23. Staminal sheath pubescence	1	1.000	1.000
24. Anterior stamens exserted/included	8	0.125	0.750
25. Posterior stamens included/exserted	4	0.250	0.667
26. Anterior filament pubescence	5	0.200	0.733
27. Posterior filament pubescence	5	0.200	0.826
28. Style pubescence	14	0.071	0.552
29. Presence of second stigmatic lobe	5	0.200	0.500
30. Seed pubescence	9	0.111	0.692
31. Seed hair type	1	1.000	-
32. Presence of a seed areola	3	0.333	0.833

FIG. 20. Strict consensus of 5439 optimal unrooted trees of length 18 from the analysis of pollen characters (characters 1-8). ◯ indicates taxa with ellipsoidal pollen; ◑ indicates taxa with spheroidal pollen. 'A' indicates the branch separating species with ellipsoidal pollen from species with spheroidal pollen; ● indicates modes also present in the strict consensus from the analysis of the total data sets. Only taxa in bold were included in the analysis. All other taxa have identical coding to one of the 15 taxa indicated in bold.

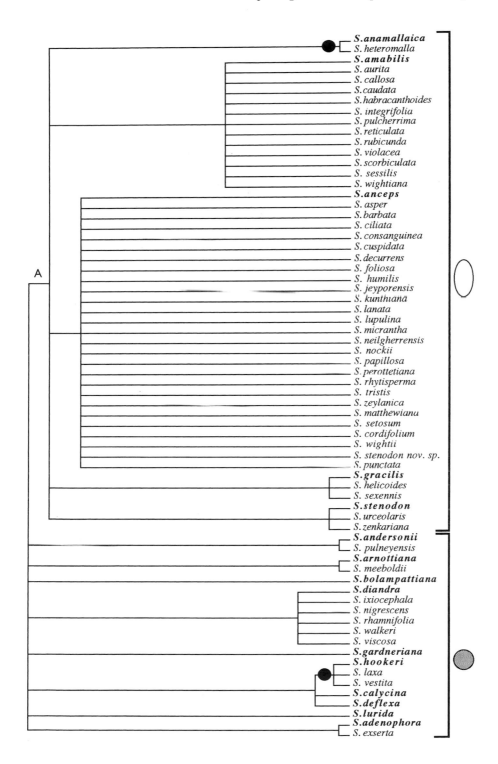

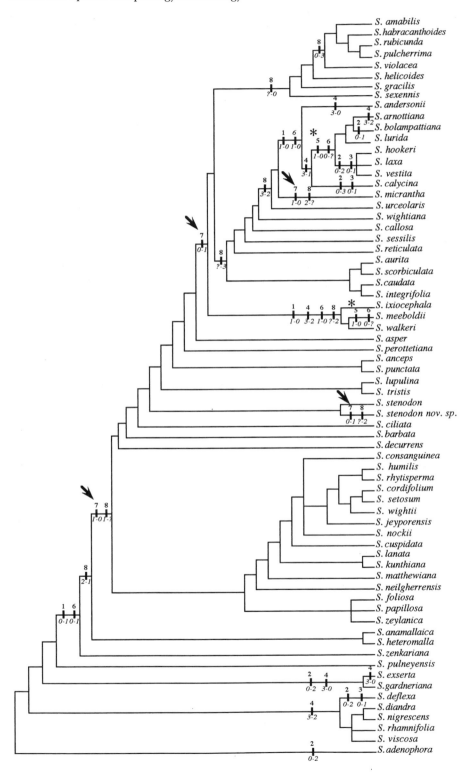

The results of the total combined analysis presented in Fig. 18, are based upon dense sampling of *Strobilanthes* from southern India and Sri Lanka (68 species out of 77). The strict consensus is well resolved, showing 77% resolution, even though 88% of the characters are homoplastic as evidenced by their c.i. values. Hillis (1996) suggested that a weak phylogenetic signal can be detected above high levels of noise (homoplasy). In our results, however, the homoplastic characters (the 'noise') are actually providing the 'phylogenetic signal'. More recently, and again in the context of molecular systematics, Hillis (1998) commented that, 'inclusion of many taxa in a densly sampled tree permits more effective use of rapidly evolving characters than in a poorly sampled tree'. Our results show that in morphological analyses, dense sampling similarly permits homoplastic characters to be used effectively in resolving species relationships.

In a cladistic analysis, the optimality criterion is parsimony. The results presented in this paper represent the optimal result for the given data, using this criterion. However, while cladistic analysis is not a statistical method, 'support' for the results of an analysis is often sought. Hillis' (1996) study was a simulation study - the true phylogeny was known. In the case of the results presented in this paper, however, it is pertinent to ask what confidence we have in the accuracy of the results.

Several approaches to measuring statistical support for cladistic analyses have been developed (see Kitching, 1998, for a review). Support for individual clades is often assessed with techniques such as bootstrap and jackknife (Felsenstein 1985, 1988). Only the clade comprising (*S. cordifolium, S. setosum, S. wightii*) was present in the 50% majority rule tree from a jackknife analysis performed on the total data set using Jac (Farris *et al.*, 1996).

In contrast to jackknife and bootstrap, the PTP test (Faith and Cranston, 1991) looks for support of the most parsimonious cladogram as a whole by comparing the length of the optimal tree against the length of trees produced using randomised data sets. A PTP-test performed on the data set showed that there was significant hierarchical structure (100 randomised matrices; $p=0.01$).

Thus, the PTP test suggests that the data set contains significant hierarchical structure. However, the jackknife analysis found only one clade present in more than 50% of the pseudoreplicates.

However, statistical methods applied to phylogeny reconstruction have theoretical and practical limitations. The lack of jackknife support for these results is not surprising, and is a function of the size of the data set and the level of homoplasy. The interpretation of a highly significant PTP test value is problematic in light of recent criticisms of the method by Carpenter *et al.* (1998) who showed that the PTP test can attribute high significance to data that support no resolved grouping. None of the statistical methods developed to assess confidence in the results of parsimony analysis can place meaningful confidence limits onto cladograms (Kitching, 1998), and meaningful support for the result presented can only come from corroboration with other data.

FIG. 21. One of 6862 optimal unrooted trees of length 148, from the analysis of the total data set showing the distribution of pollen character states. Bars represent character changes. The number above the bar is the character number. The numbers below the bar, in italics, represent the character state in taxa below that branch (first) and above that branch (second). ? = Inapplicable character. Arrows indicate character state transformations for character 7 (presence/absence of tectal ornamentation). * indicate transformations of character 5 (columella layer continuous/discontinuous).

Conclusions

Pollen characters contributed twenty five percent of the characters (8 of 32) of the total data set in this analysis of 68 species of *Strobilanthes sensu lato* from southern India and Sri Lanka.

While pollen characters provided a significant proportion of the total data set, pollen characters alone do not provide enough characters to provide a well resolved topology for the group - the 68 taxa are represented in the matrix by only 15 unique combinations of characters. The analysis of pollen data alone (characters 1-8) is also incongruent with the analysis of the total data set, as are results of the analysis of macromorphological characters (characters 9-32). Neither pollen nor macro-morphological sub-sets of characters estimate the result obtained from the analysis of the total data set.

All pollen characters in the analysis of the total data set are homoplastic. However, all pollen characters have significant levels of synapomorphy and are useful for delimiting clades as can be seen in Fig. 21, and from a comparison of c.i. and r.i. values (Table 1).

In the analysis of the total data set presented in this paper, 32 morphological characters were used to investigate the relationships of 68 species of *Strobilanthes sensu lato*. The 32 characters represent the extent of cladistically informative morphological characters in the study group. Other characters that have been used to produce classifications of *Strobilanthes sensu lato* are unsuitable for cladistic analysis, because they do not comprise discrete character states when examined across the range of taxa.

Even with a low character:taxon ratio, and high levels of homoplasy (88% of all characters are homoplastic), the strict consensus of the total data set showed 77% resolution. We suggest that dense sampling has permitted homoplastic characters to contribute effectively to the structure of the tree. Measures of statistical support do not generate meaningful confidence limits for cladograms (Kitching, 1998), and support for the results therefore must be sought by looking for corroboration with other data, for example molecular data.

The taxonomy of large, species-rich taxa, such as *Strobilanthes*, has historically proven problematic because of high levels of character conflict. However, the results presented in this paper suggest there is the potential to resolve species relationships in this group using relatively few, rigorously examined, morphological characters.

Acknowledgements

We thank the conference organisers for inviting us to speak at the meeting, Jon Bennett for checking the data matrix, Dick Olmstead for suggestions on search strategies and Dick Olmstead and Dennis Stevenson for useful reviews on the manuscript. The first author is supported by BBSRC and the second by the Royal Society.

References

Anderson, T. (1867). An enumeration of the Indian species of Acanthaceae. *Journal of the Linnean Society* 9: 425–526.

Bremekamp, C.E.B. (1944). Materials for a monograph of the Strobilanthinae (Acanthaceae). *Verhandelingen der Nederlandsche Akademie van Wetenschappen, Afdeeling Natuurkunde, Tweede Sectie* 41: 1–305.

Carine, M.A, and Scotland, R.W. (1998). Pollen morphology of *Strobilanthes* Blume (Acanthaceae) from southern India and Sri Lanka. *Review of Palaeobotany and*

Palynology 103: 143–165.

Carpenter, J.M., Goloboff, P.A. and Farris, J.S. (1998). PTP is meaningless, T-PTP is contradictory: a reply to Trueman. *Cladistics* 14: 105–116.

Catalán, P., Kellogg, E.A. and Olmstead, R.G. (1997). Phylogeny of Poaceae subfamily Pooideae based on chloroplast *ndh*F gene sequences. *Molecular Phylogenetics and Evolution* 8: 150–166.

Clarke, C.B. (1885). Acanthaceae. In: Hooker, J.D. Flora of British India, vol 4, pp 387–558. Reeve & Co., London.

Clarke, C.B. (1907). Acanthaceae. In: King, G. and Gamble, J.S. Materials for a Flora of the Malayan Peninsula, pp 838–908. West, Newman & Co., London.

Farris, J.S., Albert, V. A., Källersjö M., Lipscomb, D. and Kluge, A.G. (1995). Parsimony jackknifing outperforms neighbor-joining. *Cladistics* 12: 99–124.

Faith, D.P. and Cranston, P.S. (1991). Could a cladogram this short have arisen by chance alone? On permutation tests for cladistic structure. *Cladistics* 7: 1–28.

Felsenstein, J. (1985). Confidence limits on phylogenies: an approach using the bootstrap. *Evolution* 39: 783–791.

Felsenstein, J. (1988). Phylogenies from molecular sequences: inference and reliability. *Annual Review of Genetics* 22: 521–565.

Hawkins, J.A., Hughes, C.E. and Scotland, R.W. (1997). Primary homology assessment, characters and character states. *Cladistics* 13: 275–283.

Hillis, D.M. (1996). Inferring complex phylogenies. *Nature* 383: 130–131.

Hillis, D.M. (1998). Taxonomic sampling, phylogenetic accuracy, and investigator bias. *Systematic Biology* 47: 3–8.

Kitching, I.L. (1998). Support and confidence statistics for cladograms and groups. In: Kitching, I., Forey, P., Humphries, C. and Willliams, D. Cladistics (Second Edition), pp 118–138. Oxford Univertsity Press, Oxford.

Lindau, G. (1893). Beiträge zur systematik der Acanthaceen. *Botanische Jahrbücher* 18: 36–64.

Lindau, G. (1895). Acanthaceae. In: Engler, A. and Prantl, K. (editors), Die Naturlichen Pflanzenfamilien, vol 4, part 3b, pp 274–354. Engleman, Leipzig.

Nees von Esenbeck, C.G.D. (1832). Acanthaceae. In: Wallich, N. Plantae Asiaticae Rariores vol 3, pp 70–117. London.

Nees von Esenbeck, C.G.D. (1847). Acanthaceae. In: de Candolle, A. (Editor), Prodromus Systematis Naturalis Regni Vegetabilis, vol 11, pp 46–519. Par etc., Paris.

Radlkofer, L. (1883). Über den systematischen werth der pollen-beschaffenheit bei den acanthaceen. *Sitzungsberichte der Bayerischen Akademie der Wissenschaften zu München. Mathematisch-physikalische Klasse* 13: 256–314.

Ridley, H.N. (1923). Flora of the Malay Peninsula, vol 2. Reeve & Co., London.

Scotland, R.W. (1993). Pollen morphology of Contortae (Acanthaceae). *Botanical Journal of the Linnean Society* 111: 471–504.

Swofford, D.L. (1993). Phylogenetic analysis using parsimony, version 3.1.1. Computer programme distributed by the Illinois Natural History Survey, Champaign, Illinois.

Terao, H. (1982). Observations on the echinulate pollen of *Strobilanthes* s.l. and its allies. *Acta Phytotaxonomica et Geobotanica* 33: 371–379.

Terao, H. (1983). Taxonomic study of the genus *Strobilanthes* Bl. (Acanthaceae). Generic delimitation and infrageneric classification. Unpublished thesis submitted for the degree of PhD at Kyoto University, Kyoto, Japan.

Vishnu-Mittre and Gupta, H.P. (1966). Contribution to the pollen morphology of the genus *Strobilanthes* Blume, with remarks on its taxonomy. *Pollen et Spores* 8: 285–307.

Wood, J.R.I. (1994). Notes relating to the Flora of Bhutan XXIX: Acanthaceae, with special reference to *Strobilanthes*. *Edinburgh Journal of Botany* 51: 191–269.

APPENDIX 1. List of characters and character states

1. Pollen shape spheroidal (0); ellipsoidal (1).
2. Pollen apertures three (0); four (1); five (2); seven (3).
3. Pollen apertures equatorially distributed (0); panto-aperturate (1).
4. Ectoapertures absent (cryptoaperturate) (0); porate (1); brevicolpate (2); colpate (3).
5. Columella layer continuous (0); discontinuous (1).
6. Columella layer forming discrete islands (0); forming logitudinal rows (1); inapplicable (-).
7. Tectal ornamentation (sexine 3) absent (0); present (1).
8. Tectal ornamentation (sexine 3) comprising a solid strip (0); discontinuous strip (1); spines (2); reticulation (3); inapplicable (-).
9. Sessile gland dots on the underside of the leaf absent (0); present (1).
10. Flowers sessile (0); pedicellate (1).
11. Bracteoles absent (0); present (1).
12. Bracteoles without secondary flower buds in axils (0); bracteoles with secondary flower buds in axils (1); inapplicable (-).
13. Calyx lobes united for at least one third of their length (0); free almost basally (1).
14. Calyx lobes equal or sub-equal in length (0); calyx with one lobe longer (1).
15. Calyx venation linear (0); reticulate (1).
16. Calyx lobe apex acute (0); obtuse (1).
17. Corolla ventricose from a narrow tube (0); campanulate from a broad tube (1); curved and widening from the base (2).
18. Corolla tube not twisted at anthesis (0); corolla tube twisted at anthesis (1).
19. Corolla without hairs to retain the style (0); corolla with hairs to retain the style (1).
20. Corolla with a short row of hairs to retain style (less than half of the distance between the lip of the corolla and the stamen sheath (0); corolla with a long row of hairs to retain style (greater than 60% of the distance between the lip of the corolla and the stamen sheath) (1); inapplicable (-).
21. Hairs to retain style not borne on corolla papillae (0); hairs to retain style borne on corolla papillae (1).
22. Stamens two (0); two with two infertile staminodes (1); four (2).
23. Staminal sheath glabrous (0); pubescent (1).
24. Anterior stamens included (0); exserted (1).
25. Posterior stamens included (0); exserted (1); inapplicable (-).
26. Anterior filaments glabrous (0); pubescent (1).
27. Posterior filaments glabrous (0); pubescent (1); inapplicable (-).
28. Style without simple hairs (0); style with simple hairs (1).
29. Stigma without a short projection on the posterior surface (0); stigma with a short projection on the posterior surface (1).
30. Seeds glabrous, exareolate (0); seeds hairy, areolate (1).
31. Seed hairs not hygroscopic (0); seed hairs hygroscopic (1); inapplicable (-).
32. Seed areola absent (0); seed areola present (1); inapplicable (-).

APPENDIX 2. Data matrix

S. adenophora	02001	01200	10000	00011	02110	10??0	-1
S. amabilis	10031	11301	10110	12010	02100	10001	00
S. anamallaica	10031	11100	10001	00011	02100	1100?	??
S. anceps	10031	10-10	10100	01011	02100	11000	—
S. andersonii	00001	01200	10101	01011	020??	0000?	??
S. arnottiana	00020	-1200	10100	0100-	02011	00100	—
S. asper	10031	10-00	10101	01011	02100	11101	00
S. aurita	10031	11300	10100	01011	02100	00011	00
S. barbata	10031	10-00	10000	01011	02110	11000	—
S. bolampattiana	01010	-1200	10101	0100-	02011	00100	—
S. callosa	10031	11300	0-101	01011	02010	00001	01
S. calycina	03111	01200	10101	01011	02010	00000	—
S. caudata	10031	11300	10100	00010	02100	00111	00
S. ciliata	10031	10-00	10100	01011	02111	1100?	??
S. consanguinea	10031	10-00	10000	00110	0010-	1-?01	01
S. cuspidata	10031	10-00	10000	00010	0010-	1-001	01
S. decurrens	10031	10-00	10000	01011	02100	11000	—
S. deflexa	02121	01200	10000	00110	0010-	1-001	01
S. diandra	00021	01200	10000	00110	0011-	1-100	—
S. exserta	02001	01200	11000	00111	12110	11001	01
S. foliosa	10031	10-00	10000	00111	01100	11100	—
S. gardneriana	02031	01200	11000	00011	10110	10000	—
S. gracilis	10031	11001	10100	02010	02100	11101	00
S. habracanthoides	10031	11301	0-010	12010	02100	10011	00
S. helicoides	10031	11001	11110	12011	02100	11111	00
S. heteromalla	10031	11100	10000	00011	02100	11101	01
S. hookeri	02110	-1200	10101	11011	02011	00000	—
S. humilis	10031	10?00	11000	00110	0011-	1-101	01
S. integrifolia	10031	11300	10110	00011	02100	10111	00
S. ixiocephala	00021	01200	10101	00011	02100	11101	01
S. jeyporensis	10031	10?00	11000	00110	0010-	1-001	01
S. kunthiana	10031	10?00	10000	00010	0010-	1-101	01
S. lanata	10031	10?00	10000	00010	0010-	1-101	01
S. laxa	02110	-1200	10101	01011	02011	00000	—
S. lupulina	10031	10?00	0-100	01011	02100	11101	01
S. lurida	00010	-1200	10101	01011	02011	00000	—
S. meeboldii	00020	-1200	10101	01011	02111	1110?	??
S. micrantha	10031	10-00	10101	01011	02011	00000	—
S. neilgherrensis	10031	10-00	10000	00011	02100	10100	—
S. nigrescens	00021	01200	10000	00110	0011-	1-100	—
S. nockii	10031	10-00	10000	00110	0010-	1-000	—
S. papillosa	10031	10-00	10100	00111	02100	1110?	??
S. perrottetiana	10031	10-00	10001	01011	02100	11110	—
S. pulcherrima	10031	11301	0-110	02010	02100	10111	00
S. pulneyensis	00001	01200	10000	00111	02110	11100	—
S. punctata	10031	10-10	10100	01011	02100	1110?	??
S. reticulata	10031	11300	0?100	00011	01100	1010?	??
S. rhamnifolia	00021	01200	10000	00110	0011-	1-00?	??
S. rhytisperma	10031	10?00	0-000	00110	0011-	1-100	—
S. rubicunda	10031	11301	0-110	02010	02100	10111	00
S. scorbiculata	10031	11300	10100	01011	02100	10011	00
S. sessilis	10031	11300	0-101	01011	02100	10101	01
S. sexennis	10031	11000	10100	02011	02100	11101	00
S. stenodon	10031	11200	0-100	01010	0011-	1-100	—
S. tristis	10031	10?00	0-100	00011	02100	11100	—
S. urceolaris	10031	11200	10100	01011	02010	00100	—
S. vestita	02110	-1200	10101	0100-	02011	00000	—
S. violacea	10031	11301	10100	12010	02100	11001	00
S. viscosa	00021	01200	10000	00110	0011-	0?000	—

S. walkeri	00021	01200	10100	01011	02111	11101	10
S. wightiana	10031	11300	10101	01010	02000	00100	—
S. zeylanica	10031	10-00	10000	00111	02100	1110?	??
S. zenkariana	10031	11200	10000	00111	02110	11010	—
S. matthewiana	10031	10-00	10000	00111	02100	10101	01
S. cordifolium	10031	10-00	11000	00111	12110	10001	01
S. setosum	10031	10-00	11000	00111	11110	10001	01
S. wightii	10031	10-00	11000	00111	1011-	0-001	01
S. stenodon nov.sp.	10031	10-00	1-100	01011	1011-	1-100	—

Jamzad, Z, Harley, M.M., Ingrouille, M., Simmonds, M.S.J., and Jalili, A. (2000). Pollen exine and nutlet surface morphology of the annual species of *Nepeta* L. (Lamiaceae) in Iran. In: M.M. Harley, C.M. Morton and S. Blackmore (Editors). Pollen and Spores: Morphology and Biology, pp. 385–397. Royal Botanic Gardens, Kew.

POLLEN EXINE AND NUTLET SURFACE MORPHOLOGY OF THE ANNUAL SPECIES OF *NEPETA* L. (LAMIACEAE) IN IRAN

Z. JAMZAD[1], M.M. HARLEY[2], M. INGROUILLE[3], M.S.J. SIMMONDS[4] AND A. JALILI[1]

[1]Research Institute of Forests and Rangelands, PO Box 13185-116, Tehran, Iran
[2]The Herbarium, Royal Botanic Gardens, Kew Richmond Surrey, TW9 3AE, UK
[3]Biology Dept, Birkbeck College, University of London, WC1E 7HX
[4]The Jodrell Laboratory, Royal Botanic Gardens, Kew Richmond Surrey, TW9 3DS, UK

Abstract

The Old World genus *Nepeta* L. has about 300 species of which eight of the annual species are Iranian endemics. Existing infra-generic and specific classifications are unsatisfactory and mostly artificial. This new study aims to establish a more natural classification based on phylogenetically revealing characters. An SEM study of pollen morphology has been undertaken to evaluate the use of exine ornamentation as a taxonomic character, and as a character for later use in phylogenetic analyses of the whole genus. *Nepeta*, like other members of subfamily Nepetoideae Erdtman, has hexacolpate pollen. All the annual species studied have a bi-reticulate exine. The lumina formed by the primary muri may be polygonal, more or less rounded or elongate. Apocolpial primary lumina are smaller than those of the mesocolpia. The apocolpial exine in both groups is bi-reticulate; in one group the muri are shallow and less clearly defined than the deeper, more prominent lumina of the other group. To some extent the size of the apocolpial, and of the mesocolpial, lumina differs within species. In the pollen of *Nepeta amoena* Stapf. an unusual apomorphy occurs: three narrow mesocolpia, with a perforate-reticulate exine, alternate with three wide mesocolpia with a bi-reticulate exine.

Introduction

The genus *Nepeta* L. consists of about 300, mainly herbaceous, perennial and annual species which are distributed only in the Old World, from the Pacific Ocean to the Atlantic. Most are perennials, but there are about 30 annual species. The centre of diversity of the annual species is in Iran, with eight endemic species. Other species, although occurring in Iran, are distributed more widely, reaching into neighbouring regions such as Transcaucasia, Central Asia, Afghanistan, Pakistan and Iraq.

Annual species usually occupy foothill to alpine habitats, but rarely deserts, at an altitude of 200-3800m, mainly in the Irano-Turanian Phytogeographical Region, with a few extending to the Hyrcanian subprovince of the Euro-Siberian Region. There is no disjunction between the two sections. Species within Section *Micronepeta* Benth. are found in the Provinces of Northeast, West, Central and Southeast Iran. Species of Section *Micranthae* (Boiss.) Pojark. Occupy the same provinces, with a few also occurring in the North, Northwest and Southwest Provinces (Fig. 1).

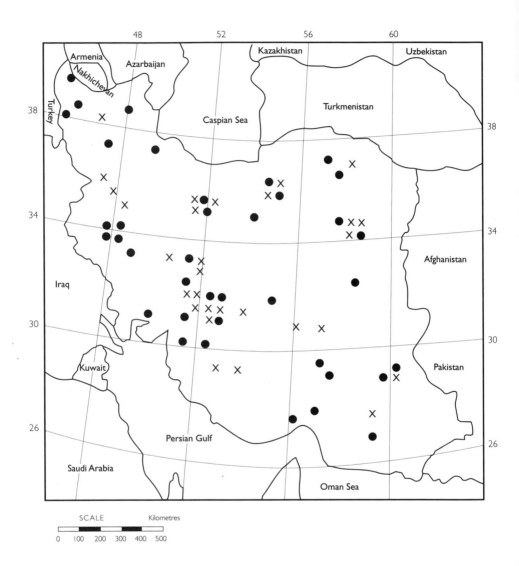

FIG. 1. Distribution map of annual species of *Nepeta* in Iran [• = sect. *Micranthae*, **X** = sect. *Micronepeta*].

Most of the existing classifications, such as those of Bentham (1834, 1848), Boissier (1879), Briquet (1895-97), Pojarkova (1954), Rechinger (1982), Hedge (1990) and Budantsev (1993), are based on morphological characters but seem artificial. Recent authors have divided the annual species into two sections (Table 1).

TABLE 1. Calyx, corolla and nutlet characters according to the sectional classification of the annual species of *Nepeta*.

	Section Micrantheae (Boiss.) Pojark.	Section Micronepeta Benth.
Calyx	With oblique throat (two-lipped)	with straight throat
Corolla	Without a swelling at the base of the middle lobe of the lower lip	Middle lobe of the lower lip with a swelling at the base
Nutlets	Surface coarsely tuberculate to finely granular	Surface smooth

Here we report the results of a study of pollen exine morphology and nutlet surface morphology in the Iranian annual species of *Nepeta*. These data will form part of a broader revision of the genus, in which the micromorphological, chemical and molecular data will be utilised, and phylogenetic relationships established using cladistic analyses.

Pollen morphological studies for Lamiaceae have been carried out by several authors, for example: Erdtman (1945), Wunderlich (1967), Henderson *et al.* (1968), Varghese (1986), Bassett and Munro (1986), Abu-Asab (1990), Abu-Asab and Cantino (1992), Harley (1992), Harley *et al.* (1992), Wagstaff (1992), Ranjbar, H., Dianat-Nejad, H. and Jamzad, Z. (1998), Khandani, S., Dianat-Nejad, H. and Jamzad, Z. (1998). As far as we are aware no previous morphological or taxonomic accounts for the pollen of *Nepeta* have been published. However, a few of authors have described the pollen of one or two species of *Nepeta*, either in general accounts of angiosperm pollen or, more specifically, of Lamiaceae pollen. These are summarised by Wunderlich (1967).

A comprehensive survey of the nutlet morphology in tribe Nepeteae was published by Budantsev and Bobova (1997).

Materials and Methods

Pollen and nutlets were removed from herbarium specimens in the collections of the Research Institute of Forests and Rangelands, Tehran (TARI), and the Royal Botanic Gardens, Kew (K). Two collections were sampled for each species. Untreated nutlets were mounted on stubs using double-sided adhesive tape, they were then sputter coated. Pollen was acetolysed and prepared for light (LM) and scanning electron microscopy (SEM) using methods fully described in Harley *et al.* (1992). For LM measurements means and ranges of 10–15 pollen grains were taken, with a few exceptions where pollen was too collapsed or damaged to find more than five measurable grains. Both nutlets and pollen were examined using an Hitachi S2400 SEM. All collections examined are listed in Appendix 1.

TABLE 2. Pollen data for the Iranian annual species of *Nepeta*. **Column 1**: Species; **Column 2**: Collector(s); **Column 3**: Polar length; **Column 4**: Equatorial width; **Column 5**: Shape (P/E); **Column 6**: Appearance of apocolpial exine; **Column 7**: Size of apocolpial primary lumina in relation to size of mesocolpial lumina; **Column 8**: Shape of lumina in mesocolpium; **Column 9**: Size comparison of mesocolpial lumina between species examined; **Column 10**: Width of the six mesocolpia.

1	2	3	4	5	6	7	8	9	10
Section Micronepeta									
Nepeta bracteata	Froughi 17276	(27-) 32.0 (-38)	(30-) 37.0 (-43)	0.86	PmP	smaller	round	similar	similar
Nepeta bracteata	Assadi & Mozaffarian 32699	(28-) 33.0 (-43)	(36-) 38.0 (-43)	0.86	PmP	smaller	round	similar	similar
Nepeta daenensis	Archibald 2758	(34-) 36.0 (-37)	(32-) 35.0 (-40)	1.02	PmP	smaller	round	similar	similar
Nepeta daenensis	Assadi & Mozaffarian 36866	(35-) 36.0 (-38)	(35-) 37.0 (-40)	0.97	PmP	smaller	round	similar	similar
Nepeta pungens	Bazargan & Arazm 24829	(26-) 28.0 (-29)	(32-) 32.0 (-32)	0.87	PmN	smaller	round	similar	similar
Nepeta pungens	Mozaffarian 64940	(35-) 37.0 (-43)	(38-) 39.0 (-40)	0.94	PmN	smaller	round	similar	similar
Nepeta satureioides	Assadi & Mozaffarian 35559	(36-) 36.0 (-36)	(36-) 36.0 (-36)	1.00	PmP	similar	round	similar	similar
Nepeta satureioides	Runemark et al. 19458	(32-) 33.0 (-34)	(36-) 36.0 (-36)	0.91	PmP	similar	round	similar	similar
Section Micrantheae									
Nepeta amoena	Wendelbo & Assadi 27882	(30-) 33.0 (-35)	(36-) 36.0 (-36)	0.91	PmN	smaller	round	similar	altern
Nepeta amoena	Sabeti 13072	(29-) 34.0 (-36)	(33-) 36.0 (-39)	0.94	PmN	smaller	round	similar	altern
Nepeta bakhtiarica	Mozaffarian 57433	(21-) 26.0 (-28)	(28-) 29.0 (-30)	0.89	PmP	smaller	round	smaller	similar
Nepeta bakhtiarica	Mozaffarian 57846	(25-) 27.0 (-29)	(29-) 30.0 (-36)	0.90	PmP	smaller	round	smaller	similar
Nepeta mirzayanii	Mozaffarian 42836	(29-) 34.0 (-39)	(29-) 34.0 (-39)	1.00	PmN	smaller	elong	similar	similar
Nepeta mirzayanii	Mozaffarian 42871	(36-) 37.0 (-43)	(36-) 37.0 (-43)	1.00	PmN	smaller	elong	similar	similar
Nepeta ispahanica	Edmonson & Miller 1534b	(29-) 36.0 (-43)	(29-) 36.0 (-43)	1.00	PmN	smaller	round	similar	similar
Nepeta aff. *ispahanica*	Assadi 29249	(30-) 34.0 (-40)	(30-) 36.0 (-42)	0.94	PmN	smaller	poly	similar	similar
Nepeta meyeri	Mozaffarian & Nowrozi 34528	(32-) 36.0 (-40)	(32-) 36.0 (-40)	1.00	PmN	smaller	poly	similar	similar
Nepeta meyeri	Mitchell, C.M. & W. 2895	(29-) 36.0 (-41)	(32-) 37.0 (-43)	0.97	PmN	smaller	poly	similar	similar
Nepeta schiraziana	Mozaffarian 59973	(31-) 33.0 (-35)	(30-) 32.0 (-35)	1.03	PmN	smaller	round	similar	similar
Nepeta schiraziana	Hewer (date 1975)	(33-) 35.0 (-40)	(30-) 32.0 (-35)	1.09	PmN	smaller	round	similar	similar
Nepeta wettsteinii	Runemark Froughi 19935	(35-) 35.0 (-35)	(35-) 35.0 (-35)	1.00	PmP	smaller	poly	similar	similar
Nepeta wettsteinii	B.S.B.E. 1474	(36-) 39.0 (-43)	(36-) 39.0 (-43)	1.00	PmP	smaller	poly	similar	similar

Abbreviations: Column 6: **PmP** = primary muri prominent; **PmN** = primary muri not prominent. Column 7: **smaller** = primary apocolpial lumina smaller than primary mesocolpial lumina; **similar** = primary apocolpial lumina similar in size to primary mesocolpial lumina. Column 8: **round** = rounded; **poly** = polygonal; **elong** = elongate. Column 9: **smaller** = primary lumina smaller in comparison to other species examined; **similar** = primary lumina similar in size to most other species examined. Column 10: **similar** = width of the six mesocolpia similar, **altern** = three wide mesocolpia alternate with three narrow mesocolpia.

388

Results and Discussion

Pollen morphology (summarised in Table 2: columns 3–7)

The pollen grains are hexacolpate (Fig. 8) and isopolar. These features are common to most pollen in species of subfamily Nepetoideae. In pollen of the annual species of *Nepeta* shape ranges from suboblate (P/E 0.86), oblate spheroidal (P/E 0.87 – 0.97), spheroidal to prolate spheroidal (P/E 1.09). Pollen size: polar axis (21–) 33.9 (–43) µm; equatorial axis (28–) 35.2 (–43) µm. In polar view the grains are circular, and the exine is bi-reticulate (see for example, Figs. 12, 16, 18, 20, 23). The exine may be simple perforate at the colpus margins and, occasionally (Fig. 22) the bi-reticulum is so subtle that, to the casual observer, the exine appears simple perforate or reticulate. The simple perforate or reticulate exine occurs in some of the perennial species of the genus (pers. obs.). In all the investigated species the colpus membrane is finely or coarsely granular (for example Fig. 10). In the apocolpium two distinct exine patterns are seen. In one group the pattern is distinctly bi-reticulate with prominent primary muri (Fig. 14), while in the other group the primary muri are less prominent (Fig. 15). Species with prominent primary muri in the apocolpium include: *N. bakhtiarica* Rech.f., *N. bracteata* Benth., *N. daenensis* Boiss., *N. satureioides* Boiss., *N. wettsteinii* H.Braun. Species with less pronounced primary muri in the apocolpium include: *N. ispahanica* Boiss., *N.* aff. *ispahanica*, *N. mirzayanii* Rech.f. & Esfand., *N. meyeri* Benth., *N. amoena* Stapf., *N. pungens* (Bunge) Benth., *N. schiraziana* Boiss.

Differences in shape, size and number of perforations in the lumina are apparent among different species. In most species studied the primary lumen size in the apocolpium is smaller than in the mesocolpium, although similar lumen size in the apocolpium and mesocolpium has been observed in *N. satureioides*. The lumina of the primary muri may be irregular to rounded (Fig. 17), polygonal (Fig. 20), or elongated (Fig. 19). A polygonal lumen shape in the primary reticulum is observed in: *N. meyeri*, *N.* aff. *ispahanica* and *N. wettsteinii*. In *N. bakhtiarica*, *N. ispahanica*, *N. pungens*, *N. schiraziana* and *N. satureioides* the lumina are rounded, and in *N. mirzayanii* they are elongated. In the pollen of *N. bakhtiarica* the lumina of the mesocolpial primary reticulum are small in comparison to other species (Fig. 17).

An interesting apomorphy for *Nepeta amoena* is observed in its pollen. Three narrow mesocolpia with a perforate-reticulate exine, alternate between three wide mesocolpia with a bi-reticulate exine (Fig. 21). The apocolpial ornamentation is similar to that of the narrow mesocolpia (Fig. 21). This type of pollen has previously been recorded in the genus *Endostemon* (Paton *et al.* 1995), although in *Endostemon* the exine ornamentation is similar for both the narrow and the wide mesocolpia.

The pollen morphology does not reflect the geographical distribution of the species within Iran (Fig. 1 and Appendix 2). Neither does the pollen show any correlation with altitude and ecological preferences of the Iranian species (Appendix 1).

Nutlet morphology (summarised in Table 2: column 8)

Three contrasting patterns of nutlet surface morphology were observed in the annual species (Figs. 2–7). In group 1 there are prominent tubercles on both ventral and dorsal surface (Figs. 2–3). In group 2 an intermediate condition is observed, with less prominent tubercles (Figs. 4–5). In group 3 the surface was more or less smooth but with a shallow reticulate pattern (Figs. 6–7). Species in groups 1 and 2 are found in Section *Micranthae*, while species in group 3 occur in Section *Micronepeta*. Full data on nutlet morphology for species of *Nepeta* that occur in Iran are in preparation (ZJ).

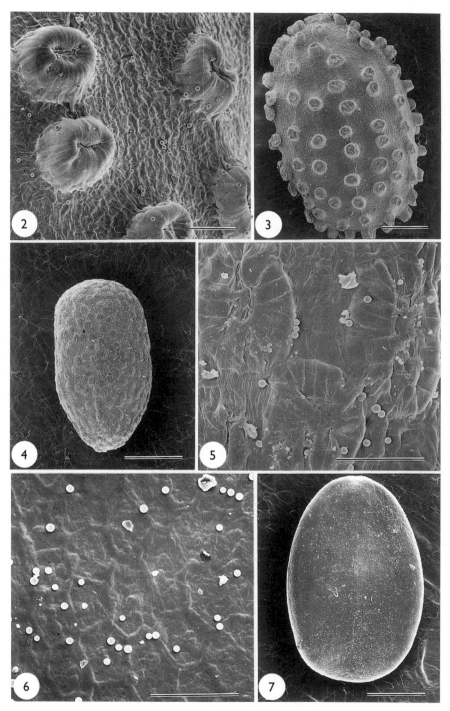

FIGS. 2–7. Nutlets of three different species of *Nepeta* showing three patterns of surface morphology: 3, 4, 7, abaxial view; 2, 5, 6, nutlet surface detail. (2-3, *N. schiraziana*, Froughi 17415; 4-5, *N. ispahanica* Assadi & Bazgosha 56572; 6-7, *N. satureioides* Jamzad & Taheri 69559. [scale bars: 2, 5-6 = 50μm; 3-4, 7 = 500μm]

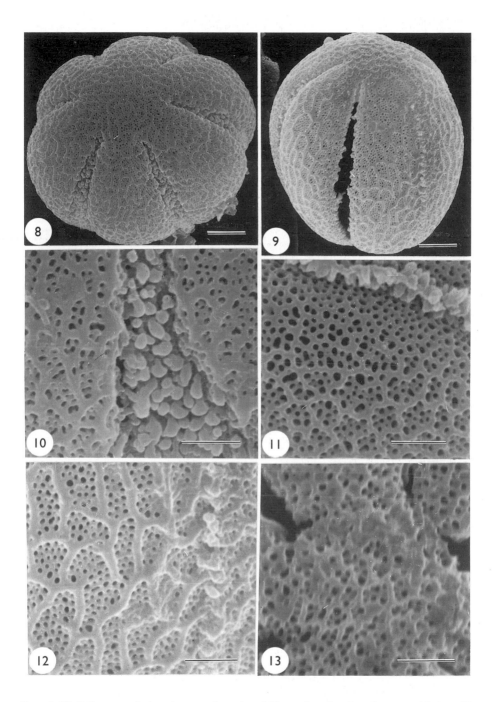

FIGS. 8–13. Pollen morphology in annual species of *Nepeta*: 8, polar view; 9, equatorial view; 10, colpus membrane; 11, mesocolpium; bi-reticulate exine, mesocolpium; 13, apocolpium. (8, 10, 13, *N. satureioides* Assadi & Mozaffarian 35559; 11, *N. bakhtiarica* Mozaffarian 57433; 9, 12. *N. wettsteinii* Runemark & Froughi 19935). [scale bars: 8-9 = 5μm; 10-13 = 2.5μm]

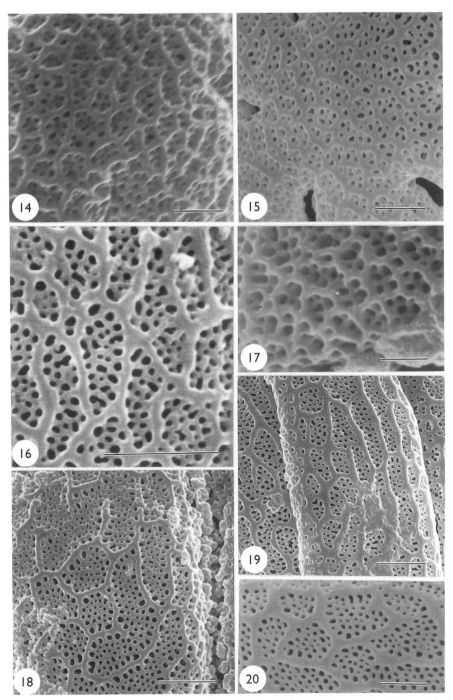

FIGS. 14–20. Apocolpium and mesocolpium of the pollen of annual species of *Nepeta*: 14-15, apocolpium; 16-20; mesocolpium. (14, *N. bracteata* Froughi 1726; 15, *N. ispahanica* Edmondson & Miller 15346; 16, *N. satureioides* Assadi & Mozaffarian 35559; 17, *N. bakhtiarica* Mozaffarian 57433; 18, *N. schiraziana* Hewer 1975; 19, *N. mirzayanii* Mozaffarian 42871; 20, *N.* aff. *ispahanica* Assadi 23249). [scale bars: 14-20 = 2.5μm]

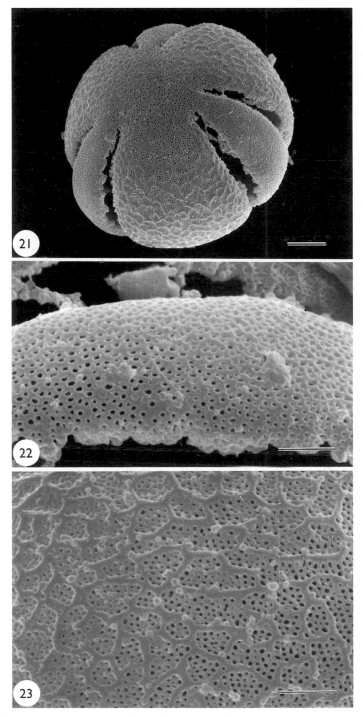

FIGS. 21–23. Pollen of *Nepeta amoena* Wendelbo & Assadi 27882: 21, polar view; 22, mesocolpium, narrow colpus; 23, mesocolpium, wide colpus. [scale bars: 21 = 5μm; 22-23 = 2.5μm]

Conclusions

This study has shown that exine morphology is not congruent with the existing sectional classification of the species. Members of both sections exhibit both types of apocolpial exine (Table 2) and, furthermore, there are other variations within and overlapping between the two sections. Pollen characteristics do not correlate with either the geographical distribution (Fig. 1), or the ecology, of the species. The existing sectional classification is congruent with the morphology of nutlet, corolla and calyx morphology. The lack of sectional correlation shown by pollen characters suggests that the pollen morphology has not evolved in parallel with the other morphological characters that have been examined. The data, particularly for *N. amoena*, indicate that wider sampling of the pollen of the annual species of *Nepeta* should be undertaken prior to cladistic analysis.

Acknowledgements

We would like to acknowledge the receipt of a grant from the Islamic Republic of Iran, Ministry of Jahad-e-sazandegi to ZJ. We also wish to thank the Research Institute of Forests and Rangelands (TARI), and the Herbarium of the Royal Botanic Gardens, Kew (K) for permission to remove pollen and nutlets from herbarium specimens and our reviewers for their helpful comments.

TABLE 3. Nutlet data for the three species examined in the present study

Species	Collection data	Nutlet length (mm)	Nutlet Width (mm)
Nepeta schiraziana	Fars, Shiraz, Dasht-e Arjan, old road to Kazeron, First Kotal, 2100m, Froughi 17415 (TARI)	2.1	0.9
N. satureioides	Fars, 5km on the road from Eghlid to Sourmagh, 2100- 2150m, Jamzad & Taheri 69559 (TARI)	1.9	1.2
N. ispahanica	Esfahan, 10km to Natanz, Ardestan (WT4), 1550m, Assadi & Bazgosha 56572 (TARI)	1.5	0.9

References

Abu-Asab, M.S. (1990). Phylogenetic implications of pollen morphology in subfamily Lamioideae (Labiatae) and related taxa. Ph.D Thesis. Ohio University, Athens, Ohio.

Abu-Asab, M.S. and Cantino, P.D. (1992). Pollen morphology in subfamily Lamioideae (Labiatae) and its phylogenetic implications. In: R.M. Harley and T. Reynolds (editors). Advances in labiate science, pp. 97–112, Royal Botanic Gardens, Kew.

Bassett, I.J. and Munro, D.B. (1986). Pollen morphology of the genus *Stachys* (Labiatae) in North America, with comparisons to some taxa from Mexico, Central and South America and Eurasia. *Pollen et Spores* **28**: 279–96.

Bentham, G. (1834). *Nepeta*. In: Labiatae genera et species, pp. 464–489. Ridgeway and Sons, London.

Bentham, G. (1848). *Nepeta*. In: A.P. de Candolle (editor). Prodromus systematis naturalis regni vegetabilis. Volume **12**, pp. 370–396. V. Masson, Paris & L. Michelsen, Leipzig.

Boissier, E. (1879). *Nepeta*. In: Flora orientalis. **4**, pp. 637–670. Genevae et Basileae.

Briquet, J. (1895-97). *Nepeta*. In: A. Engler and K. Prantl (editors). Die natürlichen pflanzenfamilien. **4.3a**, pp. 235–238. Engelman, Leipzig.

Budantsev, A.L. (1993). A synopsis of the genus *Nepeta* (Labiatae). *Bot. Zhurn.* **78**: 93–107, [In Russian].

Budantsev, A.L. and Lobova, T.A. (1997). Fruit morphology, anatomy and taxonomy of tribe Nepeteae (Labiatae). Edinburgh Journal of Botany **54,2**: 183–216.

Erdtman, G. (1945). Pollen morphology and plant taxonomy. IV. Labiatae, Verbenaceae, and Avicenniaceae. *Svensk Bot. Tidskr.* **39**: 279–285.

Harley, M.M. (1992). The potential value of pollen morphology as an additional taxonomic character in subtribe Ociminae (Ocimeae: Nepetoideae: Labiatae). In: R.M. Harley and T. Reynolds (editors). Advances in labiate science, pp. 125–138, Royal Botanic Gardens, Kew.

Harley, M.M., Paton, A., Harley, R.M. and Cade, P.G. (1992). Pollen morphological studies in tribe Ocimeae (Nepetoideae: Labiatae): I. *Ocimum* L. *Grana* **31**: 161–176.

Hedge, I.C. (1990). *Nepeta*. In: S. Ali and Y.J. Nasir (editors). Flora of Pakistan **192**, pp. 59–117. Karachi.

Henderson, D.M., Prentice, H. and Hedge, I.C. (1968). Pollen morphology of *Salvia* and some related genera. *Grana Palynologia* **8**: 70–85

Paton, A., Harley, M.M. and Weeks, S. (1995). A revision of the genus *Endostemon* N.E. Br. (Labiatae). *Kew Bulletin* **48**: 205–243.

Khandani, S., Dianat-Nejad, H. and Jamzad, Z. (1998), Taxonomic studies of the genus *Nepeta* L. sect. Stenostegiae Boiss., thesis for MS degree, Teachers Training University, Tehran (in Persian, unpublished).

Pojarkova, A.I. (1954). *Nepeta*. In: B.K. Shishkin and S.V. Yuzepchuk (editors). Flora of USSR **20**, pp. 191–293. Moscow and Leningrad.

Rechinger, K.H. (1982). *Nepeta*. In: Flora Iranica **150**, pp. 108–216. Akademische Druck-u. Verlagsanstalt, Graz, Austria.

Ranjbar,F., Dianat-Nejad, H. and Jamzad, Z. (1998). Taxonomic studies of the genus *Nepeta* L., sect. Cataria Benth., thesis for MS degree, Teachers Training University (in persian, unpublished).

Varghese, T.M., Verma, D.P.S. (1983). Pollen morphology of some Indian Labiatae. *Journal of Palynology* **4**: 77–83.

Wagstaff, S.J. (1992). A phylogenetic interpretation of pollen morphology in Tribe Mentheae (Labiatae). In: R.M. Harley and T. Reynolds (editors). Advances in labiate science, pp. 113–124. Royal Botanic Gardens, Kew.

Wunderlich, R. (1967). Ein vorschlag zu einer natürlichen gliederung der Labiaten auf grund der pollenkörner, der samenentwicklung und des reifen samens. *Österr. Bot. Z.* **114**: 383–483.

Appendix 1. Collection data and ecology of specimens examined

Species	Collection data	Ecology
Nepeta bracteata	Esfahan, Shahreza, ca. 60 km on road to Shiraz, Mt. Dombalan, 2000m, Froughi 17276 (TARI) and Tehran, Souleghan Valley, 1750m, Assadi & Mozaffarian 32699 (TARI)	Rocky stony slopes, serpentine mountains, sometimes with *Pistacia*, sandy river beds, at an altitude of 1300-3800m.
N. daenensis	Esfahan, Balehsun, between Damaneh and Khunsar, 3350m, Archibald 2758 (K) and Hamadan, ca. 8km E of Ganjnameh, 2750m, Assadi & Mozaffarian 36866 (TARI)	Rocky stony or soil slopes; in steppe vegetation, at an altitude of 2000-3350m
N. pungens	Tehran, SW of Latyan Dam, 2000m, Bazargan & Arazm 24829 (TARI) and Hamadan, Faminin, Ghorveh, Mts W. of Karafs, 2000-2600m, Mozaffarian 64940 (TARI)	Rocky stony slopes, gypsum and limestone soils, margin of salt lake and slopes of river banks at an altitude of 1200-3000m.
N. satureioides	Khorassan, 9km from Kashmar to Neyshabour, 1300m, Assadi & Mozaffarian 35559 (TARI) and Khorassan, Mts ca,. 4km S of Birjand, 1600-1900m, Runemark *et al.* 19458 (TARI)	Dry rocky stony slopes, gypsum soils at an altitude of 1060 -2200m.
N. amoena	Azarbayejan, between Meshkinshahr and Ahar, Now-Douz, 1100m, Wendelbo & Assadi 27882 (TARI) and Gilan Manjil, 1350m, Sabeti 13072 (TARI)	Stony pebbly slopes, foothills, in steppe vegetation at an altitude of 300-1900m.
N. bakhtiarica	Bakhtiari, road from Shahre Kord to Naghan, N. of Soulegan, Mt. Shapournaz, 2100m, Mozaffarian 57433 (TARI) and Bakhtiari, top region of Mt Saldaron, from Deh Cheshmeh, W slope 2200m, Mozaffarian 57846 (TARI)	Mountain slopes at an altitude of 1500-2350m.

N. mirzayanii	Balouchestan, 18km from Khash to Iranshahr, road to Irandegan,1500m, Mozaffarian 42836 (TARI) and Balouchestan, 55km from Khash to Iranshahr, 1500m, Mozaffarian 42871 (TARI)	Steep rocky slopes and open gravelly deserts at an altitude of 800-1500m.
N. ispahanica	Yazd, Darreh Bid, 30km from Mehriz to Bafgh, 2000m, Edmondson & Miller, 1534b (TARI) and Kerman, between Mahan and Kerman, 1900m, Assadi 23249 (TARI)	Stony hillsides, plain borders, flat roadside ground, deserts, at salt lake margins, at an altitude of 700-1750m
N. meyeri	Azarbayejan, 14km from Namin to Chulandareh Sofla, Anbaran, towards Germi, 1600m, Mozaffarian & Nowrozi 34528 (TARI) and Azarbayejan, Urmieh (Rezaeyeh), Mugul Dag, 20km, SE of Shapour, 1400m, Mitchell, C.M. & W. 2895 (K)	Rocky stony slopes, in steppe vegetation and in sandy gravelly cultivated plains at an altitude of 800-2250m
N. schiraziana	Mt. Zagros, 1960m, Hewer [date 1975] (K) and Bakhtiari, Sabz Kuh, 2350m, Mozaffarian 59973 (TARI)	Rocky slopes at an altitude of 1850-2600m.
N. wettsteinii	Azarbayejan, 10km along the road from Mianeh to Zanjan, 1000m, Runemark & Froughi 19935 (TARI) and Azarbayejan, 10 miles E of Khoy, 1380m, B.S. B.E 1474 (K)	Steep rocky and gravelly slopes, with *Artemisia, Astragalus, Acantholimon* and some annual grasses at an altitude of 1000-2050m.

DeVore, M.L., Zhao, Z, Jansen, R.K. and Skvarla, J. (2000). Utility of trends in pollen morphology for phylogenetic analyses: an example using subfamilies Barnadesioideae and Cichorioideae (Asteraceae). In: M.M. Harley, C.M. Morton and S. Blackmore (Editors). Pollen and Spores: Morphology and Biology, pp. 399–412. Royal Botanic Gardens, Kew.

UTILITY OF TRENDS IN POLLEN MORPHOLOGY FOR PHYLOGENETIC ANALYSES: AN EXAMPLE USING SUBFAMILIES BARNADESIOIDEAE AND CICHORIOIDEAE (ASTERACEAE)

M.L. DeVore[1], Z. Zhao[2], R.K. Jansen[2] and J. Skvarla[3]

[1]Department of Biological and Environmental Sciences, Georgia College and State University, Milledgeville, GA 31061, USA;
[2]Section of Integrative Biology, University of Texas at Austin, TX 78712, USA;
[3]Department of Botany and Microbiology, University of Oklahoma, Norman, OK 73019, USA

Abstract

The subfamily Cichorioideae was recognised as a paraphyletic group within Asteraceae consisting of Arctoteae, Cardueae, Lactuceae, Liabeae, Mutisieae, and Vernonieae. Recent efforts to clarify phylogenetic relationships of the Asteraceae using chloroplast DNA strongly suggest that Cichorioideae should be comprised of Arctoteae, Lactuceae, Liabeae, and Vernonieae. Since pollen data have proved to be congruent with molecular data within Asteraceae, we used trends in pollen morphology within tribes to assess the monophyletic nature of the Arctoteae, Lactuceae, Liabeae, and Vernonieae. Strict consensus of the most parsimonious trees obtained in the analysis produced one nested set of taxa comprised of Arctoteae, Cardueae, Lactuceae, Liabeae, and Vernonieae. The palynological data are somewhat consistent with results from a combined analysis of *ndh*F and restriction site data. Mapping of exine patterns onto the combined cpDNA restriction site and *ndh*F, tree for Arctoteae, Lactuceae, Liabeae, and Vernonieae indicate that the anthemoid pattern is derived within Cichorioideae and the ancestral lineage to the subfamily probably had caveate grains. This supports the hypothesised sister group relationship of the Cichorioideae to Asteroideae. A final point we discuss is the difficulty of using Barnadesioideae as an outgroup for palynological studies of Asteraceae. This difficulty is caused by the great amount of anagenesis within this basal lineage of Asteraceae.

Introduction

The Asteraceae have been the focus of palynological studies for more than a century. Starting with pollen wall studies at the light microscopic level by Fischer (1890), continuing with studies by Wodehouse in the 1920s and Stix (1960), and culminating with scanning and transmission electron microcope surveys of the family (Tomb, 1971; Skvarla *et al.*, 1977; Robinson and Marticorena, 1986; Hansen, 1991; Vezey *et al.*, 1994), palynologists have interpreted pollen morphology in an effort to clarify relationships among and within tribes of Asteraceae. Bolick (1978) summarised the evolutionary and functional significance of Asteraceae pollen ultrastructure and sculpture. Recently, there has been a revived interest in palynological studies of Asteraceae.

Chloroplast DNA (cpDNA) studies over the past ten years are the major reason palynology is experiencing a renaissance in synantherology. Chloroplast DNA restriction site comparisons of 361 genera from all recognised tribes within Asteraceae have been completed (Jansen and Kim, 1996). Sequences of chloroplast genes (*ndh*F and *rbc*L) have been phylogenetically analysed for numerous genera within Asteraceae (Kim *et al.*, 1992; Kim and Jansen, 1995; Bergqvist *et al.*, 1995; Jansen and Kim, 1996). Kim and Jansen (1995) combined cpDNA restriction site data and *ndh*F sequences to assess relationships among subfamilies and tribes within Asteraceae. Results of cpDNA restriction site data and sequences of chloroplast genes have produced relationships supported by pollen data (Jansen *et al.*, 1991; Jansen and Kim, 1996). Pollen data have the potential to help clarify relationships when incongruities exist among molecular and/or morphological phylogenies.

The purpose of this paper is to provide two examples of the value of palynological data in systematic studies of Asteraceae. The first example demonstrates how trends in pollen morphology and ultrastructure can be used as characters for phylogenetic analyses. This method is used to assess the delimitation of subfamily Cichorioideae Kitam. The second example, involving the basal subfamily Barnadesioideae K. Bremer & R.K. Jansen, illustrates the importance of outgroup selection for phylogenetic analyses of Asteraceae pollen data.

Subfamily relationships within Asteraceae

Three subfamilies are currently recognised within Asteraceae: Asteroideae K. Bremer, Barnadesioideae, and Cichorioideae. Barnadesioideae are the smallest subfamily (1 tribe, 9 genera, 92 species) and are found only in South America. This subfamily, which lacks a large cpDNA inversion (Jansen and Palmer, 1987), is the sister group to both the Asteroideae and Cichorioideae and is monophyletic. Pollen ultrastructural features (Skvarla *et al.*, 1977; Hansen 1992; DeVore *et al.*, unpubl.) have been said to support the basal position of the Barnadesioideae based on the similarity between the subfamily and the recognised sister group to Asteraceae, Calyceraceae. In the present paper, we show that other pollen features are highly derived.

The Asteroideae are the largest subfamily within Asteraceae (10 tribes, 1,135 genera, 16,200 species), and like Barnadesioideae, are monophyletic. The monophyly of Asteroideae is supported by numerous morphological and molecular studies. Pollen ultrastructural features, especially the presence of caveate grains, have been used to delimit the subfamily.

Subfamily Cichorioideae, unlike Asteroideae and Barnadesioideae, is paraphyletic and, in the past, has included six tribes, 391 genera, and 6,700 species (Bremer, 1994). The tribal relationships within Cichorioideae have been especially difficult to resolve because of the following four issues: 1) recognition of the sister group to the Asteroideae; 2) resolution of the "vernonioid" complex (Bremer, 1994) or recognition of a monophyletic group that could be recognised as Cichorioideae; 3) the paraphyletic nature of Mutisieae; and 4) the position of problematic genera that cannot be placed within any recognised tribe of the subfamily (Bremer, 1994; Jansen and Kim, 1996). For the focus of this paper, we would like to show how pollen data can be phylogenetically analysed and used to support the delimitiation of a monophyletic group of tribes that could be recognised as Cichorioideae. Secondly, we would like to show some caveats to using Barnadesioideae as the single outgroup for palynological studies in Asteraceae.

Selection of taxa

A phylogenetic analysis was conducted to elucidate any strongly supported monophyletic groups within the Cichorioideae. Our intent was to use pollen data as an independent means of testing morphological and molecular phylogenies for tribal relationships within Cichorioideae. We have used two approaches: 1) cladistic analysis of palynological characters and trends within Cichorioideae and 2) mapping pollen ultrastructure of tribes within Cichorioideae onto a phylogenetic tree based on combined *ndh*F and restriction site data (Jansen and Kim, 1996).

For the cladistic analysis, tribes placed within Cichorioideae, as delimited by Bremer (1994), were selected initially as the ingroup with some modifications. The Tarchonantheae, as recognised by Keeley and Jansen (1991) were added to the ingroup. Also, because the Mutisieae Cass. are paraphyletic, we included the currently recognised subtribes Mutisiinae Less. and Nassauviinae Less. in the analysis (Karis *et al.*, 1992). Subfamily Barnadesioideae was also included in the ingroup. Even though Barnadesioideae are considered the sister group to the rest of Asteraceae, their pollen morphology is highly derived and difficult to compare with pollen features in Asteroideae and Cichorioideae (Zhao *et al.*, unpubl.).

Calyceraceae and Goodeniaceae were used as outgroups for the analysis. Strong support exists for a sister group relationship of Calyceraceae and Asteraceae based on *ndh*F sequence comparisons (Kim and Jansen, 1995) and morphology (Skvarla *et al.*, 1977; Turner, 1977; Hansen, 1992, DeVore and Stuessy, 1995). Goodeniaceae were included as an outgroup since the family has also been suggested as an outgroup to Asteraceae (Michaels *et al.*, 1993; Gustafsson, 1996).

A generalised Asteroideae pollen type was coded and included in the analysis and a complementary review of pollen morphology/ultrastructure of the entire subfamily is needed. The results we present here are preliminary. Conducting a thorough phylogenetic analysis that encompasses the entire Asteraceae is a Herculean task. Clearly, such a data set, along with a revised classification of Mutisieae, will aid in clarifying the relationship of Asteroideae with the other subfamilies of Asteraceae. We are currently working on a synthesis for pollen data for the entire family.

Character Treatment and Selection

Cladistic treatment of pollen data has been difficult because of all the perceived parallelisms that exist in wall architecture. Nowicke and Skvarla (1979) eloquently addressed this issue and believed that there was indeed congruity in wall architecture and this was not a product of parallel evolution. Instead, congruity in wall architecture was a manifestation of long-standing structural similarity dictated by common ancestry. If this is indeed true, then taxa sharing a common ancestry would gain features via a similar, if not identical, series of structural modifications from one shared "ground" pollen type. To test this approach we used a trend as a character instead of viewing each structural modification as characters for phylogenetic analyses. The application of this approach is discussed below for the eight trends selected as "characters" for our analysis. All multistate "characters" are coded as ordered. The character number corresponds with the data matrix included in the appendix.

The issue of how parallelisms should be dealt with in phylogenetic analyses is controversial. Gosliner and Ghiselin (1984), in their study of opisthobranch snails, found that parallelisms provided invaluable insights to evolution of the group. Rasmussen (1983) makes a strong case for not excluding parallelisms from phylogenetic studies. In some groups rampant parallel evolution is often the rule (Gosliner and Ghiselin, 1984). We wanted to explore the use of parallelisms utilising a limited data set from Asteraceae because this appears to be the case with our data.

401

1.) Intercolpar concavities - The presence of intercolpar concavities has been suggested as a synapomorphy for Calyceraceae and Barnadesioideae. If this is the case, intercolpar concavities were lost within the Asteroideae-Cichorioideae lineage. If not, then intercolpar concavities evolved independently in Calyceraceae and subfamily Barnadesioideae. Based on Calyceraceae and Barnadesioideae pollen studies (DeVore *et al.*, unpubl.; Zhao *et al.*, unpubl.), we have found strong evidence to support the independent evolution of intercolpar concavities within Calyceraceae and Barnadesioideae. However, we can look at the trend toward intercolpar concavities within both taxa as a synapomorphy.

2.) Colpar ledges (1) - We define colpar ledges as ridges of exine that line the inner margins of the colpi. Colpar ledges are found in a number of species of Calyceraceae and have been described in the genus *Brunonia* Sm. (Skvarla *et al.*, 1977). The trend towards developing intercolpar concavities serves as a synapomorphy for Calyceraceae and Goodeniaceae.

3.) Thick foot layer (0) - The presence of a thick foot layer (at least 1/3 total exine) is a symplesiomorphy for Calyceraceae and Goodeniaceae.

4.) Thick basal columellae (0) - Thick basal columellae are found in Calyceraceae, Goodeniaceae, and all members of the ingroup with the exceptions of Tarchonantheae and Arctoteae Cass.

5.) Divided columellae (endotectum) - Some pollen types within Asteraceae have columellae divided by an endotectum. Calyceraceae and Goodeniaceae lack endotecta (0). Barnadesioideae lack endotecta, but have crosslinks among columellae (1). Tarchonantheae and Nassauviinae possess endotecta (2), however, some partial crosslinks among the fine columellae composing the upper exine suggest the initiation of a second endotectum in Mutisiinae Less. (3), Cardueae Dumort. and Arctoteae contain some pollen types appearing to lack basal columellae, but possess what we interpret to be a distinctive secondary endotectum (4). Lactuceae, Liabeae, and Vernonieae have distinct thick columellae in the upper exine (5).

6.) Reduction of basal columellae - Calyceraceae, Goodeniaceae, Nassauviinae, and Tarchonantheae have distinct, thick basal columellae. Genera within Barnadesioideae, Cardueae, Lactuceae, Liabeae, Mutisiinae, and Vernonieae exhibit some reduction in basal columellae (1). Finally, Arctoteae possess very few or usually no basal columellae and caveate pollen (2).

7.) Spinulate, spinulose-spiny, spinose, spinose and lophate - Calyceraceae, Goodeniaceae, Barnadesioideae, Nassauviinae, and Tarchonantheae all have grains with spinulose surfaces (0). Mutisiinae pollen is spinulose-spinose (1). Spinose grains are present throughout Cardueae, Arctoteae, and Liabeae (2). Tribes Lactuceae and Vernonieae are characterised by spinose and lophate pollen.

8.) Spheroidal, spheroidal-subprolate, subprolate only, spheroidal-subprolate-prolate - Spheroidal grains (0) are exclusively found in Arctoteae, Cardueae, Lactuceae Cass., Liabeae (Cass.) Rydb., and Vernonieae Cass. Spheroidal and subprolate pollen (1) has been observed in Goodeniaceae, Calyceraceae, and Barnadesioideae. Tarchonantheae possess only subprolate grains (2). Both subtribes of Mutisieae (Mutisiinae and Nassauviinae) have grains ranging from subprolate to prolate (3).

Analysis

Cladistic analysis of Barnadesioideae, Arctoteae, Cardueae, Lactuceae, Liabeae, Mutisieae (Mutisiinae and Nassauviinae), Tarchonantheae, Vernonieae, and outgroup taxa (Calyceraceae and Goodeniaceae) utilised PAUP 3.1.1 (Swofford, 1993). The heuristic search option with tree-bisection-reconnection (TBR) and mulpars (multiple parsimonious trees) options with a simple addition sequence was used to analyse the data set. A second search was completed using the branch and bound option. A strict consensus tree was calculated to elucidate any well-defined topologies.

Bootstrap re-sampling technique of Felsenstein (1985) was not utilised to assess support for particular nodes within a topology of the strict consensus tree. Our concern was identifying topologies shared by all parsimonious trees and comparing those topologies with Jansen and Kim's (1996) combined phylogeny based on restriction site variation and *ndh*F sequences.

Results

Both the branch and bound and heuristic search using Barnadesioideae, Calyceraceae and Goodeniaceae as outgroups yielded 20 trees of 23 steps (consistency index = 0.739, rescaled consistency index = 0.628, homoplasy index = 0.261, and retention index = 0.850). One major inter-nested clade was retrieved from the strict consensus tree consisting of Mutisiinae, Cardueae, Arctoteae, Lactuceae, Liabeae, and Vernonieae and generalised Asteroideae pollen type (Appendix 1).

The phylogenetic analysis based only on pollen data fails to resolve the relationship among the Lactuceae, Liabeae, and Vernonieae clade, Cardueae, and Arctoteae. This may reflect ingroup selection since the ingroup for the analysis was paraphyletic and we only included a generalised pollen type instead of all the tribes within Asteroideae. A second factor may be the uncertain circumscription of Mutisieae. In order to address trends within the subfamily, we mapped the pollen ultrastructure of tribes within Cichorioideae (Fig. 1) on a phylogenetic tree based on combined *ndh*F and restriction site data (Jansen and Kim, 1996).

Discussion

Pollen ultrastructure of Arctoteae, Lactuceae, Liabeae, and Vernonieae

The review of pollen ultrastructure is based on pollen surveys published by Tomb (1975, Lactuceae), Skvarla *et al.* (1977), Keeley (1979, Vernonieae) and Robinson and Marticorena (1986, Liabeae). Our goal is not to review pollen evolution within each tribe, but rather to look for additional pollen features useful for elucidating relationships among all four tribes.

Figure 2 shows wall architecture of tribe Arctoteae. The presence of a cavus, spine channels, and a thin foot layer are evident. Another interesting feature is the connection of the spine area to the foot layer (Fig. 2B). Skvarla *et al.* (1977) viewed the ultrastructural pattern present within the 'arctotoid' tribe as being a modification of the 'senecioid' pattern. Interestingly enough, the Senecioneae belong to one of the basal clades within the Asteroideae (Jansen and Kim, 1996). Also of interest is the presence of a cavus in Arctoteae pollen. The cavus is a feature associated with all tribes of Asteroideae with the exception of Anthemideae. At first, it appears that this feature is a synapomorphy of Arctoteae and Asteroideae. However, this is clearly a reversal since the Arctoteae are linked with Lactuceae, Liabeae, and Vernonieae. This also supports the viewpoint that Lactuceae, Liabeae, Vernonieae, along with Arctoteae, are the sister group to Asteroideae as suggested by the combined *ndh*F and restriction site phylogeny (Jansen and Kim, 1996).

Figure 3 shows a small sample of the exine types found within Lactuceae. Pollen morphology within the tribe is variable and has been the subject of several palynological surveys (Stix, 1960; Tomb, 1972, 1974, 1975; Feuer, 1974). The presence of "cavus-like areas" in Lactuceae has also been addressed (Skvarla *et al.*, 1977). Skvarla and Larson (1956) defined a cavus as being located between apertural areas. As pictured in Fig. 3, short columellae appear to interrupt the "cavus" and are joined with the foot layer. The connections are more pronounced where ridges and spines are accentuated.

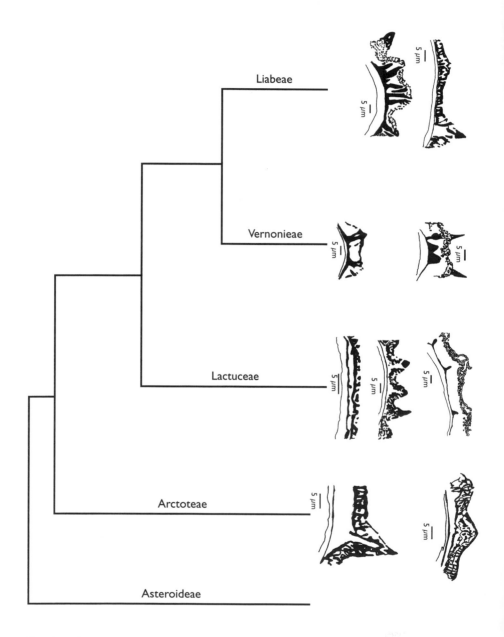

FIG. 1. Exine patterns of Arctoteae, Lactuceae, Liabeae, and Vernonieae mapped on the combined *ndh*F and chloroplast DNA phylogeny of Jansen and Kim (1996). Note that the anthemoid pattern is derived and the presence of a cavus is primitive within this clade.

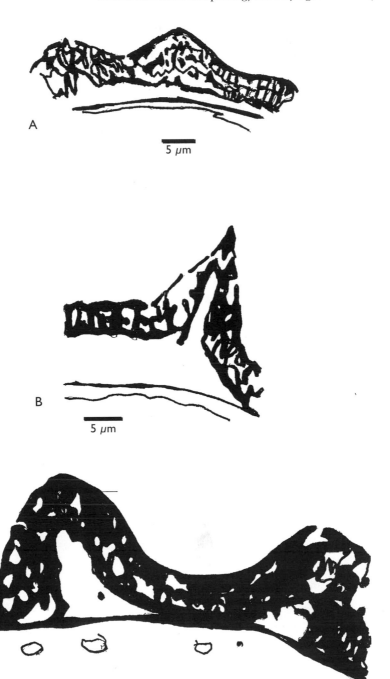

FIG. 2. Pollen exine patterns found in Arctoteae. (After Skvarla *et al.*, 1977).

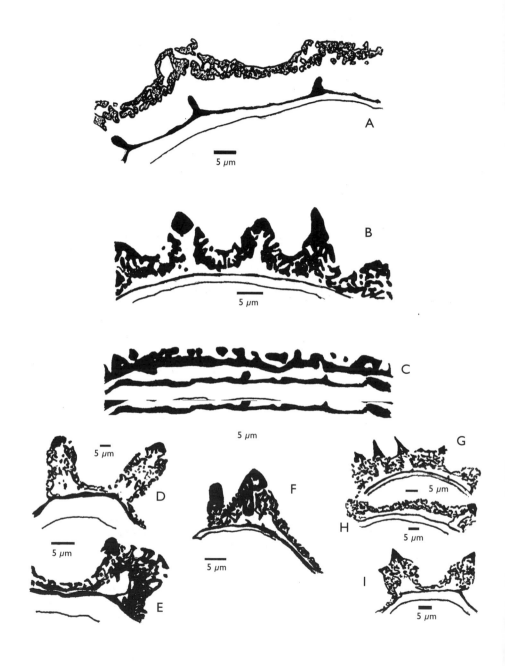

FIG. 3. Pollen exine patterns found in Lactuceae. (After Skvarla *et al.*, 1977).

Prominent columellae are present between the apertural areas of both Vernonieae (Fig. 4) and Liabeae (Fig. 5). The internal morphology, described as a modification of the 'anthemoid' pattern 'liabioid' by Skvarla et al., (1977), is similar for both tribes. The 'libioid' pattern is distinguished from the the 'anthemoid' pattern by the presence of broader columellae. In Liabeae, the lateral branches of columellae form an internal tectum, or endotectum. From the endotectum, short columellae arise and extend to a perforate tectum. Vernonieae, unlike Liabeae, possess more lateral branches connected to form anastomosing patterns.

Figure 1 shows the ultrastructural patterns of Arctoteae, Lactuceae, Liabeae, and Vernonieae mapped on the topology of the Arctoteae-Lactuceae-Liabeae-Vernonieae (A-L-L-V) clade from the phylogeny based on ndhF sequence and restriction site data (1996). If this group is sister group to subfamily Asteroideae, then the trend from caveate exines in Arctoteae to partial caveate exines in Lactuceae and distinct, non-caveate exines in Liabeae and Vernonieae fits. We would expect a caveate exine to be a synapomorphy for Asteroideae and the whole A-L-L-V clade. Within the A-L-L-V clade the character is reversed.

Non-caveate exines are characteristic of the 'anthemoid' pattern (Skvarla et al., 1977). 'anthemoid' exines consist of thick, long basal columellae that are superimposed on shorter columellae alternating with internal tecta. In the case of Lactuceae, Liabeae, and Vernonieae, the 'liabioid-anthemoid' pattern is derived from an ancestral caveate state shared with Asteroideae. The 'anthemoid' pattern evolved a second time within subfamily Asteroideae, as the name implies, in the Anthemideae.

Exine structure of subfamily Barnadesioideae and its utility as an outgroup for pollen studies

A third 'anthemoid' pattern was described for Doniophyton Wedd. in Barnadesioideae (Skvarla et al., 1977). The columellae of this genus appear similar to the complex upper columellar structures in Mutisieae. Caveate exines also evolved independently within the subfamily, where they are documented in Dasyphyllum. Kunth. It appears that caveate exines and the 'anthemoid' pattern have evolved independently in all three subfamilies (Barnadesioideae, Asteroideae, Cichorioideae) of Asteraceae.

Barnadesioideae have clearly been recognised as the sister group to the rest of Asteraceae based on the absence of a 22 kb inversion present in all members of Asteroideae and Cichorioideae (Jansen and Palmer, 1987). Members of Barnadesioideae were suggested to be most closely related to Calyceraceae based on the strong resemblance of Nastanthus Meirs (Calyceraceae) exine structure to that of Dasyphyllum and Schlechtendalia Less. (Barnadesioideae). However, this strong similarity is limited to the intercolpar depressions of all three genera. In general, the exine structure in Barnadesioideae consists of numerous fine columellae. This is in contrast to the thick basal columellae and distinct 'anthemoid' pattern found in Calyceraceae.

Bolick (1978) hypothesised, without the benefit of extensive molecular phylogenies, that thick columellae were derived from the fusion of the fine columellae present within taxa such as Dasyphyllum and Schlechtendalia. Today, in the context of molecular phylogenies, it appears that pollen morphology in Barnadesioideae is a product of anagenesis within the basal lineage of Asteraceae. Blackmore et al. (1989) suggested that a taxon will generally possess unique characters, some features homologous with related taxa, and general characters that are widespread within the group. If we look at the Barnadesioideae, at least in regard to pollen types, the subfamily possesses more unique characters than general characters shared by Asteraceae. For this reason, we suggest that Barnadesioideae alone are not a wise choice of outgroup for phylogenetic analyses of

Asteraceae pollen data sets. A far better choice would be Calyceraceae or Calyceraceae and Goodeniaceae. Better still, according to the criteria presented by Maddison and colleagues (1984), would be to use both Barnadesioideae and Calyceraceae.

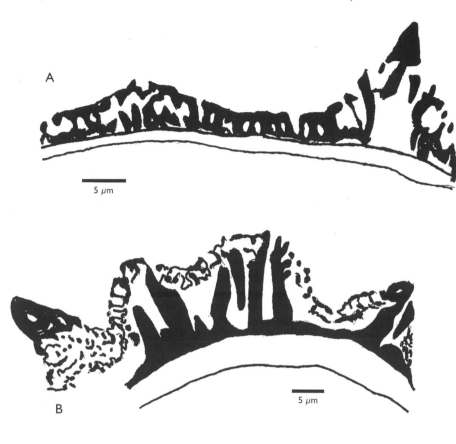

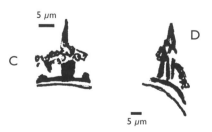

FIG. 4. Pollen exine patterns found in Liabeae. (After Skvarla *et al.*, 1977).

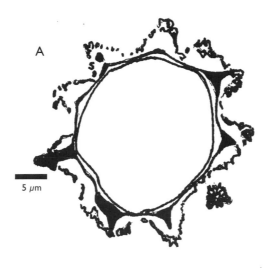

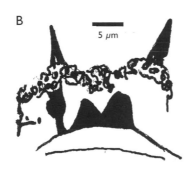

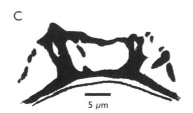

FIG. 5. Pollen exine patterns found in Vernonieae. (After Skvarla *et al.*, 1977).

Acknowledgements

We would like to thank James Doyle and our second reviewer for their constructive reviews. We thank Deborah Freile for preparation of graphics for this manuscript. Research reported in this paper was supported by NSF grant DEB-9318279 to RKJ.

References

Bergqvist, G., Bremer, B. and Bremer, K. (1995). Chloroplast DNA variation and the tribal position of *Eremothamnus* (Asteraceae). *Taxon* **44**: 341–350.

Blackmore, S., Scotland, R.W. and Stafford, P.J. (1989). The comparative method in palynology. In: K. Heine (editor). *Palaeoecology of Africa and the Surrounding Islands* **22**: 3–10. A.A. Balkema, Rotterdam.

Bolick, M.R. (1978). Taxonomic, evolutionary, and functional considerations of Compositae pollen ultrastructure and sculpture. *Plant Systematics and Evolution* **130**: 209–218.

Bremer, K. (1994). Asteraceae: Cladistics and Classification. Timber Press, Portland, Oregon.

DeVore, M.L. and Stuessy, T.F. (1995). The time and place of origin of the Asteraceae with additional comments on the Calyceraceae and Goodeniaceae. In: D.J.N Hind, C. Jeffrey, and G.V. Pope. (editors). Advances in Compositae Systematics, pp. 23–40. Royal Botanic Gardens, Kew.

Felsenstein, J. (1985). Confidence limits on phylogenies: an approach using the bootstrap. *Evolution* **39**: 783–791.

Feuer, S.M. (1974). Pollen Morphology and Ultrastructure in the Subtribe Microseridinae (Tribe Lactuceae: Family Asteraceae). MSc thesis, University of Illinois at Chicago Circle.

Fischer, H. (1890). Beiträge zur vergleichenden Morphologie der Pollenkörner. Berlin.

Gosliner, T.M. and Ghiselin, M.T. (1984). Parallel evolution in opithobranch gastropods and its implications for phylogenetic methodology. *Systematic Zoology* **33**: 255–274.

Gustafsson, M.H.G., Backlund, A. and Bremer, B. (1996). Phylogeny of the Asterales sensu lato based on *rbc*L sequences with particular reference to the Goodeniaceae. *Plant Systematics and Evolution* **119**: 217–242.

Hansen, H.V. (1991). SEM-studies and general comments on pollen in the tribe Mutisieae (Compositae) *sensu* Cabrera. *Nordic Journal of Botany* **10**: 607–623.

Hansen, H.V. (1992). Studies in the Calyceraceae with a discussion of its relationship to Compositae. *Nordic Journal of Botany* **12**: 63–75.

Jansen, R.K. and Palmer, J.D. (1987). A chloroplast DNA inversion marks an ancient evolutionary split in the sunflower family (Asteraceae). *Proceedings of the National Academy of Science U.S.A.* **84**: 5818–5822.

Jansen, R.K. and Kim, K.-J. (1996) Implications of chloroplast DNA data for the classification and phylogeny of the Asteraceae. In: D.J.N. Hind and H. Beentje (editors). Compositae: Systematics. Proceedings of the International Compositae Conference, vol. 1, pp. 317–339. Royal Botanic Gardens, Kew.

Jansen, R.K., Wallace, R.S., Kim, K.-J. and Chambers, K.L. (1991). Systematic implications of chloroplast DNA variation in the subtribe Microseridinae (Asteraceae: Lactuceae). *American Journal of Botany* **78**: 1015–1027.

Karis, P.O., M. Källersjö and K. Bremer. (1992). Phylogenetic analysis of the Cichorioideae (Asteraceae) with emphasis on the Mutisieae. *Annals of the Missouri Botanical Garden* **79**: 416–427.

Keeley, S.C. and Jones, S.B. (1979). Distribution of pollen types in *Vernonia* (Vernonieae: Compositae). *Systematic Botany* **4**: 195–202,

Keeley, S.C. and Jansen, R.K. (1991). Evidence from chloroplast DNA for the recognition of a new tribe the Tarchonantheae, and the tribal placement of *Pluchea* (Asteraceae). *Systematic Botany* **16**: 173–181.

Kim, K.-J., Jansen, R.K., Wallace, R.S., Michaels, H.J. and Palmer, J.D. (1992). Phylogenetic implications of *rbc*L sequence variation in the Asteraceae. *Annals of the Missouri Botanical Garden* **79**: 428–445.

Kim, K.-J. and Jansen, R.K. (1995). *ndh*F sequence evolution and the major clades in the sunflower family. *Proceedings of the National Academy of Science U.S.A.* **92**: 10379–10383.

Maddison, W.P, Donoghue, M.J. and Maddison, D.R. (1984). Outgroup analysis and parsimony. *Systematic Zoology* **33**: 83–103.

Michaels, H.J., Scott, K.M., Olmstead, R.O., Szaro, T., Jansen, R.K. and Palmer, J.D. (1993). Interfamilial relationships of the Asteraceae: Insights from *rbc*L sequence variation. *Annals of the Missouri Botanical Garden.* **80**: 742–751.

Nowicke, J.W. and Skvarla, J.J. (1979). Pollen morphology: the potential influence in higher order systematics. *Annals of the Missouri Botanical Garden* **66**: 633–700.

Rasmussen, F.N. (1983). On "apomorphic tendencies" and phylogenetic inference. *Systematic Botany* **8**: 334–337.

Robinson, H. and Marticorena, C. (1986). A palynological study of the Liabeae (Asteraceae). *Smithsonian Contributions to Botany* . **64**: 1–50.

Skvarla, J.J. and Larson, D.A. (1956). Interbedded exine components in some Compositae. *Southwestern Naturalist* **10**: 65–68.

Skvarla, J.J., Turner, B.L., Patel, V.V. and Tomb, A.S. (1977). Pollen morphology in the Compositae and in morphologically related families. In: V.H. Heywood, J.B. Harborne and B.L. Turner (editors). The Biology and Chemistry of Compositae. pp. 141–265. Academic Press, London.

Stix, E. (1960). Pollenmorphologische Untersuchungen und Compositen. *Grana Palynologica.* **2**: 289–297.

Swofford, D.L. (1993). PAUP: phylogenetic analysis using parsimony, version 3.1.1. Natural History Survey, Champaign, Illinois.

Tomb, A.S. (1971). Karyotypes, pollen grains and systematics in the genus *Lygodesmia* (Compositae: Cichorieae). *American Journal of Botany* **58**: 467.

Tomb, A.S. (1972). The systematic significance of pollen morphology in the family Compositae. Tribe Cichorieae. *Brittonia* **24**: 223–228.

Tomb, A.S. (1974). Pollen morphology and detailed structure in the Compositae, Tribe Cichorieae I. Subtribe Stephanomeriinae. *American Journal of Botany.* **61**: 486–498.

Tomb, A.S. (1975). Pollen morphology in tribe Lactuceae (Compositae). *Grana* **15**: 79–89.

Turner, B.L. (1977). Fossil history and geography. In: V.H. Heywood, J.B. Harborne and B.L. Turner (editors). The Biology and Chemistry of Compositae, pp. 21–39. Academic Press, London.

Vezey, E.L., Watson, L.E., Skvarla, J.J., and Estes, J.R. (1994). Plesiomorphic and apomorphic pollen structure characteristics of Anthemideae (Asteroideae: Asteraceae). *American Journal of Botany* **81**: 648–657.

Wodehouse, R. P. (1929) Pollen grains in the identification and classification of plants, IV: the Mutisieae. *American Journal of Botany* **16**: 297–313.

Appendix 1

Goodeniaceae	01110001
Calyceraceae	11110001
Barnadesioideae	10011101
Tarchonantheae	00002002
Mutisiinae	00013113
Nassauviinae	00012003
Cardueae	00014120
Arctoteae	00004220
Lactuceae	00015120
Liabeae	00015120
Vernonieae	00015130
Asteroideae	00005120

Strict consensus of 20 trees:

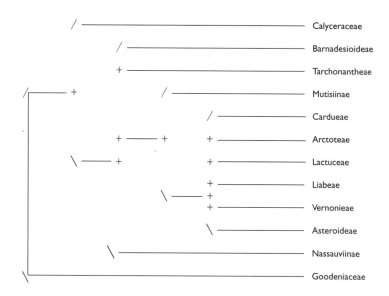

Strother, P.K. and Beck, J.H. (2000). Spore-like microfossils from Middle Cambrian strata: expanding the meaning of the term cryptospore. In: M.M. Harley, C.M. Morton and S. Blackmore (Editors). Pollen and Spores: Morphology and Biology, pp. 413–424. Royal Botanic Gardens, Kew.

SPORE-LIKE MICROFOSSILS FROM MIDDLE CAMBRIAN STRATA: EXPANDING THE MEANING OF THE TERM CRYPTOSPORE

PAUL KLEE STROTHER AND JOHN H. BECK

Paleobotanical Laboratory, Weston Observatory of Boston College, Department of Geology & Geophysics, 381 Concord Road, Weston Massachussetts 02193, USA

Abstract

Recent findings of spore-like microfossils from Middle Cambrian deposits lead us to propose a liberal interpretation of the term, cryptospore, to include all spore-like remains of non-marine origin from the lower Paleozoic. Assemblages from the Middle Cambrian Bright Angel Shale contain tetrads and dyads in addition to polyads, but the irregular configurations of these forms do not indicate a direct affinity to the embryophytes. Cryptospore dyads now appear to be quite ancient although their relations to higher plants remain problematical. The cryptospores in this expanded sense should now include algal ancestors to the embryophytes as well as spores of problematic and extinct forms of early cryptogams.

Introduction

Reports of Ordovician cryptospores by Vavrdová (1984, 1990), Richardson (1988), Wellman (1996) and Strother *et al.* (1996) have established that spores with apparent embryophytic affinity existed throughout the Middle and Late Ordovician. This palynological record implies that plants of bryophyte grade lived in moist habitats on the terrestrial surface beginning in the Llanvirn. It seems reasonable to assume that there was a prior existence of photosynthetic organisms, perhaps at an algal grade of organisation, which would have initially colonised both freshwater (aquatic) and moist habitats on the terrestrial surface. Indeed, Retallack (1985) has already designated labels for the series of transitions that should characterise this sequence of the evolution of soil type as an expression of plant development in the terrestrial realm.

The palaeontological record of pre-Devonian terrestrial remains is exceedingly sparse, however, there is very little direct evidence for any sort of terrestrial organisms existing before the Llanvirn. Yet, there are a few claims of terrestrial algae/cyanobacteria from the Precambrian extending as far back as the 1 Ga Nonesuch Shale (Strother, 1994). The cyanobacteria today are principally a freshwater group, with provision for UV protection that may have enhanced their ability to colonise moist surficial habitats even in atmospheres of lower pO_2. Since cyanobacteria are known to form stromatolites in shallow marine habitats, their presence in a palynological sample does not imply a non-marine source. Cyanobacterial sheaths are found with some regularity in Proterozoic and Lower Cambrian deposits, but they are generally not reported from younger units.

There are no reports of non-marine palynological assemblages from Cambrian and lower Ordovician strata. Vavrdová (1984) and Combaz (1968) found isolated spore-like microfossils, *Attritasporites* Combaz and *Virgatasporites* Combaz from Tremadocian to Arenig age sediments. These few occurrences remain problematic, in part because there is little stratigraphic continuity for these forms which are not seen elsewhere in the column. Nevertheless, if *Attritasporites* and *Virgatasporites* are really of terrestrial origin, as was strongly argued by Combaz (1968), then they represent the only record of non-marine "spores" of this age. These forms are spore-like in character, discoidal in shape with moderately thick walls, and some retain a sculpture of radially aligned striations on one surface. Yet, without a trilete scar or preservation in a tetrahedral tetrad, no spore-like microfossil can convincingly be ascribed to an embryophytic origin. These spores could have been derived from freshwater or subaerial algae, and, given the general paucity of such spore-like morphologies in the marine realm, this is a likely possibility. Palynologists have long recognised that non-marine palynomorphs (organic-walled microfossils) are typically transported into shallow marine settings (Muller, 1959). Therefore, it is reasonable to assume that lower Paleozoic palynological assemblages from near-shore marine deposits should include a sporomorph component that was derived from subaerial sources.

Terrestrial deposits of pre-Devonian age are quite rare, one must look to shallow marine and nearshore deposits to recover a palaeontological record of early land-dwelling organisms. We have used the lower Silurian, Tuscarora Formation which is characterised by mixed sandstones and shales thought to be of nearshore marine deposition (Cotter, 1983), as a model lithofacies for discovering older mixed clastic deposits containing shales with a non-marine palynological signature. These sandstone-shale interbeds are replete with trace fossils including *Phycodes* Richter, *Skolithos* Haldeman, *Paleophycus* Hall, *Cruziana* d'Orbigny and *Rusophycus* Hall. Within the Tuscarora, the shales yield a completely non-marine assemblage of cryptospores with almost no record of marine palynomorphs (Gray and Boucot, 1971; Pratt *et al.*, 1978; Strother and Traverse, 1979; Johnson, 1985). The Bright Angel Shale appears to be analogous to the Tuscarora in its sedimentological, palynological and trace fossil content, along with its general lack of normal marine body fossils.

The Bright Angel Shale is a proximal marine deposit that is transitional from the coarse Tapeats Sandstone to the overlying Muav Limestone. It represents the middle component of a classic Cambrian transgression onto the North American Platform (McKee and Resser, 1945). The original topographic expression of this transgressed surface can be observed throughout the Canyon, and at some localities, the Bright Angel shales rest directly on the nonconformity. Typically, however, the base of the Bright Angel is transitional with underlying coarse, crossbedded Tapeats sandstones rich in *Skolithos*. Ferruginous sandstones decrease in abundance and thickness up section, giving way to about 100 m of olive-green to greenish gray shales interbedded with decimeter scale medium to fine grained sandstones.

Traditional interpretations of the deposition environment for the Bright Angel generally call for open shelf, below wave base deposition (McKee and Resser, 1945). Middleton and Elliot (1990) characterise more recent sedimentological studies as having documented shallow water deposits, including tidally-influenced sedimentation. We have found wrinkle structures at 96 m at Sumner Butte which have been compared to Recent supratidal microbial mats by Hagadorn and Bottjer (1997). Similar features were interpreted as interference ripples by McKee (1969, Fig. 22 B). *Lingula* sp. occurs at 27 m at Sumner Butte from shale samples which are palyniferous. Thus, there is evidence for tidal to estuarine or restricted marine settings within the Bright Angel Shale. These settings are consistent with both the palynological signature and the lack of normal marine body fossils within the section.

The age of the Bright Angel Shale has long been considered to be Middle Cambrian based on trilobite biozonation (McKee and Resser, 1945). At Sumner Butte, the *Alokistocare-Glossopleura* horizon occurs at about 30 above the base of the formation. This places the Bright Angel Shale within the Delamaran Stage of the Lincolnian Series using the terminology proposed recently for the Cambrian of Laurentia (Palmer, 1998).

We have now examined three sequences within the Middle Cambrian Bright Angel Shale that have overall lithic similarity to the Tuscarora Formation. The first section is from Deer Creek drainage, the second is from Tapeats Creek drainage, and the third sequence comes from a complete section through the Bright Angel Shale at Sumner Butte which is between Bright Angel Creek and Clear Creek drainage in the eastern part of the Grand Canyon. Work on these deposits is in progress and only preliminary results that have an effect on the meaning of cryptospores will be discussed here.

Materials and Methods

Shale samples were processed according to conventional palynological techniques. This consisted of first macerating about 20 grams of sample in HF followed by water washing. Next the organic fraction was separated from the remaining insoluble mineral fraction by a heavy liquid flotation using $ZnCl_2$. The resultant organic fraction was then oxidised in Schultze's solution (HNO_3 and $KClO_3$) if needed. Residues were washed and mounted in glycerin jelly on glass slides for observation under the light microscope or rinsed and dried onto stubs for observation with the Scanning Electron Microscope (SEM).

Results

Sections measured at Deer Creek (Strother and Baldwin, unpublished) and at Sumner Butte (Strother, Baldwin and Skilliter, unpublished), were complemented by additional spot sampling in the Tapeats Creek area. Approximately 70 samples have been examined to date, about a quarter of which are palyniferous.

The assemblages contain populations of membrane-enclosed spore-like cells or cysts. They are spherical to collapsed cells, often enclosed within a membranous envelope (Figs. 2–8). Cells are typically clustered into groups of four or more but they may be solitary (Figs. 1 and 9) or in dyads (Figs. 2 and 7). The clusters are not multicellular, but appear to be individual spherical to sub-spherical cells, each representing a separate spore-like body (Figs. 3, 4, 8, 10). In some cases the arrangement of cells into tetrads appears quite regular, but in general, there is very little resemblance to the perfectly tetrahedral arrangements seen in the cryptospore tetrads from Ordovician and Silurian strata. Individual naked cells are generally unornamented; but membranous coverings are either smooth (Figs. 5, 7, 8) or ornamented with small granules (Figs. 1, 2, 6), forming a scabrate to granulate sculpture. Many forms are ornamented by thickened knobs, several microns in diameter, which appear as darkened knobs on otherwise translucent spore walls (Fig. 8). These features appear to be discrete thickened portions of the spore wall. Sizes range from about 10 to 40µm; these are well within the normal range for fossil spores of eukaryotes.

415

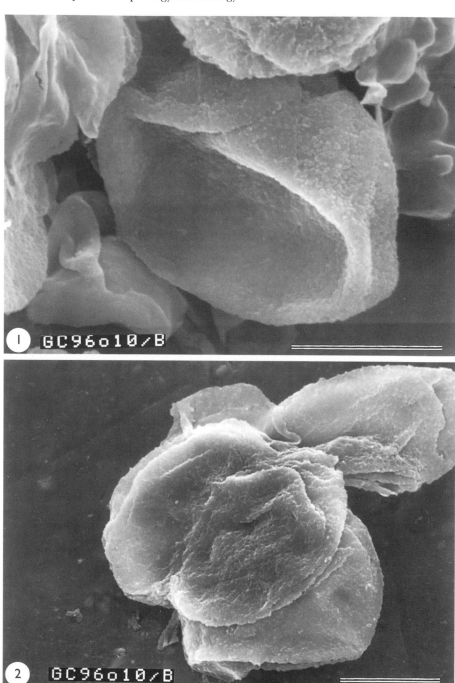

FIGS. 1–2. SEM images of cryptospores from the Bright Angel Shale, Thunder River locality in Tapeats drainage area, sample no. GC9610/B. Scale bar = 10 μm. Fig. 1. Monad with scabrate sculpture. Note invaginated surface and lack of surficial folds that would typically characterise thinner-walled leiospheres. Fig. 2. Dyad surrounded by enclosing envelope with scabrate sculpture.

Tetrads of cells may exist completely unsheathed (Fig. 3) or they may appear as distinct, separate forms although completely encased within a membranous sac (Figs. 5-6). Many tetrads appear to be dyad pairs, in fact, many of the polyads seem to be built on different combinations of paired cell sets. There is no clear ordering of tetrads into regular geometric arrangements, although planar tetrads form the most typical examples (Figs. 3, 5-6, 8). Given that some of these tetrads are completely ensheathed and that cell sizes within clusters are very similar, it is unlikely that cell clusters represent random associations of planktonic cells.

Discussion

Expanding the definition of Cryptosporites

In typical marine shale of Middle Cambrian age, palynological assemblages would be dominated by acanthomorphic acritarchs, algal cysts characterised by radial processes such as *Asteridium* Moczydłowska, *Comasphaeridium* Staplin *et al.* and *Skiagia* Downie (for example Moczydłowska, 1991). However, in spite of excellent preservation, the Bright Angel suite, has yet to reveal even a single acanthomorphic acritarch. Individual shale samples in the Bright Angel that contain *Lingula*, a brackish water indicator, also contain well-preserved sporomorphs. Thus, the spore-like clusters are likely to be the remains of freshwater or brackish water algal species that were transported into a shallow, proximal marine setting.

There is a possible modern analogue to the above-mentioned condition occurring when freshwater cyanobacteria, chlorophytes, and other algae are washed into marine waters (Foerster, 1973). *Oöcystis* Näg. loses its shape to become more like *Tetraëdron* Kütz., which consists of spore-like cells enclosed in an organic membrane (Foerster, 1971). Although four cells can make up the *Tetraëdron* cluster, other combinations of numbers of encysted cells are possible. *Scenedesmus obliquus* (Turp.) Kütz balls up to form autospores when grown in salinities of 10 to 30 ‰, and these morphs have been documented *in situ* from actual plankton tows Foerster (1973). *Chlorella*-like cells and other cyst-like forms that were collected from shallow marine phytoplankton tows off Long Island by Foerster (1973), were viable when grown on freshwater media. Since *Chlorella* Beyerinck is known to form resistant walls, its documentation in shallow marine waters is significant as a possible analogue to the occurrence of terrestrial algae in the Bright Angel Shale.

There are no spores in the Bright Angel assemblage that are convincingly like spores from any known embryophytes. Most of the tetrads are either planar or simply irregular in their arrangement. It is possible that the dyads are merely examples of two cells stuck together, but in the majority of cases, individuals of a dyad pair are quite similar morphologically and are most likely division pairs. Such dyads and tetrads, if found in younger strata, would be placed into the cryptospores. In addition monads, such as those figured in Figs. 1 and 9, would readily be placed into the cryptospores if found in younger units. It is now necessary to consider whether these kinds of lower Paleozoic "spores" should be classified as cryptospores or if they should remain as acritarchs.

Although the term acritarch is correctly applied to cysts and other organic walled microfossils whose affinities with extant organisms are unknown, the acritarchs are considered by most authors as representative of marine environments. This parallels what is known in the extant marine microbiota with cyst-forming dinoflagellates representing the closest ecological analogue to Paleozoic acritarch species. In the case of the Bright Angel assemblage, there is no marine signal at all in the palynoflora; these forms are very likely to be non-marine in origin. The cryptospores, as a group, need not represent only the products of embryophytes. *Quadrisporites* Hennelley is a planar tetrad which is included in the cryptospores. However, its topology alone does not constitute proof of embryophyte affinity. Likewise, individual monads such as *Rugosphaera* Strother & Traverse and *Qualisaspora* Richardson, Ford & Parker have no direct morphological connection to the embryophyte lineage, yet such forms are now accommodated by turma Cryptosporites.

Thus it seems reasonable that spore-like, organic-walled microfossils of non-marine origin be accommodated by the term cryptospore. Belonging to the cryptospores does not demonstrate any special relation to the embryophytes, individual morphology must be used to make such a determination. A spore-like microfossil is usually discoidal in shape rather than being fully spherical. Leiospheres derived from spherical cells are characterised by arcuate folds in their walls that are formed by compression during diagenesis. Spore-like microfossils are often thicker walled than their marine counterparts. This is an important distinction in the Llanvirnian assemblage from Saudi Arabia which contains both acritarchs and cryptospores (Strother *et al.*, 1996). The distinction is most difficult when comparing simple leiospheres with smooth-walled spores and there may always be individual specimens that cannot be placed unambiguously into either class.

The condition of "non-marine" origin is an even less robust distinguishing characteristic as it is divorced from morphology. Perhaps the best argument for its inclusion in defining the cryptospores as a class is simply that this distinction is needed to exclude such fossils from the acritarchs which have a strong marine connotation. All of these characteristics are subjective and are dependent upon the overall circumstances that characterise individual palynological assemblages. It is necessary to view spore-like microfossils within the context of non-marine and proximal marine assemblages in order to consider them as cryptospores.

These arguments lead to the inclusion of decidedly non-embryophytic spores as belonging to the cryptospores. There is a good reason for this as it helps to focus attention on terrestrial remains from lowermost Paleozoic strata. These remains should include chlorophytes and ancestral streptophytes that were capable of producing sporopollenin or other organic polymers as a resistant spore coating. There is no *a priori* need to limit the cryptospores to only those spores which derive from embryophytes; it is perhaps more realistic to consider them as being derived from the cryptogams which would then include extinct sources as well as lineages that did not give rise to the embryophytes.

Stratigraphic record and evolution of Cryptosporites

Expanding the meaning of Cryptosporites to include generally non-marine palynomorphs of lower Paleozoic age necessarily creates a heterogeneous group. This makes discussion of phylogenetic relations even more tenuous than they already are. Obviously, the best examples of phylogenetic analysis are based on cryptospores found *in situ* in sporangia of known botanical affinity (for example, Fanning *et al.*, 1991; Shute *et al.*, 1996; Edwards, 1998; Wellman, 1998). It is important not to become too constrained by potential associations with known fossil and extant plant types. The cryptospores represent a probable fossil record of the evolutionary transition that gave rise to the higher land plants. We see the result of that transition culminating in the origin of the tracheophytes, but the record of steps taken within that transition are poorly constrained by fossil data to date.

The transition from a Cambrian terrestrial/aquatic world to a late Silurian world of tracheophytes can be modelled after cladistic views of tracheophyte origins (for example, Bremer *et al.* 1987), which propose a progressive accumulation of plant-like characters in a chlorophyte to charophyte to embryophyte lineage. Since sporopollenin production is known for extant chlorophytes (Atkinson *et al.*, 1972; see references in Butterfield *et al.*, 1994), this is a primitive condition that could have existed throughout the lineage. Presumably, extant charophytes have lost the ability to produce sporopollenin. The transition from chlorophyte to embryophyte could well be represented in a cryptospore record from the Middle Cambrian to the Middle Ordovician if the Llanvirnian tetrads reported by Strother *et al.* (1996) represent early embryophytes. Subsequent cryptospore evolution throughout the remainder of the Ordovician seems to be marked by stasis (Wellman, 1996). This could represent a period of time during which the bryophyte groups differentiated, but it appears that the evolution of spore morphology was not a significant component of evolutionary change at this time.

The cryptospore record does not tell us when the tracheophytes subsequently evolved. For one thing, the line between cryptospore and trilete spore becomes blurred with forms such as *Ambitisporites? Vavrdovae* Richardson being considered to be the result of physical breakage of cryptospore tetrads rather than an independent trilete spore produced under the genetic control of a tracheid-containing plant (Wellman, 1996). The burst of sculptural evolution seen in the Homerian (Wenlock) by Burgess and Richardson (1995), Dufka (1995) and Beck (1998) may well represent the initial evolutionary diversification of the tracheophytes. Certainly a Wenlockian origin would fit the evidence from macrofossil remains (Edwards *et al.*, 1983).

Cryptospore dyads remain a curious puzzle because they are so rare in the extant *sporae dispersae*, yet they range very far back into the fossil record. In addition to their occurrence in the Bright Angel Shale, dyads occur in the Upper Cambrian of northwest Spain (Strother and Baldwin, unpublished) and Vavrdová (1990) has reported dyads from the Arenig/Llanvirn boundary in the Prague Basin. Dyads are easily recognised in their dispersed form in Silurian monads of *Hispanaediscus* (Cramer) Burgess & Richardson, *Confossuspora* Strother, *Laevolancis* Burgess & Richardson and *Artemopyra* Burgess & Richardson. Some of the late Ordovician and early Silurian dyads are clearly related on the basis of common sculpture. This can be seen in the rugose forms of Johnson (1985) and Wellman (1996). Brown and Lemmon (1991) present a possible clue to dyad affinity in demonstrating that the second meiotic division in *Isoëtes* microsporogenesis takes place after cell plate and wall formation. This plant may have retained a developmentally primitive state in sporogenesis relating to dyad production by completion of cytokinesis after division I. Combined with the Silurian record of the lycophyte *Baragwanathia* Lang & Cookson, the antiquity of cryptospore dyads could be evidence of an early origin of the lycophytes.

Conclusions

We are moving to a condition wherein non-marine problematica can be accommodated by the taxon cryptospores, much in the same way that marine problematica are accommodated by the term acritarch. These spore-like microfossils were derived from terrestrial cryptogams that included sporopollenin-producing chlorophytes and ancestral streptophytes in addition to cryptogams from extinct lineages. It is important to emphasise this heterogeneity of origin, for there is no direct evidence that the fossils from the Bright Angel Shale represent either embryophytes or their evolutionary ancestors. However, they do represent a rare glimpse of a record of either freshwater aquatic or, perhaps, subaerial cryptogams colonising a very early terrestrial landscape. Seen in this light, the stratigraphic record of the cryptospores is consistent with current views of plant origins based on phylogenetic systematics.

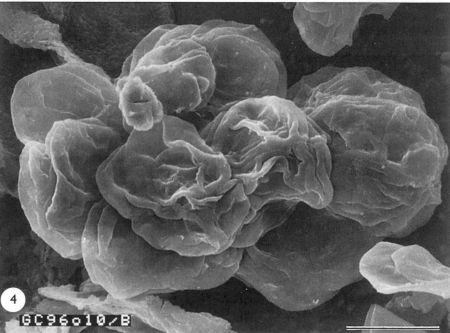

FIGS. 3–4. SEM images of smooth-walled tetrads and polyads from the Bright Angel Shale, Thunder River locality in Tapeats drainage area, sample no. GC9610/B. Scale bar equals 10 μm. Fig. 3. Cryptospore tetrad in planar configuration in which individual spores display sutures clearly. Fig. 4. Typical polyad showing the generally irregular nature of attachment with folded envelopes surrounding individual cells.

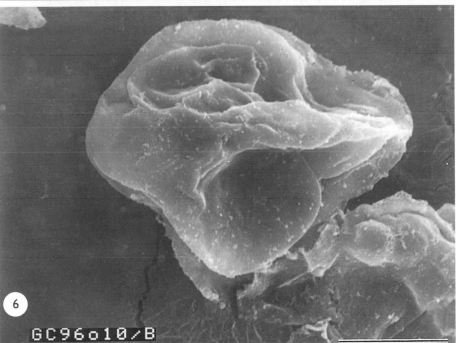

FIGS. 5–6. SEM images of cryptospore tetrads from the Bright Angel Shale, Thunder River locality in Tapeats drainage area, sample no. GC9610/B. Scale bar equals 10 μm. Fig. 5. Planar tetrad with smooth-walled envelope enclosing entire tetrad. Fig. 6. Planar tetrad with lightly scabrate envelope enclosing entire tetrad.

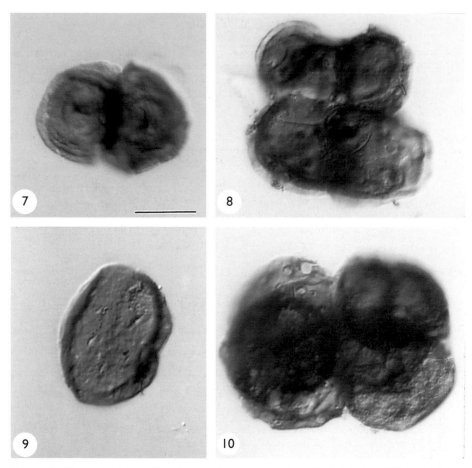

FIGS. 7–10. Transmitted green light images of cryptospores from the Bright Angel Shale. Scale bar = 15 μm. Fig. 7. Dyad in which each sporomorph appears to be enclosed by an individual envelope. Slide no. GC96-9/A1, Bright Angel Shale at Deer Creek. Fig. 8. Planar tetrad of cryptospores enclosed within an envelope. Sculptural elements (knobs) appear as circular patches seen on the lower two sporomorphs. Slide no. GC96-9/A1, Bright Angel Shale at Deer Creek. Fig. 9. Distinct, smooth-walled monad with marginal thickenings. This form would be classified as *Laevolancis*, if it had been found in deposits of Silurian age. Slide no. GC98-6, Bright Angel Shale at Sumner Butte. Fig. 10. Irregular cryptospore tetrad. Slide no. GC98-6, Bright Angel Shale at Sumner Butte.

Textbooks of Earth history are replete with tales of trilobites and other interesting creatures that inhabited the Cambrian seas, yet we know literally nothing of the inhabitants of the continental surface at this time. There can be little doubt that cyanobacteria and various eukaryotic algae could have inhabited aquatic and marginal aquatic habitats during the lowermost Paleozoic. The recovery of cryptospores from the Middle Cambrian does not extend the stratigraphic range of embryophytes, but it does give us our first glimpse into the terrestrial cryptogamic world at that time. How extensive such cryptogamic floras were in creating soils and sequestering carbon remains to be seen, but it is possible that the atmospheric drawdown of CO_2 began as a result of the expansion of these early terrestrial communities.

Acknowledgements

This research was supported in part by grants from the National Science Foundation (EAR-9526562) and the National Geographic Society (5908-79). Deborah Skilliter assisted in the Grand Canyon project and was responsible for processing. We received field assistance from Christopher Baldwin, Cecilia Lenk, Laurence Marschall, Judith Baldwin, Kevin Pyle, Peter Brown and John Strother.

References

Atkinson, A.W., Gunning, B.E.S. and John, P.C.L. (1972). Sporopollenin in the cell wall of *Chlorella* and other algae: Ultrastructure, chemistry, and incorporation of $_{14}$C-acetate, studied in synchronous cultures. *Planta* 107: 1–32

Beck, J.H. (1998). Paleopalynology of the Silurian Arisaig Group, Nova Scotia. Unpublished Ph.D. dissertation, Boston University, pp. 300.

Bremer, K., Humphries, C.J., Mishler, B.D. and Churchill, P. (1987). On cladistic relationships in green plants. *Taxon* 36: 339–349.

Brown, R.C. and Lemmon, B.E. (1991). Sporogenesis in simple land plants. In: S. Blackmore and S.H. Barnes (editors). Pollen and spores: patterns of diversification, pp. 9–24. Oxford University Press, Oxford.

Burgess, N.D. and Richardson, J.B. (1995). Late Wenlock to early Přídolí cryptospores and miospores from south and southwest Wales, Great Britain. *Palaeontographica B* 236: 1–44.

Butterfield, N.J., Knoll, A.H. and Swett, K. (1994). Paleobiology of the NeoProterozoic Svanbergfjellet Formation, Spitsbergen. *Fossil & Strata* 34: 1–84.

Combaz, A. (1968). Un microbios du Trémadocien dans un sondage d'Hassi-Messaoud. *Actes de la Société Linnéenne de Bordeaux* 104B 29: 4–26.

Cotter, E. (1983). Shelf, paralic, and fluvial environments and eustatic sea level fluctuations in the origin of the Tuscarora Formation (Lower Silurian) of central Pennsylvania. *Journal of Sedimentary Petrology* 53: 25–49.

Dufka, P. (1995). Upper Wenlock miospores and cryptospores derived from a Silurian volcanic island in the Prague Basin (Barrandian area, Bohemia). *Journal of Micropalaeontology* 14: 67–79.

Edwards, D (1998). *In situ* spores of early land plants. Pollen and Spores: Morphology and Biology. July 6–9, Royal Botanic Gardens, Kew. Abstracts, p. 6.

Edwards, D., Feehan, J. and Smith, D.G. (1983). A late Wenlock flora from Co. Tipperary, Ireland. *Botanical Journal of the Linnean Society* 86: 19–36.

Fanning, U., Richardson, J.B. and Edwards, D. (1991). A review of *in situ* spores in Silurian land plants. In: S. Blackmore and S.H. Barnes (editors). Pollen and spores: patterns of diversification, pp. 25–47. Oxford University Press, Oxford.

Foerster, J. W. (1971). Environmentally induced morphological changes in *Oöcystis lacustris* Chodat (Chlorophyta). *Bulletin of the Torrey Botanical Club* 98: 225–227.

Foerster, J. W. (1973). The fate of freshwater algae entering an estuary. In: L. H. Stevenson and R. R. Colwell (editors). Estuarine Microbial Ecology, pp.387–420. University of South Carolina Press, Columbia, South Carolina.

Gray, J. and Boucot, A. (1971). Early Silurian spore tetrads from New York: Earliest New World evidence for vascular plants? *Science* 173: 918–921.

Hagadorn, J. W. and Bottjer, D. J. (1997). Wrinkle structures: Microbially mediated sedimentary structures common in subtidal siliciclastic settings at the Proterozoic-Phanerozoic transition. *Geology* 25(11): 1047–1050.

Johnson, N.G. (1985). Early Silurian palynomorphs from the Tuscarora Formation in central Pennsylvania and their paleobotanical and geological significance. *Review of Palaeobotany and Palynology* 45 :307–360.

McKee, E. D. (1969). Stratified rocks of the Grand Canyon. *USGS Professional Paper* 669: 23–58.

McKee, E. D. and Resser, C. E. (1945). Cambrian history of the Grand Canyon Region. *Carnegie Institution of Washington*, Publication 563: 232 p.

Middleton, L. T. and Elliot, D. K. (1990). Tonto Group. In: S. S. Beus and M. Morales (editors). Grand Canyon Geology, pp. 83–106. Oxford University Press and Museum of Northern Arizona Press, Oxford.

Moczydłowska, M. (1991). Acritarch biostratigraphy of the Lower Cambrian and the Precambrian-Cambrian Boundary in southeastern Poland. *Fossils & Strata* 29: 1–127.

Muller, J. (1959). Palynology of Recent Orinoco delta and shelf sediments. *Micropaleontology* 5: 1–32.

Palmer, A. R. (1998). A proposed nomenclature for stage and series for the Cambrian of Laurentia. *Canadian Journal of Earth Sciences* 35: 323–328.

Pratt L.M., Phillips, T.L. and Dennison J.M. (1978). Evidence of non-vascular land plants from the early Silurian (Llandoverian) of Virginia, U.S.A. *Review of Palaeobotany and Palynology* 25: 121–149.

Retallack, G.J. (1985). Fossil soils as grounds for interpreting the advent of large plants and animals on land. In: W.G Chaloner and J. D. Lawson (editors). Evolution and environment in the late Silurian and early Devonian. *Philosophical Transactions of the Royal Society of London* B 309: 105–140.

Richardson, J. B. (1988) Late Ordovician and Early Silurian cryptospores and miospores from northeast Libya. In: A. El-Arnauti *et al.* (editors). Subsurface Palynostratigraphy of Northeast Libya, pp. 89–109. Garyounis University, Benghazi, Libya.

Shute, C.H., Hemsley, A.R. and Strother, P.K. (1996). Reassement of dyads contained in a late Silurian rhyniophytoid sporangium. *Special Papers in Palaeontology* 55: 137–145.

Strother, P.K. (1994). Sedimentation of palynomorphs in rocks of pre-Devonian age. In: A. Traverse (editor). Sedimentation of Organic Particles, pp. 489–502. Cambridge University Press, Cambridge.

Strother, P.K. and A. Traverse. (1979). Plant microfossils from Llandoverian and Wenlockian rocks of Pennsylvania. *Palynology* 3: 1–21.

Strother, P.K., Al-Hajri, S. and Traverse, A. (1996). New evidence for land plants from the lower Middle Ordovician of Saudi Arabia. *Geology* 24(1): 55–58.

Vavrdová, M. (1984). Some plant microfossils of possible terrestrial origin from the Ordovician of central Bohemia. *Věstnik Ústředníko ústavu geologického* 59: 165–170.

Vavrdová, M. (1990). Coenobial acritarchs and other palynomorphs from the Arenig/Llanvirn boundary, Prague basin. *Věstnik Ústředníko ústavu geologického* 65(4): 237–242.

Wellman, C. (1996). Cryptospores from the type area of the Caradoc Series in southern Britain. *Special Papers in Palaeontology* 55: 103–136.

Wellman, C. (1998). Morphology and wall ultrastructure of early land plant spores (Cryptospores and miospores). Pollen and Spores: Morphology and Biology. July 6–9, Royal Botanic Gardens, Kew, London. Abstracts, p. 23.

Taylor, W.A. (2000). Spore wall development in the earliest land plants. In: M.M. Harley, C.M. Morton and S. Blackmore (Editors). Pollen and Spores: Morphology and Biology, pp. 425–434. Royal Botanic Gardens, Kew.

SPORE WALL DEVELOPMENT IN THE EARLIEST LAND PLANTS

WILSON A. TAYLOR

**Department of Biology, University of Wisconsin-Eau Claire
Eau Claire, Wisconsin, 54701 (USA)**

Abstract

By combining careful analysis of the mature ultrastructure of fossil cryptospores with information on sporoderm development in modern systems, it is possible to construct hypothetical developmental pathways for the spores of extinct plants. Following a brief review of the wall ultrastructure of the enclosed (i.e., having a common covering surrounding all spore bodies) dyads *Dyadospora murusdensa* and *D. murusattenuata*, as well as the tetrad *Velatitetras reticulata* (which has the same ultrastructure as the dyad *Abditusdyadus histosus*), developmental scenarios are proposed that would result in the production of these various complex walls. These scenarios are plausible, given what is known of development in modern systems, and differ primarily with respect to the timing of meiosis and the deposition of the various wall layers.

Introduction

Development is an active process. Its elucidation in extinct systems relies on the coordinated and parallel approaches of careful, detailed morphological analysis of mature form in the fossil systems, (and the occasional fortuitous analysis of immature fossil specimens) and developmental analysis of living (and, optimally, related and homologous) systems/structures. This paper is a consideration of the inferences that can be drawn about the developmental processes that may have operated to produce spore walls in the earliest land plants, and the taxonomic implications of those inferences - specifically, how these spore walls (including lamellae, globules and the envelope) might have been produced using developmental systems in modern plants as analogues.

Despite the paucity of terrestrial plant megafossils in the Ordovician and Silurian, there is an extensive microfossil record. In the absence of information on the megafossils that produced these spores, they are referred to as cryptospores (Gensel, *et al.*, 1990; Richardson, 1996). Cryptospores occur as single spores (monads), pairs of spores – tightly adhered (permanent) or separated to varying degrees (true dyads) – and tetrads. Any of these types can be naked or covered by a common covering (envelope – enclosed).

Materials and Methods

All specimens figured in this report were collected from the Upper Ordovician (Ashgill) Drakes Formation, or the Lower Silurian (Rhuddanian) Centerville Formation from

Adams County, Ohio, USA (for complete locality information, see Taylor, 1997). Preparation for TEM involved standard techniques (Taylor, 1990). Sections were triple stained with potassium permanganate, uranyl acetate and lead citrate. Micrographs were taken on a Hitachi HS-9 transmission electron microscope.

Results

Since this paper focuses on spore wall development, results will be limited to cross sections produced using TEM showing the various layers seen in these mature dispersed cryptospores. For detailed information on surface sculpture, spore size ranges, etc., please refer to Taylor (1995, 1996, 1997).

Dyadospora murusdensa Strother and Traverse emend. Burgess and Richardson, 1991
The original diagnosis of the genus *Dyadospora* was:
"Palynomorphs consisting of two inaperturate spore or spore-like palynomorphs occurring in a dyad configuration; individual spore spherical to subspherical in outline; walls psilate; overall length of flattened dyad body 25 to 50 μm."
The emended diagnosis of Burgess and Richardson is as follows:
"Dyads composed of two laevigate hilate cryptospores. Individual spores not strongly mutually attached, a clear line of separation is always seen; in equatorial compression the two spores are often partially separated."
The emended diagnosis of *D. murusdensa* is:
"A *Dyadospora* with unfolded walls." (Burgess and Richardson, 1991).

In the light microscope, this specimen was 53 μm in length (perpendicular to the contact face), with thick (dark) unfolded walls. There were clearly two separate spores joined along a contact face, but with a clear line of separation.
The wall of this cryptospore consists of two principal layers (Fig. 1). The outer layer is approximately 1 μm thick, has an undulatory inner surface, and appears homogeneous in cross section. This outer layer is present over the entire non-contact face surface of the dyad members, and protrudes inward to cover a portion of the contact face. The inner layers of the adjacent members are in contact in the centre of the contact surface. The inner layer consists of 10-15 separate lamellae that are each approximately 0.1 μm thick. The entire dyad appears to be surrounded by a very thin envelope, but its thinness and fragmentary nature render these results somewhat equivocal (Taylor, 1996).

Dyadospora murusattenuata Strother and Traverse emend. Burgess and Richardson, 1991
The emended diagnosis of this species is:
"A *Dyadospora* with thin folded walls." (Burgess and Richardson, 1991).

In the light microscope, this dyad is 40 μm in length, and has folded walls.

The wall of this cryptospore is also two layered (Fig. 2). The outer layer is homogeneous and approximately 1 μm thick. It grades inward to a coarsely lamellated layer, also approximately 1 μm thick. The innermost layer consists of smaller granular units that coalesce to form inwardly directed rounded protrusions. The outer layer is distributed over the individual dyad members as described for *D. murusdensa* above. *D. murusattenuata* dyads are surrounded by a slightly darker staining envelope approximately 100nm thick (Taylor, 1997).

Velatitetras reticulata Burgess, 1991 (enclosed tetrad) and *Abditusdyadus histosus* Wellman and Richardson, 1996 (enclosed dyad)

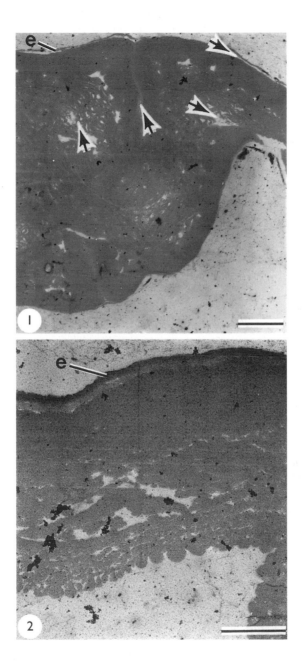

FIGS. 1–2. Transmission electron micrographs of cryptospore wall cross-sections. e = envelope; Fig. 1. *Dyadospora murusdensa*. Contact face between the dyad members. The walls between the similarly oriented pairs of arrows mark contact face (left hand dyad member) and non-contact face (right hand member) wall cross-sections. Bar = 2 μm. Fig. 2. *Dyadospora murusattenuata*. Non-contact face wall. Bar = 1 μm.

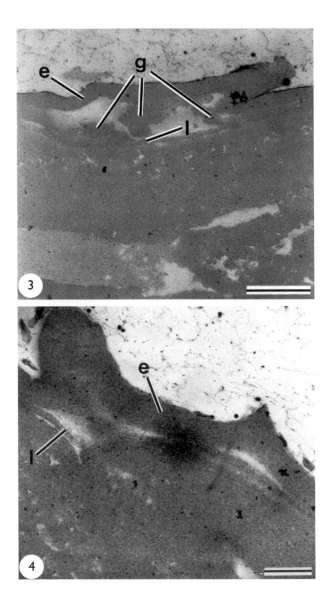

FIGS. 3–4. Transmission electron micrographs of wall cross-sections of cryptospore *Velatitetras reticulata.* e = envelope; l = lamella; g = globules. Fig. 3. Non-contact face wall with globules between the envelope and the lamella. Bar = 1 *μ*m Fig. 4. Non-contact face wall. Bar = 0.5 *μ*m.

Both of these taxa are recognisable at the level of the light microscope by reticulate surface ornamentation that is formed by the enclosing envelope.

The ultrastructure of these two taxa is identical. The main spore wall is densely spongy and approximately 1.2 μm thick, with no evidence of lamellar construction (Fig. 3, 4). The entire tetrad/dyad is enclosed by a single lamella, approximately 0.1 μm in thickness (a thickness comparable to that of the individual lamellae in *Dyadospora murusdensa*), which is overlain by a thicker envelope (150-300nm). The lamella is most apparent beneath folds in the envelope, but is traceable and not fused to the envelope in places between folds. The lamella is not discernible where the envelope and the individual cryptospore walls are tightly appressed, but this is interpreted as being due to preservational compression. Ornamentation is produced by folds in the envelope (Fig. 4, left hand projection) and peaked thickenings (Fig. 4, right hand projection). In places, globular units occur between the lamella and the envelope (Fig. 3; Taylor, 1996).

Discussion

Blackmore and Barnes (1987) recognised four modes of sporopollenin deposition in extant embryophytes: 1) accumulation on tripartite lamellae (in which mature walls may be homogeneous or lamellate), 2) deposition from surrounding cells of the sporangium onto previously formed layers (this may produce mature walls that are homogeneous, and/or structures known as globules, orbicules or Ubisch bodies), 3) accumulation within a pre-patterned cell surface glycocalyx (primexine; this may produce walls that are homogeneous, or differentiated into distinct strata) and 4) centripetal deposition to existing sporoderm layers without the involvement of tripartite lamellae, giving rise to amorphous sporopollenin accumulations on the inner surface of the sporoderm.

Which of these mechanisms might be implicated to produce the structures/layers present in cryptospore walls?

Relevant structures that occur in cryptospores include: lamellae (multiple in *Dyadospora murusdensa*, and *D. murusattenuata*; single in *Abditusdyadus histosus* and *Velatitetras reticulata*) and the outer envelope ("membrane") that surrounds many cryptospores (the term enclosed is used for these). The walls of many cryptospores are homogeneous (for example, *in situ* dyads described in Wellman, *et al.*, 1998a; *in situ* laevigate hilate cryptospores, Wellman, *et al.*, 1998b; *Pseudodyadospora* sp., Taylor, 1996; some *Dyadospora murusattenuata*, Taylor, 1997). Unfortunately, this structure can be produced in one of several ways, and the final product leaves little evidence of its mode of formation. Therefore, little can be discerned about developmental processes of strictly homogeneous walls.

Lamellae
Nearly all modern plants that have been examined produce spores whose resistant walls are lamellated at some point in development. Lamellae are produced outside the plasma membrane by the spore cytoplasm following meiotic cytokinesis. They may be occluded late in development by the addition of sporopollenin from cells of the sporangium wall (the tapetum, which may remain in place around the perimeter of the sporangium locule [secretory or cellular tapetum], or enter the sporangium locule to surround the developing spores [plasmodial tapetum]; Brown and Lemmon, 1990; Lugardon, 1990; Kurmann, 1990; Blackmore and Barnes, 1990).

429

Essentially all liverworts produce spores with persistent lamellae (and without a perispore), since there is little to no sporangial involvement in sporopollenin production in this group (Blackmore and Barnes, 1987; Brown and Lemmon, 1990). Even in layers of the mature spore wall that do not look especially lamellated in liverworts, closer inspection reveals a substructure that began as separate lamellae. A tapetum has been documented in some liverworts, (for example, *Sphaerocarpos*, Kelley and Doyle, 1975), and may be present in all embryophytes (Pacini, 1990), but its activity is probably limited to spore nutrition and possible production of sporopollenin precursors, but not the deposition of homogeneous sporopollenin that occurs in many other plant groups. In mosses and many pteridophytes, the tripartite lamellae are occluded by sporangial sporopollenin, and a perispore is present (Neidhart, 1979; Lugardon, 1990).

Wellman, *et al.* (1998), in addition to three different homogeneous walled monads, found lamellations in the inner walls of two different types of cryptospore monads. The lamellations differed between the two specimens – concentric continuous lamellae in one and laterally discontinuous, overlapping and irregularly spaced lamellae with structure typical of white line centred lamellae, in the other. Fossil lamellae have also been reported in the Devonian fossils *Protosalvinia* (Taylor and Taylor, 1987) and *Parka* (Hemsley, 1989). There are distinct differences (thickness, lateral extent, etc.) between all of these fossil lamellae and those reported here. Some of these differences could be due to various diagenetic histories, but some exist between specimens at the same locality. A wide range of lamella morphology also exists in modern plants – especially within the liverworts (Brown and Lemmon, 1990). This is not unexpected given the probable antiquity of this group (Qiu, *et al.*, 1998).

Envelope

Many cryptospores of all types are surrounded by a common covering - an envelope. Johnson (1985) argued that the term perispore is not appropriate for this layer since, based on developmental patterns in modern plants, it would not be expected to produce a layer that surrounded the *entire* tetrad/dyad. Gray (1991) defends the use of the term as applied in the original sense of Bower (1959) - i.e., conspicuous, loosely fitting, readily detachable, ornamented or often more or less unornamented or membranous layer - and argues that the morphological analogue between the perispore and the cryptospore envelope only seems poor due to the lack of any modern perispore producing cryptogams which have mature spores that remain united in dyads or tetrads. The only modern group of free-sporing plants which contains members that routinely produce adherent tetrads (no modern plants normally produce dyads) is the hepatics, but they produce no perispore.

Johnson (1985) suggested that the envelope may represent a "persistent mother cell wall," similar to that seen surrounding zygotes of extant *Coleochaete*. The sporopollenin containing layer surrounding zygotes of *Coleochaete* is a single darkly staining layer approximately 100nm in thickness (Graham, 1990). This ultrastructure is similar to that of the envelope which surrounds many cryptospores (for example, *Dyadospora murusattenuata*, Taylor, 1997). The development of the sporopollenin wall surrounding zygotes of *Coleochaete* is not known, but there is obviously no involvement of sporangial cells since the zygote represents the entire sporophyte generation.

In many preparations, cryptospore envelopes (especially those of *Abditusdyadus histosus* and *Velatitetras reticulata*) stain with a different intensity from the underlying spore wall. This implies a different chemical composition, which is consistent with the interpretation of the envelope as a tapetal product (perispore). This difference in staining is also seen in modern bryophyte (Brown and Lemmon, 1990) and pteridophyte (Lugardon, 1990) spores.

Development of complex lamellated walls of *Dyadospora murusdensa*, and *D. murusattenuata*

The two, parted walls of *Dyadospora murusdensa*, and *D. murusattenuata* most resemble the spore walls of some modern liverworts in the Sphaerocarpales - being composed of an inner polylamellated layer and an outer, less lamellated, layer. In some modern sphaerocarpaleans (for example, *Sphaerocarpos cristatus*, *S. drewei*, *Geothallus tuberosus*, Steinkamp, 1973) the outer layer recognisably consists (at least partially) of fused lamellae. This is not the case with the fossil dyads where the outer layer appears homogeneous. The outer layer in the dyads must have either 1) lost its lamellated substructure due to preservational changes (unlikely since the inner lamellae are so well preserved), 2) lost its lamellated substructure (occlusion) due to developmental changes (which would require a degree of sporangial involvement in sporopollenin deposition that is unknown in modern liverworts), or 3) been deposited without the involvement of lamellae either a) by the accumulation of sporangially derived sporopollenin on a pre-existing template ("primexine"), or b) by the accumulation of non-lamellated sporopollenin from the spore cytoplasm (documented in only one hepatic [*Corsinia*; Wiermann and Weinert, 1969], one moss [*Andreaea*; Brown and Lemmon, 1984] and the hornwort *Anthoceros* [Ridgway, 1965]).

Most of the modern liverworts that produce permanent tetrads are also members of the Sphaerocarpales. In these permanent tetrads the outer, less conspicuously lamellated layer, does not protrude between adjacent spores, and this results in the adpression of the inner lamellated layer of adjacent tetrad members over most of the contact face. In the case of the lamellated fossil dyads, the outer layer protrudes inward to cover part of the contact face, resulting in contact of the inner lamellated layer of the adjacent spores of the dyad in the centre of the contact face. In the living taxa, the lamellae of adjacent tetrad members show some degree of fusion, while in the fossils, there appears to be little evidence of lamellae fusion.

The final point of distinction between modern sphaerocarpalean liverworts and these fossil dyads (aside from their dyadic nature) is the presence of an envelope. If these fossil dyads were produced in a sporangium with some tapetum-like activity, it could explain both the developmental occlusion of an original lamellate construction of the outer layer and the subsequent deposition of a perispore. This is unlike the situation in modern liverworts. However, the absence of a perispore in modern liverworts does not preclude its presence in an ancestor of that lineage.

One possible scenario for the deposition of the complex lamellate wall of *Dyadospora murusdensa* is: 1) differentiation of spore mother cells in a sporangium; 2) formation of a special mother cell wall around the separated products of the first meiotic division (which may then serve as a primexine-like template for the outer unlamellated half of the spore wall); 3a) second meiotic division and subsequent deposition of the inner, lamellated spore wall layer by the newly formed spore cytoplasm; 3b) deposition of sporangially derived sporopollenin on the outer wall template (3a and 3b could be in any order or simultaneous); 4) deposition of a very thin perispore that comes to surround the entire persistent dyad. *D. murusattenuata* could have developed in the same way, except that the individual dyad members deposited their sporopollenin (step 3a) initially as thick lamellae (or lamellae that become thicker than those of *D. murusdensa*) and finally as globular units. The perispore (envelope) of *D. murusattenuata* is also much thicker than that of *D. murusdensa*. The plausibility of this scenario is based on the fact that it relies on developmental mechanisms known in modern plants and requires only changes in timing of meiosis and spore wall deposition (heterochrony) of a type that has been previously suggested (Hemsley, 1994; Strother, 1991; Gray, 1993). The origin of dyads by separation of the products of meiosis I was favoured by Fanning, *et al.* (1991).

Development of cryptospore walls of *Abditusdyadus histosus* and *Velatitetras reticulata*

Enclosed tetrads (*V. reticulata*) are occasionally found in homogeneous clumps (i.e., all tetrads) suggestive of their production in a sporangium. The lack of overall lamellate construction in the mature spore wall (unlike modern liverworts) suggests the involvement of the sporangial wall in sporopollenin deposition. While it is not impossible that the spore cytoplasm alone deposited the spongy spore wall, it would be highly unusual, relative to modern developmental systems, since the spore cytoplasm usually produces sporopollenin on tripartite lamellae. As such, one possible source of the single lamella would be the surface of the spore mother cell prior to meiosis. The other possible source is that it represents the innermost portion of the perispore deposited by cells of the sporangial wall. Several ferns have exospores surrounded by thin lamellae overlain by uniformly thick perispores (for example, *Gleichenia bankrostii*, *G. oceanica* and *Anemia phyllitidis*; Lugardon, 1971), but the perispore and lamellae surround individual spores in these taxa, not the entire tetrad.

One possible scenario of the development of the walls of *Abditusdyadus histosus* and *Velatitetras reticulata* is: 1) separation of the products of meiosis I followed by meiosis II where the products remain joined (dyads) *or* completion of meiosis I and II where the four products remain joined (tetrads) in a sporangium; 2) cytoplasmic deposition by the products of meiosis II of granular/globular structural elements that fuse to form a tightly spongy spore wall; 3) deposition of a two-parted perispore by the cells of the sporangial wall.

Conclusion

It is difficult to envision a reasonable developmental pathway for producing these complex cryptospore walls that does *not* involve tapetal deposition of sporopollenin and/or the envelope (perispore). This is *unlike* modern liverworts. Therefore, the lack of tapetal involvement in sporopollenin deposition that we see in modern liverworts may be secondarily derived from some ancestor that occluded its lamellae by tapetal sporopollenin production.

Cryptospore wall ultrastructure is diverse. By the late Ordovician, land plants had already evolved the basic structural units (and the developmental patterns to produce those units) that make up the sporoderm of most modern land plants. Whether this diversity represents multiple groups making the initial move to land, rapid evolution of new developmental patterns from a common ancestor, or a much earlier origin for terrestrial plants, or some combination of these, remains to be seen.

Acknowledgements

Acknowledgment is made to the donors of The Petroleum Research Fund, administered by the ACS, for partial support of this research.

References

Blackmore, S. and Barnes, S.H. (1987). Embryophyte spore walls: origin, development and homologies. *Cladistics* 3: 185–195.

Blackmore, S. and Barnes, S.H. (1990). Pollen wall development in angiosperms. In: S. Blackmore and R.B. Knox (editors). Microspores: evolution and ontogeny. pp. 173–192. Academic Press, London.

Bower, F.O. (1959) Primitive land plants: also known as the Archegoniatae. Hafner, New York, NY.

Brown, R.C. and Lemmon, B.E. (1984). Spore wall development in *Andreaea* (Musci: Andreaeopsida). *American Journal of Botany* 71: 412–420.

Brown, R.C. and Lemmon, B.E. (1990). Sporogenesis in bryophytes. In: S. Blackmore and R.B. Knox (editors). Microspores: evolution and ontogeny. pp. 55–94. Academic Press, London.

Burgess, N.D. (1991). Silurian cryptospores and miospores from the type Llandovery area, south-west Wales. *Palaeontology* 34: 575–599.

Burgess, N.D. and Richardson, J.B. (1991). Silurian cryptospores and miospores from the type Wenlock area, Shropshire, England. *Palaeontology* 34: 601–628.

Fanning, U., Richardson, J.B. and Edwards, D. (1991). A review of *in situ* spores in Silurian land plants. In: S. Blackmore and S.H. Barnes (editors). Pollen and spores: patterns of diversification, pp. 25–47. Clarendon Press, Oxford.

Gensel, P.G., Johnson, N.G. and Strother, P.K. (1990). Early land plant debris (Hooker's "Waifs and strays"?). *Palaios* 5: 520–547.

Graham, L. (1990). Meiospore formation in charophycean algae. In: S. Blackmore and R.B. Knox (editors). Microspores: evolution and ontogeny. pp. 43–54. Academic Press, London.

Gray, J. (1991). *Tetrahedraletes, Nodospora*, and the "cross" tetrad: an accretion of myth. In: S. Blackmore and S.H. Barnes (editors). Pollen and spores: patterns of diversification, pp. 49–87. Clarendon Press, Oxford.

Gray, J. (1993). Major Paleozoic land plant evolutionary bio-events. *Palaeogeography, Palaeoclimatology, Palaeoecology* 104: 153–169.

Hemsley, A.R. (1989). The ultrastructure of the spores of the Devonian plant *Parka decipiens*. *Annals of Botany* 64: 359–367.

Hemsley, A.R. (1994). The origin of the land plant sporophyte: an interpolational scenario. *Biological Review* 69: 263–273.

Johnson, N.G. (1985). Early Silurian palynomorphs from the Tuscarora Formation in central Pennsylvania and their paleobotanical and geological significance. *Review of Palaeobotany and Palynology* 45: 307–360.

Kelley, C.B. and Doyle, W.T. (1975). Differentiation of intracapsular cells in the sporophyte of *Sphaerocarpos donnellii*. *American Journal of Botany* 62: 547–59.

Kurmann, M.H. (1990). Exine ontogeny in conifers. In: S. Blackmore and R.B. Knox (editors). Microspores: evolution and ontogeny. pp. 158–172. Academic Press, London.

Lugardon, B. (1971). Contribution à la connaissance de la morphogenèse et de la structure des parois sporales chez les Filicinées isosporées. Thèse, Université Paul Sabatier, Toulouse.

Lugardon, B. (1990). Pteridophyte sporogenesis: a survey of spore wall ontogeny and fine structure in a polyphyletic plant group. In: S. Blackmore and R.B. Knox (editors). Microspores: evolution and ontogeny. pp. 95–120. Academic Press, London.

Neidhart, H.V. (1979). Comparative studies of sporogenesis in bryophytes. In: G.C.S. Clarke and J.G. Duckett (editors). Bryophyte systematics, pp. 251–280. Academic Press, London.

Pacini, E. (1990). Tapetum and microspore function. In: S. Blackmore and R.B. Knox (editors). Microspores: evolution and ontogeny. pp. 213–237. Academic Press, London.

Qiu, Yin-Long, Cho, Yangrae, Cox, J.C. and Palmer, J.D. (1998). The gain of three mitochondrial introns identifies the liverworts as the earliest land plants. *Nature* 394: 671–674.

Richardson, J.B. (1996). Lower and middle Palaeozoic records of terrestrial palynomorphs. In: J. Jansonius and D.C. McGregor (editors). Palynology: principles and applications. pp. 555–574. American Association of Stratigraphic Palynologists Foundation, Vol. 2.

Ridgway, J.E. (1965). Some aspects of morphogenesis, biotic coaction, and ultrastructure in the genus *Anthoceros*. Ph.D. dissertation, University of Texas, Austin.

Steinkamp, M.P. (1973). Spore wall ultrastructure in the Hepaticae with special reference to the Sphaerocarpales. Ph.D. Thesis, University of California, Santa Cruz. 302 pp.

Strother, P.K. (1991). A classification schema for the cryptospores. *Palynology* 15: 219–236.

Taylor, W.A. and Taylor, T.N. (1987). Spore wall ultrastructure of *Protosalvinia*. *American Journal of Botany* 74: 437–443.

Taylor, W.A. (1990). Comparative analysis of megaspore ultrastructure in Pennsylvanian lycophytes. *Review of Palaeobotany and Palynology* 62: 65–78.

Taylor, W.A. (1995). Ultrastructure of *Tetrahedraletes medinensis* (Strother and Traverse) Wellman and Richardson, from the Upper Ordovician of southern Ohio. *Review of Palaeobotany and Palynology* 85: 183–187.

Taylor, W.A. (1996). Ultrastructure of Lower Paleozoic dyads from southern Ohio. *Review of Palaeobotany and Palynology* 92: 269–279.

Taylor, W.A. (1997). Ultrastructure of lower Paleozoic dyads from southern Ohio II: *Dyadospora murusattenuata*, functional and evolutionary considerations. *Review of Palaeobotany and Palynology* 97: 1–8.

Wellman, C.H. and Richardson, J.B. (1996). Sporomorph assemblages from the "Lower Old Red Sandstone" of Lorne, Scotland. *Special Papers in Palaeontology* 55: 41–101.

Wellman, C.H., Edwards, D. and Axe, L. (1998a). Permanent dyads in sporangia and spore masses from the Lower Devonian of the Welsh Borderland. *Botanical Journal of the Linnean Society* 127: 117–147.

Wellman, C.H., Edwards, D. and Axe, L. (1998b). Ultrastructure of laevigate hilate cryptospores in sporangia and spore masses from the Upper Silurian and Lower Devonian of the Welsh Borderland. *Philosophical Transactions of the Royal Society of London*. 353B: 1983-2004.

Wiermann, R. and Weinert, H. (1969). Untersuchungen zur Sporodermentwicklung bei *Corsinia coriandrina* (Sprengl) Lindb. *Berichte der Deutschen Botanischen Gesellschaft* 82: 175–82.

van Konijnenburg-van Cittert, J.H.A. (2000). Osmundaceous spores throughout time. In: M.M. Harley, C.M. Morton and S. Blackmore (Editors). Pollen and Spores: Morphology and Biology, pp. 435–449. Royal Botanic Gardens, Kew.

OSMUNDACEOUS SPORES THROUGHOUT TIME

JOHANNA H. A. VAN KONIJNENBURG-VAN CITTERT

Laboratory of Palaeobotany and Palynology, University of Utrecht, Budapestlaan 4, 3584 CD Utrecht, The Netherlands

Abstract

The living fern family Osmundaceae comprises three genera: *Osmunda*, *Todea* and *Leptopteris*. The spores of these three genera are described, including the presence of spores with double laesurae (bipolar apertures): a 'normal' triradiate scar together with a clear monolete mark with a well-defined margo that might also have served as a way for germination. The first fertile osmundaceous fossils with *in situ* spores originate from the Late Triassic. Especially from the Jurassic and Lower Cretaceous, many osmundaceous fossils have been recorded which yielded *in situ* spores. Evolutionary trends in osmundaceous spores are indicated including number of spores per sporangium, spore size and shape, laesurae, exospore ornamentation and perispore absence or presence and ornamentation.

Introduction

The fern family Osmundaceae is characterised by a typical stem anatomy (leaf traces and the vascular bundle of the rachis are C-shaped), and simultaneous development of large sporangia with a shield-shaped annulus (Fig. 1) that are either arranged along the veins on the lower pinnule surface, or along the veins in modified leaves that have lost their lamina. The extant Osmundaceae comprise three genera: *Osmunda*, *Todea* and *Leptopteris*. *Osmunda* is the largest genus, and is a cosmopolite with ca. 10 species; *Todea* is a monotypic genus found in South Africa, Australia and New Zealand, and *Leptopteris* is a South Pacific genus with six species. In *Osmunda* the sporangia are arranged along the veins in modified leaves (or occasionally only pinnae in a normal sterile leaf) that have lost their lamina. In *Todea* and *Leptopteris* the fertile leaves do not differ in general from the sterile leaves; the sporangia are arranged in distinct or indistinct sori along the veins on the abaxial side of the leaf. Spores from the three genera are remarkably similar: they are produced in large numbers, and are globose, usually trilete and possess an exospore covered with granulae or baculae that may or may not be fused in small groups forming irregular patterns. The thin perispore which follows the exospore closely is covered, over the sculptural elements, with small spines (for example, Tryon and Lugardon, 1991). The spores of the living Osmundaceae are discussed in detail (pp. 437–441).

The family can be traced back to the Upper Permian where a number of structurally preserved stem genera possess typical osmundaceous features. The first fertile osmundaceous fossils with *in situ* spores originate from the Late Triassic, while many osmundaceous fossils have been recorded with *in situ* spores, especially from the Jurassic. The *in situ* spores are briefly described and discussed (pp. 443–445).

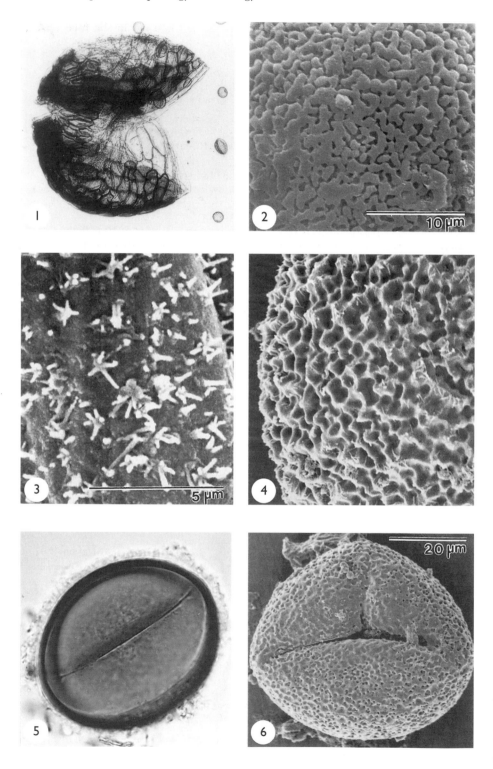

Evolutionary trends in osmundaceous spores will be discussed in the final section, including number of spores per sporangium, spore size and shape, laesurae, exospore ornamentation and perispore absence or presence and ornamentation.

Spores of the living Osmundaceae

As mentioned in the introduction, spores from the three genera are remarkably similar: they are produced in large numbers, they have been recorded as being globose, trilete and possessing an exospore covered with granulae or baculae that may or may not be fused in small groups forming irregular patterns (Fig. 2). Over the sculptural elements the thin perispore which surrounds the exospore is covered with small, narrow spines (Fig. 3) but it is often eroded in older spores (Fig. 4) (for example, Hanks and Fairbrothers, 1981; Tryon and Lugardon, 1991; Large and Braggins, 1991).

In most descriptions of extant osmundaceous spores, they are recorded as being trilete. However, our study of *Osmunda regalis* L. spores proved that not all spores are trilete, a percentage of the spores appears to be monolete (Fig. 5) or incompletely trilete (Fig. 6). This percentage varies considerably in every sample; percentages of 0–20% have been found, but 4–5% is average. The sample with 0% originated from the Isle of Man, while the samples with the highest percentage of monolete and incompletely trilete spores came from Denmark and England (Norfolk). Monolete spores have also been found in some other *Osmunda* species (for example, *O. japonica* Thunb. and *O. claytoniana* L.) but in very small quantities. Monolete spores have been described by Bobrov (1966), who also described and figured spores with double lacsurac ("bipolar apcrturcs") in *O. regalis* and *O. cinnamomea* L.: a 'normal' triradiatc scar together with a clear monolete mark with a well-defined margo that might also have served as a way for germination. None of the later authors have seen this feature or commented on it, but the present study revealed this character in some of the samples of *O. regalis*: in 5 out of 12 samples at least one spore with a double laesura was observed (Figs. 7, 8). The feature seems to be unrelated to the percentage of monolete/imperfectly trilete spores, as spores with a double laesura were not found in any of the samples with a high percentage of monolete/imperfectly trilete spores; but they occurred in samples with 4–5% of these spores. An explanation of the feature cannot be given at the moment.

Osmunda

Based on anatomical differences, Miller (1971) divided the species of *Osmunda* into three groups (or subgenera): the *Osmunda* group, with *O. regalis*, *O. japonica*, *O. claytoniana* and *O. lancea* Thunb., the *Osmundastrum* group with *O. cinnamomea*, and the *Plenasium* group with *O. banksiifolia* (Presl) Kuhn, *O. javanica* Bl., *O. bromeliaefolia* (Presl) Cop. and *O. vachelii* Hook. Although some authors use the group names *Osmundastrum* and *Plenasium* at generic level (for example, Bobrov, 1967), most use them at subgeneric level.

Captions – Unless stated differently or indicated by bar, magnifications are × 1000.

FIGS. 1–6. *Osmunda*. Fig. 1. Dehisced sporangium of *Osmunda claytoniana*; note the shield-shaped annulus (arrow); × 100. Fig. 2. *Osmunda regalis* exospore ornamentation; × 2000. Fig. 3. *Osmunda cinnamomea* perispore with echinate sculpture; × 10,000. Fig. 4. *Osmunda regalis* eroded perispore; × 2000. Fig. 5. *Osmunda regalis* monolete spore. Fig. 6. *Osmunda regalis* imperfectly trilete spore.

Spores have been studied from each of the three groups to see if spore morphological features coincide with this division on anatomical features. From the *Osmunda* group, *O. regalis* (samples from several localities in Europe), *O. japonica* (Japan) and *O. claytoniana* (samples from the Himalaya and China) were examined. From the *Osmundastrum* group *O. cinnamomea* (Parona), and from the *Plenasium* group *O. banksiifolia* (Taipei) and *O. vachelii* (Hongkong). As the osmundaceous spores of all species are quite similar, differences might be found in spore size, exospore thickness (the perispore being uniformly thin, with small spines over the apices of the sculptural elements), and especially the variation in exosculptural elements and the degree in which they are fused to each other.

Study of the literature revealed that the three groups can probably not be distinguished on the basis of their spores: for example Hanks and Fairbrothers (1981) stated that the spores of *O. cinnamomea* (*Osmundastrum*) and *O. claytoniana* (*Osmunda*) have greater similarity to each other than either has to spores of *O. regalis* (*Osmunda*). The same was indicated by Zhang *et al.* (1976).

Osmunda regalis is a widespread species in the Northern Hemisphere and the spores have been described by various authors (for example Hanks and Fairbrothers, 1981; Bobrov *et al.*, 1983; Large and Braggins, 1991; Tryon and Lugardon, 1991). The spores are quite variable in size, even within one sporangium, with an average diameter of ca. 55–60 μm. The laesura is fine, usually extending $^3/_4$ of the spore radius. The exospore is 2–3 μm thick. The exospore sculpture is usually described as verrucate, although the terms tuberculate and baculate are sometimes used as well. The sculptural elements are often coalesced into small groups (Fig. 9), sometimes so much that they form an irregular reticulum (Fig. 2, 6); see also Tryon and Lugardon (1991).

Osmunda japonica is quite similar to *O. regalis* in its spores, although they are usually slightly smaller (40–50 μm) and the laesural arms are $^7/_8$ of the spore radius and more pronounced than in *O. regalis*. The exospore is again 2–3 μm thick with a baculate to verrucate ornamentation of which the elements are partly coalesced (Fig. 10).

Osmunda claytoniana has quite large spores (55–65 μm) with an exospore up to 4 μm thick. The laesural arms almost reach the equator. The exospore has rather coarse verrucate elements that are usually not fused (Fig. 11) (Hanks and Fairbrothers, 1981; Tryon and Lugardon, 1991). The spores of *O. claytoniana* differ in this respect from others of the *Osmunda* group. Monolete spores occur occasionally (Fig. 12).

Osmunda cinnamomea (the only species in *Osmundastrum*) has quite large spores (usually 60–70 μm) with thin laesural arms that reach almost to the equator. The exospore, as in *O. claytoniana*, is ca. 4 μm thick and usually displays baculate elements that are not fused (Fig. 13). Again monolete spores occur occasionally (Fig. 14). One spore (Fig. 16) was found with a tri- and a monolete mark, as recorded by Bobrov (1966).

Osmunda banksiifolia (*Plenasium*) has large spores (65–80 μm) with laesural arms $^3/_4 - ^7/_8$ of the spore radius. The exospore is ca. 4 μm thick and covered with baculate to finely verrucate elements that are occasionally coalesced (Fig. 15).

FIGS. 7–12. *Osmunda*. Fig. 7. *Osmunda regalis* spore with bipolar apertures, focus on trilete mark. Fig. 8. Same spore, focus on monolete mark. Fig. 9. *Osmunda regalis* exosporal elements fused into small groups. Fig. 10. *Osmunda japonica*. Fig. 11. *Osmunda claytoniana* exospore ornamentation. Fig. 12. *Osmunda claytoniana* monolete spore.

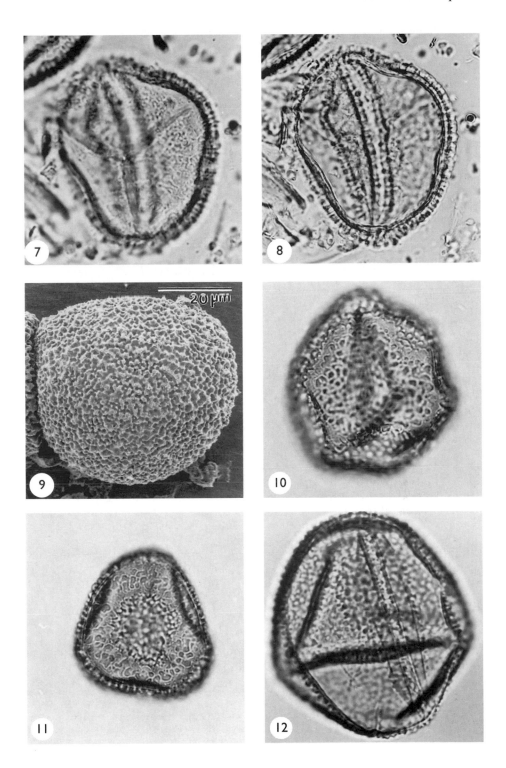

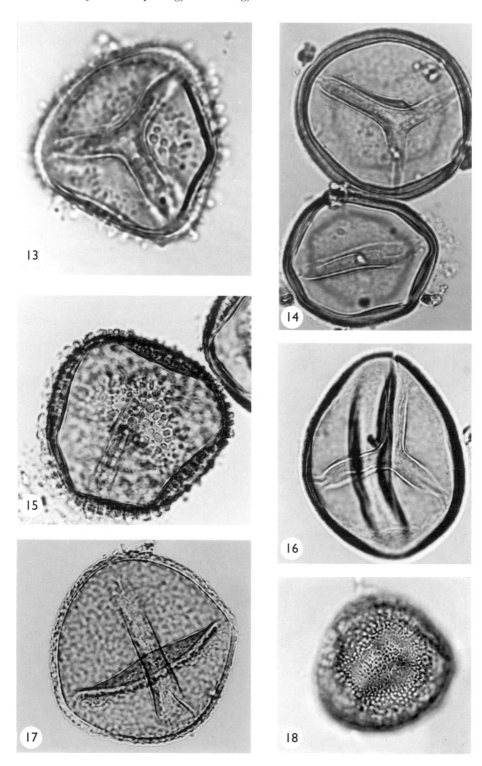

Osmunda vachelii has spores 40–60 μm, with broad laesural arms that almost reach the equator. The exospore is 3.5–4 μm thick and covered with baculae that are usually fused basally. In older, eroded, spores an irregular reticulum can sometimes be seen (Fig. 18), but in younger spores the elements seem to be separate.

One might conclude that there is no spore morphological support for the division into the three subgenera *Osmunda, Osmundastrum* and *Plenasium*. Coalescent elements (sometimes forming an irregular reticulum) occur in the subgenera *Osmunda* and *Plenasium* (*O. vachelii*) while, as stated above, *O. cinnamomea* (*Osmundastrum*) and *O. claytoniana* (*Osmunda*) are so much alike that *O. claytoniana* might better be placed in *Osmundastrum*.

Todea

Todea barbara (L.) Moore (sample from New South Wales) has trilete, globose spores, but occasionally monolete spores occur (Fig. 17). They are quite variable in size (usually 50–75 μm but up to 90 μm has been recorded; Tryon and Lugardon, 1991). The laesural arms reach only ½ of the spore radius according to Tryon and Lugardon (1991), but Harris (1955), Large and Braggins (1991) and the present study reveal that they may be up to ¾ of the spore radius (Fig. 19). The ca. 2 μm thick exospore is covered by coarse baculae and/or tuberculae (Fig. 20). The perispore is, just as in all extant Osmundaceae, a thin continuous layer following the exospore closely, with echinate elements above the baculae/tuberculae (Tryon and Lugardon, 1991).

Leptopteris

Leptopteris spores are the smallest in the Osmundaceae: 30–45 μm is the usual size. Four species have been examined: *L. superba* (Col.) Presl (sample from New Zealand), *L. hymenophylloides* (A.Rich.) (from New Zealand), *L. wilkesiana* (Brack.) Christ. (from Samoa) and *L. fraseri* (Hook. et Grev.) Presl (from New South Wales). The spores are quite uniform: globose, trilete with laesural arms usually ¾ of the spore radius. Alete spores have been recorded from *L. hymenophylloides* (Large and Braggins, 1991). Incomplete trilete to monolete spores occur occasionally in *L. superba* (Fig. 21) but have not been recorded from any of the other species. The exospore is ca. 1–2 μm thick, with tuberculae, baculae or verrucae that are not fused (Figs. 22, 23). The perispore is, just as in *Osmunda* and *Todea*, a thin continuous layer following the exospore closely, with echinate elements above the baculae, tuberculae or verrucae (Fig. 23) (Tryon and Lugardon, 1991).

Spores of the fossil Osmundaceae

The Osmundaceae have an extensive fossil record from the Permian onwards, mainly based on permineralised stems (Miller, 1971; Tidwell and Ash, 1994; Collinson, 1996). These include records of all three subgenera of the modern genus *Osmunda* by the Palaeocene (Miller, 1971). At least three specialised lines of stems appeared in the Paleozoic and Mesozoic (Collinson, 1996) but these will not be discussed here.

FIGS. 13–18. *Osmunda* and *Todea*. Fig. 13. *Osmunda cinnamomea* exosporal elements. Fig. 14. *Osmunda cinnamomea* trilete and monolete spore. Fig. 15. *Osmunda banksiifolia*. Fig. 16. *Osmunda cinnamomea* spore with bipolar apertures. Fig. 17. *Todea barbara* monolete spore. Fig. 18. *Osmunda vachelii*.

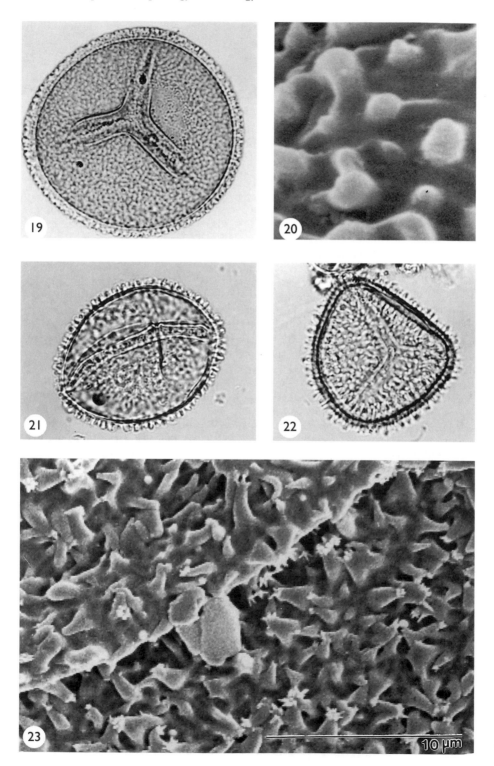

Fossil foliage assigned to the Osmundaceae includes forms such as *Cladophlebis* for sterile leaves (but not all *Cladophlebis* species are Osmundaceae!; for example see Harris, 1931, 1961; Collinson, 1996), and *Todites* for fertile isomorphic leaves that look like living *Todea* and *Leptopteris*. *Todites* is especially common in the latest Triassic and throughout the whole of the Jurassic and the lowermost Cretaceous. In the Jurassic *Osmundopsis* occurs as well, with reduced fertile foliage like living *Osmunda*. This genus has also been recorded from the Early Cretaceous but without spores, while species from the Jurassic - Tertiary have been assigned to the extant genus *Osmunda* (Barthel, 1976; Krassilov, 1978; Zhang Yulong *et al.*, 1976). There is no fossil record of *Todea*, and that of *Leptopteris* consists of a few doubtful Tertiary petiole bases (Collinson, 1996).

The first *in situ* spores were recorded from the Late Triassic. Van Konijnenburg-van Cittert (1978) gave a review of the morphology of the majority of osmundaceous *in situ* spores, and Balme (1995) describes later material. The various descriptions will not be repeated here. Since then, Van Konijnenburg-van Cittert (1996) described a new *Osmundopsis* species with *in situ* spores and Schweitzer *et al.* (1997) described the spores of *Todites crenatus*, *T. nebbensis* and *Osmundopsis sturii*.

In general, there are three different types of *in situ* fossil osmundaceous spores: the *Osmundopsis*-type (Figs. 24, 25, 27) with granulate or baculate elements on the exospore that are usually partly fused, the *Todites hartzu*-type (Fig. 26) with granules that are not fused, and the *Todites williamsonii*-type (Fig. 28) with an almost smooth exospore. All fossil osmundaceous spores are globose and trilete. No monolete or imperfectly trilete spores have been recorded and none of the *in situ* fossil spores demonstrate a perispore. There are some differences in size, but in general the spores are between 30 and 60 μm in diameter. In the following treatment, the three types of fossil spores will be discussed and the mean size of each species will be mentioned.

Todites princeps (Presl) Gothan (Late Triassic and Jurassic) is unique in its anadromic branching of pinnules and veins (Harris, 1961; Van Konijnenburg-van Cittert, 1978) – it is the only osmundaceous species (living or fossil) that displays this feature. Its spores (Fig. 29) are 35 μm in diameter and, as the exospore is scabrate, they should fall into the *Todites williamsonii*-type

The spores of most (if not all, see below) *Osmundopsis* and fossil *Osmunda* species belong to the *Osmundopsis*-type: *Osmundopsis sturii* (Raciborski) Harris – 48 μm, Middle Jurassic (Van Konijnenburg-van Cittert, 1978; Schweitzer *et al.*, 1997; Figs. 24, 25), *O. plectrophora* Harris – 50 μm, lowermost Jurassic (Van Konijnenburg-van Cittert, 1978; Fig. 27), *Osmunda diamensis* (Seward) Krassilov – 60 μm, Jurassic (Krassilov, 1978; Balme, 1995) and *O. lignitus* (Giebel) Stur – 55 μm, Tertiary (Barthel, 1976). Furthermore, the spores of two Late Triassic *Todites* species belong to this group: *T. nebbensis* (Brongniart) Kilpper – 32 μm (Schweitzer *et al.*, 1997) and *T. crenatus* Barnard – also 32 μm (Schweitzer *et al.*, 1997). The spores of the latter species have an exospore with widely spaced baculae, granules or spines that are not fused, contrary to all other spores from the *Osmundopsis*-type where the exosporal elements are always irregularly fused in small groups (Fig. 25). However, only a few spores of *T. crenatus* were recovered and, therefore, these results may not be representative for the whole species.

FIGS. 19–23. *Todea* and *Leptopteris*. Fig. 19. *Todea barbara*. Fig. 20. *Todea barbara* exosporal elements; × 10,000. Fig. 21. *Leptopteris superba* monolete spore. Fig. 22. *Leptopteris superba* echinate spore. Fig. 23. *Leptopteris wilkesiana* conate exospore with echinate perispore × 5,000.

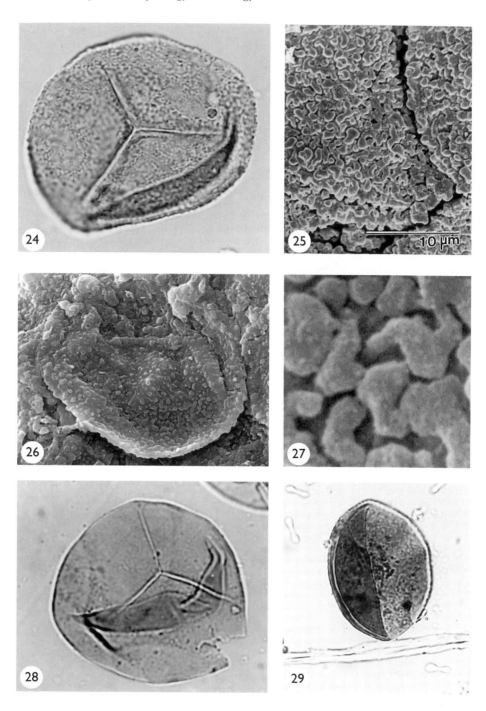

FIGS. 24–29. *Osmundopsis* and *Todites*. Fig. 24. *Osmundopsis sturii*. Fig. 25. *Osmundopsis sturii* partly fused exosporal elements. Fig. 26. *Todites hartzii*. Fig. 27. *Osmundopsis plectrophora* × 10,000. Fig. 28. *Todites williamsonii*. Fig. 29. *Todites princeps*.

The *Todites hartzii*-type comprises many *Todites* species: *Todites fragilis* Daugherty – 35 μm, Late Triassic (Litwin, 1985), *T. hartzii* Harris – 46 μm, uppermost Triassic/lowermost Jurassic (Van Konijnenburg-van Cittert, 1978; Balme, 1995; Fig. 26), *T. recurvatus* Harris – 38 μm, uppermost Triassic/lowermost Jurassic (Harris, 1931; Balme, 1995), *T. thomasii* Harris – 47 μm, Middle Jurassic (Van Konijnenburg-van Cittert, 1978) and *T. denticulatus* (Brongniart) Krasser – 50 μm, Middle Jurassic (Van Konijnenburg-van Cittert, 1978). The spores of *Cacumen expansa* Cantrill et Webb – 20–50 μm, Cretaceous (Cantrill and Webb, 1987; Balme, 1995) also belong to this type. The spores of *Osmundopsis hillii* Van Konijnenburg-van Cittert as described by Van Konijnenburg-van Cittert, 1996 might also belong to this type, and not to the *Osmundopsis*-type. However, just as in the case of *Todites crenatus*, only a few spores were recovered from the fossil material, and therefore, the exospore sculpture may be imperfectly known and the attribution to the *Todites hartzii*-type or the *Osmundopsis*-type is not clear.

Apart from possibly *T. princeps*, only two species belong to the *Todites williamsonii*-type: *T. williamsonii* (Brongniart) Seward – 62 μm, Late Triassic – Late Jurassic (Van Konijnenburg-van Cittert, 1978; Balme, 1995; Fig. 28) and *T. goeppertianus* (Muenster) Krasser – 50 μm, uppermost Triassic/lowermost Jurassic (Harris, 1931; Van Konijnenburg-van Cittert, 1978; Balme, 1995). All these results are summarised in Table 1.

Some *in situ* osmundaceous spores that are not listed above, *Osmundopsis prynadae* and *Todites pseudoraciborskii*, probably had immature spores which did not show many details (Balme, 1995), while Krassilov (1969) briefly discussed the spores of *Osmundopsis kugartensis* and *Cladotheca kazachstanica* without figuring them (and they have not been figured elsewhere) so comparison cannot be made.

TABLE 1. Comparison of fossil osmundaceous spores

Species name	Spore type	Size in μm	Age
Osmundopsis sturii	Osmundopsis	48	Middle Jurassic
Osmundopsis plectrophora	Osmundopsis	50	Early Jurassic
Osmunda diamensis	Osmundopsis	60	Jurassic
Osmunda lignitus	Osmundopsis	55	Tertiary
Todites nebbensis	Osmundopsis	32	Late Triassic
?*Todites crenatus*	Osmundopsis	32	Late Triassic
Todites fragilis	T. hartzii	35	Late Triassic
Todites hartzii	T. hartzii	46	Early Jurassic
Todites recurvatus	T. hartzii	38	Early Jurassic
Todites thomasii	T. hartzii	47	Middle Jurassic
Todites denticulatus	T. hartzii	50	Middle Jurassic
Cacumen expansa	T. hartzii	20–50	Cretaceous
Osmundopsis hillii	?T. hartzii	42	Middle Jurassic
Todites goeppertianus	T. williamsonii	50	Early Jurassic
Todites williamsonii	T. williamsonii	62	Late Triassic–Late Jurassic
Todites princeps	? T. williamsonii	35	Late Triassic–Jurassic

Evolutionary trends in the Osmundaceae

A. Number of spores per sporangium

Bower (1923) was the first to deal with the number of spores per sporangium. The number 64 is certainly the most widespread among the ferns nowadays, but eusporangiate families, such as the Marattiaceae and the Ophioglossaceae, have much higher numbers (where 256 spores per sporangium is normal), and the Osmundaceae usually have 128 or sometimes 256 spores in their large sporangia. This is also the case in the fossil Osmundaceae (Harris, 1961; Van Konijnenburg-van Cittert, 1996).

B. Spore size

Spore size in extant Osmundaceae can vary considerably. This is not the case in most of the fossil material. There the size of the spores is around 30-60 μm (see Table 1) with a mean variation of ca. 20% within one species. The size variation in some of the living species may be up to almost 100%. However, some populations of *Osmunda regalis* showed a rather constant spore size (for example on the Isle of Man), while others were highly variable. Large and variable spore size has been correlated with polyploidy (for example in the Ophioglossaceae; Wagner, 1974). This may be the case here as well in the cosmopolitan genus *Osmunda*, although polyploidy has not been reported from *Osmunda*.

Judging from Table 1, the smallest fossil spores (30–35 μm) are those found in the Late Triassic but from the Jurassic onwards they are usually between 40 and 60 μm. Extant osmundaceous spores vary in size, from 30–45 μm in *Leptopteris*, via 40–80 μm in *Osmunda* to 50–90 μm in *Todea*.

In general, it seems that spore size in the Osmundaceae has increased in the course of evolution. It is not clear if the rather small size of the present *Leptopteris* spores is a primary or a secondary feature, as there is no fossil record of *Leptopteris*.

C. Overall spore shape and laesurae

Spore shape and aperture type (trilete or monolete) reflect the alignment of spores in the tetrad. Trilete spores, arranged tetrahedrally, are generally believed to be primitive because they occur early in the fossil record, and are dominant in most primitive fern families such as the Marattiaceae, Ophioglossaceae and Osmundaceae. Bilateral spores with a monolete aperture are dominant in some of the more specialised families such as the Polypodiaceae *sensu lato* or the Dipteridaceae. At one time it was generally believed that the living Osmundaceae comprised only trilete spores, but Bobrov (1966) and the present study have demonstrated that occasionally monolete and incompletely trilete spores occur, and also spores with double laesurae ("bipolar apertures"). All fossil osmundaceous spores described so far are trilete; no trace of monolete spores was found although they were looked for.

Two basic spore shapes occur within the large group of fern taxa with tetrahedral, trilete spores. These are globose spores with a more or less circular equatorial outline and spores with a (rounded) triangular outline. Of these two types, the globose shape is probably more primitive as it occurs both in the eusporangiate families (Marattiaceae and Ophioglossaceae) and in the protoleptosporangiate Osmundaceae (Van Konijnenburg-van Cittert, 1975, 1978, 1996). The globose shape of osmundaceous spores does not seem to have been affected during evolution as the oldest recorded *in situ* spores had this shape, and all the living spores, except occasional monolete ones, still have it.

D. Exosore thickness and sculpture

The exospore is considered to be extremely important for establishing evolutionary relationships. Exospore architecture is very diverse. Pettitt (1966) was the first to study exospore structure in detail, as revealed by light and electron microscopy. Since then, numerous scientists have dealt with the subject, culminating in the work of Tryon and Lugardon (1991) which describes pteridophyte spores from all over the world with the aid of SEM and TEM. The exospore will be discussed only in very general terms, considering possible evolutionary trends in exospore architecture.

Exospore thickness

Thin exospores (1–2 μm) are considered to be primitive, while exospores from 2-6 μm are probably derived (Wagner, 1974). However, Tryon (1990) records that in some of the more primitive families, for example Marattiaceae, Ophioglossaceae and Osmundaceae, the exospore is relatively thick (2–3 μm) and is overlaid by a thin, conforming perispore (see paragraph E for the perispore). In her opinion this is the primitive wall structure. It seems certain that exospore thicknesses exceeding 6 μm is a derived feature. This may be an adaptation to varying environmental circumstances, such as hot days and cold nights, or extreme droughts, and may protect the spores against drying. In the case of the Osmundaceae, there is only a slight variation in exospore thickness, both in fossil and living species, and no evolutionary trend could be detected.

Exospore sculpture

The general concept is that primitive exospore sculpture is either almost smooth or finely ornamented, and that massive, coarse sculpture is derived (Wagner, 1974). The relatively simply ornamented exospores of the Osmundaceae are only slightly derived, because the sculptural elements are not granulae but baculae or verrucae, while, the fusion of the elements in some extant osmundaceous species is derived. In the fossil material the exospore is either almost smooth (*Todites williamsonii*-type) or granulate (*Todites hartzii*-type) to baculate (*Osmundopsis*-type) occasionally with fused elements. Here, an evolutionary trend is clearly visible.

E. Perispore

Perispore presence or absence

Although many fossil *in situ* spores with perispores have been recovered, the majority of fossil spores lack a perispore, and this is the case in all osmundaceous fossil spores that have so far been found *in situ*. It is not clear whether they did not possess a perispore, or if the perispore was so thin that it was destroyed either during fossilisation or during spore preparation.

Perispore thickness and sculpture

A primitive perispore is a thin layer that overlies the exospore and follows its sculpture closely. These thin perispores consist of one layer only (Tryon, 1990), and have been recorded from Carboniferous marattiaceous spores (Millay, 1979). Nowadays they are found in the Marattiaceae and Ophioglossaceae. Their sculpture is usually smooth or scabrate. Perispores in extant Osmundaceae are also thin, but they are echinate and, in this feature, are unique in the pteridophytes (Tryon, 1990; Tryon and Lugardon, 1991). As mentioned above, none of the osmundaceous fossil spores has demonstrated a perispore.

Conclusion

To conclude, one might say firstly that the spore features of the Osmundaceae clearly reflect the rather primitive position of the family (Protoleptosporangiatae) and, secondly, that the family appears to be homogeneous, well-delimited and stable with regard to evolution from the Triassic onwards.

References

Balme, B.E. (1995). Fossil *in situ* spores and pollen grains: an annotated catalogue. *Review of Palaeobotany and Palynology* 87: 81–323.

Barthel, M. (1976). Eozäne Floren des Geiseltales. Farne und Cycadeen. *Abhandlungen des Zentralen Geologischen Instituts* 26: 439–490.

Bobrov, A.E. (1966). A contribution to the spore morphology of the Osmundaceae. *Botaniscesky Zhurnal* 51(10): 1452–1460.

Bobrov, A.E. (1967). The family Osmundaceae (R.Br.) Kaulf., its taxonomy and geography. *Botanical Journal Academy of Sciences USSR* 52: 1600–1610.

Bobrov, A.E. Kuprianova, L.A., Litvintseva, M.V. and Tarasevich, V.F. (1983). Sporae Pteridophytorum et Pollen Gymnospermarum Monocotyledonearumque Florae Partis Europaeae URSS. Acad. Sci. URSS Komarovii Inst. Bot., Moscow.

Bower, F.O. (1923). The Ferns. Vol. I. The criteria of comparison. Cambridge University Press.

Cantrill, D.V. and Webb, J.G. (1987). A reappraisal of *Phyllopteroides* Medwell (Osmundaceae) and its stratigraphic significance in the Lower Cretaceous of eastern Australia. *Alcheringa* 11: 59–85.

Collinson, M.E. (1996). "What use are fossil ferns?" – 20 years on: with a review of the fossil history of extant Pteridophyte families and genera. In: Camus, J.M. and Johns, R.J. (eds.): Pteridology in perspective, pp. 349–394. Royal Botanic gardens, Kew.

Hanks, S.L. and Fairbrothers, D.E. (1981). A palynological investigation of three species of *Osmunda*. *Bulletin of the Torrey Botanical Club* 108(1): 1–6.

Harris, T.M. (1931). The fossil flora of Scoresby Sound, East Greenland. Part 1. Cryptogams. *Meddelelser om Grønland* 85(2): 1–102.

Harris, T.M. (1961). The Yorkshire Jurassic Flora. I. Thallophyta–Pteridophyta. British Museum (Natural History), London.

Harris, W.F. (1955). A manual of the spores of the New Zealand Pteridophyta. *New Zealand Department of Scientific Industrial Research Bulletin* 116: 1–186.

Krassilov, V.A. (1969). A critical review of the taxonomy of fossil ferns of the USSR with spores *in situ*. *Fossil fauna and flora of the Far East* 1: 117–128.

Krassilov, V.A. (1978). Mesozoic lycopods and ferns from the Bureja Basin. *Palaeontographica* Abt. B 166: 16–29.

Large, M.F. and Braggins, J.E. (1991). Fern spore atlas. *New Zealand Journal of Botany* 29: 1–167.

Litwin, R.J. (1985). Fertile organs and *in situ* spores of ferns from the Late Triassic Chinle Formation of Arizona and New Mexico, with discussion of associated dispersed spores. *Review of Palaeobotany and Palynology* 44: 101–146.

Millay, M.A. (1979). Studies of Palaeozoic marattialeans: a monograph of *Scolecopteris*. *Palaeontographica* B 169: 1–169.

Miller, C.N. (1971). Evolution of the fern family Osmundaceae based on anatomical studies. *University of Michigan Publications, Contribution Museum Palaeontology* 23: 105–169.

Pettitt, J.M. (1966). Exine structure in some fossil and recent spores and pollen as revealed by light and electron microscopy. *Bulletin British Museum (Natural History) Geology* 13: 221–257.

Schweitzer, H.-J., Van Konijnenburg - van Cittert, J.H.A. and Van der Burgh, J. (1997). The Rhaeto-Jurassic flora of Iran and Afghanistan. 10. Bryophyta, Lycophyta, Sphenophyta, Pterophyta-Eusporangiatae and Protoleptosporangiatae. *Palaeontographica* B 243: 103–192.

Tidwell, W.D. and Ash, S.R. (1994). A review of selected Triassic to Early Cretaceous ferns. *Journal Plant Research* 107: 417–442.

Tryon, A.F. (1990). Fern spores: evolutionary levels and ecological differentiation. *Plant Systematic Evolution* Suppl. 5: 71–79.

Tryon, A.F. and Lugardon, B. (1991). Spores of the Pteridophyta. Springer Verlag, New York.

Van Konijnenburg-van Cittert, J.H.A. (1975). Some notes on *Marattia anglica* from the Jurassic of Yorkshire. *Review Palaeobotany Palynology* 20: 205–214.

Van Konijnenburg-van Cittert, J.H.A. (1978). Osmundaceous spores *in situ* from the Jurassic of Yorkshire, England. *Review Palaeobotany Palynology* 26: 125–141.

Van Konijnenburg-van Cittert, J.H.A. (1996). Two *Osmundopsis* species and their sterile foliage from the Middle Jurassic of Yorkshire. *Palaeontology* 39: 719–731.

Wagner, W.H. (1974). Structure of spores in relation to fern phylogeny. *Annals Missouri Botanical Garden* 61: 332–353.

Zhang Yulong, Xi Yizhen, Zhang Jintan, Gao Guizhen, Dui Naiqiu, Sun Xiangjun and Kong Zhaochen (1976). Spore Morphology of Chinese Pteridophytes. Science Press, Beijing, China.

Srivastava, S.K. (2000). Palaeogeography of some Neocomian-Albian pollen and their significance in the evolution of phytogeoprovinces around the Early Cretaceous Atlantic Ocean. In: M.M. Harley, C.M. Morton and S. Blackmore (Editors). Pollen and Spores: Morphology and Biology, pp. 451–466. Royal Botanic Gardens, Kew.

PALAEOGEOGRAPHY OF SOME NEOCOMIAN-ALBIAN POLLEN AND THEIR SIGNIFICANCE IN THE EVOLUTION OF PHYTOGEOPROVINCES AROUND THE EARLY CRETACEOUS ATLANTIC OCEAN

SATISH K. SRIVASTAVA

3054 Blandford Drive, Rowland Heights, California 91748-4825, USA

Abstract

Ephedroid monosulcate *Jugella* has an affinity with extant Araceae pollen. It was first described from the Neocomian-Barremian of the Ural Mountains and northwestern Siberia. It occurs in the *Retimonocolpites-Afropollis* assemblage from Hauterivian sediments of southern England and in the coeval *Dicheiropollis etruscus* assemblage from the offshore Canary Islands. The widespread occurrence of *Jugella* indicates that continents around the North Atlantic were linked by land during the Neocomian. The equatorial *D. etruscus* phytogeoprovince spanned southern Italy, north Africa, northern South America, and southern North America. A further northward extension of *D. etruscus* was restricted by environmental barriers. Similarly, palaeoecological barriers restricted the entrance of the *Retimonocolpites-Afropollis* spp. complex into the equatorial phytogeoprovince until the Aptian. Aptian-Albian occurrences of elater-bearing pollen indicate that the Afro-South American landmass was connected in the northwest with the Ozarks of North America and in the northeast with Bohemian land in Europe through the Southern Alps. An ancestral *Nypa*-like pollen appears during the Albian in the equatorial phytogeoprovince. The *Nypa*-like pollen occurrence became widespread in Early Cenozoic sediments of southern Europe.

Introduction

A *Retimonocolpites* Pierce - *Afropollis* Doyle, Jardiné & Doerenkamp - spp. complex in association with *Jugella* Mchedlishvili & Shakhmundes spp. occurs in various Hauterivian core samples from southeastern England (Hughes, 1994). The genus *Jugella* was not recognised from the Neocomian equatorial *Dicheiropollis* phytogeoprovince (Hughes, 1994) that included Africa, South America and southern North America (Srivastava, 1994). On the basis of the recently reported occurrence of elater-bearing pollen in the Bohemian Cenomanian (Pacltová and Svobodová, 1992), the Aptian-Albian boundary of the equatorial phytogeoprovince needs to be adjusted further northward. The earliest occurrence of *Nypa*-like pollen grains was found in the Albian of the Arabian Gulf that was part of the Aptian-Albian equatorial phytogeoprovince. In this paper, the Early Cretaceous phytogeography of *Retimonocolpites-Afropollis* spp., ephedroid monosulcate *Jugella* spp., *Dicheiropollis etruscus* Trevisan, elater-bearing palynomorphs, and *Nypa*-like pollen is discussed.

Material and Methodology

Material

Dicheiropollis etruscus and associated ephedroid monosulcate *Jugella* spp. were recovered from a Hauterivian sample of DSDP Leg 47A, Hole 397A (Core 47, Section 1, #17-21cm), offshore Canary Islands. Elater-bearing and *Nypa*-like pollen were recovered from Albian sub-bottom core samples from the Arabian Gulf. For SEM photomicrographs, acetolysed pollen of extant *Nypa fruticans* van Wurmb and *Spathiphyllum* Schott. sp. were used from the A. R Loeblich, Jr., collection at the former Chevron Oil Field Research Company (now Chevron Petroleum Technology Company), La Habra, California.

Methodology

The edges of a landmass are major repositories of its organic matter that is transported with sediments to the ocean. Pollen assemblages from such sediments indicate the floral composition and prevailing environments of the respective landmasses. However, when a landmass breaks up, its fragments disperse and join other landmasses, enabling an intermixing of flora from one geoprovince to another. Inland seaways of these landmasses may act as physical barriers for advancing floras and define the boundaries of phytogeoprovinces (Srivastava, 1978). Plant migration patterns can be explained either by the transgression/regression history of seaways (Srivastava, 1981) or by plate tectonic data (Srivastava, 1988). Such an approach and concept has been used here in applying the palaeogeography of some Neocomian-Albian pollen to determine the evolution of phytogeoprovincial boundaries around the Early Cretaceous Atlantic Ocean.

Neocomian-Barremian phytogeoprovinces

Early angiosperm pollen from southern England

The genus *Afropollis* Doyle, Jardiné & Doerenkamp (Figs. 1–2) is very similar to *Retimonocolpites* Pierce but has been considered distinct in being an inaperturate or zonasulcate pollen (Doyle *et al.*, 1982). *Retimonocolpites peroreticulatus* Brenner and *R. reticulatus* (Brenner) Doyle in Doyle, Van Campo & Lugardon are small coarsely reticulate monocolpate pollen (Fig. 3). The coarse reticulum is easily detached from the foot layer of the sexine (Doyle *et al.*, 1975). The earliest occurrence of both these genera is in the Hauterivian subsurface core samples of southern England (Hughes, 1994). They became widespread during Aptian-Albian time in the boreal and equatorial phytogeoprovinces (Doyle *et al.*, 1982).

Ephedroid monosulcate pollen also occur in the Hauterivian *Retimonocolpites-Afropollis* spp. assemblages of southern England (Hughes, 1994; fig. 9.19). The genus *Jugella* Mchedlishvili & Shakhmundes was proposed for longitudinally ribbed monosulcate pollen with ephedroid exine from the Neocomian and Barremian of the Ural Mountains

---

FIGS. 1–2. *Afropollis jardinus* (Brenner) Doyle, Jardiné & Doerenkamp, Cenomanian, the Arabian Gulf. Fig. 1. Complete specimen (scale = 10 μm). Fig. 2. Detail of the exine (scale = 1 μm). **FIG. 3.** *Retimonocolpites peroreticulatus* (Brenner) Doyle in Doyle, Van Campo & Lugardon, Albian, the Arabian Gulf (scale = 10 μm).

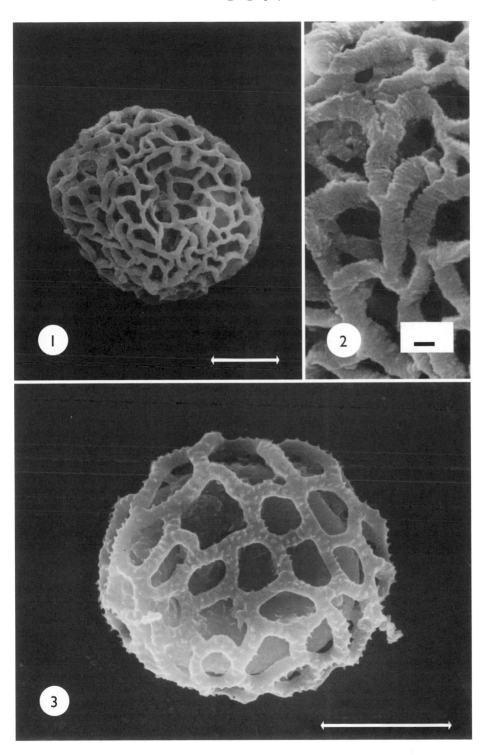

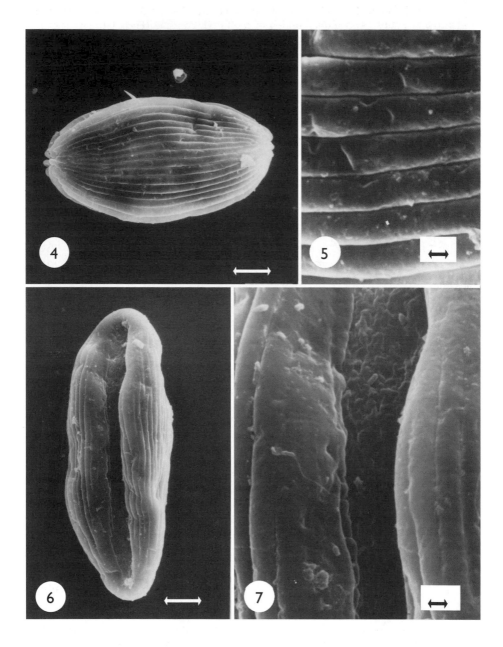

FIGS. 4–5. *Jugella* Mchedlishvili & Shakhmundes sp., Hauterivian, offshore Canary Islands; Fig. 4. A complete specimen in distal view (scale = 10 μm). Fig. 5. Detail of the ribbed exine (scale = 1 μm). FIGS. 6–7. *Jugella* Mchedlishvili & Shakhmundes sp., Hauterivian, offshore Canary Islands; Fig. 6. A complete specimen in a proximal view showing the sulcus (scale = 10 μm). Fig. 7. Detail of the sulcus and ribbed exine (scale = 1 μm).

and northwestern Siberia (Mchedlishvili and Shakhmundes, 1973) and the pollen were compared to extant *Spathiphyllum* pollen of the family Araceae (Figs. 4–6; Figs. 12–14). Trevisan (1980) examined ultrathin-sections of *Ephedripites* Bolkhovitina sp. and extant *Spathiphyllum wallisii* Regel pollen and found that the exine of some *Ephedripites* sp. specimens is comparable to that of *Spathiphyllum* pollen. Berriasian -Valanginian records of *Jugella* spp. from the Ural Mountains predate southern England records. *Equisetosporites multicostatus* (Brenner) S. K. Srivastava reported from the Neocomian section of the DSDP Site 105, western North Atlantic (Habib, 1977) and from the lower Barremian of France (Srivastava, 1984a; pl. 38, fig. 11) may also belong to the genus *Jugella*. The Hauterivian ephedroid monosulcate pollen were not recognised in the coeval equatorial area, designated as the *Dicheiropollis* phytogeoprovince (Srivastava, 1978). Thus, Hughes (1994) considered them restricted to the boreal phytogeoprovince.

Jugella Mchedlishvili & Shakhmundes occurrence in the Neocomian equatorial phytogeoprovince

Dicheiropollis Trevisan (Figs. 8–11), a gymnospermous dyad pollen (Trevisan, 1972), is restricted to the Neocomian. It has been reported from the Neocomian of southern Italy, northern Africa, South America (Jardiné *et al.*, 1974) and the southeastern part of North America (Srivastava, 1994). Its reported rare occurrences in early Barremian are not reliable.

Documented Neocomian *Dicheiropollis* assemblages are sparse, hence very little is known about palynomorphs associated with this assemblage. Müller's (1966) *Classopollis* Pflug forms without equatorial striations from the Neocomian of Brazil have been accepted as *Dicheiropollis* sp. (Jardiné *et al.*, 1974). His "Polyplicate (predominantly *Ephedripites*)" forms could be *Jugella* spp. *Jugella* spp. (Figs. 4–7) occur in the Hauterivian *Dicheiropollis etruscus* assemblage of a DSDP core from offshore Canary Islands (Map 1). The Canary Islands specimens are similar to those recorded from southern England (Hughes, 1994), the Ural Mountains and western Siberia (Mchedlishvili and Shakhmundes, 1973). The data support that *Jugella* spp. occurred in the boreal as well as the equatorial phytogeoprovince during the Neocomian.

Neocomian palynomorph palaeogeography in the boreal and equatorial phytogeoprovinces (Map 1)

North America and Africa had a land connection through South America during the Late Jurassic (Kimmeridgian) (Moore *et al.*, 1995). However, Neocomian tectonic activity in the Caribbean area and the Gulf of Mexico obscured any firm evidence for the existence of a land connection between the North American and Afro-South American continents. An island arc type bridge has been proposed for Neocomian-Cenomanian time that facilitated biotic migration between the two continental blocks (Pitman III *et al.*, 1993). The presence of *Dicheiropollis* sp. in the Neocomian subsurface samples of Texas (Srivastava, 1994) supports such a land connection.

Although *Jugella* spp. occurred in all the continents around the North Atlantic during the Neocomian, *Dicheiropollis* did not enter the boreal phytogeoprovince (Map 1). Neocomian deposits do not occur on the Atlantic coast of North America (Williams and Stelck, 1975) which may account for the lack of a *Dicheiropollis* record. The *Retimonocolpites-Afropollis* spp. complex did not enter the equatorial phytogeo-province during the same time. Palaeoecological barriers such as the Appalachian Mountains could be one of the causes for the restriction of floras to respective areas during the Neocomian.

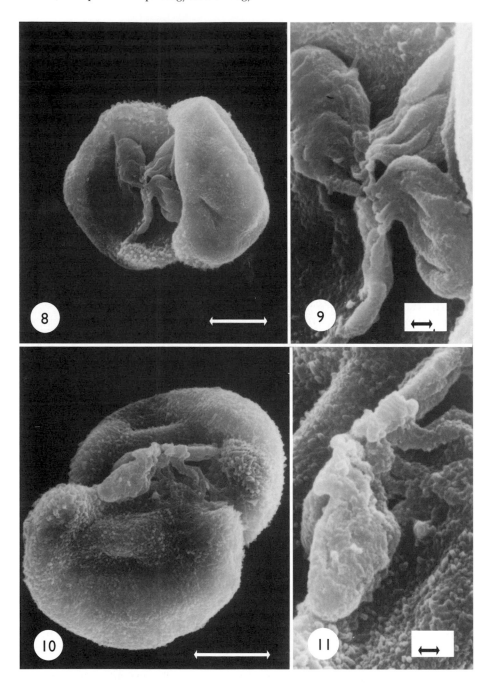

FIGS. 8–11. *Dicheiropollis etruscus* Trevisan, Hauterivian, offshore Canary Islands. Fig. 8. A complete dyad showing connecting strands (scale = 10 μm). Fig. 9. Detail view of connecting strands (scale = 1 μm). Fig. 10. Another specimen of *D. etruscus* (scale = 10 μm). Fig. 11. Details of connecting strands and a portion of the exine (scale = 1 μm).

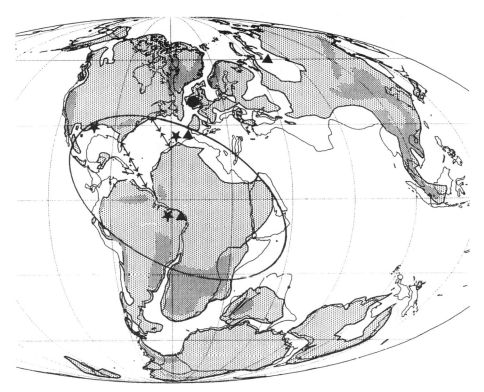

MAP 1. World map showing Neocomian land and sea distribution and the boundary of the equatorial phytogeoprovince. • Earliest occurrence of *Retimonocolpites-Afropollis* spp. complex in association with *Jugella* spp.; ▲ Neocomian *Jugella* spp. occurrences; ★ Neocomian occurrence of *Dicheiropollis etruscus*; ↔↔ possible Neocomian land connection between Afro-South American continent and North America; encircled area with solid line is the possible equatorial phytogeoprovince during the Neocomian. Modified on a Hauterivian map after Smith *et al.*, 1994.

Aptian-Albian phytogeoprovinces

Elater-bearing pollen from the equatorial phytogeoprovince

Elaterosporites Jardiné, *Elaterocolpites* Jardiné & Magloire emend. Jardiné, *Elateroplicites* Herngreen, and *Pentarhethus* S.K. Srivastava (Figs. 15–18) are the main elater-bearing genera (Srivastava, 1984b) that defined the boundaries of the equatorial phytogeoprovince during Aptian-Albian time. Re-worked pollen occurrences such as *Fustispollenites* Tschudy & Pakiser from the Late Cretaceous of Kentucky (Tschudy and Pakiser, 1967) and *Elaterosporites klaszii* (Jardiné & Magloire) Jardiné from the Paleocene-Eocene Wilcox Group of Louisiana (Gregory and Hart, 1992) indicate that the distribution of plants that produced elater-bearing pollen extended into the North American southeastern interior during the Aptian-Albian. These plants may have occupied the Ozarks that remained above water throughout the Cretaceous. Local Aptian-Albian deposits could be the source of re-worked elater-bearing pollen that eroded either primarily or subsequently into the Paleocene Wilcox Group. Elater-bearing pollen do not occur in the Aptian-Albian sediments of North America east of

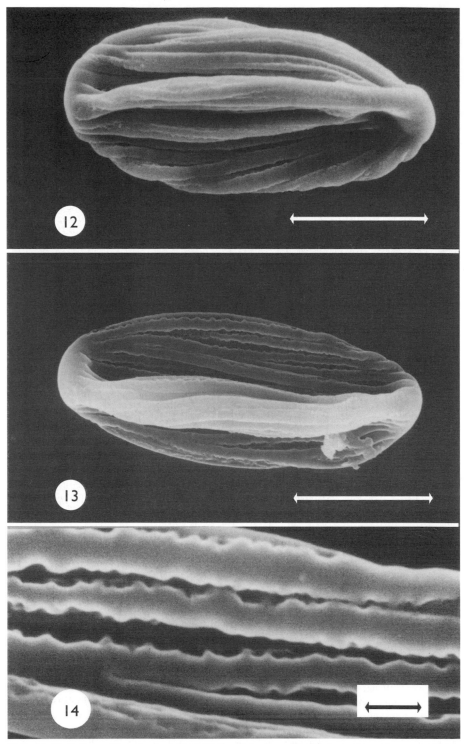

FIGS. 12–14. Pollen grains of extant *Spathiphyllum* Schott. sp. Fig. 12. A specimen in proximal view (scale = 10 μm). Fig. 13. Another specimen in proximal view (scale = 10 μm). Fig. 14. details of the ribbed exine (scale = 1 μm).

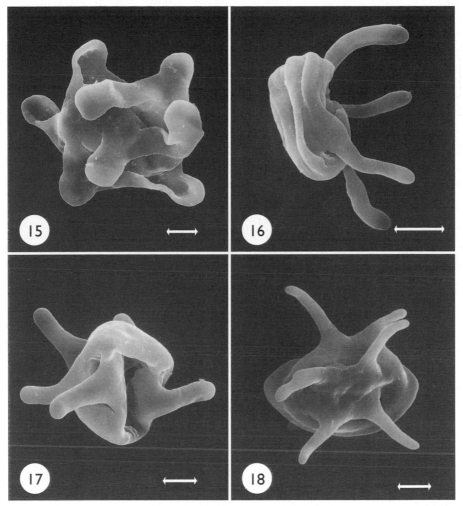

FIG. 15. *Elaterocolpites castelainii* Jardiné & Magloire, late Albian, the Arabian Gulf. FIG. 16. *Elateroplicites africaensis* Herngreen, Cenomanian, the Arabian Gulf. FIG. 17. *Elaterocolpites pentarhethus* S. K. Srivastava, late Albian, the Arabian Gulf. FIG. 18. *Elaterosporites klaszii* (Jardiné & Magloire) Jardiné, Albian, the Arabian Gulf (scale, Figs. 15–18 = 10 μm).

the Appalachian Mountains (Brenner, 1963; Doyle and Robbins, 1977). The Appalachian Mountains existed throughout the Cretaceous Period (Williams and Stelck, 1975) and may have formed a palaeoecologic barrier for further spread of plants that produced elater-bearing pollen.

The Bohemian Cenomanian Peruc-Korycany Formation overlies unconformably on an erosional Carboniferous surface (Čech *et al.*, 1997). Rare elater-bearing pollen occur in the basal (Pacltová and Svobodová, 1992) and the upper claystone beds of the Peruc Member (Pacltová, 1990). On the basis of pollen assemblages, Pacltová (1971) and Svobodová (1991) correlated the lower-upper claystone beds of the Peruc Member with the upper Albian of Hungary and the Patapsco Formation of the Atlantic Coastal Plain in North America. Konzolova (pers. comm.) confirmed that the elater-bearing

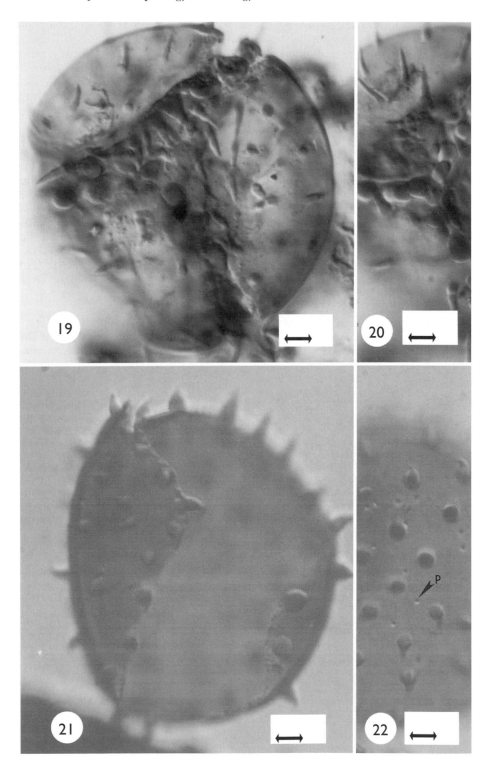

pollen beds of the Peruc member are now considered to be of Albian age. Czech Mid-Cretaceous palynomorphs have been considered as having affinity with both Tethyan and boreal bioprovinces (Svobodová, 1997). The Bohemian Cretaceous basin is filled with Albian transgressive facies followed by the regressive Cenomanian and younger Late Cretaceous facies (Čech, 1989; Pacltová and Svobodová, 1992). Regressive facies receive eroded sediments from the proximal older sediments. Thus, the rare occurrences of poorly preserved elater-bearing pollen appear to be the examples from the eroded Aptian-Albian deposits of the Bohemian basin.

Plants producing elater-bearing pollen had reached the northern Italy and southern Switzerland area during the Aptian-Albian (Hochuli, 1981). Proximal land areas of central European and Bohemian massifs were separated only by deltaic shallow marine sands during the Aptian-Albian (Ziegler, 1988). Elater-bearing pollen producing plants must have reached the Bohemian land during this time when the possibility of a land contact between southern European land areas was much greater. Thus, rare and poorly preserved Bohemian elater-bearing pollen may represent reworked examples from the eroded pre-Peruc beds. Bohemian records of *Elaterocolpites* spp. (Pacltová and Svobodová, 1992) appear to indicate a northeastward extension of the land connection between the Afro-South American and European continents up to the Bohemian land through the Alps during the Aptian-Albian time (Map 2).

Albian *Nypa*-like pollen in the equatorial phytogeoprovince

Spinose zonasulcate pollen with sparsely punctate exine occur in the Albian of the Arabian Gulf. These pollen grains often split along the sulcus and each half appears as a spinose monosulcate pollen grain (Figs. 19–22). These pollen differ from the genuinely monosulcate genus *Echimonocolpites* van der Hammen & Garcia de Mutis in being zonasulcate but with the dehisced halves appearing to be monosulcate. The genus *Spinizonocolpites* Muller could well accommodate these Albian pollen grains due to the presence of a zonate sulcus, spines, and sparsely punctate exine surface. However, the type species, *S. echinatus* Muller, has a baculate infratectal exine with a retipilate tectal surface. Although Harley *et al.* (1991) noted several distinctions in the spines and exine of Senonian *Spinizonocolpites* spp., they still compared the species with pollen of the extant monotypic genus *Nypa* Steck. (Figs. 23–26) on the premise that the fossil history of *Nypa* must have been more diverse than its modern representation. Considering the zonasulcate nature of the pollen, the shape and size of the spines, and the sparsely punctate exine surface, the Albian pollen (Figs. 19–22) could represent ancestral forms of a *Nypa* lineage. The sparsely punctate exine surface of the Albian pollen evolved into the well-defined baculate infratectal and retipilate tectal sculpture of Senonian *Spinizonocolpites* spp. Albian *Nypa*-like pollen could have reached southern Europe along with elater-bearing pollen from the equatorial phytogeoprovince during Albian time. They occurred widely in the Early Cenozoic of southern Europe, the Middle East, northern Africa, and northern South America. The distribution of modern monotypic *Nypa* Steck. is restricted to the Indo-Malayan and Austro-Malayan regions (Uhl and Dransfield, 1987; Srivastava and Binda, 1991). Thus *Nypa fruticans* Van Wurmb may be a relict species of a previously diverse genus in fossil history (Harley *et al.*, 1991).

FIGS. 19–22. Albian *Nypa*-like pollen grains from the Arabian Gulf; Figs. 19–20. A complete trichotomosulcate spinose specimen and details of its exine. Figs. 21–22. A dehisced part of a specimen and details of its exine showing sparsely scattered punctae (indicated by P at the arrow) (scale, Figs. 19–22 = 10 μm).

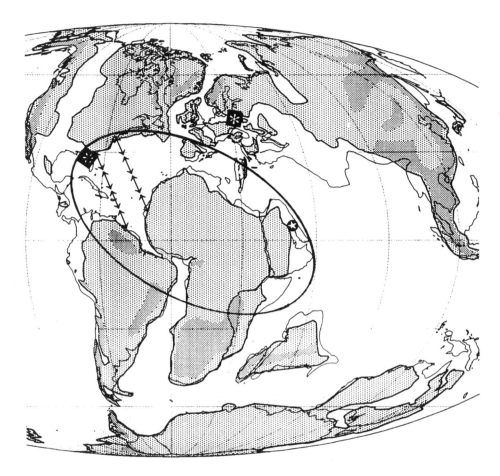

MAP 2. World map showing Aptian-Albian land and sea distribution and the boundary of the equatorial phytogeoprovince. ✳️Reworked elater-bearing pollen record near Ozarks in North America and on the Bohemian land in Europe; ↔↔ possible Aptian-Albian land connection between Afro-South American continent and North America; encircled area with solid line is the possible equatorial phytogeoprovince during Aptian-Albian time; ✪ Albian *Nypa*-like pollen occurrence in the Arabian Gulf. Modified on an Albian map after Smith *et al.*, 1994.

Conclusions

1. *Jugella* occurred in the Neocomian of the boreal and equatorial phytogeoprovinces indicating land connections that maintained plant migratory paths between the two provinces during the early Neocomian. The *Retimonocolpites-Afropollis* spp. complex and *Dicheiropollis etruscus* remained restricted to the boreal and equatorial phytogeoprovinces respectively, possibly due to palaeoecological barriers, such as the Neocomian Appalachian Mountains.

2. The *Retimonocolpites-Afropollis* spp. complex reached the equatorial phytogeoprovince in Aptian-Albian time due to the break down of Neocomian palaeoecologic barriers. However, the Albian occurrence of elater-bearing pollen

remained restricted to the equatorial phytogeoprovince. The area of the equatorial phytogeoprovince increased more during Aptian-Albian time than during the Neocomian. This is indicated by the extended distribution of elater-bearing pollen towards the northwest in southeastern North America and towards the northeast in southern Europe up to the Bohemian land.

3. An ancestral *Nypa*-like pollen occurred in the equatorial phytogeoprovince during the Albian. *Nypa*-like pollen became common in the Early Cenozoic of southern Europe and the equatorial areas.

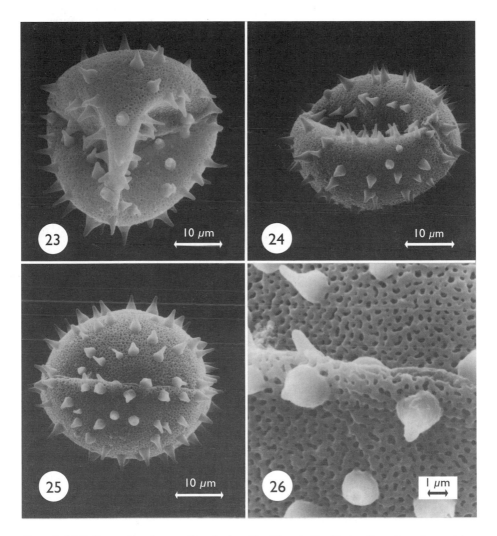

FIGS. 23–26. Pollen grains of extant *Nypa fruticans* Van Wurmb; Fig. 23. A collapsed specimen giving a false appearance of trichotomosulcate nature of the grain (scale = 10 μm). Fig. 24. A specimen showing zonasulcate nature of the grain (specimen is slightly collapsed) (scale = 10 μm). Figs. 25–26. A specimen showing zonasulcate nature and details of retipilate exine with swollen-based spines (scale, Fig. 25 = 10 μm; Fig. 26 = 1 μm).

463

Acknowledgements

Rosalind A. Srivastava read and improved the manuscript. Michael C. Boulter, Madeline Harley, and G. F. Waldemar Herngreen critically reviewed the manuscript and provided constructive suggestions.

References

Brenner, G.J. (1963). The spores and pollen of the Potomac Group of Maryland. *Bulletin Maryland Department of Geology, Mines and Water Resources* 27: 1–215.

Čech, S., (1989). Bohemian Cretaceous basin In: Bůzek, Č., Dvořák, Z., Čech, S., Knobloch, E., Kvaček, Z., and Prokeš, M. (editors). *Excursion guide to International Symposium "Palaeofloristic and Palaeoclimatic changes in the Cretaceous and Tertiary", Prague.* pp. 13–14.

Čech, S., Uličný, D., and Zitt, J. (1997). Cretaceous: Cenomanian/ Turonian events - Transgressive tract of the Cenomanian-Turonian, anoxic event - Pecínov quarry near Nové Strašecí; In: Čejchan, P. and Hladil, J. (editors). UNESCO - IGCP Project #335 "Biotic recoveries from mass extinctions", Recoveries '97 Conference Field Trip Book, pp. 1–4. Eurocongress Centre, Praha 4.

Doyle, J.A., Van Campo, M. and Lugardon, B. (1975). Observations on exine structure of *Eucommiidites* and Lower Cretaceous angiosperm pollen. *Pollen et Spores* 17: 429–486.

Doyle, J.A., Jardiné, S. and Doerenkamp, A. (1982). *Afropollis,* a new genus of early angiosperm pollen, with notes on the Cretaceous palynostratigraphy and paleoenvironments of northern Gondwana. *Bulletin des Centres de Recherches Exploration - Production Elf-Aquitaine* 6: 39–117.

Doyle, J.A. and Robbins, E.I. (1977). Angiosperm pollen zonation of the continental Cretaceous of the Atlantic Coastal Plain and its application to deep wells in the Salisbury Embayment. *Palynology* 1: 43–78.

Gregory, W.A. and Hart, G.F. (1992). Towards a predictive model for the palynologic response to sea-level changes. *Palaios* 7: 3–33.

Habib, D. (1977). Comparison of Lower and Middle Cretaceous palynostratigraphic zonations in the western North Atlantic. In: F.M. Swain (editor). Stratigraphic micropaleontology of Atlantic basin and borderlands, pp. 341–467. Elsevier Scientific Publishing Company, Amsterdam - Oxford - New York.

Harley, M.M., Kurmann, M.H., and Ferguson, I.K. (1991). Systematic implications of comparative morphology in selected Tertiary and extant pollen from the Palmae and the Sapotaceae. In: S. Blackmore and S.H. Barnes (editors). Pollen and spores: patterns of diversification. pp. 225–238. Clarendon Press, Oxford.

Hochuli, P.A. (1981). North Gondwana floral elements in Lower to Middle Cretaceous sediments of the Southern Alps (southern Switzerland, northern Italy). *Review of Palaeobotany and Palynology,* 35: 337–358.

Hughes, N.F. (1994). The enigma of angiosperm origins. Cambridge University Press, Cambridge.

Jardiné, S., Doerenkamp, A. and Biens, P. (1974). *Dicheiropollis etruscus,* un pollen caractéristique du Crétacé inférieur Afro-Sudaméricain: Conséquences pour l'évaluation des unités climatiques et implications dans la dérive des continents. *Science Geologie Bulletin* 27: 87–100.

Mchedlishvili, N.D. and Shakhmundes, V.A. (1973). Occurrence of Araceae pollen in the Lower Cretaceous sediments. *Trudy 3rd Mezhdunarodnoi Palinologicheskoi Konferéntsii SSSR, Novosibirsk, 1971,* pp. 137–142. Nauka, Moskva. (in Russian with English summary).

Moore, G.T., Barron, E.J., and Hayashida, D.N. (1995). Kimmeridgian (Late Jurassic) general lithostratigraphy and source rock quality for the western Tethys Sea inferred from paleoclimate results using a general circulation model. *AAPG Studies in Geology* 40: 157–172.

Müller, H. (1966). Palynological investigations of Cretaceous sediments in northeastern Brazil. In: J.E. van Hinte (editor). *Proceedings of the 2nd West African Micropaleontological Colloquium (Ibadan, 1965)*, pp. 123–136. E. J. Brill, Leiden.

Pacltová, B. (1971). Palynological study of Angiospermae from the Peruc Formation (? Albian - Lower Cenomanian) of Bohemia. *Sbornik Geologickych ved, paleontologie*, Section P, 13: 105–141.

Pacltová, B. (1990). Marginal facies of the Bohemian Upper Cretaceous (palynological study). *Proceedings of the Symposium "Paleofloristic and paleoclimatic changes in the Cretaceous and Tertiary" 1989, Prague*, pp. 47–52.

Pacltová, B. and Svobodová, M. (1992). Facial characteristics from the palynological point of view in the area of the Bohemian Cenomanian. *Proceedings of the Symposium "Paleofloristic and Paleoclimatic changes in the Cretaceous and Tertiary"*, September 14–20, 1992, Bratislava, pp. 17–21.

Pitman III, W.C., Cande, S., LaBrecque, J., and Pindell, J. (1993). Fragmentation of Gondwana: the separation of Africa from South America. In: P. Goldblatt (editor). Biological relationships between Africa and South America, pp. 15–34. Yale University Press, New Haven and London.

Smith, A.G., Smith, D.G. and Funnell, B.M. (1994). Atlas of Mesozoic and Cenozoic coastlines. Cambridge University Press, Cambridge.

Srivastava, S.K. (1978). Cretaceous spore-pollen floras - a global evaluation. *Biological Memoirs* 3: 1–130.

Srivastava, S.K. (1981). Fossil pollen genus *Kurtzipites* Anderson. *Journal of Paleontology* 55: 868–879.

Srivastava, S.K. (1984a). Barremian dinoflagellate cysts from southeastern France. *Cahiers de Micropaléontologie* 2–1984: 5–90.

Srivastava, S.K. (1984b). A new elater-bearing late Albian pollen species from offshore eastern Saudi Arabia. *Botanical Journal of the Linnean Society* 89: 231–238.

Srivastava, S.K. (1988). *Ctenolophon* and *Sclerosperma* paleogeography and Senonian Indian plate position. *Journal of Palynology* 23/24: 239–253.

Srivastava, S.K. (1994). Evolution of Cretaceous phytogeoprovinces, continents and climates. *Review of Palaeobotany and Palynology* 82: 197–224.

Srivastava, S.K. and Binda, P.L. (1991). Depositional history of the early Eocene Shumaysi Formation, Saudi Arabia. *Palynology* 15: 47–61.

Svobodová, M. (1991). Earliest Upper Cretaceous palynomorphs of basal (transgressive) strata in the Blansko Graben, Monrovia, Czechoslovakia. *Proceedings of the Pan-European Palaeobotanical conference, Vienna, 19-23 September, 1991*; p. 313–320.

Svobodová, M. (1997). Mid-cretaceous palynomorphs from the Blansko Graben (Czech Republic): affinities to both Tethyan and Boreal bioprovinces. *Proceedings 4th EEPC; Mededelingen Nederlands Instituut voor Toegepaste Geowetenschappen TNO*, no. 58: 149–155.

Trevisan, L. (1972). *Dicheiropollis*, a pollen type from Lower Cretaceous of southern Tuscany (Italy). *Pollen et Spores* 13: 561–596.

Trevisan, L. (1980). Ultrastructural notes and considerations on *Ephedripites*, *Eucommiidites* and *Monosulcites* pollen grains from Lower Cretaceous sediments of southern Tuscany (Italy). *Pollen et Spores* 22: 85–132.

Tschudy, R.H. and Pakiser, H.M. (1967). *Fustispollenites*, a new Late Cretaceous genus from Kentucky. *U.S. Geological Survey Professional Paper* 575-B: B54–B56.

Uhl, N.W. and Dransfield, J. (1987). Genera Plantarum. A classification of the palms based on the work of H.E. Moore, Jr. The L.H. Bailey Hortorium and International Palm Society; 610 pp. Allen Press, Lawrence, Kansas.

Williams, G.D. and Stelck, C.R. (1975). Speculations on the Cretaceous paleo-geography of North America. *The Geological Association of Canada Special Paper* Number 13: 1–20.

Ziegler, P.A. (1988). Evolution of the Arctic-North Atlantic and the western Tethys. *AAPG Memoir* 43: 198 pp.

Marchant, R. and Taylor, D. (2000). Modern pollen - vegetation relationships in montane rainforest of Uganda: an aid to interpretation of fossil sequences from central Africa. In: M.M. Harley, C.M. Morton and S. Blackmore (Editors). Pollen and Spores: Morphology and Biology, pp. 467–480. Royal Botanic Gardens, Kew.

MODERN POLLEN - VEGETATION RELATIONSHIPS IN MONTANE RAINFOREST OF UGANDA: AN AID TO INTERPRETATION OF FOSSIL SEQUENCES FROM CENTRAL AFRICA

ROBERT MARCHANT[1] AND DAVID TAYLOR[2]

[1]Hugo de Vries-Laboratory, Department of Palynology and Paleo/Actuo-Ecology, University of Amsterdam, Kruislaan 318, 1098 SM Amsterdam, The Netherlands
[2]Department of Geography, National University of Singapore, 10 Kent Ridge Crescent, Singapore, 119260

Abstract

The arboreal composition of montane rainforest within the Mubwindi Swamp catchment in southwest Uganda is compared to pollen spectra recovered from the swamp surface. For the majority of the arboreal taxa the results indicate that (in the case of the Mubwindi Swamp catchment) the proportions of pollen from arboreal sources found in surface samples match those of the assumed parent taxa in the surrounding montane rainforest. The surface pollen samples coincide with the location of several fossil pollen sequences. Pollen data from a 5 m sedimentary sequence with four radiocarbon dates are presented. Suggestions are made regarding the robustness of fossil pollen sequences as an indicator of late Holocene montane forest dynamics in south-west Uganda and inferences drawn that are applicable throughout central Africa for the interpretation of fossil pollen sequences. For example, *Celtis*, Ericaceae, *Faurea*, *Hagenia*, *Nuxia*, *Olea*, *Podocarpus*, *Polyscias*, *Prunus* and *Schefflera* although being important components of the non-local pollen sum are likely to be significantly less important within the surrounding vegetation over the late Holocene than indicated by their pollen. *Alchornea*, *Anthocleista*, *Croton*, *Dombeya*, *Ilex*, Myrtaceae, *Newtonia*, *Rapanea* and *Zanthoxylum* pollen types are shown to representatively record the abundance of their parent taxa within the surrounding vegetation. However, *Chrysophyllum*, *Drypetes* and *Tabernaemontana* are not recorded by their pollen, but are numerically important components of the surrounding vegetation. This omission renders pollen-based reconstructions of montane rainforest only partial. Although it is only possible to base vegetation reconstructions on the pollen data that are present within fossil sediments, future reconstructions of montane rainforest should take into account the omissions of these taxa from fossil pollen sequences.

Introduction

Southwest Uganda has been a focus of pollen-based palaeoecological research in central Africa for four decades (Morrison, 1961; Hamilton, 1982; Taylor 1990; 1993; Taylor and Marchant, 1996; Marchant *et al.*, 1997; Marchant and Taylor, 1998). As a result of this work there have been significant interpretations made on the response of vegetation to climate forcing over the late Quaternary period; such as the response of vegetation to late glacial climates (Jolly *et al.*, 1997). More recently, the impact of human activity over the past 4000 years has been recognised (Taylor and Marchant, 1996). For example, one finding of this research indicates that although montane rainforest is regarded as climax vegetation type for much of south-west Uganda, this vegetation has been extensively modified by human activity (Taylor, 1990, 1993; Taylor and Marchant, 1996). Despite this modification, a significant remnant of natural montane rainforest still remains; Bwindi-Impenetrable Forest, within which Mubwindi Swamp is located, is one such remnant (Fig. 1). The hydrological catchment of Mubwindi Swamp supports a dense cover of montane forest.

The relationship between present-day trees and the deposition of their pollen (for example, Hamilton, 1972; Flenley, 1973; Meadows, 1984, 1989; Crowley *et al.*, 1994; Islebe and Hooghiemstra, 1995; Rodgers and Horn, 1996; Vincens *et al.*, 1997) is central to the reconstruction of vegetation history in the tropics. However, in most cases such relationships have a relatively limited utility as the modern pollen samples are from quite different locations and environments to that from which the fossil pollen sequence has been obtained. For example, modern pollen samples are often collected from soils underneath a range of vegetation types, along altitudinal transects (Bonnefille and Riollet, 1988; Mancini, 1993; Bonnefille *et al.*, 1993; Islebe and Hooghiemstra, 1995; Vincens *et al.*, 1997), whereas fossil pollen sequences are taken from large sedimentary basins, such as lakes or swamps. The sedimentary processes, and the impact that these may impart on the incorporation of pollen within the fossil sequence, are quite different to those that have taken place within surface sample sites.

Site Description

Mubwindi Swamp (1°05'S, 29°45'E) is located within Bwindi-Impenetrable Forest, southwest Uganda (Fig. 1). The hillsides adjacent to Mubwindi Swamp support a mosaic of moist lower montane rainforest at different stages of regeneration (Marchant and Taylor, 1998). Some areas of the forest are thought to be undisturbed by human activity (Howard, 1991). *Neoboutonia macrocalyx* Pax and *Syzygium cordatum* Hochst. ex Krauss are common on the lower slopes of valleys; at mid-altitude *Chrysophyllum albidum* G.Don and *C. gorungosanum* Engl. are abundant in association with *Cassipourea ruwenzorensis* (Engl.) Alston, *Drypetes gerrardii* Dg and *Strombosia scheffleri* Engl. At higher altitudes taxa such as *Faurea saligna* Engl., *Hagenia abyssinica* Pax. and *Nuxia congesta* Engl. become common. *Podocarpus falcatus* (Thunb.) Engl. and *Olea capensis* subsp. *welwitschii* ((Knobl.) I. Friis & P.S. Green, previously common on ridge and hill top locations, have now been mostly removed by pit-sawyers (Hamilton, 1969). Where the forest has been degraded by human activity, large open gaps occur. Within these *Alchornea hirtella* Benth., *Macaranga kilimandscharica* Pax., *Neoboutonia macrocalyx* and *Polyscias fulva* (Hiern) Harms. can be found.

Materials and Methods

Vegetation surveys

Forest on the hillsides that delimit the northeast part of Mubwindi Swamp was surveyed using a belt transect approach. Trees with a trunk diameter of > 15 cm at 1.3 m above the ground surface (dbh) were enumerated along 11, 5 m-wide, belt transects (Fig. 2). The length of these transects varied from 150 to 900 m and were fixed at a width of 5 m. The location of the transects followed a path from the swamp margin to the surrounding hill crests. The locations of the transects, where the cores containing fossil sediments were obtained, were fixed so that the vegetation in the area surrounding was surveyed. These vegetation transects account for a total area of 22,500 m² (2.8 % of the total area of the catchment).

Results from each transect were plotted on scaled figures so as to represent changes in slope angle and tree height (Fig 3). Trees from the vegetation survey are recorded at species level, and pollen grains from the surface sediments commonly to generic level. To make the two data sets directly comparable, species were summed to the same taxonomic level as achieved from the pollen analysis. Each taxon encountered was calculated as a percentage value of the total number of specimens recorded within the entire sample area (Table 1). This provided results that were compatible with the pollen data (in that the abundance of any given taxon was dependent on all the other taxa recorded).

TABLE 1. Percentage pollen and forest data summed from 24 surface samples and 11 belt transects respectively.

TAXA	% SURFACE SEDIMENT	% VEGETATION SURVEY
Alchornea	12.1	13.6
Anthocleista	1.6	1.8
Cassipourea	1	11.5
Croton	1.6	2.4
Cyathea	1	1
Dombeya	0.8	0.9
Ilex	2.1	1
Macaranga	6	5.1
Myrtaceae	12.5	11.2
Neoboutonia	5.8	15.5
Newtonia	0.6	1.2
Olea	10.8	0.8
Podocarpus	7.8	0.9
Polyscias	3.8	1.2
Prunus	3	1.2
Rapanea	2.8	1.4
Strombosia	1.4	11.3
Zanthoxylum	1.9	2
Chrysophyllum	0	5.7
Drypetes	0	4.7
Tabernaemontana	0	5.6
Celtis	4.1	0
Ericaceae	7.5	0
Faurea	4.8	0
Hagenia	1.7	0
Nuxia	5.2	0
Schefflera	1	0

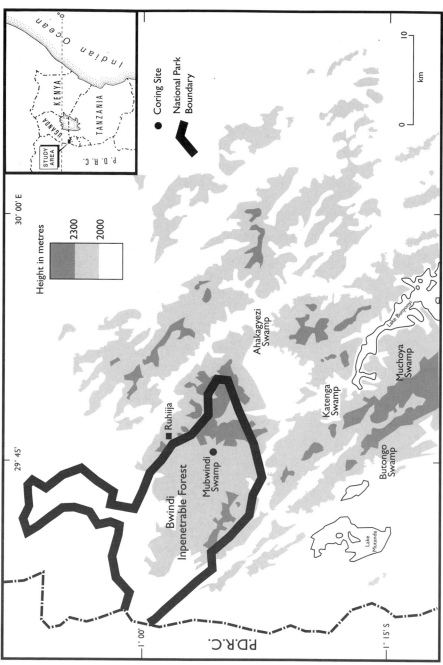

FIG. 1. Location of Bwindi-Impenetrable Forest National Park showing regional relief and the swamps that have yielded pollen-based records of vegetation history.

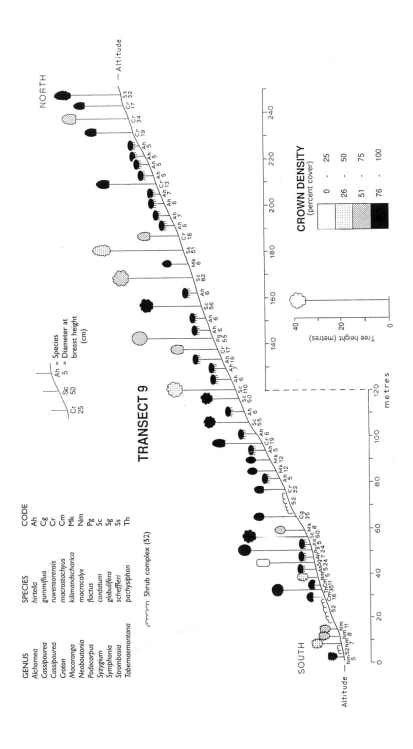

FIG. 2. Vegetation recorded along transect 9 from the Mubwindi Swamp catchment.

Pollen analysis of surface sediments

Surface sediment samples were obtained by first removing any vegetation on the swamp surface and then collecting a scrape of the uppermost peat. 24 surface samples were collected in this way from the northeast part of the swamp. Pollen in the surface sediment samples was concentrated at the University of Hull, using the standard preparation procedure of Faegri and Iversen (1989). Due to the fibrous nature of the surface sediment, the samples were left in KOH for up to 30 minutes to allow for easier sieving of the material, and hence release of pollen for subsequent concentration. Throughout the preparation, a standard amount of sample (1 cm³), and reagent (5 ml) were used. Reaction and centrifuge times were also kept constant.

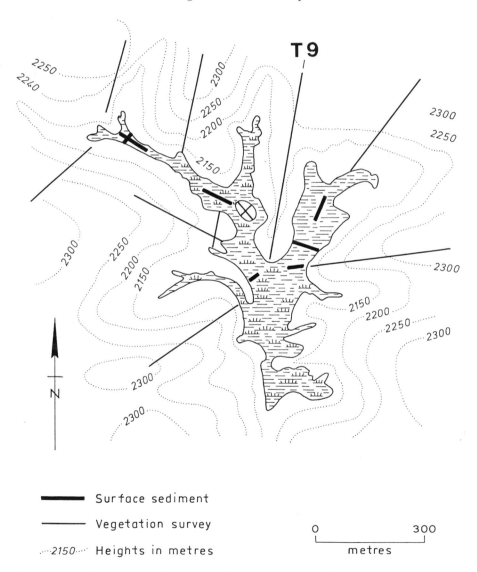

Surface sediment

Vegetation survey

····2150···· Heights in metres

0 ⊢————————⊣ 300

metres

FIG. 3. Mubwindi Swamp, showing topography and the location of transects for vegetation and surface sediment sample surveys.

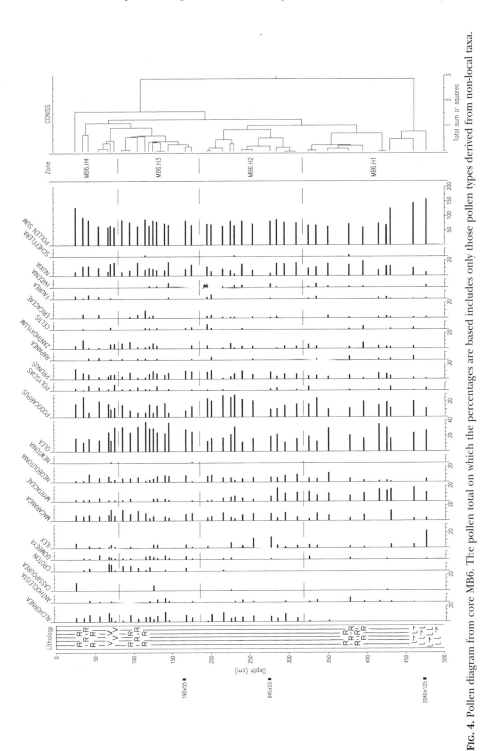

FIG. 4. Pollen diagram from core MB6. The pollen total on which the percentages are based includes only those pollen types derived from non-local taxa.

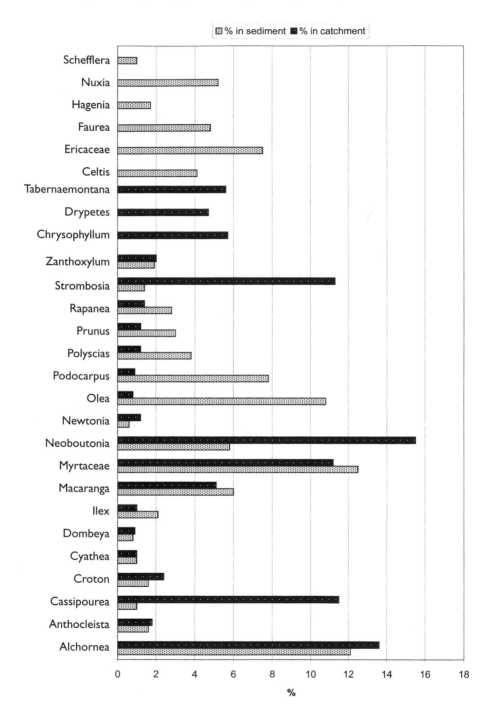

FIG. 5. Percentages of parent vegetation and pollen taxa present in the surrounding catchment and surface sample respectively.

Pollen analysis was carried out as detailed in Marchant *et al.*, (1997). To enable the pollen spectra to be compared to the surrounding vegetation pollen counts for individual taxa were converted to a percentage value after having first removed pollen derived from taxa growing on the swamp itself. This 'non-local' pollen sum varied from 64 to 151 pollen grains per sample. Pollen data from the twenty-four surface samples were summed to give a non-local pollen data set of 2652 pollen grains; from this a percentage value was computed for the individual pollen types (Table 1).

The relationship between the percentage of a taxon's pollen derived from the surface sediment, and the abundance of the same taxon in the surrounding vegetation was investigated. Pollen types and tree species were separately ranked in order of decreasing importance and Spearman Rank Correlation analysis applied to determine the degree of correlation between the ordering of the two data sets. The taxa are arranged to represent: those represented both in the vegetation survey and in the surface samples, those represented within the vegetation survey but not by their pollen, and those that are represented by their pollen but are not present within the surrounding vegetation (Fig. 4).

Pollen analysis of fossil sediments

Twenty three sediment cores have been extracted from Mubwindi Swamp using a combination of Russian (D-section) and Hiller corers. One of these cores (MB6) provided sediments in the uppermost five metres that have dating control in the form of two AMS and two standard radiocarbon ages and provide a record of late Holocene montane rainforest dynamics (Marchant and Taylor, 1998). Two of these dates are from the same level; these were averaged to give a single date. The relatively large standard deviation of this date reflects this process.

Data from the analysis of fossil pollen are portrayed in percentage form (Fig. 5). Fig. 5 was plotted using the PC-based software TILIA and TILIAGRAPH (Grimm, 1991). The pollen and spore data were classified into four pollen zones (MB6.H1 - MB6.H4) according to a stratigraphically constrained, numerical clustering package within TILIA. The pollen taxa were divided into two categories; one category comprised taxa that were recorded both within the vegetation survey and the surface sediments, a second category included only taxa that were recorded from pollen in the surface sediments, but were not present within the vegetation survey.

Results

Mubwindi Swamp vegetation

Results from transect 9 (Fig. 3) highlight the open nature of the forest within the Mubwindi Swamp catchment. When data from all the transects are summed, three associations of taxa can be determined. These are forest fringing the swamp, a mid slope association and an association that is confined to the hilltops. The forest fringing the Mubwindi Swamp is dominated by *Syzygium cordatum*, which accounts for around 12% of the total number of specimens recorded. *Anthocleista zambesiaca* Baker was also recorded in this position close to the swamp margin. Both species rarely extended to higher altitudes away from the swamp margin, possibly being maintained by the locally wetter environmental conditions associated with a swamp-edge habitat.

The mid-slope association was dominated by *Neoboutonia macrocalyx*; this taxon accounting for 16% of all specimens recorded. The high percentage of this taxon, being characteristic of open montane forest, highlights the open nature of the montane rainforest around Mubwindi Swamp, and thus the extent of recent disturbance. Other taxa commonly associated with disturbed forest, such as *Croton macrostachyus* Hochst., *Dombeya goetzenii*, K. Schum., *Macaranga kilimandscharica* and *Polyscias fulva* were all present, *M. kilimandscharica* being the most common of these. *Cassipourea ruwensorensis* and *Strombosia scheffleri* both accounted for approximately 12% of the total number of specimens recorded. These two species are often found in close association, particularly on mid-altitude sites, where they form the dominant forest type along with occasional species of *Drypetes* spp. and *Tabernaemontana pachysiphon* Stapf. Taxa indicative of more established montane forest, such as *Ilex mitis* (L.) Radlk., *Podocarpus* spp., *Olea* spp., *Prunus africana* (Hook.f.) Kalkm., *Rapanea* spp. and *Zanthoxylum gilletii* (De Wild.) Waterman were all present in low quantities. At the highest altitudes *Chrysophyllum albidum* was the most common species, dominating the vegetation at the highest altitudes within the catchment, often with an under-story of *Drypetes* spp.

Pollen within surface sediments

Myrtaceae, *Alchornea*, and *Olea* dominated the non-local pollen sum, each accounting for approximately 12% of the pollen sum. *Macaranga* and *Podocarpus* pollen each accounted for about 8% of non-local pollen. Ericaceae, *Neoboutonia*, *Polyscias* and *Prunus* pollen all contributed approximately 4% to the non-local pollen sum. There were low amounts of *Celtis*, *Faurea*, *Ilex* and *Nuxia* pollen present. Other pollen types recorded were present at levels of 2% or lower.

An aid to the interpretation of fossil pollen sequences

The zonation of the pollen data presented in Fig. 5 is similar to that derived from all the non-local pollen taxa presented in Marchant and Taylor (1998) - four separate pollen zones are identified (MB6.H1 - MB6.H2) with the boundaries located in similar stratigraphic positions. The pollen taxa have been arranged in two groups; the first includes all taxa recorded in the survey of pollen and vegetation, the second group includes taxa that were only recorded by their pollen. The sequence represents a transition from relatively dense mixed montane rainforest to a more open forest. This is characterised initially by high amounts of *Alchornea* and *Neoboutonia*, and latterly by *Croton* and *Dombeya*; all these taxa are relatively common in forest gaps (Marchant *et al.*, 1997).

Discussion

How representative is pollen data of the surrounding vegetation?

The close relationship between the abundance of many pollen types and their parent taxa within vegetation on hillsides adjacent to Mubwindi Swamp suggests that fossil pollen from the site provide a robust record of vegetation history, although with some notable caveats. *Croton*, *Dombeya* and *Newtonia* pollen have been thought of as being under-representative of their parent taxa in past studies of modern pollen based in Uganda (Hamilton, 1972). Thus, these pollen types would be expected to be under-represented by their pollen in this analysis. However, the results obtained suggest otherwise. In the case of Mubwindi Swamp sediments, pollen from these taxa are an accurate reflection of the abundances of the parent taxa in the surrounding vegetation. By comparison, *Macaranga* pollen has been viewed as being over-representative of the parent taxa (Hamilton, 1972; Bonnefille *et al.*, 1993). However, in the case of the Mubwindi Swamp the amount of pollen was found to closely reflect the density of the parent taxa within the surrounding vegetation.

Although common in the vegetation surrounding Mubwindi Swamp, *Cassipourea ruwensorensis*, *Neoboutonia macrocalyx* and *Strombosia scheffleri* were poorly represented by their pollen in the surface sediments. Hamilton (1972) has classified all these pollen taxa as being under-representative of their parent taxa and this status is confirmed by the present analysis. Species of *Chrysophyllum*, *Drypetes* and *Tabernaemontana* are relatively common components of the vegetation surrounding Mubwindi Swamp, but are not recorded by their pollen in the surface sediments. Species of *Chrysophyllum* and *Tabernaemontana* were mainly recorded as mature specimens, and were locally dominant, the former forming a discrete type of vegetation on the ridge-tops surrounding Mubwindi Swamp. The absolute lack of pollen from the parent taxa indicates that the species are very under-represented by their pollen.

Olea, *Podocarpus*, *Polyscias* and *Prunus* pollen types had abundance in surface sediment samples that was much higher than recorded by the vegetation survey. This finding is in agreement with Hamilton's (1972) classification of these pollen types and the findings of Meadows (1984) and Bonnefille *et al.* (1993). Even so, particularly for *Olea*, the high amount of pollen present in the surface samples is quite remarkable, given the low presence of the parent taxa recorded from the vegetation survey.

Celtis, Ericaceae, *Faurea*, *Hagenia*, *Nuxia* and *Schefflera* pollen are recorded at low percentages in the surface sediments but not by their parent taxa within the surrounding vegetation. Consequently, these pollen types must be viewed as being over-representative of their parent taxa. Within this group there are two major differences from Hamilton's (1972) classification. Firstly *Faurea* pollen, which is classed as being under-representative of its parent taxa, was recorded at relatively high levels. Given the absence of the parent taxa in vegetation surrounding the swamp it is suggested that *Faurea* should be re-classified as a pollen type that is over-representative relative to the number of the parent taxa in surrounding vegetation. Secondly, Ericaceae pollen has been traditionally classed as a moderately well-dispersed type (Hamilton, 1972). Due to the relatively high percentage of Ericaceae within the surface sediments, it should be re-classified as being over representative of its parent taxa. This re-classification would also place less confidence on the assumption that Ericaceae within fossil sequences is always derived from relatively high altitude vegetation (Bonnefille *et al.*, 1993) growing close to the site of deposition.

An aid to the interpretation of fossil pollen sequences

There are two main ways in which this study contributes to an improved interpretation of fossil pollen data. First, through narrowing-down the possible sources of pollen. Second, by determining the potential accuracy of vegetation reconstruction based upon counts of fossil pollen.

Source of pollen

Pollen within surface and accumulated sediments from Mubwindi Swamp has commonly been identified to either generic or family taxonomic level (Marchant *et al.*, 1997; Marchant and Taylor, 1998). By carrying out the survey of the surrounding vegetation it is possible to indicate, particularly over the late Holocene time period, which species are likely to form the specific parent taxa of the pollen identified within the Mubwindi Swamp sediments. For example, *Syzygium cordatum*, the only member of the Myrtaceae found within the catchment, can be identified as the parent taxa responsible for the pollen identified as Myrtaceae within fossil sediments. Similarly, the pollen identified as *Alchornea* is derived from *Alchornea hirtella*, rather than *Alchornea cordatum* Benth. These specific relationships allow for greater precision in palaeoecological reconstruction of the vegetation; the ecology of single species being more restricted than that for taxa at generic or family level. The degree to which this specific modern relationship can be applied to fossil pollen sequences will diminish through time (sediment depth) as different species to those identified within the modern data set may have been present within the catchment.

How representative of the vegetation is fossil pollen data ?

Notwithstanding the factors highlighted here, the data presented from the surface pollen spectra and vegetation surveys indicate that non-local pollen in samples from Mubwindi Swamp sediments reflects to a good degree of accuracy the non-local vegetation growing within the immediate catchment. This relationship is particularly relevant as the fossil pollen to modern analogue comparison is subjected to less error than would be the case if the modern pollen samples been collected from a different sedimentary environment, such as forest hollows (Calcote, 1995), moss polsters (Spieksma *et al.*, 1994; Kershaw and Bulman, 1994; Ritchie, 1974), pollen traps (Davis and Thompson Webb III, 1975; Meadows, 1984; Hamilton and Perrot, 1980) and soils beneath forests (Bonnefille *et al.*, 1993; Jolly *et al.*, 1996). Where the relationship does not hold, for taxa such as *Celtis, Chrysophyllum, Drypetes,* Ericaceae, *Faurea, Hagenia, Nuxia, Schefflera* and *Tabernaemontana* allowances can be made in the interpretation of fossil sequences.

Olea, Podocarpus, Polyscias and *Prunus,* although being numerically important in the pollen sum would have been significantly less important components of the surrounding vegetation over the late Holocene than indicated by their pollen. Similarly, Ericaceae, *Celtis, Faurea, Hagenia, Nuxia* and *Schefflera* would have been less important components of the surrounding vegetation over the late Holocene than indicated by their pollen. This is particularly the case for *Nuxia,* the pollen of which accounted for greater than 10% of the non-local pollen sum in samples from throughout the sedimentary sequence considered here. It is thought much of this fossil pollen came from sources outside the immediate catchment. Indeed, *Nuxia congesta* was noted to be a locally important component of the forest at higher altitudes than those found within the Mubwindi Swamp catchment. Abundance of Myrtaceae, *Alchornea, Anthocleista, Croton, Dombeya, Ilex, Newtonia, Rapanea* and *Zanthoxylum* pollen types are shown to provide a representative record of the density of their parent taxa within the surrounding vegetation. The closeness of these relationships is particularly relevant for the reconstruction of vegetation history as they indicate that the majority of the taxa that make up fossil pollen spectra in sediments from Mubwindi Swamp are an accurate reflection of the parent taxa found within the confines of the immediate catchment. Furthermore, members of the genus *Anthocleista* and the family Myrtaceae are components of swamp forest elsewhere in south-west Uganda, and have in the past been classed as local taxa for the calculation of pollen sums (Taylor, 1990, 1993). The close relationship between the density of the parent taxa in non-swamp vegetation, and the pollen in surface samples, ratifies the incorporation of these taxa in the non-local pollen sum (Marchant *et al.*, 1997).

One of the key findings is that *Chrysophyllum, Drypetes* and *Tabernaemontana* are not recorded by their pollen, but are numerically important components of the surrounding vegetation. This omission renders past reconstructions of montane rainforest only partial. Although it is only possible to base vegetation reconstructions on the pollen data that is present within fossil sediments, future reconstructions of montane rainforest should take into account the omissions of these taxa from fossil pollen sequences. In some cases, as at Mubwindi Swamp, these omissions can render pollen-based vegetation reconstructions only partial. Indeed, future work should highlight that some taxa, not reflected by the analysis of fossil pollen sequences, are likely to have been present in the surrounding vegetation.

Conclusions

Pollen accumulating within the surface sediments of Mubwindi Swamp are derived largely from plants growing within the immediate catchment. The modern pollen-vegetation relationship constructed is probably unique to the Mubwindi Swamp catchment due to the individual characteristics of the catchment. Indeed, the results present here, although relevant to other sites in central Africa, should be applied, and be used by other palaeoecologists, with an understanding of the possible influence that the individual catchment may impart on the pollen accumulating within the sedimentary environment. The results have implications for the interpretation of fossil pollen data from central Africa.

Acknowledgements

The work presented here was supported by the University of Hull, the Bill Bishop Memorial Fund, the British Institute in East Africa and Natural Environment Research Council. Thanks are due the Government of Uganda for research permission, to Myooba Godfy for his invaluable help and company during the summer of 1994, and to the staff and students of the Institute of Tropical Forest Conservation field station at Ruhiija. We are indebted to Keith Scurr for production of the figures. The manuscript for this work was improved by Anne-Marie Lézine and an anonymous reviewer, their comments have been most useful.

References

Bonnefille, R. and Riollet, R. (1988). The Kashiru pollen sequence (Burundi). Palaeoclimatic implications for the last 40,000 years in tropical Africa. *Quaternary Research* **30**: 19–35.

Bonnefille, R., Buchet, G., Friis, I., Kelbessa, E. and Mohammed, M.U. (1993). Modern pollen rain on an altitudinal range of forests and woodlands in Southwest Ethiopia. *Opera Botanica* **121**: 71–84.

Calcote, R. (1995). Pollen source area and pollen productivity, evidence from forest hollows. *Journal of Ecology* **83**: 591–602.

Crowley, G.M., Grindrod, J, Kershaw, A.P. (1994). Modern pollen deposition in the tropical lowlands of north-east Queensland, Australia. *Review of Palaeobotany and Palynology* **83**: 299–327.

Davis, R.B. Webb, T. (1975). The contemporary distribution of pollen in eastern North America: a comparison with the vegetation. *Quaternary Research* **5**: 395–434.

Faegri, K, and Iversen, J. (1989). Textbook of pollen analysis. Fourth edition, John Wiley, Chichester.

Flenley, J.R. (1973). The use of modern pollen rain samples in the study of the vegetation history of tropical regions. In: Birks H.J.B. and West R.G. (editors). Quaternary Plant Ecology, pp. 131–141, Blackwell, Oxford.

Grimm, E.C. (1991). Tilia Graph 1.18. Springfield: Illinois State Museum.

Hamilton, A.C. (1969). The vegetation of south-west Kigezi. *Uganda Journal* **33**: 175–199.

Hamilton, A.C. (1972). The interpretation of pollen diagrams from highland Uganda. *Palaeoecology of Africa* **7**: 45–149.

Hamilton, A.C. (1982). *The Environmental History of East Africa.* Academic Press.

Hamilton, A.C. and Perrot, R.A. (1980). Modern pollen deposition on a tropical African mountain. *Pollen et Spores* **22**: 437–468.

Howard, P.C. (1991). Nature conservation in Ugandaís Forest Reserves. The I.U.C.N. Tropical Forest Program. Gland. Cambridge.

Islebe, G.A. and Hooghiemstra, H. (1995). Recent pollen spectra of highland Guatemala. *Journal of Biogeography* **22**: 1091–1099.

Jolly, D., Bonnefille, R., Burcq, S. and Roux, M. (1996). Représentation pollinique de forêt dense humide du Gabon, tests statistiques. *C.R.* Académie Sciences Paris 322, série II a: 63–70.

Jolly, D., Taylor, D., Marchant, R., Hamilton, A.C., Bonnefille, R., Buchet, G. and Riollet, G. (1997). Vegetation dynamics in central Africa since 18,000 yr BP: pollen records from the interlacustrine highlands of Burundi, Rwanda and western Uganda. *Journal of Biogeography* **24**: 495–512.

Kershaw, A.P. and Bullman, D. (1994). The relationship between modern pollen samples and environment in the humid tropic region of north-eastern Australia. *Review of Palaeobotany and Palynology* **83**: 83–96.

Mancini, M.V. (1993). Recent pollen spectra from forest and steppe of South Argentina: a comparison with vegetation and climate data. *Review of Palaeobotany and Palynology* **77**: 129–142.

Marchant, R.A., Taylor, D.M. and Hamilton, A.C. (1997). Late Pleistocene and Holocene history of Mubwindi Swamp, south-west Uganda. *Quaternary Research* **47**: 316–328.

Marchant, R.A. and Taylor, D.M. (1998). A late Holocene record of montane forest dynamics from south-west Uganda. *The Holocene* **8**: 375–381.

Meadows, M.E. (1984). Contemporary pollen spectra and vegetation of the Nyika Plateau, Malawi. *Journal of Biogeography* **11**: 223-233.

Meadows, M.E. (1989). Contemporary pollen rain studies in southern Africa. *Paleoecology of Africa* **20**: 155–162.

Morrison, M.E.S. (1961). Pollen analysis in Uganda. *Nature* **190**: 383–386.

Ritchie, J.C. (1974). Modern pollen assemblages near the arctic tree line, Mackenzie Delta region, Northwest Territories. *Canadian Journal of Botany* **52**: 381–396.

Rodgers, J.C. and Horn, S.P. (1996). Modern pollen spectra from Costa Rica. *Palaeogeography, Palaeoclimatology, Palaeoecology* **124**: 53–71.

Spieksma, F.T.M., Nikkels, B.H. and Bottema, S. (1994). Relationship between recent pollen deposition and airborne pollen concentration. *Review of Palaeobotany and Palynology* **82**: 141–145.

Taylor, D.M. (1990). Late Quaternary pollen records from two Ugandan mires: evidence for environmental change in the Rukiga highlands of south-west Uganda. *Palaeogeography, Palaeoclimatology, Palaeoecology* **80**: 283–300.

Taylor, D.M. (1993). Environmental change in montane south-west Uganda: a pollen record for the Holocene from Ahakagyezi Swamp. *The Holocene* **3**: 324–332.

Taylor, D.M. and Marchant, R. (1996). Human impact in the lacustrine region: long-term pollen records from the Rukiga highlands. In: Sutton J. (editor). *The growth of farming communities in Africa from the equator southwards*, The British Institute in Eastern Africa.

Vincens, A., Ssemmanda, I., Roux, M. and Jolly, D. (1997). Study of modern pollen rain in western Uganda with a numerical approach. *Review of Palaeobotany and Palynology* **96**: 145–168.

Long, D.J., Tipping, R., Carter, S., Davidson, D., Boag, B. and Tyler, A. (2000). The replication of pollen stratigraphies in soil pollen profiles: a test. In: M.M. Harley, C.M. Morton and S. Blackmore (Editors). Pollen and Spores: Morphology and Biology, pp. 481–497. Royal Botanic Gardens, Kew.

THE REPLICATION OF POLLEN STRATIGRAPHIES IN SOIL POLLEN PROFILES: A TEST

DEBORAH. J. LONG[1], RICHARD TIPPING[1], STEPHEN CARTER[2], DONALD DAVIDSON[1], BRIAN BOAG[3] AND ANDREW TYLER[1]

[1]Department of Environmental Science, University of Stirling, Stirling, FK9 4LA
[2]Headland Archaeology Ltd, Edinburgh
[3]Scottish Crop Research Institute, Invergowrie, Dundee

Abstract

Pollen profiles from soils can potentially be used to represent changes in local vegetation patterns. However, differential rates of biological, chemical and physical processes in different soil types may lead to systematic distortions in how vegetation histories are recorded in pollen analyses. We report here the main conclusions of work that attempts to test the replication of a uniform vegetation history from different soil types against a secure stratigraphic peat sequence.

Five different soil types and a small peat hollow, c. 2 m diameter, were excavated within a small area of a treeless, upland pasture in southern Scotland. Soil types varied in hydrological status and, to a lesser degree, pH. Arboreal pollen arrives at these sites from regional sources and is assumed to represent a uniform input to each soil type. Dating controls in the peat allow us to understand the temporal sequence of arboreal pollen and organic and chemical pollutants to all soil types over the last c. 150 years. Given uniform input of particles to these soils, the resultant pollen diagrams should be comparable. We test this using correspondence analysis and conclude that no soil profile contains a pollen stratigraphy comparable to that of the peat sequence. No soil profile appears to provide a stratigraphy that can be interpreted with respect to vegetation history.

Reasons for this include contrasts in the rates and character of bioturbation and differential pollen decay and loss. The implications of our interpretations for archaeopalynology are briefly discussed.

Introduction

Pollen analysis of buried soils is frequently employed to examine past vegetation patterns, often at archaeological sites. The most common application of the technique in Britain derives from the approach of Dimbleby (1985), in which a pollen stratigraphy is constructed and changes in vegetation through time are interpreted.

481

However, the interpretation of pollen from soils as temporal sequences requires the use of certain assumptions, defined by Dimbleby in 1985. These assumptions are:

- pollen assemblages are deposited through time in stratigraphic sequence
- pollen stratigraphies are maintained and can be interpreted chronologically
- pollen spectra can be recognised as coherent assemblages despite the occurrence of differential decay of pollen types.

This project was set up to test these assumptions by studying the pollen record from different soil profiles at a site where known-age microfossil markers, particularly pollen, have been recorded. Pollen records were compared to a control profile against which the loss of stratigraphic integrity and the occurrence of differential decay can be measured.

Correspondence analysis is then used to compare the regional pollen records at each sampling site and to test for similarities in the pollen records. Where the pollen stratigraphy of a soil is similar to a control sequence, then that soil type can be said to be stratigraphically secure and able to provide a pollen record of vegetation history. Pedological characteristics are used to identify the processes responsible for altering the pollen stratigraphy of different soil types. Tipping *et al.* (1997) provide the background to the project and Carter *et al.* (1997) discuss the recent history of the study site and surrounding area, stressing that all soil types have evolved naturally. Detailed discussion of the methods employed and the interpretation of the pollen data, with regard to stratigraphic integrity at each site, are given in Long *et al.* (in press), while the interpretation of data from podzol profiles, in particular, is discussed in Tipping *et al.* (in press). Soil micromorphological descriptions and interpretations are discussed in Davidson *et al.* (1999).

The study site

The study site in the Scottish Borders, near Peebles (Fig. 1) is an area of rough grazing supporting *Calluna, Pteridium* and grassland communities. The hillside lies to the south-east, and downwind of Dawyck Botanic Gardens, the oldest arboretum in southern Scotland (Bown 1992) (Fig. 1). This supplies a number of recognisable non-native tree pollen types (Tipping *et al.* 1997). The six sampling sites lie within 200 metres of each other. The nearest source of arboreal pollen to the sampling sites are the spruce plantations at Lour Wood and the High Wood of Posso (Fig. 1), *c.* 300 metres away at the nearest point. All arboreal pollen received at the sampling sites is regional (*sensu* Jacobson and Bradshaw 1981).

Each sampling site has received the following dated inputs:

- Non-native tree pollen, including *Pinus* diploxylon and haploxylon types; *Ulmus* species; *Fagus* species; *Abies* species; *Picea* species and *Larix* species (Tipping *et al.* 1997), recorded in the pollen diagrams from Lour and probably originating from Dawyck Botanic Gardens, although lags between planting at Dawyck and representation in the pollen record at Lour, and differences in the numbers and location of the trees planted, have resulted in the identification of a single injection of non-native tree pollen in the control peat profile (see below). This injection has been dated by ^{210}Pb dating to *c.* 1840 (Carter *et al.* 1997; Tipping *et al.* 1997).
- Spheroidal carbonaceous particles (SCPs) derived from fossil-fuel combustion provide an additional dated microfossil marker at Lour. Rose *et al.* (1995) have shown that SCP input to lake sediments in southern Scotland and northern England can be reliably dated to post 1900.

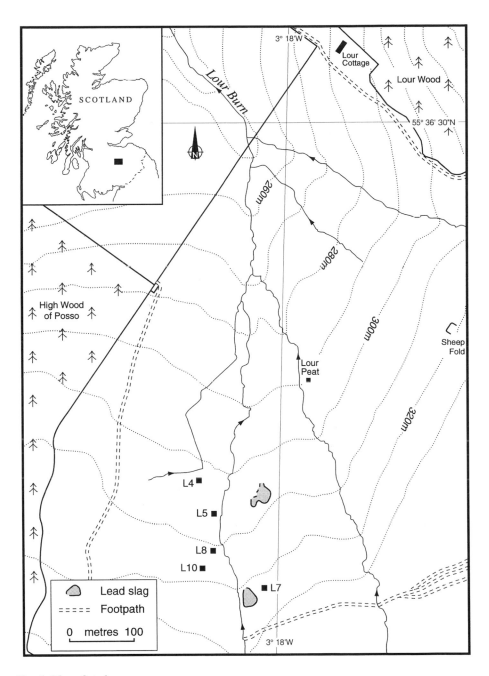

FIG. 1. Map of study area

The soil types

A control profile from a small diameter (*c.* 2 m) peat deposit (LP) at Lour, located adjacent to the soil profiles sampled (Fig. 1) is used to define a palynological sequence at Lour typical of an undisturbed profile with high stratigraphic integrity. Fig. 2 presents arboreal taxa in stratigraphic sequence in this profile from *c.* AD1800 to the present; a full diagram from this site is given in Carter *et al.* (1997). This pollen stratigraphy should be found at the soil profiles sampled.

The five soil profiles studied here (Fig. 1 and Table 1) are distributed mainly along a hydrological gradient with pH as a secondary axis. The soil profiles sampled have had most variables held constant, while the main differentiating variable, the hydrological regime, was measured over a two year period. Table 1 provides details of the soil profile characteristics of the uppermost horizons.

Differences between the soil types sampled at Lour are for the most part a function of hydrological regime, governed by location on the hillslope. The soils are developed on colluvium from acid Silurian mudstones: all profiles are acidic and the particle size distribution of the mineral component is comparable between sites. Furthermore, the study area, including the six sampling sites, has been demonstrated to have been subject to the same land management strategies (Carter *et al.* 1997).

Methods

Details of the field and laboratory techniques used are provided in Long *et al.* (in press). Pollen data used here (Figs. 2 – 7) are restricted to arboreal pollen (AP) taxa (counted to a total of 200 AP except at L10 where this sum could not be achieved). Pollen zones have been defined objectively using constrained cluster analysis (CONISS; Grimm 1987). Pollen taxonomy follows Bennett (1994), although "type" is used where a pollen type includes non-native species, that cannot be identified to species level. Non-native pollen types include all of the non-native tree types recorded at Lour (Tipping *et al.* 1997; Long *et al.* in press).

The arboreal pollen data were subjected to detrended correspondence analysis (DECORANA) (Hill and Gauch 1980) to test inter-site dissimilarity. Percentage arboreal pollen data were used in the analysis and were subjected to square-root transformation, with no downweighting of rare taxa. Data were detrended using 2nd-order polynomials (Jongman *et al.* 1995). Average zone scores were calculated as a mean of the DECORANA sample scores within these pollen zones. Linear regression was used to identify any relationship between axis 1, identified in DECORANA as accounting for the greatest proportion of variation within the dataset, with independent environmental variables.

Canonical correspondence analysis (CANOCO) (Ter Braak 1987) was used to see if any of the measured environmental variables at each site (pH, dominant moisture regime (measured as the percentage of time the water table was recorded at a depth within 10% of the base of the profile) and percentage organic matter) could account for this variation. The presence of soil fauna was predicted to be important but could not be used as an independent variable, as this is strongly dependent on the dominant moisture regime and pH (Boag *et al.* 1997; Lee 1985).

Other quantitative measures used in this paper are the percentage total pollen loss and taxonomic ratios at each sampling site. Percentage total pollen loss uses a total pollen loading at each site, which is the total non-native pollen concentrations per cm^3 of sediment in the uppermost pollen zones (defined by the base of the rational pollen limit for *Pinus diploxylon* type). The total pollen loss from each soil profile is then calculated as a percentage of the total pollen loading at LP (assuming no loss from the peat profile). This assumption that the peat profile has preserved a record of pollen influx over time that has not been subject to pollen loss is unlikely. Pollen preservation at LP in these upper pollen zones is imperfect: 60% of total pollen and spores have deteriorated to some degree, although this includes crumpling and breakage (Carter *et al.* 1997). However, intense deterioration is a minor component and LP has lower rates of pollen destruction than the soil profiles (currently unpublished data). Taxonomic ratios are used as measures of arboreal species richness and are calculated from the number of arboreal pollen types recorded at each sampling site as a ratio of the number recorded at LP, again assuming no loss of pollen information from LP.

Results and Interpretation

The DECORANA plot (Fig. 8) is used here to demonstrate that the non-native pollen curves at the soil sites are not similar to LP. Fig. 8 shows the mean and range of sample scores at each site (*cf.* Birks and Berglund 1979), along axes 1 and 2, accounting for 34.2% of total variation. Axes 3 and 4, accounting for 13% of total variation, are not shown. DECORANA uses the presence or absence of non-native tree pollen types and the differences in the proportions of these taxa between sites to compare the similarity of the post-1840 pollen assemblages identified at each sampling site. Fig. 8 can be interpreted in two ways:

1. there is no uniform input of regionally-derived non-native tree pollen types to the profiles at Lour
2. arboreal pollen proportions and stratigraphies are different between sites because of differences in soil-forming processes and differential pollen destruction

The first interpretation cannot be dismissed but for the purposes of this paper, the assumption is maintained that arboreal pollen input is uniform between sites. The concept of a uniform regional pollen input has been illustrated in the USA for lakes and small forest hollows by Sugita (1994) and Calcote (1995). However, this assumption may not be valid for treeless areas where sediment profiles are capped by vegetation cover of different types and densities. An investigation at the University of Stirling has tested this assumption, using pollen trap data and vegetation mapping techniques, although results are as yet unpublished.

If the assumption that arboreal pollen is uniform between sites is maintained (and this has been tested in a follow-up project at the University of Stirling, unpublished to date), the second interpretation is preferred. If this is the case, soil-forming processes in each profile are critical to the preservation of a pollen stratigraphy. Of the soil variables measured (Table 1), pH and moisture regime are independent variables: particle size data show little variation between sampling sites owing to the homogenous parent material. Other variables, including soil faunal activity, are dependent on moisture regime and pH. Regression of the DECORANA scores for axis 1 (accounting for 24% of total variance) with pH and moisture regime indicates that moisture regime is the more important variable in explaining variation along axis 1 (Table 2). Regression of axis 1 against pH shows little correlation, although the sites sampled were of a very limited pH range (Table 1). Monte Carlo permutation tests used in conjunction with CANOCO indicate that axis 1 is significant and accounts for 20.4% of total variation. However, correlation coefficients show that pH and moisture are both closely negatively correlated with axis 1 (Table 2).

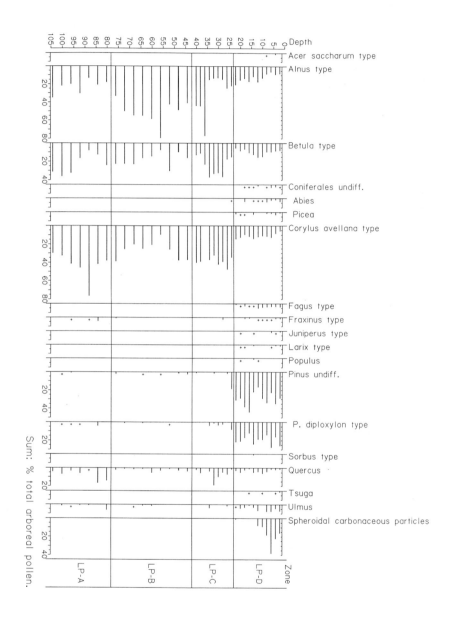

FIG. 2. Non-native arboreal pollen input to Lour peat (LP).

TABLE 1. Summary of soil profile characteristics at Lour. The soil type and site name are given in column 1 and these are used throughout the text. Column 2 gives the depth of the soil horizons, identified in the field and using soil micromorphological analyses and column 3 the depth of the pollen zones, identified using constrained cluster analysis. pH ranges for each soil horizon are given in column 4. Moisture regime in column 5 is a measure of the length of time that a profile remained waterlogged, expressed as the percentage of time that the water table was located within 10% of the depth of the profile. Column 6 is a measure of water table variability, which is a measure of fluctuations in water table height, expressed as the distance from soil surface as a percentage of the total profile depth. Mean organic content was measured as a percentage weight loss, following loss-on-ignition at 1 cm intervals throughout the profile, and expressed as a mean value for each soil horizon in column 7. Column 8 indicates the presence of soil macrofauna excrement as a subjective measure, derived from soil micromorphological analyses.

Site and soil type	Uppermost soil horizons & depths (cms)	Uppermost pollen zones & depths (cms)	Mean pH (CaCl₂)	Moisture regime	Water table variability	Mean organic content (% loi)	Presence of soil macrofauna excrement
Peat (**LP**)	0-24.3	D 0-18.3 C 18.3-24.3	-	0.0	0.03	74.96	0
Humic orthic gley (**L4**)	L/F: 0-2 Oh: 2-11	E 0-4.3 D 4.3-7.9	2.97	8.6	0.18	78.48	0
Groundwater orthic gley (**L5**)	Ahg: 0-23 Bg: 23-46	B 0-28.1	4.68 4.29	16.7	0.22	12.36	very dominant very dominant
Cambic stagnogley (**L8**)	A(g): 0-20 B(g): 20-34	B 0-25	3.99 4.26	68.3	0.38	13.13	very dominant very dominant
Typical orthic brown soil (**L10**)	Ah: 1-13 Bwl: 13.22	B 0-18.3	4.82 4.84	70.6	0.2	24.06	very dominant very dominant
Ferric podzol (**L7**)	L/F: 0-4 H: 4-10	C 0-6.2	2.70 2.89	98.3	0.07	76.11	very dominant dominant

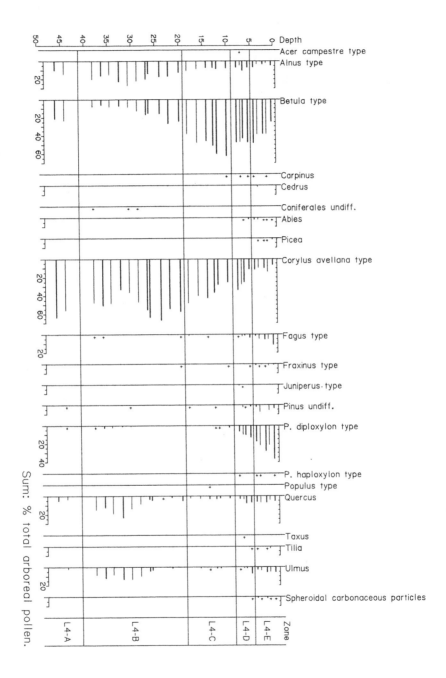

FIG. 3. Non-native arboreal pollen input to Lour 4 (L4).

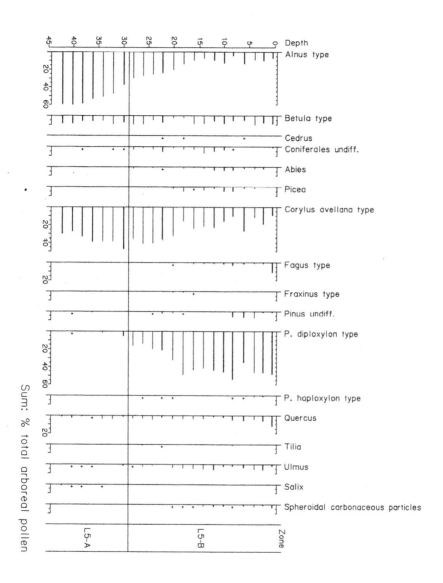

FIG. 4. Non-native arboreal pollen input to Lour 5 (L5).

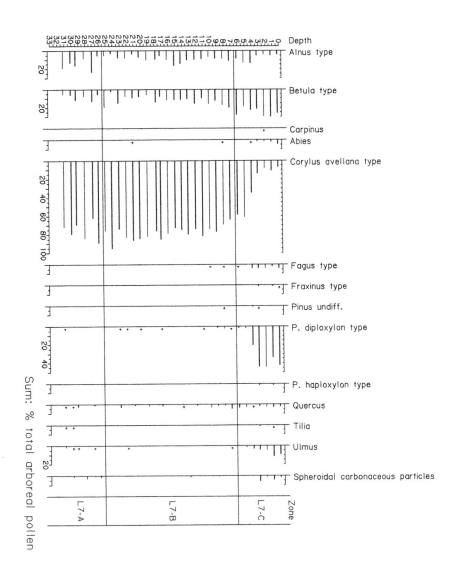

FIG. 5. Non-native arboreal pollen input to Lour 7 (L7).

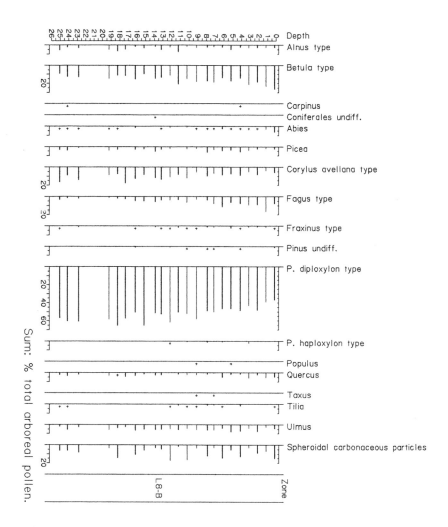

FIG. 6. Non-native arboreal pollen input to Lour 8 (L8).

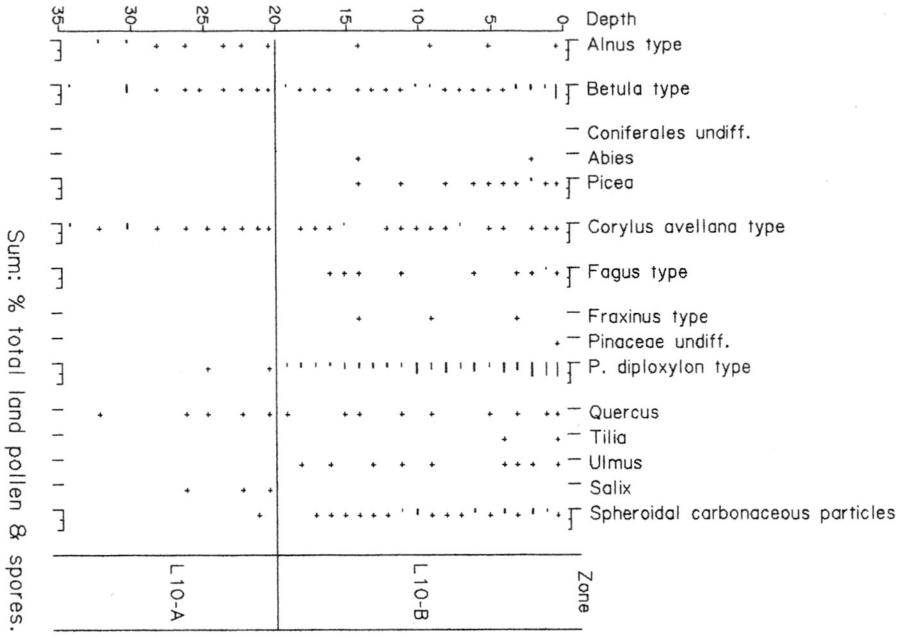

FIG. 7. Non-native arboreal pollen input to Lour 10 (L10).

TABLE 2. Correlation coefficients and R² values.

Axis 1	R²	Regression correlation coefficient	CANOCO correlation coefficient
moisture regime	53.3%	-.0175	-0.718
pH	15.7%	-0.165	-0.829

If moisture regime can be used to explain variation in the pollen record between sites at Lour, other variables dependent on the moisture regime, such as the presence of macrofauna and pollen preservation, are also likely to be important. Table 3 illustrates the relationship between the order of sites along axis 1, the moisture regime and the presence of macrofauna. A measure of differential pollen decay has not been calculated directly, although percentage pollen loss is used as an estimate of the relative importance of pollen decay in each profile, whether through mechanical or corrosion damage (Havinga 1964). Pollen loss is not necessarily equal to differential pollen decay, as pollen may be lost through downwashing through soil voids or extreme crumpling, which renders the pollen unrecognisable. In addition, this may affect some taxa more than others. Downwashing through void space has been rejected as an important mechanism for the redistribution of pollen in a soil profile (Davidson et al. 1999) and where crumpling obscured grains, percentages of crumpled unidentifiable grains were low: this suggests that differential pollen decay is probably the most important component of pollen loss in soil profiles. Table 3 indicates that axis 1 sorts the sites in terms of moisture regime, with the exception of L7, which although well-drained, is located midway between the well- and poorly-drained sites. This ordering of the sites accords very well with the dependent variables of the presence of macrofauna and percentage pollen loss, although L5 is an exception. This profile is adjacent to a small stream and is periodically flooded. Earthworms are recorded in this profile and it is likely that they adapt to the changes in the moisture regime by non-permanent use. Periodic flooding of this profile could also explain the lower than expected percentage value for pollen loss by inhibiting microbial activity and hence pollen decomposition. This will be tested and presented elsewhere using taxon-specific pollen preservation analyses. Table 3 suggests that moisture regime can account for inter-site dissimilarities through its control on macrofaunal activity and the rate of pollen decay.

TABLE 3. Relationship between axis 1 (DECORANA) and the independent and dependent environmental variables.

Axis 1	Moisture regime	Presence of macrofauna	% pollen loss since c. 1840
L10	70.6	Y	69
L8	68.3	Y	33
L7	98.3	Y	56
L4	8.6	N	24
L5	16.7	Y	31
LP	0	N	0

TABLE 4. Stratigraphic comparisons between LP and the soil sites.

Site	Maximum depth of SCPs (cm)	Maximum depth of non-native pollen (<2%TAP) (cm)	Maximum depth of macrofaunal excrement (cm)	Taxonomic ratio
LP	12.3	24.3	-	1.00
L4	6.55	10.0	-	0.81
L5	18.1	28.1	46	0.68
L8	25.1	25.1	34	0.68
L10	17.2	18.3	56	0.62
L7	3.25	5.2	10	0.56

Table 4 provides some additional semi-quantitative information about the movement of pollen and SCPs. In L10, L8 and L5, the dated markers reach similar depths in the stratigraphies and the curves are elongated throughout the A and into the B horizons to depths of 28 cm in L5. Macrofaunal activity extends well below these depths (Table 4). This suggests that the presence of earthworms could be responsible for the loss of stratigraphic integrity within the A horizon. Loss of pollen information from these profiles is indicated by pollen loss values of between 30% and 60% (Table 3) and also by the low arboreal pollen taxonomic ratios for these sites in comparison to LP (Table 4). Mechanisms for this include ingestion-excretion by earthworms and dry conditions favouring microbial activity.

In L4, the depth to which SCPs are recorded lies above the maximum depth of the non-native pollen markers, suggesting some degree of stratigraphic integrity in the uppermost pollen horizons. Macrofauna are not recorded at this site (Tables 1 and 4). The stratigraphic integrity of the L/F and Oh horizons has probably been maintained through the inhibition of earthworm activity, owing to the low pH and the predominantly wet moisture regime (Table 1), which has also favoured the accumulation of organic matter. The L/F and Oh horizons at this sampling site therefore have been accumulating in a stratigraphic sequence that has not been subjected to post-depositional mixing since about 1840. Percentage pollen loss (Table 3) and the taxonomic ratio at L4 suggest that pollen information loss has occurred but is less severe than at the other soil sites. Both of these measures show that L4 has suffered a loss of up to 24% of pollen information.

Podzols have been used as stratigraphic sequences by many workers by assuming that mor humus horizons maintain some sort of stratigraphic integrity (for example, Stockmarr 1975). However, evidence from L7 suggests that, at this site, mixing has occurred within the mor humus horizons (the L/F and H horizons in this profile), with some loss of pollen information (Table 3). The SCPs reach 4 cm, and the non-native pollen markers reach 5 cm, 1 cm below the base of the L/F horizon (Tables 1 and 4), suggesting complete mixing to this level. The top 4 cm of this profile is subject to surface feeding invertebrates (Tables 1 and 4) and these could be removing the stratigraphic integrity of this horizon (Table 1). The very small proportion of markers penetrating through to the top of the H horizon, below 4 cm, suggests that the mixing mechanism is limited to the L/F horizon. If the mixing mechanism at this site is these surface-feeding invertebrates, then a limited mixing range would be consistent with the activities of these invertebrates. High values for percentage pollen loss (Table 3) and a low taxonomic ratio (Table 4) indicate a significant loss of pollen information from L7. This profile is discussed in detail in Tipping *et al.* (in press).

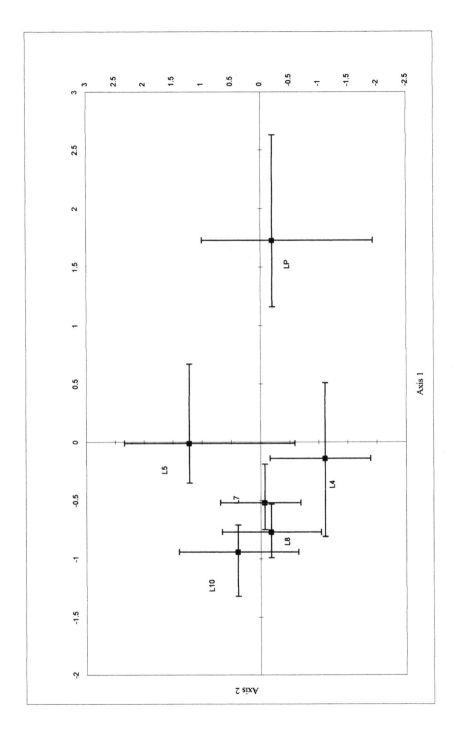

FIG. 8. DECORANA plot: average zone scores.

The problems of a loss of stratigraphic integrity in soil profiles inhabited by soil macrofauna and the loss of pollen information are sufficient in all of the soil profiles studied here to obscure the pollen record and render it uninterpretable for environmental reconstructions. None of the soil profiles here were able to reproduce the pollen record at the control site (LP). This suggests that the use of soil pollen analysis to produce temporal sequences of vegetation history is invalid in the soil types studied here. A more useful approach to environmental reconstruction using soil pollen analysis would be the use of thin samples from buried soil surfaces. Nevertheless, the limitations of mixing by surface-feeding invertebrates acting to blur the temporal resolution of the surface samples, and the possibility of differential pollen decay removing parts of the record must be recognised.

Conclusions

Soil profiles that are subject to mixing by invertebrates, particularly earthworms but also surface feeding invertebrates, do not preserve stratigraphic integrity and pollen loss from these systems is likely to be considerable. Soil profiles with pH and moisture regimes within the tolerance range of earthworms will be subject to stratigraphic mixing and the selective removal of pollen data. Even when these conditions are achieved only periodically, stratigraphic mixing can be severe. Where soil environmental conditions have changed to conditions inhibiting earthworm activity, past invertebrate activity will have removed the stratigraphic integrity of that part of the record. Only where these soil macrofauna are absent, and have been absent for the period under study, is soil stratigraphic integrity preserved. However, even where this is the case, a significant proportion of pollen information can be lost. This has been demonstrated for the upper organic horizons of the peaty gley and podzol in this study. In addition to this, the organic top to the podzol profile studied here (L7) has been subject to complete reworking and mixing by surface-feeding invertebrates for at least 150 years, removing stratigraphic integrity over this period. Only where the organic horizons can be demonstrated to have accumulated sufficiently quickly to provide a useful temporal resolution and where invertebrate activity has been inhibited can these soil horizons preserve stratigraphic integrity. We would suggest that to demonstrate a useful temporal resolution and a sufficiently fast accumulation rate would be difficult to achieve in most soil profiles. Even if this can be demonstrated, loss of pollen information can be sufficient to thwart archaeopalynological interpretations.

Acknowledgements

This research was funded by a NERC award (GR/J73421). Pollen samples were prepared by Richard Kynoch. We would like to thank David Mann and David Knott for their support and help at Dawyck and the landowner at Lour, Mr Balfour, for access.

References

Bennett, K.D. (1994). An annotated catalogue of pollen and pteridophyte spore types of the British Isles. Unpublished manuscript. University of Cambridge. Cambridge. http://www.kv.geo.uu.se/pc-intro.html

Birks, H.J.B. and Berglund, B.E. (1979). Holocene pollen stratigraphy of southern Sweden: a reappraisal using numerical methods. *Boreas* **8**: 257–279.

Boag, B., Palmer, L.F., Neilson, R., Legg, R. and Chambers, J. (1997) Distribution, prevalence and intensity of earthworm populations in arable land and grassland in Scotland. *Annals of Applied Biology* **130**: 153–165.

Bown, D. (1992). Four gardens in one: the Royal Botanic Garden, Edinburgh. HMSO, Edinburgh.

Calcote, R. (1995). Pollen source area and pollen productivity: evidence from forest hollows. *Journal of Ecology* **83**: 591–602.

Carter, S., Tipping, R., Davidson, D., Long, D. and Tyler, A. (1997). A multiproxy approach to the function of postmedieval ridge-and-furrow cultivation in upland northern Britain. *The Holocene* **7**: 447–456.

Davidson, D., Carter, S., Boag, B., Long, D., Tipping, R. and Tyler, A. (1999). Analysis of pollen in soils: processes of incorporation and redistribution of pollen in five soil profile types. *Soil Biology and Biochemistry* **31**: 643–653.

Dimbleby, G.W. (1985). The palynology of archaeological sites. Academic Press, London.

Grimm, E.C. (1987). CONISS: a FORTRAN 77 program for stratigraphically constrained cluster analysis by the method of incremental sum of squares. *Computers and Geoscience* **13**: 13–35.

Havinga, A.J. (1964). An investigation into the differential corrosion susceptibility of pollen and spores. *Pollen et Spores* **6**: 621–635.

Hill, M.O., and Gauch, H.G. (1980). Detrended correspondence analysis: an improved ordination technique. *Vegetatio* **42**: 47–58.

Jacobson, G.L. and Bradshaw, R.H.W. (1981). The selection of sites for palaeovegetational studies. *Quaternary Research* **16**: 80–96.

Jongman, R.H.G., Ter Braak, C.J.F. and van Tongeren, O.F.R. (1995). Data analysis in community and landscape ecology. Cambridge University Press, Cambridge.

Lee, K.E. (1985). Earthworms: their ecology and relationships with soils and land-use. Academic Press, Sydney.

Long, D.J., Carter, S., Davidson, D., Tipping, R., Boag, B. and Tyler, A. (in press). Testing the assumptions behind soil pollen analysis and the use of soil pollen analyses in environmental reconstruction. *Journal of Archaeological Science*.

Rose, N. L., Harlock, S., Appleby, P.G. and Battarby, R.W. (1995). Dating of recent lake sediments in the United Kingdom and Ireland using spheroidal carbonaceous particle (SCP) concentration profiles. *The Holocene* **5**: 328–335.

Stockmarr, J. (1975). Retrogressive forest development, as reflected in a mor pollen diagram from Mantingerbos, Drenthe, the Netherlands. *Palaeohistoria* **17**: 38–51.

Sugita, S. (1994). Pollen representation of vegetation in Quaternary sediments: theory and method in patchy vegetation. *Journal of Ecology* **82**: 881–897.

Ter Braak, C.F.G. (1987). CANOCO - a FORTRAN program for canonical community ordination by [partial] [detrended] [canonical] correspondence analysis, principal components analysis and redundancy analysis (version 2.1). Agricultural Mathematics Group, Wageningen.

Tipping, R., Carter, S., Davidson, D., Long, D. and Tyler, A. (1997). Soil pollen analysis: a new approach to understanding the stratigraphic integrity of data. In: A. Sinclair (editor). Archaeological Sciences 1995. Liverpool University Press, Liverpool.

Tipping, R., Long, D., Carter, S., Davidson, D., Tyler, A. and Boag, B. (1999). Testing the potential of soil stratigraphic palynology in podsols. *Journal of the Geological Society of London*, Special Publication **165**, pp. 79–90.

SUBJECT INDEX

aberrant grains 152, 287, 351
aborted or shrivelled spores 134
absolute volume 11
ACCTRAN 76, 81, 356
accumulation rate 496
aceto-soluble micro-fibrillar layer 313
acetolysis-resistant 229–230, 232, 234, 236–237, 241–242
acetolysis-susceptible 229, 233, 236
acritarchs 417–419
actin cytoskeleton 68
adaptations 74, 222, 447
adaptive advantages 221
adhesion of pollen 345
adsorbing surfaces 38
aerodynamic 74, 152
Afghanistan 385
Africa 133, 144, 193, 241, 261, 451, 455, 461
African 144–145, 168, 245
Afro-Madagascan 57
Afro-South American 451, 455, 461–462
agar medium 59
aggregation 36, 294–295
air-borne pollen 154
air-filled sacci 160
air-water interphase 153, 159
alate spores 134–136
Albian 452, 457, 459, 461–462
alete spores 441
algae 417
algal 413, 417
algorithm 275
allele 289
Alokistocare-Glossopleura horizon 415
alpine habitats 385
Alps 461
altitude 389, 476, 478
altitudinal transects 468
alveolar 174, 181
alveolate 148, 150, 155, 157, 160
ambavioid 259, 261, 263, 266, 270, 274–276, 278
ambophilous species 75
American 235, 259, 285–286
2-amino-ethanol 3
ammonium persulphate 33, 35–36
amoeboid 107, 242
amorphous sporopollenin accumulations 429
amyl acetate 327
amylase 91–92

anadromic branching 443
anagenesis 399, 407
ancestor 157, 356
ancestral 148–150, 264, 276, 326, 344, 399, 401, 418–419, 451, 461
androecium 289, 325–326, 343
anemophilous 73–75, 78–82, 84–85
anemophily 73, 78
angiosperms 118, 154, 160–161, 164, 169, 180–181, 193, 195, 205–206, 227–228, 237, 253, 259–261, 270, 285–286, 292
Angola 140
animal pollinated species 343
annonoids 261, 265, 267, 270, 272–273, 276
annual 385, 387–388
annulate 114
annulus 46
anomalous behaviour 160
ANOVA 92
Antarctica 187, 189, 198
anthemoid 404, 407
anther 325, 328, 330, 341, 343–345, 354
anther glands 337–338, 342–343
anthesis 325
anthophyte 163–164, 174, 180–181
antibodies 92
antigen 92
apertural 26, 28, 54, 57, 63–64, 68–69, 122, 167–168, 171, 173, 303, 316–317, 319, 323
aperturate pollen 250, 302
aperture 77–78, 84, 122, 205, 307, 313–314, 317–318, 323, 446
aperture absence 249
aperture evolution 266
aperture number 323
aperture position 323
aperture sites 9
aperture symmetry 206
apetalous 301–303
apical articulation 274
apiculus 275
apocolpium 305–307, 310, 323, 394
apomorphic 264, 301, 318–319
apomyctic 134
Appalachian Mountains 455, 459, 462
Aptian 451, 457
Aptian-Albian 451, 457, 463
aquatic 249–250, 254, 413, 422
aqueous 31, 99–101, 103, 105–107, 159

TAXONOMIC INDEX